ELECTRONIC EXCITATION ENERGY TRANSFER IN CONDENSED MATTER

MODERN PROBLEMS IN CONDENSED MATTER SCIENCES

Volume 3

Series editors

V.M. AGRANOVICH

Moscow, USSR

A.A. MARADUDIN

Irvine, California, USA

NORTH-HOLLAND PUBLISHING COMPANY
AMSTERDAM · NEW YORK · OXFORD

ELECTRONIC EXCITATION ENERGY TRANSFER IN CONDENSED MATTER

V.M. AGRANOVICH

Institute of Spectroscopy, USSR Academy of Sciences
Troitsk, Moscow ob. 142092
USSR.

M.D. GALANIN

Lebedev Physical Institute, USSR Academy of Sciences
Leninski Prospect 53, Moscow 117924
USSR

Translated from the Russian
by
O. GLEBOV

1982

NORTH-HOLLAND PUBLISHING COMPANY.
AMSTERDAM · NEW YORK · OXFORD

ISBN 0444 86335 4

PUBLISHERS:
NORTH-HOLLAND PUBLISHING COMPANY
AMSTERDAM · NEW YORK · OXFORD

SOLE DISTRIBUTORS FOR THE USA AND CANADA:
ELSEVIER SCIENCE PUBLISHING COMPANY, INC.
52 VANDERBILT AVENUE
NEW YORK, N.Y. 10017

Library of Congress Cataloging in Publication Data

Agranovich, V. M. (Vladimir Moiseevich), 1929-
Electronic excitation energy transfer in condensed matter.

(Modern problems in condensed matter sciences)
Translation of: Perenos ėnergii ėlektronnogo vozbuzhdenii͡a v kondensirovannykh sredakh.
Bibliography: p.
Includes index.
1. Condensed matter. 2. Electronic excitation. 3. Energy transfer. I. Galanin, M. D. II. Title. III. Series.
QC173.4.C65A3713 1982 530.4 82-8081
ISBN 0-444-86335-4 AACR2

PRINTED IN THE NETHERLANDS

MODERN PROBLEMS IN CONDENSED MATTER SCIENCES

Vol. 1. SURFACE POLARITONS
V.M. Agranovich and D.L. Mills, *editors*

Vol. 2. EXCITONS
E.I. Rashba and M.D. Sturge, *editors*

Vol. 3. ELECTRONIC EXCITATION ENERGY TRANSFER IN CONDENSED MATTER
V.M. Agranovich and M.D. Galanin

Vol. 4. SPECTROSCOPY AND EXCITATION DYNAMICS OF CONDENSED MOLECULAR SYSTEMS
V.M. Agranovich and R.M. Hochstrasser, *editors*

Oh, how many of them there
are in the fields!
But each flowers in its
own way—
In this is the highest achievement
of a flower!

Matsuo Bashó
1644–1694

PREFACE TO THE SERIES

"Modern Problems in Condensed Matter Sciences" is a series of contributed volumes and monographs on condensed matter science that is published by North-Holland Publishing Company. This vast area of physics is developing rapidly at the present time, and the numerous fundamental results in it define to a significant degree the face of contemporary science. This being so, it is clear that the most important results and directions for future developments can only be covered by an international group of authors working in cooperation.

Both Soviet and Western scholars are taking part in the series, and each contributed volume has, correspondingly, two editors. Furthermore, it is intended that the volumes in the series will be published subsequently in Russian by the publishing house "Nauka".

The idea for the series and for its present structure was born during discussions that took place in the USSR and the USA between the former President of North-Holland Publishing Company, Drs. W.H. Wimmers, and the General Editors.

The establishment of this series of books, which should become a distinguished encyclopedia of condensed matter science, is not the only important outcome of these discussions. A significant development is also the emergence of a rather interesting and fruitful form of collaboration among scholars from different countries. We are deeply convinced that such international collaboration in the spheres of science and art, as well as other socially useful spheres of human activity, will assist in the establishment of a climate of confidence and peace.

The General Editors of the Series,

V.M. Agranovich A.A. Maradudin

PREFACE TO THE ENGLISH TRANSLATION

We are glad that this translation of our book into English will help a wider range of readers to get acquainted with various problems encountered in the studies of electronic excitation energy transfer in condensed matter. Owing to the wide use of lasers, this research field encompassing problems in optics, solid state physics and physics of liquids continues to develop rapidly.

Two years have passed since the Russian edition of this book. During this period a number of interesting theoretical and experimental studies have been published and we have made a considerable number of changes and additions in the current text. These changes concern primarily the new results on processes occurring in molecular crystals at high excitation levels, on polariton luminescence, reabsorption and metallic quenching of excitons. Moreover, a section on optical properties of mixed crystalline solutions has been added to Chapter 3.

The English translation of the book was carefully read by Prof. R. Hochstrasser and we are deeply grateful to him for his remarks reflected in this edition.

V.M. Agranovich and M.D. Galanin

April 1981

PREFACE

Electronic excitation energy transfer in liquids and solids is one of the fundamental problems in modern physics of condensed matter. This is a fairly universal problem since transfer of electronic excitation energy is an intermediate process between the primary event of electronic excitation and the final processes that utilize the energy of electrons. Therefore, the analysis of the mechanisms of electronic energy transfer and, in particular, the analysis of their efficiency are indispensable in the studies of the interaction between various types of radiation and matter when what is studied is not just absorption of radiation energy but also the accompanying effects on the absorbing medium.

Electronic excitation energy transfer plays an essential role in such modern research fields as luminescence, radiation physics, radiation chemistry, photosynthesis, biochemistry, bioenergetics, to name but a few.

In the last thirty years especially intense efforts were aimed at studying electronic excitation energy transfer in crystals, apart from that in liquid solutions. This is due not only to general development of experimental and theoretical solid state physics but also to numerous practical applications. For instance, electronic excitation energy transfer in crystals is a very important feature for development of various luminophors, scintillators and materials for quantum electronics (laser crystals and solutions). Wide-ranging studies of electronic excitation energy transfer in organic crystals are stimulated, among other factors, by the fact that they possess sensitized luminescence and by the role which energy migration can play in biological systems. These studies aim at finding an answer to the fundamental question of modern bioenergetics which, according to Szent-Györgyi (1957) is: how energy controls life and how it activates the living machine?

Electronic excitation energy transfer is such an extensive field now that it would be difficult to deal with all its aspects in a single book, and, naturally, we have not attempted to do that. The selection of material in the book is, to a certain extent, subjective. Apart from fundamentals of the theory, we tried to treat those subjects which are best understood at present and which are frequently used for interpreting experimental results. On the other hand, we attempted to demonstrate the limits of applicability of the theory and to note the problems to be solved in the future. Among important subjects not dealt with directly in the text are the effects of a magnetic field on energy transfer involving triplet levels of organic molecules, biological energy transfer and other specialized applications. Readers are referred to the original papers in these instances.

Molecular liquids and molecular crystals as well as crystals activated with rare-earth elements are perfect model systems for analysis of energy transfer. The main features of electronic excitation energy transfer which are known now have been discovered in the studies of optical and photoelectric properties of these systems.

This book discusses the theoretical results from these studies and experimental material is used only to demonstrate agreement with theoretical predictions. A more detailed discussion of experimental studies is given by Bodunov et al. (1977), Wolf (1967), Widsor (1965), Rice and Jortner (1967), Avakian and Merrifield (1968), Avakian (1974), Powell and Soos (1975).

Amorphous and crystalline materials are analyzed in connection with their optical properties, in particular, the effect of energy transfer on their luminescence parameters. The book treats various mechanisms of energy transfer (resonance transfer of excitations, excitons, radiative transfer) and specific features of energy transfer at high excitation levels.

Many questions discussed in the book have practically never been treated in monographs and, therefore, the book may contain some inaccuracies. We shall be thankful for any critical suggestions which we hope to make use of in the future.

We are especially thankful to those Soviet and foreign scientists who have sent us reprints of their papers. We are grateful to P.P. Feofilov for his suggestions which we took into account in the final draft of the manuscript.

V.M. Agranovich and M.D. Galanin

INTRODUCTION

Energy transfers from excited atoms, ions, or molecules to unexcited particles frequently occur in nature. We frequently encounter this phenomenon when the concentration of the interacting particles and the lifetime of the excited state are high enough. The simplest visible manifestation of energy transfer is the quenching of luminescence due to interaction between excited and unexcited molecules or, conversely, sensitization of luminescence, that is, development of luminescence of molecules which are unexcited earlier.

The phenomena associated with energy transfer and deactivation of excited molecules were found in the early studies of fluorescence of organic dye solutions. It was these studies that led to the first observations of depolarization and concentration quenching of luminescence. These occur when the concentrations of solutions are increased, as well as by the quenching of luminescence by foreign materials.

Initially, attempts were made to analyze the quenching of luminescence in liquid solutions by analogy with the treatment of interactions between molecules or atoms in gases based on the concept of "second-order collisions". This concept was introduced at first for describing the collisions of excited atoms with electrons accompanied by conversion of the energy of excited atoms into the kinetic energy of electrons; later, it was applied to all processes resulting in deactivation of the excited state (Mitchell and Zemansky 1934, 1961).

Vavilov (1925, 1929, 1954) was the first to attempt explaining quenching of fluorescence of liquid solutions by means of collisions between excited molecules and quenching molecules. This approach proved to be successful for analyzing quenching of fluorescence of solutions by foreign colourless quenchers, particularly, when the motion of molecules in liquids had been taken into account. We know that in ideal solutions the number of collisions is determined by the gas-kinetics equation and does not depend on the viscosity. However, the statistics of collisions in gases differ from the statistics of collisions in liquids in that collisions in liquids are not distributed uniformly but occur in series. Since even the first collision (the so-called encounter) typically results in quenching, the kinetics of quenching is determined by diffusion and, therefore, is strongly determined on the solution viscosity. Vavilov was the first to note this fact. Vavilov (1928, 1936, 1954) Sveshnikov (1935, 1937, 1938) developed the diffusion theory of luminescence quenching by foreign colourless quenchers proceeding from the coagulation theory of Smoluchowski which was in a good agreement with the experimental data.

However, the attempts to apply similar concepts to concentration quenching and depolarization due to the interactions between identical molecules proved to be unsuccessful. The weak dependence of these effects on the solution viscosity indicated that under such conditions it was insufficient to take into account only the short-range interaction as had been done by Vavilov and Sveshnikov, and that an essentially new approach was needed.

A new idea first put forward by J. Perrin (1924) and later elaborated by F. Perrin (1929) was that the concentration phenomena in luminescent solutions could be explained by "molecular induction", that is, by the long-range electromagnetic interaction between the excited molecule and the unexcited one, similar to the interaction between the classical electronic dipole oscillators. This attractive and fundamentally correct idea did not, however, take into consideration the relaxation effects and, thus, yielded quantitatively wrong results.

Owing to the lack of a satisfactory theory of energy transfer Vavilov in his later studies on the concentration phenomena attempted to develop a phenomenological theory (Vavilov and Feofilov 1942, Vavilov 1942, 1943, 1947, 1954). He defined the transfer probability as an empirical quantity proportional to the concentration. Vavilov proceeded from the idea that the concentration phenomena, that is, quenching and depolarization, are caused by the same process of energy transfer. It was assumed that quenching occurred because some of the transfer events were accompanied by deactivation of both interacting molecules ("transfers with quenching"). For some time such a process appeared to be possible since the fluorescent solutions exhibited a drop in fluorescence yield with anti-Stokes excitation (Galanin 1951, 1960).

Later, however, this treatment of concentration depolarization and concentration quenching as the same process was shown to be invalid. Levshin and Baranova (1959), Levshin and Grineva (1968) demonstrated that typical cases of concentration quenching were caused by energy transfer from excited molecules to nonluminescent (or luminescent in other spectral regions) dimers. Nevertheless, Vavilov's ideas on the role of "inductive resonance" in energy transfer proved to be fruitful and strongly stimulated experimental and theoretical work which later yielded many important results on energy transfer. For instance, quenching of luminescence by absorbers was shown to be related to the extent of spectral overlap (Vavilov and Galanin 1949). Feofilov and Sveshnikov (1940) established the relationship between quenching and the lifetime of the excited state, etc.

Förster (1948, 1949, 1959) was the first to put forward a satisfactory theory of energy transfer for molecules having broad spectra in a condensed medium. This theory was based on pertrubation theory in the adiabatic approximation. The theory assumes that energy transfer occurs owing to the weak dipole–dipole interaction between molecules. The interaction is assumed to be so weak that it does not change the initial optical spectra of the molecules. Förster has shown that under these conditions the probability of energy transfer can, indeed, be expressed in terms of the integral of the overlap of the luminescence and absorption spectra of the interacting molecules.

Moreover, Förster was the first to average the probability of transfer over the molecules in the solid solution obtaining relationships which could be compared with the experimental data. The Förster's theory is still quite useful since its results are in a good agreement with the experimental data when the conditions for its applicability are met.

Later, Dexter (1953) generalized Förster's theory and applied it to the cases of multipole and exchange interactions. Subsequently, the theory was developed to take into account various complicating factors such as the diffusion of molecules during the lifetime of the excited state.

Since Förster's theory is an approximate one the limits of its applicability were repeatedly discussed. For instance, Galanin (1951, 1960) derived Förster's formula in the framework of the classical model of two dipole oscillators with friction only under the assumption that the acceptor excitation lifetime is much shorter than the time of the reverse transfer. Robinson and Frosch (1962, 1963) obtained a similar result for a quantum mechanical model. Förster himself discussed the conditions for applicability of his theory (Förster 1965). He developed a classification of interaction types according to which his initial theory corresponds to the case of the "very weak coupling".

The limits of applicability of Förster's theory were also discussed in numerous papers proceeding from the density matrix approach. It is just these papers that led to the best understanding of the role played by the relaxation processes in the transfer phenomena, and formulated clear-cut criteria of applicability of various approximations (see ch. 1).

Thus, the "induced resonance" energy transfer is regarded as a result of weak interaction between two molecules – the energy donor and the energy acceptor. The experimentally observed effects of energy transfer such as sensitization or quenching of luminescence can be regarded as a result of independent elementary transfer events. This approach is correct for sufficiently diluted solutions in optically neutral transparent solvents.

Typically, the impurity molecules in solution form a spatially disordered system. Therefore, the excitation energy of the impurity and the energy of its resonance interaction with the surrounding molecules prove to be random quantities which are, generally, different for different impurity centres. This fact considerably complicates analysis of the transfer of the electronic excitation energy via the impurity molecules so that the problem is essentially similar to that of electric conductivity of spatially disordered media (Mott and Davis 1971).

This aspect of energy transfer via impurity molecules in crystals is especially important at sufficiently low temperatures and in cases when the inhomogeneous line width for impurity absorption is not only large compared with the homogeneous line width but is also larger than the energy of the resonance interaction between the impurity molecules at the mean distance between them. In particular, a significant part is played in this case by the nonresonance transfer processes accompanied with production or disappearance of the lattice phonons or other elementary lattice excitations. It is just the processes of this kind that destroy localization of states in disordered media predicted by Anderson in 1958 and make possible energy transfer via the impurity molecules for fairly low impurity concentrations (see ch. 5 sect. 9).

The physical picture of electronic excitation energy transfer is essentially different in crystals where any of the unit cells or any of the molecules comprising the crystal can be excited owing to the translational symmetry. It is well known that energy transfer in crystals can be due to motion of quasiparticles known as excitons. The fundamental concept of excitons put forward by Frenkel in (1931), which is of major importance for modern solid state physics, is directly applicable to the mechanism of energy transfer in crystals. This fact was particularly well understood as early as 1938 when Frank and Teller published their paper "Migration and photochemical effect of the excitation energy in crystals" (Frank and Teller 1938) in which they were the first to discuss how the energy absorbed at a certain unit cell could migrate in the crystal and later be utilized for photochemical or other process at any sufficiently distant point of the crystal. In this connection, Frank and Teller (1938) wrote that: "A crystal consists of identical elementary cells, which may be atoms, molecules or groups of atoms, molecules or ions. If a crystal is excited, the excitation may be localized in any one of these elementary cells. Such a localized excitation will not represent, however, a stationary state. The coupling between the particles in the crystal and the resonance caused by the identity of the crystal cells has the consequence that the excitation energy will be transferred from one crystal cell to another". The qualitative analysis of Frank and Teller based on the results of Frenkel (1931) and Peierls (1932) is, to a varying degree, relevant to the current problems of radiationless transfer of electronic excitation energy in crystals. Frank and Teller discussed energy transfer by wave packets as well as energy transfer by localized (incoherent) excitons. In the latter case the causes of exciton migration were given as thermoactivated hops and the tunnel effect (at low temperatures). Though Frank and Teller used the term "diffusion process" to describe migration of only incoherent excitons, they treated the migration of wave packets using the mean free path concept and assumed that scattering of excitons by phonons involved sharp change of the direction of motion of excitons (but not their energy). They also made a number of remarks on the mechanism of interaction between excitons and crystal defects or crystal surface, in particular, on the mechanism of exciton capture by foreign molecules. In fact, Frank and Teller formulated a consistent program of theoretical studies.

Experimental studies of energy transfer in crystals were started approximately in the same period. In particular, Bowen (1938), Bowen and Mikiewicz 1947) were the first to carry out an extensive investigation of the effects of low-concentration impurities on luminescence of molecular crystals. He found that the greatest effect on luminescence of crystals is produced by those impurities that have a high light absorption in the spectral region of luminescence of the pure crystal. The presence of such impurities decreases the intensity of luminescence of the crystal material and this intensity decrease is accompanied with a decrease in the duration of decay of the crystal luminescence. Bowen emphasized that the main part in the process was played by a radiationless transfer mechanism, that is, a mechanism which was not a result of the production of photons corresponding to the luminescence of the host material.

Apker and Taft (1951) were the first to carry out experimental studies of energy transfer by excitons in alkali halide crystals. They observed energy transfer from excitons to F centers. (For a discussion of later experiments with alkali halide crystals, see Knox (1963), Luschik et al. (1977), Kuusman et al. (1976), Kink et al. (1969)). The recent results on exciton mobility in crystals of inert gases (Ne, Ar, Xe) are reported by Himsel et al. (1975), Ophir et al. (1975).

Owing to small effective masses and comparatively weak exciton–phonon interaction, the mobility of excitons (the Wannier–Mott excitons) in semiconductor crystals proved to be so high (in some crystals the diffusion coefficient of excitons can be as high as 10^3 cm^2/s; for comparison, note that the diffusion coefficient of the singlet excitons in anthracene is about 3×10^{-3} cm^2/s) that special experimental procedures had to be designed (Pokrovsky 1972, Pokrovsky et al. 1970). The diffusion coefficients of excitons in semiconductors were successfully measured only quite recently in the studies of luminescence with high exciton concentrations.

In semiconductors energy is transferred by coherent excitons (diffusion of wave packets). Owing to large (macroscopic) Bohr radii of excitons in these crystals ($r_B \approx 50 - 100$ Å) a variety of new phenomena (biexcitons, electron–hole drops) are observed at exciton concentrations as low as $N \lesssim r_B^{-3} \approx 10^{18}\,\mathrm{cm}^{-3}$ and the study of them comprises a special field of modern semiconductor physics (Haken and Nikitine 1975). For instance, analysis of these phenomena must take into account not only the mobility of individual excitons but also the mobilities of various exciton complexes (such as exciton–hole drops), and the interactions between these complexes and other effects (Keldysh 1970). Some of the results discussed in this book can be used to describe the properties of the Wannier–Mott excitons (for instance, the temperature dependence of the diffusion coefficient or specific problems of kinetics and luminescence of excitons). Moreover, the more extensive discussion of the growth kinetics of the electron–hole drops in germanium given in ch. 6 illustrates the nature of the relevant problems. However, a more or less full discussion of the collective properties of excitons, biexcitons, electron–hole drops and other possible exciton complexes is outside the scope of this book and should be given in specialized reviews and monographs.

Turning back to the theory of exciton energy transfer in molecular and ionic crystals note that, at first, this theory developed to a certain extent independently of the work on energy transfer between molecules in solutions. Only later, at the end of the fifties, did the intimate links between these theories become appreciated.

The concept of the diffusional motion of excitons was first used for calculating the intensity of surface annihilation of excitons in Cu_2O as a function of the absorption coefficient k for the exciting light (Karkhanin and Lashkarev 1955), for calculating the rate of exciton capture as a function of k in molecular crystals (Faidysh 1955, Agranovich et al. 1956, 1957), and for estimating the rate of exciton capture by naphthacene impurity controlled by exciton diffusion (Galanin and Chizhikova 1956, see also ch. 7). Later (Agranovich and Faidysh 1956, Agranovich 1957) a more general integro-differential equation for the exciton concentration was derived separately taking into account the variation of the exciton concentration due to reabsorption of luminescent light.

The concept of exciton diffusion proved to be very fruitful and the diffusion approximation in most cases gives a valid picture of exciton migration. In spite of the difference between the transfer mechanisms in solutions and crystals the similarity between them is perceived much more clearly now. The similarity lies in not only that the elementary transfer event in both cases is often due to interaction of the same type (for instance, the dipole–dipole interaction) but also in that there is a continuous transition from donor molecule – acceptor molecules transfer in solutions to energy transfers via identical molecules in solutions and, finally, to migration of excitons in crystals. The diffusion approach can be employed to take into account not only the diffusion of molecules in solutions but also the diffusion of excitations in macromolecules and the diffusion of excitons. Finally, there is a certain correspondence between the cases of energy transfer by coherent and incoherent excitons and the cases (according to Förster) of strong and weak interaction between molecules.

The first two chapters of this book deal mainly with the theory of energy transfer in solutions. In particular, the first chapter discusses the elementary event of energy transfer between two interacting molecular systems at a given distance from each other. In this and some other chapters we tried to make the discussion as simple as possible in order to present a clear picture of the physical fundamentals of the theory. We thought it worthwhile to introduce the concept of the density matrix for the two-level system, that is, we used the approach that gained popularity with development of quantum electronics. Though the basic model is primitive this approach contributes to the understanding of the physical meaning of various theoretical approximations and yields rough but realistic order-of-magnitude estimates of the energy transfer probability. The first chapter also describes Förster's theory which, as mentioned above, played an essential part in the understanding of energy transfer problems.

The second chapter discusses methods and results of averaging the energy transfer probability over the ensemble of donors and acceptors of energy. The Förster's result and some modifications of it are given in this chapter. The chapter presents also a description of the theory that includes the motion of molecules in liquid solution. The great variety of experimental studies of energy transfer in solutions are illustrated here by only a few results demonstrating the good agreement between the theory and the experiments provided that the conditions of applicability of the theory have been met and the relevant experimental precautions have been taken.

Chapters 3–8 deal with energy transfer in crystals. These chapters discuss the relationships between the experimentally measured physical quantities and those microscopic crystal parameters which ultimately determine the magnitudes of the processes of electronic excitation energy transfer in crystals.

How can experimental data on energy transfer in crystals be used to determine what type of excitons (coherent or incoherent) make the main contribution to energy transfer? How can we determine the diffusion coefficient of excitons in a given crystal? What is the mechanism of interaction between excitons and impurities or interfaces? How do excitons interact with each other? What is the contribution of radiative energy transfer? What are the specific features of energy transfer between impurity molecules in crystals and what part is played by the host? This is an incomplete list of questions discussed in chs. 3–8.

In particular, the acting (local) field method is used in ch. 3 to analyze the effect of the host crystal on the energy of the resonance interaction between the impurity molecules which results in the radiationless transfer of the electronic excitation energy between them. If the distance between the impurity molecules is large compared with the size of the unit cell the effect of the host crystal on the resonance interaction energy is determined, primarily, by the dielectric constant of the medium. Therefore, the calculations of the energy of the resonance interaction between the impurity molecules prove to be closely linked to the calculations of the dielectric constant tensor of the crystal matrix: this matter determines the contents of ch. 3. On the other hand, the dielectric constant tensor completely determines the optical properties of the medium and, thus, such an approach makes possible a consistent treatment of the relationship between the rate of radiationless energy transfer between the impurity molecules, the structure of exciton spectra, and these optical properties.

Chapter 4 presents those results of the exciton theory (for instance, concerning the exciton–phonon interaction) that are used in ch. 5 for calculating the exciton kinetic parameters which determine the energy transfer rate. Chapter 4 also discusses the properties of biphonons and biexcitons which can, generally speaking, affect the kinetics of collisions between quasiparticles (and, in particular, the rate of bimolecular annihilation of excitons). Moreover, ch. 4 presents a consistent treatment of the microscopic theory of electronic excitation energy transfer between the impurity molecules which reveals the effects of virtual excitons on the transfer rate and takes into account the effects of resonance interaction between the impurity molecules and the crystal matrix.

In many crystals, in particular in many luminescent crystals in contrast to gases, the interaction between the electronic excitations (excitons) and the electromagnetic field quanta is so strong that at sufficiently low temperatures a correct description of their properties (mobility, luminescence and absorption spectra) should be based on the concept of polaritons, that is, the quasiparticles formed at strong exciton–photon interaction, rather than individual photons and excitons. Chapter 4 shows how these concepts can be used to describe within a unified physical framework a fairly wide variety of phenomena, such as radiationless and radiative transfer of the electronic excitation energy, exciton absorption and luminescence, etc.

Chapter 5 presents the theory of exciton mobility and the theory of the interactions of excitons with impurity molecules and with each other. This chapter describes also electronic excitation energy transfer via the impurity molecules, the Anderson localization of excitations and nonresonance processes. In ch. 5 attention is focussed on calculations of the kinetic parameters that are assumed to be given in ch. 6 which presents the phenomenological theory of exciton migration. The results derived from the phenomenological theory are usually compared with the experimental data to find the kinetic parameters. In this connection, ch. 6 presents calculations of experimentally measured parameters such as the quantum yield of luminescence, the decay time of luminescence, the quantum yield of sensitized luminescence, and so on.

Chapter 7 reviews the results of experimental studies of energy transfer by excitons. The main attention is given here to comparing the predictions of the theories of radiative and radiationless energy transfer with the experimental data which can be used for estimating the kinetic parameters, in particular, the diffusion coefficient of excitons and its temperature dependence. It is just the nature of this dependence (see ch. 5) that shows whether coherent or incoherent ("localized") excitons transfer energy in a given crystal or, in other words, whether the process involves diffusion of wave packets or diffusion corresponding to the model of random hops of excitation along the lattice sites.

Chapter 8 discusses the physical phenomena that take place when excitons interact at the interface between crystal and a metal surface quenching excitons. These problems are largely similar to those discussed in ch. 5 where the mechanism of interaction between excitons and a quenching impurity is discussed; the parallels can be drawn between excitons localized at the impurity and excitons localized at the interface (the surface excitons) and so on. Moreover, the analysis of metal quenching opens up a number of interesting possibilities. For instance, the rate of metal quenching of excitons describes not only the diffusion coefficient of excitons and the character of their migration but also electronic restructuring at the insulator–metal boundary which can result in formation of a "dead zone", etc. The studies of this type of exciton energy transfer (energy transfer to the surface) are just beginning to develop and we hope that our discussion of the relevant problems will stimulate further work in this field.

CONTENTS

CHAPTER 1

Elementary Energy Transfer Events

1. *Two-level system approximation*

Let us start by considering the elementary event of energy transfer between two fixed molecules at a given distance from each other. If the molecules interact in some way the most general considerations show that this results in energy transfer between them. This phenomenon is, typically, illustrated by the classical model of oscillations in the system of two coupled harmonic oscillators (for instance, two coupled pendulums) with identical fundamental frequencies ω_0. The interaction is known to produce frequency splitting so that the fundamental frequencies of the system are $\omega_0+\frac{1}{2}\Omega$ and $\omega_0-\frac{1}{2}\Omega$ where $\Omega=\beta/\omega_0$ and β is the coupling coefficient. If one of the oscillators is initially excited then energy is transferred to the second oscillator and the oscillators exchange energy with the frequency Ω until the oscillations decay completely.

A similar phenomenon can be treated in the framework of quantum mechanics using the two-level system approximation. First, let us solve the following elementary quantum mechanical problem. Consider a system of two identical atoms a and b each with two energy levels. We assume that the other energy levels are sufficiently high so that their effect can be neglected. Let the atomic states be described by the wave functions φ_a and φ_b for the lower state and φ_a' and φ_b' for the upper state. In the absence of interaction the state of the system in which one of the atoms is excited is degenerate and the respective wave functions are:

$$\Psi_1=\varphi_a'\varphi_b,\quad \Psi_2=\varphi_a\varphi_b'. \tag{1.1}$$

Now introduce interaction between the atoms described by the Hamiltonian V which is independent of time. Then the degeneracy disappears and the states Ψ_1 and Ψ_2 are no longer stationary states of the system. Assume that interaction is absent when both atoms are in the lower or upper state, that is, only the excited and unexcited atoms can interact. Using the perturbation theory we can write the wave function of the system in the following form:

$$\Psi=a_1(t)\Psi_1+a_2(t)\Psi_2. \tag{1.2}$$

It is well-known that the factors a_1 and a_2 can be found from the following equations (Landau and Lifshits 1965, sect. 38):

$$i\hbar\frac{da_1}{dt} = V_{12}a_2, \quad i\hbar\frac{da_2}{dt} = V_{21}a_1, \tag{1.3}$$

Here V_{12} is the matrix element of the interaction energy. If we eliminate, for instance, a_2 from eqs. (1.3) we obtain a simple equation of oscillation in $a_1(t)$. We have to find $|a_1|^2$ and $|a_2|^2$, that is, the probabilities that the system is in the states Ψ_1 or Ψ_2. If the initial conditions are $|a_1(0)|^2 = 1$ and $|a_2(0)|^2 = 0$ we obtain:

$$|a_1(t)| = \tfrac{1}{2}(1 + \cos \Omega t), \quad |a_2(t)|^2 = \tfrac{1}{2}(1 - \cos \Omega t), \tag{1.4}$$

where $\Omega = 2|V_{12}|/\hbar$. Thus, in such a quantum mechanical system we have transfer and exchange of energy with the frequency Ω which is proportional to the interaction energy in full analogy with the classical model of coupled oscillators.

This energy exchange is fully reversible and caused by the nonstationary character of the initial state. Generally speaking, such an initial state can be realized and then energy exchange will, indeed, take place.

However, we shall use the term "energy transfer" only for those cases of energy transfer which involve essential irreversibility and, hence, some relaxation processes. In treating such cases we should take into account the fact that the dynamic electron system under consideration is linked to a dissipative system.

This can be done in the framework of two-level system approximation if the density matrix approach is used. When we describe the state of a quantum mechanical system by the density matrix, rather than the wave function, we can phenomenologically take into account the relaxation processes caused both by spontaneous transitions and the interaction of the system with its environment which acts, for instance, as a thermostat whose exact state need not be known (see sect. 14, Landau and Lifshits 1965). For the two-level system two relaxation times are usually defined phenomenologically – the so-called "longitudinal" relaxation time (T_1) and the "transverse" relaxation time (T_2). The longitudinal time describes relaxation of the diagonal elements of the density matrix owing to radiative and radiationless transitions between the system levels (the diagonal elements of the density matrix give the relative populations of the levels). The transverse relaxation time describes relaxation of the nondiagonal elements, that is, the decay time of the phase relationships between the states. It will be shown below, that this relaxation time ("the phase memory time") is significant for energy transfer.

The equations for the density matrix replacing the Schrödinger equation for the wave function have the following form (Landau and Lifshits 1965, Pantell and Puthoff 1969):

$$\dot{\rho}_{jj} = \frac{1}{i\hbar}[V, \rho]_{jj} + \frac{1}{T_1}(\rho^{e}_{jj} - \rho_{jj}),$$

$$\dot{\rho}_{ij} = \frac{1}{i\hbar}(E_i - E_j)\rho_{ij} + \frac{1}{i\hbar}[V, \rho]_{ij} - \frac{1}{T_2}\rho_{ij} \quad (i \neq j). \tag{1.5}$$

Here V is the interaction, the square brackets denote the commutator, and ρ^{e}_{jj} is the equilibrium diagonal element of the density matrix.

Now let us write down the density matrix for the following three states of our two-level systems (the prime denotes the upper states):

$$|1\rangle = \varphi'_a\varphi_b, \qquad |2\rangle = \varphi_a\varphi'_b, \qquad |3\rangle = \varphi_a\varphi_b. \tag{1.6}$$

Assume that the excitation energy is sufficiently large so that $(E' - E) \gg kT$ where T is the absolute temperature. Then, we can assume for the equilibrium diagonal elements of the density matrix $\rho^{e}_{11} = \rho^{e}_{22} = 0$, $\rho^{e}_{33} = 1$. Denote the transverse relaxation times by τ_a and τ_b assuming that they are due to spontaneous transitions between levels. The transverse relaxation time for the elements ρ_{12} is denoted by T_2. This time is related to the level full-widths at half maximum by $2/T_2 = \Delta\omega_a + \Delta\omega_b$ (typically, $T_2 \ll \tau_a, \tau_b$). Then eqs. (1.5) yield:

$$\dot{\rho}_{11} = \frac{1}{i\hbar}(V_{12}\rho_{21} - V_{21}\rho_{12}) - \frac{\rho_{11}}{\tau_a},$$

$$\dot{\rho}_{22} = \frac{1}{i\hbar}(V_{21}\rho_{12} - V_{12}\rho_{21}) - \frac{\rho_{22}}{\tau_b},$$

$$\dot{\rho}_{12} = \frac{1}{i\hbar}V_{12}(\rho_{22} - \rho_{11}) - \frac{\rho_{12}}{T_2} + \frac{\Delta E}{i\hbar}\rho_{12}, \tag{1.7}$$

$$\dot{\rho}_{21} = \frac{1}{i\hbar}V_{21}(\rho_{11} - \rho_{22}) - \frac{\rho_{21}}{T_2} - \frac{\Delta E}{i\hbar}\rho_{21}.$$

Here $\Delta E = (E'_a - E_a) - (E'_b - E_b)$ and we take into account only those matrix elements of the interaction energy which correspond to transfer of excitation from the system a to the system b and back. The state $|3\rangle$ in which both systems are in the lower level is introduced to conserve the normalization condition $\rho_{11} + \rho_{22} + \rho_{33} = 1$.

The general solution of eqs. (1.7) is rather cumbersome; to illustrate the physical meaning it is better to analyze several specific cases (Burshtein 1965, Gamurar et al. 1969, Trifonov and Shekhtman 1970, 1972, Agabekyan and Melyakyan 1970, Kustov and Surogin 1970).

(1) The case of strong interaction corresponds to the conditions

$$\Omega = \frac{2|V_{12}|}{\hbar} \gg \frac{1}{\tau_a}, \frac{1}{\tau_b}, \frac{1}{T_2}. \qquad (1.8)$$

In this case eqs. (1.7) yield (for $\Delta E = 0$):

$$\ddot{\rho}_{11} + \Omega^2 \rho_{11} = \text{const}, \qquad (1.9)$$

that is, here we obtain oscillations of a population corresponding to the purely dynamic case described above. These oscillations can, in principle, be observed if the time resolution is sufficiently good and if, what is very significant, the initial condition $\rho_{11}(0) = 1$ is satisfied. For instance, we obtain then for the intensity of spontaneous radiation of the system a: $I_a(t) = (n_a/\tau_a)\rho_{11}(t)$ where n_a is the total number of the systems a.

(2) The case of identical systems ($\tau_a = \tau_b = \tau$, $\Delta E = 0$) and short transverse relaxation time $T_2 \ll \tau$. Here we obtain the equation

$$\ddot{\rho}_{11} + \left(\frac{2}{\tau} + \Omega^2 T_2\right)\dot{\rho}_{11} + \left(\frac{1}{\tau^2} + \Omega^2 \frac{T_2}{\tau}\right)\rho_{11} = 0. \qquad (1.10)$$

For the initial condition $\rho_{11}(0) = 1$ its solution is

$$\rho_{11}(t) = \tfrac{1}{2}\exp(-t/\tau)[1 + \exp(-\Omega^2 T_2 t)]. \qquad (1.11)$$

It should be noted that for $\Omega^2 T_2 \gg 1/\tau$ the values of ρ_{11} and ρ_{22} approach the limit $\frac{1}{2}$, that is, the populations of the states $|1\rangle$ and $|2\rangle$ become equal and then decrease according to the spontaneous decay law as $\exp(-t/\tau)$.

(3) The case of short T_2 and short lifetime for the initially excited system $T_2 \ll \tau_a \ll \tau_b$. Here we obtain:

$$\dot{\rho}_{11} + \left(\frac{1}{\tau_a} + \frac{\Omega^2 T_2/2}{1 + (\Delta E T_2/\hbar)^2}\right)\rho_{11} = 0, \qquad (1.12)$$

or

$$\rho_{11}(t) = \exp(-t/\tau_a - Wt), \qquad (1.13)$$

where

$$W = \frac{2|V_{12}|^2 T_2/\hbar^2}{1 + (T_2 \Delta E/\hbar)^2}. \qquad (1.14)$$

This is the case corresponding to the above-mentioned concept of energy transfer as an irreversible process.

The quantity W whose dimension is s^{-1} is known as the transfer rate or probability of transfer per unit time. It is proportional to the squared matrix element of the interaction energy. It should be noted that eq. (1.14) can be derived from the classical model of two coupled oscillators if damping of one of the oscillators (b) is assumed to be very strong (Galanin

1951, 1960). However, the classical model does not distinguish between T_2 and τ_b since the classical oscillator is described by only one decay time.

The general solution of the system of eqs. (1.7) has the character of damped oscillations. Bypassing explicit solution of eqs. (1.7) we can find the quantity

$$\bar{W} = -\frac{1}{\tau_a} + \left[\int_0^\infty \rho_{11}(t)\,\mathrm{d}t\right]^{-1}, \tag{1.15}$$

which can be regarded as a generalized transfer probability (if $\rho_{11}(t)$ decreases exponentially we have $\bar{W} = W$). According to Konyshev and Burshtein (1968) this can be done by applying the Laplace transformation to the variables in eqs. (1.7):

$$\mathscr{L}(\rho_{ij}) = \int_0^\infty \mathrm{e}^{-st}\rho_{ij}(t)\,\mathrm{d}t, \qquad \mathscr{L}(\dot{\rho}_{ij}) = s\mathscr{L}(\rho_{ij}) - \rho_{ij}(0). \tag{1.16}$$

Taking $s = 0$ we obtain now a system of algebraic equations instead of the system of differential equations. For the initial conditions, $\rho_{11}(0) = 1$, $\rho_{22}(0) = \rho_{12}(0) = 0$, the solution of this system of equations is,

$$\left[\int_0^\infty \rho_{11}(t)\,\mathrm{d}t\right]^{-1} = \frac{1}{\tau_a} + \frac{2|V_{12}|^2 T_2/\hbar^2}{1 + (T_2\Delta E/\hbar)^2 + (2|V_{12}|^2/\hbar^2)T_2\tau_b},$$

or

$$\bar{W} = \frac{2|V_{12}|^2 T_2/\hbar^2}{1 + (T_2\Delta E/\hbar)^2 + (2|V_{12}|^2/\hbar^2)T_2\tau_b}. \tag{1.17}$$

In the case of weak interaction when $(2|V_{12}|^2/\hbar^2)T_2\tau_b \ll 1$ we again obtain $\bar{W} = W$. In the case of very strong interaction the generalized transfer probability $\bar{W}$ tends to $1/\tau_b$, that is, the transfer rate is determined by the rate of transition of the system b from the excited state to the ground state and is independent of the interaction magnitude. It should be noted that in most experiments the time τ_b corresponds not to the radiation lifetime of the transition from the excited electronic state, but to the vibrational relaxation time which is usually much shorter. Below, we shall make quantitative estimates for the case of dipole–dipole interaction.

In conclusion, note that various approximations in the above theory of energy transfer between two two-level systems are analogous to various cases of resonance interaction of radiation with a two-level system (Allen and Eberly 1975). This is explained by the fact that the effect of the resonance field on a two-level system is, in some respects, similar to the effect of a perturbation which is constant in time (Landau and Lifshits

1965, sect. 40). In the framework of this analogy the case of strong interaction and long transverse relaxation time corresponds to coherent interaction and the case of weak interaction or short time T_2 corresponds to incoherent interaction. Indeed, in case of coherent interaction the resonance radiation can fully transfer the population from the lower to the upper level (the so-called π pulse) or back to the lower level (2π pulse) and so on. This situation corresponds to oscillations in the transfer theory. In the case of incoherent interaction the resonance radiation of a sufficiently high intensity can, in the limit, only make the populations of the levels equal. This corresponds to the second case discussed above in which the populations of the upper and lower levels tend to become equal as a result of interaction.

2. *Dipole–dipole interaction*

The excitation energy is usually transferred at distances much smaller than the wavelength of radiation corresponding to the given transition. Therefore, energy transfer is due to the Coulomb interaction between molecules. If the transition is allowed the dipole–dipole interaction plays a major part. It should be borne in mind that in the case of energy transfer the parameter for expansion in multipoles is a/R where a is the size of the molecule and R is the distance between molecules, rather than a/λ as in the case of interaction with radiation. Since $R \ll \lambda$ the former parameter is considerably larger than the latter one. The multipole and exchange interactions can play a part when the dipole transitions are forbidden and at small distances (see sect. 5).

The energy of the dipole–dipole interaction is

$$M = \frac{1}{R^3}\left\{(\boldsymbol{p}_a\boldsymbol{p}_b) - \frac{3}{R^2}(\boldsymbol{p}_a\boldsymbol{R})(\boldsymbol{p}_b\boldsymbol{R})\right\}. \tag{1.18}$$

where $\boldsymbol{p}_a$ and $\boldsymbol{p}_b$ are the dipole moments of the interacting systems, and R is the distance between them (the contribution of the dielectric constant of the medium is discussed in sect. 4). Equation (1.18) gives the following dependence of the interaction energy on the angles describing the relative orientation of the dipoles:

$$\chi(\vartheta_1, \vartheta_2, \varphi) = \sin\vartheta_1 \sin\vartheta_2 \cos\varphi - 2\cos\vartheta_1 \cos\vartheta_2. \tag{1.19}$$

Here we employ a polar system of coordinates in which $\boldsymbol{R}$ is the polar axis, ϑ_1 and ϑ_2 are the angles between the dipoles and this axis, and φ is the difference between the azimuthal angles. If all the relative orientations are equally probable we obtain the following mean value:

$$\bar{\chi}^2 = 2/3\,. \tag{1.20}$$

Let us make a rough estimate of the energy transfer rate in the case of dipole–dipole interaction between two-level systems. Assume that $\chi^2 = \frac{2}{3}$. Then the squared matrix element of the interaction energy is

$$M_{12}^2 = \frac{2}{3}\frac{|p_a|^2|p_b|^2}{R^6}. \quad (1.21)$$

The matrix elements of the dipole moments $|p_a|$ and $|p_b|$ can be expressed in terms of the oscillator strengths f_a and f_b. For instance,

$$p_a^2 = \frac{\hbar e^2}{2m\omega} f_a. \quad (1.22)$$

In the absence of radiationless transitions the lifetime τ_a can be related to the oscillator strength by

$$\frac{1}{\tau_a} = \frac{2}{3}\frac{e^2\omega^2}{mc^3} f_a = \frac{f_a}{\tau_{cl}}, \quad (1.23)$$

where $\tau_{cl} = \frac{3}{2}(mc^3/e^2\omega^2)$ is the decay time for the classical electronic oscillator.

Now we can estimate the dimensionless parameter,

$$W\tau_a = \frac{2|M_{12}|^2}{\hbar^2} T_2\tau_a, \quad (1.24)$$

which determines the transfer rate.

Substitution of eqs. (1.21) and (1.23) into eq. (1.24) yields:

$$W\tau_a = \frac{3}{4} f_b \left(\frac{\lambda\!\!\!^{-}}{R}\right)^6 \frac{T_2}{\tau_{cl}}, \quad (1.25)$$

where $\lambda\!\!\!^{-} = \lambda/2\pi = c/\omega$.

For instance, if the line width is $\Delta\tilde{\nu} = 100\ \mathrm{cm}^{-1}$ we obtain $T_2 = 0.5 \times 10^{-13}$ s. Since $\tau_{cl} \approx 10^{-8}$ s then for $f_b = 1$ the characteristic distance R_0 for which $W\tau_a = 1$ is estimated as $R_0 \approx 0.02\lambda$. Note that according to eq. (1.25) $W\tau_a$ is independent of f_a.

Let us now estimate the distance at which the interaction can be regarded as strong, that is, for which the quantity $(2|V_{12}|^2/\hbar^2)T_2\tau_b$ is of the order of unity. Clearly, this distance is $R \approx R_0(\tau_b/\tau_a)^{1/6}$. Therefore, in the two-level approximation if $\tau_b \sim \tau_a$ this distance is comparable to R_0. Indeed, as mentioned above transfer often involves vibrational sublevels and then τ_b is the vibrational relaxation time which is, typically, of the order of 10^{-12}–10^{-13} s, that is, $\tau_b \sim T_2$. In this case the distance of strong interaction is smaller than R_0 by a factor of 5–10. The typical experimental values of R_0 are about 50 Å. Therefore, it should be borne in mind that the approximation of weak interaction can be invalid at distances of the order of the molecule's diameter in solutions or the lattice parameter in crystals.

If $\tau_b \sim T_2$ the condition of weak interaction can be written as $\Omega \ll \Delta\omega$, that is, the splitting due to the interaction must be much smaller than the line widths.

3. *Transitions in the quasicontinuous spectrum; Förster's theory*

The two-level model enables us to employ the density matrix approach without special difficulties and to analyze various approximations in the theory of energy transfer without using the perturbation theory. The density matrix method can be extended to more complicated models, for instance, those that take into account vibrational states (Davydov and Serikov 1972).

However, another approach, first put forward by Förster in 1948 (Förster 1948, 1949, 1951, 1959) proves to be more fruitful for the case of wide spectra when the transverse relaxation time is short or when the two-level system approximation is not valid.

Förster suggested treating wide spectra as continuous ones and using the well-known equation of perturbation theory for the transition probability in the continuous spectrum (Landau and Lifshits 1965, sect. 43):

$$dW = \frac{2\pi}{\hbar} |V_{12}|^2 \delta(\Delta E), \tag{1.26}$$

Here $\Delta E = E_a' - E_a - (E_b' - E_b)$ as shown in fig. 1.1. Equation (1.26) implies that the transfer probability during the transverse relaxation time is small corresponding to the "case of very weak interaction" according to Förster (Förster 1965, Dexter and Förster 1969), so that the coherent effects cannot

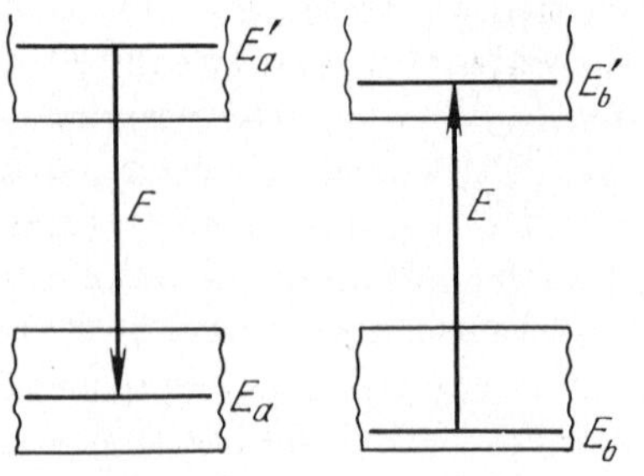

Fig. 1.1. Energy levels in Förster's theory.

be taken into account. Equation (1.26) implies also that very fast relaxation (for instance, via the vibrational states) occurs in the final state so that the population of the final state of transfer can be ignored. This means that the reverse transfer, if it is energetically feasible, must be regarded as being independent of the direct transfer. It should be emphasized that eq. (1.26) is fully applicable to transfers both between identical and different molecules.

Equation (1.26) implies integration over energies of the initial and final states under the condition $\Delta E = 0$, that is, the condition of energy conservation.

Using Förster's approach we shall assume that the wave functions of the initial and final states can be written as

$$\Psi_1 = \varphi'_a \varphi_b \Phi'_a(E'_a)\Phi_b(E_b),$$
$$\Psi_2 = \varphi_a \varphi'_b \Phi_a(E_a)\Phi'_b(E'_b), \qquad (1.27)$$

where φ are the electronic wave functions, and Φ are the vibrational wave functions. Then the matrix element of the interaction energy is

$$V_{12} = \langle \Psi_1 | V | \Psi_2 \rangle = M_{ab} \langle \Phi'_a(E'_a) | \Phi_a(E_a) \rangle \langle \Phi'(E'_b) | \Phi(E_b) \rangle$$
$$= M_{ab} S_a(E'_a, E_a) S_b(E_b, E'_b), \qquad (1.28)$$

where

$$M_{ab} = \chi |\boldsymbol{p}_a||\boldsymbol{p}_b|/R^3, \qquad (1.29)$$

is the matrix element of the dipole–dipole interaction, $|\boldsymbol{p}_a|$ and $|\boldsymbol{p}_b|$ are the matrix elements of the dipole moment, and S_a and S_b are the overlapping integrals of the vibrational wave functions.

The condition $\Delta E = 0$ yields

$$E'_a - E_a = E'_b - E_b = E.$$

Assume now that $g'(E'_a)$ is the energy distribution of the molecules a in the excited state and $g(E_b)$ is the energy distribution of the molecules b in the lower state. For instance, these may be equilibrium distributions. In order to find the total transfer probability W we have to integrate eq. (1.26) over these distributions and the energy difference $E = \hbar\omega$. Substitution of eq. (1.28) yields:

$$W = \frac{2\pi}{\hbar} \int M^2_{ab} \int g'(E'_a) S^2_a(E'_a, E'_a - E)\, dE'_a$$
$$\times \int g(E_b) S^2_b(E_b, E_b + E)\, dE_b\, dE$$
$$= \frac{2\pi\chi^2}{R^6} \int \left[p^2_a \int g'(E'_a) S^2_a(E'_a, E_a - E)\, dE'_a \right]$$
$$\times \left[p^2_b \int g(E_b) S^2_b(E_b, E_b + E)\, dE_b \right] d\omega. \qquad (1.30)$$

The remarkable fact making it possible to express the transfer probability in terms of experimentally measured quantities is that the terms in the square brackets in eq. (1.30) are proportional, respectively, to the spectra of radiation of the molecule a (the energy donor) and absorption of the molecule b (the energy acceptor). Indeed, these terms can be expressed as distributions over the transition frequencies $\omega = E/\hbar$:

$$p_a^2 \int g'(E'_a) S_a^2(E'_a, E'_a - E)\, dE'_a = \frac{1}{\hbar} p_a^2 G_a(\omega),$$

$$p_b^2 \int g(E_b) S_b^2(E_b, E_b + E)\, dE_b = \frac{1}{\hbar} p_b^2 G_b(\omega). \tag{1.31}$$

Here the normalization condition is $\int G_a\, d\omega = \int G_b\, d\omega = 1$. On the other hand, we have the well-known relationships between the lifetime τ, the absorption index $\mu(\omega)$ and the matrix element of the dipole moment of the transition (see, for instance, Pantell and Puthoff 1969, El'yashevich 1962):

$$F_a(\omega) = \frac{4\omega^3}{3\hbar c^3} p_a^2 \tau_a G_a(\omega),$$

$$\mu_b(\omega) = \frac{4\pi^2 \omega}{3\hbar c} N_b p_b^2 G_b(\omega). \tag{1.32}$$

where $F_a(\omega)$ is the normalized radiation spectrum expressed as the number of quanta per unit frequency range (that is, $\int F_a(\omega)\, d\omega = 1$), $\mu_b(\omega)$ is the absorption coefficient in cm^{-1}, that is, the absorption index in the Beer's law $I(z) = I_0 \exp(-\mu z)$. N_b is the number of the acceptor molecules per cm^3. (Note that the first of eqs. (1.32) can be rewritten as:

$$\frac{1}{\tau_a} = \frac{4p_a^2}{3\hbar c^3} \int F_a(\omega)\, d\omega \Big/ \int \omega^{-3} F_a(\omega)\, d\omega,$$

so that it is identical to the equation derived by Strickler and Berg (1962) not only for narrow but also for wide spectra.)

Substitution of eqs. (1.32) into eq. (1.30) yields

$$W = \frac{9\chi^2 c^4}{8\pi N_b \tau_a R^6} \int F_a(\omega) \mu_b(\omega) \frac{d\omega}{\omega^4}. \tag{1.33}$$

The characteristic distance R_0 at which the transfer probability equals the probability of spontaneous transition in the donor molecule is given by $W\tau_a = 1$. Equation (1.33) yields

$$R_0^6 = \frac{9\chi^2 c^4}{8\pi N_b} \int F_a(\omega) \mu_b(\omega) \frac{d\omega}{\omega^4}. \tag{1.34}$$

To find an order-of-magnitude of R_0 assume that the molecules a and b are identical and the spectral band corresponding to the given transition is homogeneously broadened, that is, it has a Lorentz shape with full width at half maximum $\Delta\omega = 2\Gamma$ and no Stokes shift. This estimate will, clearly, give the maximum transfer probability. We obtain

$$G_a(\omega) = G_b(\omega) = \frac{1}{\pi}\frac{\Gamma}{(\omega - \omega_0)^2 + \Gamma^2}. \tag{1.35}$$

Assume that $\Gamma \ll \omega_0$. Then, when integrating we can take out ω^4 from under the integral. Since in this case $\int G_a^2(\omega)\, d\omega = 1/2\pi\Gamma = T_2/\pi$ we obtain, assuming $\chi^2 = \frac{2}{3}$,

$$R_0^6 = \frac{3}{4}\lambda^6 \frac{T_2}{\tau_{cl}} f_b,$$

that is, exactly the same result as eq. (1.25) derived for the two-level system model. It is only natural that for the Lorentz line shape the results given by the Förster model and the two-level model are identical (Trifonov 1971). However, the Förster theory can be also applied to bands which do not have the Lorentz shape, for instance, to wide spectra having Stokes shifts. This theory is applicable under the conditions of weak interaction and fast relaxation as discussed above. Moreover, it should be borne in mind that the Förster's theory implies independence of the vibrational wave functions of the interacting molecules. This condition may be violated for molecules which are sufficiently close that their vibrational states are coupled (see ch. 4 sect. 10).

4. *The effect of the polarizability of the medium*

Up to now we assumed that the interacting molecules were in vacuum. In fact, the theory can be applied to molecules in a condensed medium. The role played by the medium is discussed in detail in chs. 3 and 4. Here we shall make only preliminary notes needed to derive equations suitable for comparing with the experimental data.

If the distance between molecules is considerably larger than the intermolecular distances and the medium is transparent in the given spectral region we can take into account the role of the medium by introducing into the interaction energy (1.18) the dielectric constant ϵ for the transition frequency, that is, the squared index of refraction. But the electric field acting on a molecule in the medium differs from the mean macroscopic field (the "effective (local) field"). The well-known Lorentz model gives the following relationship between the effective and the mean field:

$$\mathbf{E}_{eff} = \mathbf{E}(\epsilon + 2)/3.$$

The effective field will be discussed in more detail in ch. 3; here we shall just use a simple model to show that the medium can be taken into account by introducing $\epsilon^2 = n^4$ into the denominator of eq. (1.33), that is, the interaction is weakened by a factor of n^2, so that eq. (1.33) does not explicitly reflect any other influence of the effective field.

Let us consider the interaction between two dipoles. One dipole with the moment $\boldsymbol{p}_1$ is in the medium with the dielectric constant ϵ and corresponds to the Lorentz model while the other dipole with the moment $\boldsymbol{p}_2$ is in the centre of a spherical cavity of the radius R_1 in the medium. The second dipole will serve as an auxiliary model for which we can easily find accurate expressions both for the "acting field" and the field generated by the dipole in the medium.

Indeed, we can readily see that the field of the second dipole outside the spherical cavity, that is, in the medium with the dielectric constant ϵ, is a field of an effective dipole with the moment $\boldsymbol{p}_2^{\text{eff}}$. In order to find the ratio $\boldsymbol{p}_2^{\text{eff}}/\boldsymbol{p}_2$ we have to solve the appropriate electrostatic problem, that is, to find the fields satisfying the boundary conditions at the spherical surface. In doing this we must take into account, apart from the field of the dipole $\boldsymbol{p}_2$, the uniform field in the cavity produced by the polarization charge on the spherical surface. Thus, the potential outside the sphere is,

$$\varphi_{\text{e}}(\boldsymbol{R}) = \boldsymbol{p}_2^{\text{eff}}\boldsymbol{R}/R^3,$$

and the potential inside the sphere is,

$$\varphi_{\text{i}}(\boldsymbol{R}) = \boldsymbol{p}_2\boldsymbol{R}/R^3 - B(\boldsymbol{p}_2\boldsymbol{R}),$$

where B is a constant describing the uniform field. We have the following conditions of continuity for the potential and the normal component of the electric flux density at the spherical surface:

$$\varphi_{\text{e}}(R_1) = \varphi_{\text{i}}(R_1),$$

$$\epsilon \left.\frac{\partial \varphi_{\text{e}}}{\partial R}\right|_{R=R_1} = \left.\frac{\partial \varphi_{\text{i}}}{\partial R}\right|_{R=R_1}.$$

Eliminating B from these equations we obtain

$$p_2^{\text{eff}} = p_2 \frac{3\epsilon}{2\epsilon + 1}$$

In a similar way, we find that the external uniform field E produces inside a hollow sphere the uniform field,

$$E_{\text{sph}} = E \frac{3\epsilon}{2\epsilon + 1}$$

(see, for instance, Landau and Lifshits 1962). Thus, we have a relationship between the effective dipole p_2^{eff} and the field inside the spherical cavity E_{sph} which can be regarded as the acting field in this model.

Let us show that a similar relationship can be found for the dipole $\boldsymbol{p}_1$ for which the acting field is expressed according to Lorentz. According to the general electrodynamic principle of reciprocity following from the Maxwell equations for any two dipoles we have the relationship

$$\boldsymbol{p}_1\boldsymbol{E}_{12} = \boldsymbol{p}_2\boldsymbol{E}_{21},$$

where $\boldsymbol{E}_{12}$ and $\boldsymbol{E}_{21}$ are, respectively, the fields produced by the second dipole at the point of the first dipole and vice versa. Evidently, these fields should be regarded as corresponding to the effective fields (though $\boldsymbol{p}_1$ and $\boldsymbol{p}_2$ are, of course, the actual, rather than effective, moments). Assuming that the distance between the dipoles is $R \gg R_1$ we can take

$$E_{12}^{\text{eff}} \sim p_2^{\text{eff}} \quad \text{and} \quad E_{21}^{\text{eff}} \sim p_1^{\text{eff}}.$$

Assuming $p_1^{\text{eff}} = xp_1$ we can write the reciprocity principle in the form:

$$p_1p_2\frac{\epsilon+2}{3}\frac{3\epsilon}{2\epsilon+1} = p_2p_1x\frac{3\epsilon}{2\epsilon+1}$$

(here we have taken into account that $E_{12}^{\text{eff}} \sim \frac{1}{3}(\epsilon+2)p_2^{\text{eff}}$ and $E_{21}^{\text{eff}} \sim [3\epsilon/(2\epsilon+1)]p_1^{\text{eff}}$). Hence, we have $x = p_1^{\text{eff}}/p_1 = \frac{1}{3}(\epsilon+2)$. Thus, for the dipole $\boldsymbol{p}_1$ corresponding to the more realistic Lorentz model we obtained the same relationship between the effective dipole and the effective field as that derived for the above auxiliary model.

Now let us consider the interaction between two dipoles with the effective fields corresponding to the Lorentz model. Then the squared matrix element of the interaction contains the factor,

$$\frac{1}{\epsilon^2}\left(\frac{\epsilon+2}{3}\right)^4 = \frac{1}{n^4}\left(\frac{n^2+2}{3}\right)^4.$$

However, according to eq. (1.32), the experimental values of the radiative lifetime τ_a and the absorption index μ_b enter eq. (1.33). The relationship between these parameters and the matrix element of the dipole moment depends also on the index of refraction (Pantell and Puthoff 1969, El'yashevich 1962). The first of the eqs. (1.32) must contain the factor $n[(n^2+2)/3]^2$ and the second equation must contain the factor $(1/n)\times[(n^2+2)/3].^2$. Finally, when we substitute eqs. (1.32) into eq. (1.30) only n^4 remains in the denominator of eq. (1.33).

We should stress once more that this result is due to the use of experimental values of τ and $\mu(\omega)$ in eq. (1.33). If we tried to use a theoretical model for the molecular parameters then the transfer probability would, certainly, depend on the effective field via the factor $(1/\epsilon^2)[(\epsilon+2)/3]^4$.

Let us now write down the equations of the Förster theory which are suitable for comparison with experimental results. Introduce into eq. (1.33) the experimental value of the donor lifetime τ_0. This lifetime can differ from the radiative lifetime τ_a if the quantum yield η_0 of the donor luminescence differs from unity. In this case we have $\tau_a = \tau_0/\eta_0$. Another useful parameter is the effective cross section of acceptor absorption $\sigma(\omega) = \mu_b(\omega)/N_b$ with the dimension cm^2. Finally, we can rewrite eqs. (1.33) and (1.34) in the form:

$$W = \frac{3}{2}\chi^2 \frac{R_0^6}{\tau_0 R^6}, \tag{1.36}$$

$$R_0^6 = \frac{3}{2(2\pi)^5} \frac{\eta_0}{n^4} \int F(\tilde{\nu})\sigma(\tilde{\nu}) \frac{d\tilde{\nu}}{\tilde{\nu}^4}, \tag{1.37}$$

where $\tilde{\nu} = \omega/2\pi c$ is the frequency in cm^{-1} and R_0 is the distance at which $W\tau_0 = 1$ for $\chi^2 = \frac{2}{3}$ (if the dispersion of the index of refraction cannot be neglected in the given spectral region the factor $n^{-4}(\omega)$ should be placed under the integral).

Equation (1.37) is sometimes written in the form

$$R_0^6 = 5.86 \times 10^{-25} \frac{\eta_0}{n^4} \int F(\tilde{\nu})\epsilon(\tilde{\nu}) \frac{d\tilde{\nu}}{\tilde{\nu}^4}, \tag{1.38}$$

where $\epsilon(\tilde{\nu})$ is the molar extinction coefficient.

5. *Dipole–quadrupole and exchange interactions*

In the majority of the known cases energy transfer is due to the dipole–dipole interaction. If one or both of the dipole transitions are forbidden, energy transfer can be caused by the interaction of higher transition multipoles or by the exchange interaction. For instance, energy transfer involving triplet states of molecules is frequently caused by the exchange interaction determined by overlapping of the wave functions.

To illustrate interaction between multipoles we shall discuss the dipole–quadrupole interaction when a dipole transition occurs in the donor and a quadrupole transition in the acceptor (Dexter 1953).

The electrostatic potential of the quadrupole can be expressed in terms of the quadrupole moment tensor $D_{\alpha\beta}$ as,

$$\varphi^{(2)} = D_{\alpha\beta} e_\alpha e_\beta / 2R^3,$$

where $\boldsymbol{e} = \boldsymbol{R}/R$ is the unit radius vector and summation over repeated subscripts is implied (Landau and Lifshits 1962, sect. 41). Hence, the components of the electric field of the quadrupole are

$$E_i = -\frac{1}{R^4}(\tfrac{5}{2}D_{\alpha\beta}e_\alpha e_\beta e_i - D_{i\alpha}e_\alpha).$$

The energy of the interaction between the dipole and the quadrupole is,

$$V_{\rm dq} = -p_\alpha E_\alpha = \frac{1}{R^4}(\tfrac{5}{2}D_{\alpha\beta}e_\alpha e_\beta p_\gamma e_\gamma - D_{\alpha\beta}p_\beta e_\alpha). \tag{1.39}$$

The tensor $D_{\alpha\beta}$ is symmetric and the sum of its diagonal elements is zero (Landau and Lifshits 1962).

Selecting the system of coordinates in order to obtain the diagonal form of the tensor and axial symmetry (Landau and Lifshits 1962) we have,

$$D_{zz} = -2D_{xx} = -2D_{yy}.$$

We shall limit ourselves to this case and denote $D_{zz} = D$. Then the energy of the dipole–quadrupole interaction can be written as,

$$V_{\rm dq} = \chi_{\rm dq}pD/R^4. \tag{1.40}$$

where $\chi_{\rm dq}$ depends on the orientation of the dipole with respect to the chosen system of coordinates. Averaging over various orientations yields $\chi^2_{\rm pq} = \frac{1}{4}$.

Now we can proceed with calculations similar to those made above to derive eq. (1.33) from eq. (1.26) for the dipole–dipole interaction. However, it is better here to replace the absorption index with the radiative lifetime $\tau_{\rm b}^{(\rm q)}$ of the acceptor which is easier to measure experimentally than the very weak quadrupole absorption. This lifetime is related to the quadrupole moment by

$$\frac{1}{\tau_{\rm b}^{(\rm q)}} = \frac{1}{90\hbar}\frac{\omega^5}{c^5}D^2, \tag{1.41}$$

(see, Landau and Lifshits 1962, El'yashevich 1962). We obtain

$$F_{\rm b}(\omega) = \frac{1}{90\hbar}\frac{\omega^5}{c^5}D^2\tau_{\rm b}^{(\rm q)}G_{\rm b}(\omega), \tag{1.42}$$

where $F_{\rm b}(\omega)$ is the normalized absorption spectrum of the acceptor, that is, $\int F_{\rm b}(\omega)\,{\rm d}\omega = 1$. Hence, the transfer probability averaged over orientations is

$$W_{\rm dq} = \frac{135\pi\hbar c^8}{4\tau_{\rm a}\tau_{\rm b}^{(\rm q)}R^8}\int F_{\rm a}(\omega)F_{\rm b}(\omega)\frac{{\rm d}\omega}{\omega^8}. \tag{1.43}$$

A similar expression for the transfer probability for the dipole–dipole interaction is

$$W_{dd} = \frac{3\pi\hbar c^6}{4\tau_a\tau_b R^6} \int F_a(\omega)F_b(\omega)\frac{d\omega}{\omega^6}.$$

Now we can write the ratio,

$$\frac{W_{dq}}{W_{dd}} \approx \frac{45}{4\pi^2}\left(\frac{\lambda}{R}\right)^2 \frac{\tau_b}{\tau_b^{(q)}}. \tag{1.44}$$

The ratio between the lifetimes of the dipole and quadrupole is

$$\frac{\tau_b}{\tau_b^{(q)}} \sim 3\left(\frac{a}{\lambda}\right)^2,$$

where a is of the order of the atomic size (El'yashevich 1962). Hence, we obtain the ratio,

$$W_{dq}/W_{dd} \sim 3(a/R)^2. \tag{1.45}$$

This result agrees with the remarks made at the beginning of sect. 2 on the parameter for expanding the interaction in multipoles in the case of energy transfer.

In a similar way we can find the expression for the quadrupole–quadrupole interaction. Dexter (1953) (see also, Axe and Weller 1964, Brown et al. 1965, Imbush 1967) has shown that the quadrupole–quadrupole interaction is the dominant interaction in some media. The energy of this interaction is $V_{qq} \sim 1/R^5$ so that the respective transfer probability is $W_{qq} \sim 1/R^{10}$. If we take into account the mechanisms of virtual excitons and phonons we obtain resonance interaction of an even shorter range (see ch. 4 sect. 7). According to Orbach and Tachiki (1967) in ruby the energy transfer via the Cr^{3+} ions (R_1 emission line) for the nearest neighbours is determined by the interaction due to exchange of virtual phonons, rather than quadrupole–quadrupole interaction.

Let us consider now the expression for the probability of transfer due to exchange interaction which can be derived from the weak interaction theory. In this case the probability of the exchange transfer can also be expressed in terms of the integral of the spectral overlap (Dexter 1953). However, in contrast to the multipole interaction case, the transfer probability cannot be obtained from the optical properties of the interacting molecules. The probability of transfer due to exchange interaction can be written as (Dexter 1953):

$$W_{exc} = \frac{2\pi}{\hbar} Z^2 \int F_a(\omega)F_b(\omega)\, d\omega.$$

It is often assumed that the parameter Z determining the transfer probability is exponentially dependent on the distance between molecules since it is related to the overlapping of the wave functions of the interacting molecules (Dexter 1953).

In conclusion, it should be noted that the remarks on the effect of the index of refraction of the medium given in sect. 4 cannot be applied to the multipole and exchange interactions since the derivation is exclusively based on dipole–dipole interaction. We could extend these remarks to the cases of the multipole and exchange interactions but it is doubtful whether a treatment based on the concept of continuous medium would yield any meaningful results for these interactions which are only significant at small distances. A better approach is to use the microscopic theory for these cases (see ch. 3).

CHAPTER 2

Energy Transfer in Solutions

1. Dipole–dipole transfer in rigid solutions

1.1. Kinetics of decay of the number of donors.

To find the macroscopic effects of energy transfer we must average the transfer probability over the totality of molecules donating and accepting energy. The results of such averaging depend on the assumptions on the relative positions of the molecules and their variation during the lifetime of the excited state of the donor molecules.

Consider a solution in a solvent transparent in the given spectral range, and assume that the distribution of the energy donors and acceptors in the solution is statistically uniform. If the viscosity of the solvent is sufficiently high neither the distances between the molecules nor their orientations will change during the lifetime of the excited state. These solutions are referred to usually as rigid solutions; they can be both crystalline and amorphous.

In most cases the concentration of the excited donors can be assumed to be very low, that is, the excitation source is sufficiently weak. Then we can consider each excited molecule independently of the other molecules and averaging reduces to taking into account various concentrations of acceptors with respect to donors.

As for the concentration of the acceptor molecules, in general, it is not necessarily low. The only restriction applied to the acceptor concentration in the following discussion is due to the fact that we ignore the interaction between the acceptor molecules and their finite volume and, hence, assume that the presence of one acceptor molecule near a donor molecule does not prevent the presence of another acceptor. In other words, we do not take into account that the number of positions in the nearest coordination spheres around the donor is finite. Under these assumptions the donor–acceptor interaction is not limited to binary interaction of each donor with only one acceptor.

Therefore, the averaging of the transfer rate over the totality of the donor and acceptor molecules has some special features. Indeed, an increase in the number of acceptors in the vicinity of a given donor implies the necessity to sum the transfer rate over the acceptors. On the other hand, we must average the rate of energy transfer from various donors and, in doing so, carry out the averaging over various configurations of acceptors with respect to donors. This means averaging over various exponential dependences describing the decrease of the number of donors with various surrounding acceptor configurations, rather than averaging over the transfer rates. A procedure first suggested by Förster (1948, 1949, 1959) and its modifications (Galanin 1955, Rozman 1958) makes it possible to take into account both these effects by carrying out summation and averaging prior to integration of the kinetic equations. We shall follow the approach suggested by Samson (1962) and Agrest et. al. (1969).

Assume that at first we have one donor and one acceptor. The probability that by the moment t no transfer has occurred (if we ignore spontaneous decay for the present, so that we refer to the probability that the donor molecule is still excited) and the acceptor is in the volume element $\mathrm{d}V$ at the distance R from the donor, is

$$\rho_1(R, t)\frac{\mathrm{d}V}{V}, \tag{2.1}$$

where V is the volume under consideration. In this chapter we shall always assume that the conditions of applicability of the Förster theory are satisfied. Therefore, for the stationary molecules we have

$$\rho_1(R, t) = \exp[-W(R)t], \tag{2.2}$$

where $W(R)$ is defined for the dipole–dipole transfer according to eq. (1.36).

Let us now consider one donor and N_A acceptors which are in the volume elements $\mathrm{d}V_1, \mathrm{d}V_2, \ldots, \mathrm{d}V_i, \ldots, \mathrm{d}V_{N_A}$. Then the respective probability is the product of probabilities if we assume that the acceptors do not interact with each other (this corresponds to summation of the transfer rates over the acceptors):

$$\prod_{i=1}^{N_A} \rho_1(R_i, t)\frac{\mathrm{d}\,V_i}{V^{N_A}}. \tag{2.3}$$

Now we have to average this probability over the totality of the donors. If we assume random distribution of the acceptors with respect to the donors and ignore the finite volume of the acceptors then the averaging over a large number of donors can be reduced to integration over all $\mathrm{d}V_i$ since the difference between donors is determined only by the relative arrangements of acceptors around them. All these integrals will be identical since the presence of one acceptor does not influence the presence of another acceptor according to the above assumptions. Hence, the mean probability that no transfer has occurred is

$$\rho(t) = \left(\frac{1}{V}\int_V \rho_1(R, t)\,\mathrm{d}V\right)^{N_A}.$$

When $V \to \infty$ this expression can be transformed by making use of the fact that $\rho_1(R, t)$ is smaller than unity only in a small region around $R = 0$. Denoting

$$x = 1 - \frac{1}{V}\int_V \rho_1(R, t)\,\mathrm{d}V,$$

we have $x \ll 1$ for $V \to \infty$ and, hence, $1 - x \approx \mathrm{e}^{-x}$. Now we obtain,

$$\rho(t) = \left[\exp\left(\frac{1}{V}\int \rho_1\,\mathrm{d}V - 1\right)\right]^{N_A} = \exp\left[\frac{N_A}{V}\left(\int \rho_1\,\mathrm{d}V - V\right)\right] = \mathrm{e}^{-H(t)},$$

where,

$$H(t) = c_A \int (1 - \rho_1)\,\mathrm{d}V. \tag{2.4}$$

Here $c_A = N_A/V$ is the number of acceptors in the unit volume.

The function $\rho_1(R, t)$ satisfies the equation

$$\frac{\partial \rho_1}{\partial t} = -W(R)\rho_1,$$

and is the normalized acceptor concentration under the assumption that all donors are placed at the origin of coordinates. It should be stressed that the spatial distribution of acceptors in such a "superposed" configuration is, generally speaking, not uniform. The distribution is uniform only at the initial moment after the instantaneous excitation impulse (δ excitation) and is not uniform under the steady-state conditions. Indeed, immediately after a transfer to an acceptor it should disappear from the superposed configuration since the corresponding donor in the real configuration has lost the excitation.

Evidently, the configurations in which the acceptors are closer to the origin (that is, the donor–acceptor distance is shorter) and their orientation is more favourable will be the first to be eliminated. This "depletion effect" results in a nonexponential decay of the number of donors and is determined by the character of excitation. It is similar to that suggested at first in the theory of phosphorescence of crystal phosphors (Antonov-Romanovsky 1936, 1966). For instance, the decay after δ excitation differs from the decay after the cut-off of a steady-state excitation (Galanin 1951, 1960).

Now, if we take into account that the number of donors decreases owing to not only energy transfer but also by spontaneous emission we can write:

$$n(t) = n_0 \exp(-t/\tau_0)\rho(t) = n_0 \exp[-t/\tau_0 - H(t)]. \tag{2.5}$$

When we calculate $H(t)$ from eq. (2.4) we must bear in mind that the transfer rate depends not only on the distance R but also on the relative orientation of the dipole moments. We can assume that the dipole moment of the donor is parallel to the z-axis and carry out averaging over orientations of the acceptors.

Let us now calculate the function,

$$H(t) = c_A \int\int (1 - e^{-Wt})\, dV\, d\Omega,$$

where $d\Omega$ is the element of the solid angle determining the direction of the dipole moment of the acceptor and the integral is taken over the total volume and all directions of the dipole moment and radius vector. If the transfer probability is given by eq. (1.36) we obtain:

$$H(t) = \frac{c_A}{6}\sqrt{\frac{3}{2}}R_0^3\sqrt{\frac{t}{\tau_0}}\int_0^\infty (1 - e^{-y})\, y^{-3/2}\, dy \int \sqrt{\chi^2}\, d\Omega. \tag{2.6}$$

Here we made the substitution

$$y = \tfrac{3}{2}\chi^2 t R_0^6/\tau_0 R^6,$$

and the integration limit $y = \infty$ (or $R = 0$) implies that the radii of the molecules can be ignored (see sect. 1.5).

The integral over y is $2\sqrt{\pi}$ and integration over angles should be carried out in two stages – at first, we average the projection of the acceptor dipole on the electric field of the donor (see fig. 2.1), and then we average the angular dependence of the dipole field ($\sqrt{\chi^2} = \sqrt{3\cos^2\vartheta_1 + 1}$) over the

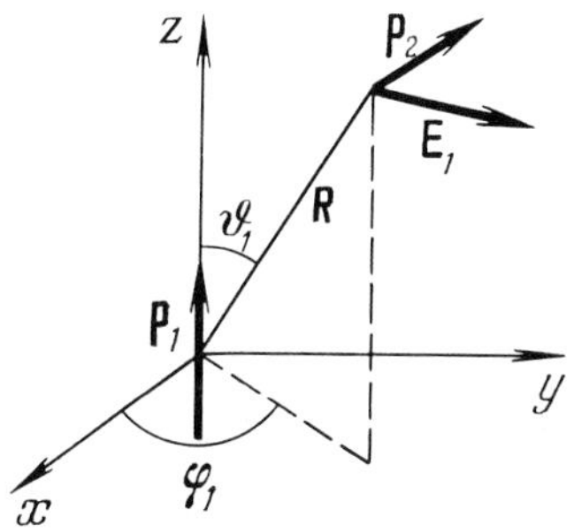

Fig. 2.1. Scheme of averaging of the dipoles p_1 and p_2 over their mutual orientation.

direction of the radius vector:

$$(1)\quad \frac{1}{4\pi}\int_{-\pi/2}^{\pi/2}\int_{0}^{2\pi} \cos\vartheta_2 \sin\vartheta_2 \, d\vartheta_2 \, d\varphi_2 = \tfrac{1}{2};$$

$$(2)\quad \int_{-\pi/2}^{\pi/2}\int_{0}^{2\pi} \sqrt{3\cos^2\vartheta_1 + 1}\, \sin\vartheta_1 \, d\vartheta_1 \, d\varphi_1 = 2\left[1 + \frac{1}{2\sqrt{3}} \ln(2+\sqrt{3})\right] 2\pi.$$

The numerical value of the term in square brackets is 1.38. Hence, we have:

$$H(t) = \frac{c_A}{3}\sqrt{\frac{3}{2}} R_0^3 \sqrt{\frac{t}{\tau_0}} \sqrt{\pi} \cdot 2\pi \cdot 1.38 = 2q_1 \sqrt{\frac{t}{\tau_0}},$$

where

$$q_1 = \frac{1.38}{\sqrt{6}} \pi^{3/2} R_0^3 c_A \approx 3.14\, R_0^3 c_A, \tag{2.7}$$

Thus, we obtain the following equation for the decay of the number of donors:

$$n(t) = n_0 \exp(-t/\tau_0 - 2q_1\sqrt{t/\tau_0}). \tag{2.8}$$

(The respective equation for the multipole interaction is

$$n(t) = n_0 \exp[-t/\tau_0 - 2s(t/\tau_0)^{3/m}],$$

where $s = \Gamma(1-3/m)\,\frac{2}{3}\pi R_0^3 c_A$ and $m = 8$ or 10 for the dipole–quadrupole or quadrupole–quadrupole interaction (Sveshnikov and Shirokov 1962).)

In some cases the orientational motion can be considerably faster than the translational motion so that averaging over orientations can occur during the lifetime of the excited state. In such cases we should take $\chi^2 = \frac{2}{3}$ at the beginning of calculations. Then we obtain eq. (2.8) in which q_1 is replaced by (Maksimov and Rozman 1962):

$$q = \frac{q_1}{0.845} = \tfrac{2}{3}\pi^{3/2} R_0^3 c_A \approx 3.71\, R_0^3 c_A. \tag{2.9}$$

Equation (2.8) was derived first by Förster (1948, 1949, 1959) and later by other authors (Galanin 1955, Rozman 1958). If we know the decay dependence for the δ excitation we can find the decay dependence for any other type of excitation. For instance, Galanin (1951, 1960) calculated this dependence for steady-state excitation and the result proved to be much closer to an exponential dependence than eq. (2.8).

1.2. Quenching kinetics for the acceptor

In the cases when the acceptor molecules exhibit luminescence it is interesting to consider the kinetics of variation of the number of the excited acceptor molecules, that is, the kinetics of sensitized luminescence.

Assume that the excitation impulse is very short (δ excitation) and the acceptor molecules acquire energy only by means of energy transfer (actually, energy transport is frequently due to reabsorption of donor radiation as discussed in ch. 6 sect. 5). Then we have the following kinetic equations for the number n_1 of the excited donor molecules and the number n_2 of the excited acceptor molecules:

$$\frac{dn_1}{dt} = -\frac{n_1}{\tau_1} - wn_1, \qquad \frac{dn_2}{dt} = -\frac{n_2}{\tau_2} + wn_1, \tag{2.10}$$

Here τ_1 and τ_2 are the decay times for the donor and acceptor taking into account both radiative and radiationless transitions.

If the transfer probability w is independent of time the solutions of these equations for $\tau_2 > \tau_1$ and the initial conditions $n_1(0) = n_{10}$ and $n_2(0) = 0$ are,

$$n_1 = n_{10} \exp\left(-\frac{1}{\tau_1} - w\right)t, \tag{2.11}$$

$$n_2 = \frac{n_{10} w \tau_1 \tau_2}{\tau_2 - \tau_1 + w\tau_1\tau_2}\left[\exp\left(-\frac{t}{\tau_2}\right) - \exp\left(-\frac{1}{\tau_1} - w\right)t\right].$$

However, according to eq. (2.8) for the dipole–dipole mechanism of transfer between stationary molecules the transfer probability depends on time as,

$$w(t) = \frac{dH}{dt} = (\tau_1 t)^{-1/2} q. \tag{2.12}$$

In this case (see, Galanin 1951, 1960; Mikhelashvili et al. 1974) we obtain:

$$n_2(t) = \sqrt{\pi}\, q_2\, e^{-q_2^2}\left\{\Phi\left(q_2 + \sqrt{\left(\frac{1}{\tau_1} - \frac{1}{\tau_2}\right)t}\right) - \Phi(q_2)\right\} e^{-t/\tau_2}, \tag{2.13}$$

where

$$q_2 = q\left(1 - \frac{\tau_1}{\tau_2}\right)^{-1/2}, \qquad \Phi(q) = \frac{2}{\sqrt{\pi}}\int_0^q e^{-x^2}\, dx.$$

The acceptor decay plots are given by Galanin (1951, 1960) and Mikhelashvili et al. (1974).

1.3. *Luminescence yield and mean donor lifetime; comparison with experimental data*

Energy transfer decreases the yield of luminescence of the donor molecules and the mean lifetime of their excited state. This effect is usually referred to as resonance quenching. Characteristically, the variation of the luminescence yield is not proportional to the variation of the mean lifetime owing to the nonexponential decrease of the number of donors with time.

If we know the decay dependence for δ excitation we can find the relative quantum yield of luminescence and its mean duration from:

$$\frac{\eta}{\eta_0} = \frac{1}{\tau_0 n_0}\int_0^\infty n(t)\,\mathrm{d}t, \qquad \tau = \int_0^\infty t n(t)\,\mathrm{d}t \Big/ \int_0^\infty n(t)\,\mathrm{d}t, \tag{2.14}$$

where η_0 and τ_0 are the quantum yield and the duration of luminescence in the absence of energy transfer.

Indeed, the number of quanta emitted per unit time is the product of the radiation probability $1/\tau_a$ and the number $n(t)$ of the excited molecules. In the absence of energy transfer the number of excited molecules decreases as $n_0 \exp(-t/\tau_0)$. Therefore, we have

$$\eta_0 = \frac{1}{\tau_a}\int_0^\infty \mathrm{e}^{-t/\tau_0}\,\mathrm{d}t = \frac{\tau_0}{\tau_a}, \qquad \eta = \frac{1}{n_0 \tau_a}\int_0^\infty n(t)\,\mathrm{d}t.$$

In order to find the mean lifetime we should bear in mind that the number of quanta emitted in the time interval from t to $t+\mathrm{d}t$ is

$$\mathrm{d}n = \frac{n(t)}{\tau_a}\,\mathrm{d}t.$$

Therefore, the mean duration of luminescence is

$$\tau = \frac{\int t\,\mathrm{d}n}{\int \mathrm{d}n} = \frac{\int_0^\infty t n(t)\,\mathrm{d}t}{\int_0^\infty n(t)\,\mathrm{d}t}.$$

Using eq. (2.8) for $n(t)$ we obtain further,

$$\frac{\eta}{\eta_0} = 1 - 2q\,\mathrm{e}^{q^2}\int_q^\infty \mathrm{e}^{-x^2}\,\mathrm{d}x,$$

$$\frac{\tau}{\tau_0} = \frac{1 + q^2 - (3 + 2q^2)q\,\mathrm{e}^{q^2}\int_q^\infty \mathrm{e}^{-x_3}\,\mathrm{d}x}{1 - 2q\,\mathrm{e}^{q^2}\int_q^\infty \mathrm{e}^{-x_2}\,\mathrm{d}x}. \tag{2.15}$$

The last equation can be rewritten as:

$$\frac{\tau}{\tau_0} = \frac{3}{2} + q^2 - \frac{\eta_0}{2\eta}.$$

For low acceptor concentrations when $q \ll 1$ we can make the approximations

$$\frac{\eta}{\eta_0} = 1 - \sqrt{\pi} q, \qquad \frac{\tau}{\tau_0} = 1 - \frac{\sqrt{\pi}}{2} q, \tag{2.16}$$

showing that at low concentrations the rate of decrease of the luminscence yield is twice that of the luminescence duration. It should be remembered that q is proportional to the acceptor concentration and the third power of the characteristic distance R_0. The proportionality factor is found from eqs. (2.7) or (2.9) according to the assumptions about the orientational motion.

Since R_0 is found from the optical characteristics of the donor and acceptor molecules we can compare theoretical predictions with experiment without introducing any adjustable parameters. Such comparisons were made a number of times. Now it can be assumed that the theory and the experimental results are in complete agreement if the conditions of applicability of the theory are satisfied to a sufficient extent. The most reliable data have been obtained for two-component solutions of organic compounds of the luminescent dye type.

Figure 2.2 shows the results obtained by Galanin (1951) for resonance quenching in glycerol solutions of 3-aminophthalimide and chrysoidine. The glycerol viscosity is usually sufficiently high that displacements of

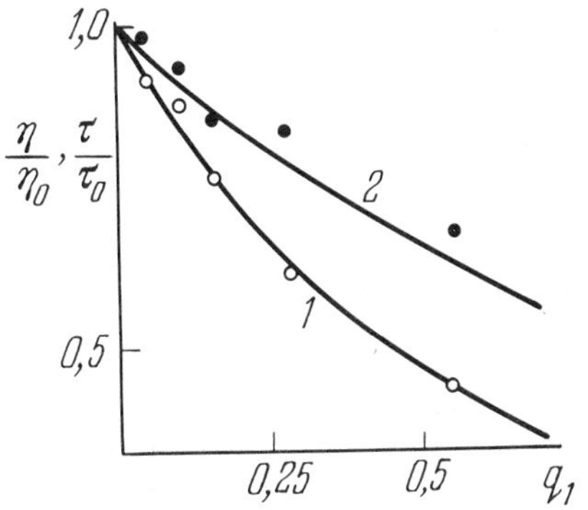

Fig. 2.2. Resonance quenching in solutions of 3-aminophthalimide and chrysoidin in glycerin as a function of the parameter q_1 (see eq. (2.7)). Solid plots present the results calculated from eq. (2.15). Plot (1) – the relative yield η/η_0. Plot (2) – relative mean duration τ/τ_0. The experimental points represent the results of Galanin (1951, 1960).

molecules are negligible during the lifetime of excited state. The characteristic distance R_0 of 43 Å was calculated from the overlapping of the luminescence spectrum of 3-aminophthalimide and the absorption spectrum of chrysoidine. Though measurements were made with solutions of comparatively low concentrations we see a good agreement between theoretical predictions and experimental data both for the relative yield and the mean lifetime (which was measured with a phase fluorimeter).

Rozman and co-workers carried out the most extensive studies of energy transfer between complex organic molecules in rigid solutions (Rozman 1961; Andreeshchev et al. 1963a, b, c). They studied solutions in plastics and carefully took into account and eliminated the energy transfer owing to reabsorption of radiation. It should be stressed that radiative transfer inevitably accompanies resonance radiationless transfer since overlapping of spectra is significant for both effects. However, the contribution of radiative transfer depends on geometric factors can be reduced, for instance, by carrying out measurments in thin layers. The contribution of radiative transfer can also be taken into account theoretically (see ch. 6 sect 5).

At first, Rozman et al. (1961) reported that the experimental transfer rates were higher by a factor of 1.8 than the predicted values (Rozman 1961, Andreeshchev et al. 1963a, b, c). Later this discrepancy was attributed to a mistake in calculations in which the factor $\sqrt{\pi} \approx 1.77$ had been omitted (Terskoi et al. 1968) which presents also a summary of other measurements.

To illustrate these results fig. 2.3 shows the yield of the donor luminescence (1, 1′, 5, 5′-tetraphenyl–3, 3′-dipyrazoline) as a function of the acceptor

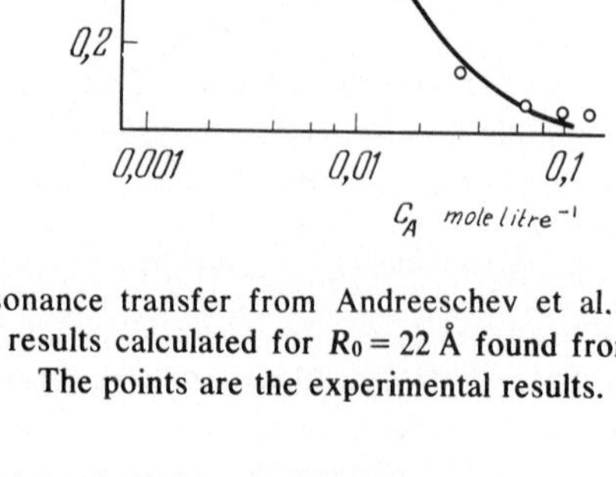

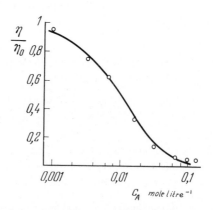

Fig. 2.3. A case of resonance transfer from Andreeschev et al. (1963c). The solid curve presents the theoretical results calculated for $R_0 = 22$ Å found from overlapping of spectra. The points are the experimental results.

concentration (1, 3, 5-triphenyl–Δ^2-pyrazoline). The solvent was polysterene. The figure shows a good agreement with theoretical predictions for variation of the acceptor concentration over two orders of magnitude. The relationship between the luminescence yield and the lifetime was also shown to be in good agreement with theory (see, Andreeshchev et al. 1963c). The acceptors used by Rozman (1961) and Andreeshchev et al. (1963a, b, c) exhibited their own luminescence making it possible to measure sensitized acceptor luminescence, apart from quenching of donor luminescence. The transfer rate found from the intensity of sensitized luminescence was also in a good agreement with the theory. The small discrepancies observed in some cases can, apparently, be attributed to the difficulties in determining accurately the contribution of the radiative transfer.

Thus, we can assume that experimental data support the Förster theory for the rigid solutions of organic compounds. This demonstrates the validity of the above assumptions, that is, that the conditions of the weak dipole–dipole interaction are met, the "forbidden" volume can be ignored (see, sect. 1.5), the molecules are stationary and their distribution is random, and so on. It can be noted also that some of these assumptions were directly verified by Stryer and Haugland (1967) and Haugland et al. (1969) for intramolecular energy transfer: They found the R^{-6} law and the dependence of transfer on the spectral overlap integral. Weakly coupled parts of a large molecule served as the donor and acceptor and distance between these parts and the R^{-6} dependence could be found from the structural considerations. In other experiments the distance between the donor and the acceptor was constant but the emission spectrum of the donor was varied by changing the solvent. When the spectral overlap integral was changed by a factor of 40 the corresponding variation of the energy transfer probability was found.

1.4. *Effect of energy transfer on polarization of luminescence (concentration depolarization of luminescence)*

If the donors and acceptors are identical molecules then energy transfer does not affect the yield of luminescence. However, energy transfer between identical molecules directly influences the degree of polarization of luminescence since the anisotropy produced by excitation rapidly disappears owing to energy transfer.

Concentration depolarization of luminescence is especially noticeable when the dipole moment of electronic transition has the same direction for absorption and emission and its relation to the molecular skeleton is fixed. This is the case in many complex organic molecules for transitions between the nondegenerate ground electronic state and the first excited state. The degree of polarization of luminescence is known in such cases to reach up to 50% in solutions with sufficiently high viscosity and low concentration.

Let us calculate polarization of luminescence after energy transfer (Galanin 1950). Assume that luminescence is excited by the polarized light with the electric vector parallel to the z-axis. Denote by I_z and I_x the intensities of luminescence with the electric vector parallel to the z- or x-axis respectively (as seen in the direction of the y-axis). A convenient parameter for describing the anisotropy of luminescence in this case is

$$r = \frac{I_z - I_x}{I_z + 2I_x} \tag{2.17}$$

(the degree of polarization $P = (I_z - I_x)/(I_z + I_x)$ is related to r by $r = 2P/(3-P)$; the parameter r is more convenient to use than the degree of polarization since r is the ratio of the polarized part of the intensity to the total intensity taking into account all three components and, therefore, it is additive, that is $r = \Sigma\, r_i I_i$ where I_i is the intensity with the anisotropy r_i). We assume that the orientation of molecules in the rigid solution is random and the orientation of the dipole moments is not changed during the lifetime of the excited state.

Since the electric vector of the exciting light is parallel to the z-axis the directions of both dipoles $\boldsymbol{p}$ and $\boldsymbol{p}'$ (donor and acceptor) should be assumed to be arbitrary (see fig. 2.4). The radius-vector $\boldsymbol{R}$ from one dipole to another is $\boldsymbol{R} = R\boldsymbol{e}$ where the components of $\boldsymbol{e}$ are $e_x = \sin\vartheta\cos\varphi$, $e_y = \sin\vartheta\sin\varphi$ and $e_z = \cos\vartheta$. According to eq. (1.33) the transfer probability is proportional to the orientation factor

$$|\chi|^2 \sim [3(\boldsymbol{pe})(\boldsymbol{p'e}) - \boldsymbol{pp'}]^2.$$

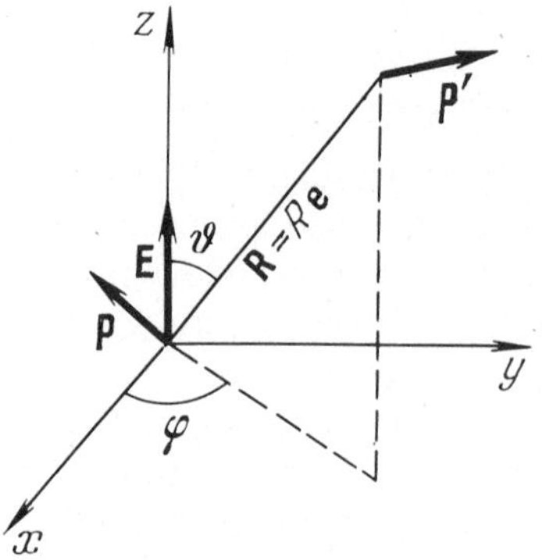

Fig. 2.4. The calculation of the effect of energy transfer on anisotropy of luminescence.

The intensity components of the luminescence after one transfer event are

$$I_z \sim \overline{p_z^2 p_z'^2 \chi^2}, \qquad I_x \sim \overline{p_z^2 p_z'^2 \chi^2}, \tag{2.18}$$

where the bar over symbols denotes averaging over all angles.

The presence of three factors in eq. (2.18) can be explained as follows. The amplitude of oscillations of the dipole $\boldsymbol{p}$ is proportional to its projection on the exciting electric field $E = E_z$, that is, p_z. The component of the electric field of the dipole $\boldsymbol{p}$ parallel to the dipole $\boldsymbol{p}'$ is proportional to $|\chi|$. Finally, the components of the amplitude of the electric field generated by the dipole $\boldsymbol{p}'$ are proportional to the projections of $\boldsymbol{p}'$ on the z- and x-axis, respectively. Replacing amplitudes with intensities we obtain eq. (2.18). Now, we have to average the expression

$$I_z \sim p_z^2 p_z'^2[3(p_x \sin\vartheta \cos\varphi + p_y \sin\vartheta \sin\varphi + p_z \cos\vartheta)$$
$$\times (p_x' \sin\vartheta \cos\varphi + p_y' \sin\vartheta \sin\varphi + p_z' \cos\vartheta) - (p_x p_x' + p_y p_y' + p_z p_z')]^2,$$

and a similar expression for I_x.

Averaging over the directions of the radius vector and the dipoles can be done independently. In doing so we should use the following table of mean values (all the expressions containing first powers of sines and cosines vanish):

$$\overline{\cos^2\vartheta} = 1/3, \qquad \overline{\cos^4\vartheta} = 1/5, \qquad \overline{\sin^2\varphi \cos^2\varphi} = 1/8,$$
$$\overline{\sin^2\vartheta} = 2/3, \qquad \overline{\sin^4\vartheta} = 8/15, \qquad \overline{\sin^2\vartheta \cos^2\vartheta} = 2/15,$$
$$\overline{\sin^2\varphi} = \overline{\cos^2\varphi} = 1/2, \qquad \overline{\sin^4\varphi} = 3/8,$$
$$\overline{p_x^2 p_x'^2 p_z^2 p_z'^2} = p^8/225, \qquad \overline{p_x^2 p_x'^2 p_z^4} = p^8/75, \qquad \overline{p_z^4 p_z'^4} = p^8/25.$$

Calculations yield,

$$I_z \sim \frac{1}{5}\left(\frac{14}{225} + \frac{12}{75} + \frac{4}{25}\right) = \frac{86}{1125},$$
$$I_x \sim \frac{1}{5}\left(\frac{13}{225} + \frac{14}{75} + \frac{3}{25}\right) = \frac{82}{1125},$$

or

$$r = \frac{I_z - I_x}{I_z + 2I_x} = \frac{2}{125} = 0.016.$$

Prior to energy transfer the intensity components are $I_{z0} \sim \overline{p_z^4}$ and $I_{x0} \sim \overline{p_z^2 p_x^2}$, so that the anisotropy is $r_0 = 0.4$. Thus, even one transfer event reduces anisotropy of luminescence by a factor of 25. Therefore, the theory of concentration depolarization typically assumes that radiation is fully depolarized even after one transfer event.

Despite this simplification, the theory of concentration depolarization is fairly complicated since it must take into account not only the energy transfer from the initially excited molecules but also the reverse transfer. In the case of Förster's mechanism the reverse transfer can be regarded as being independent of the direct transfer. The averaging procedure in this case is complicated by the fact that excitation can return to the initially excited molecule after a number of migration steps in the system of molecules.

Many authors analyzed various transfer models involving depolarization (Jablonski 1955, 1958, Ore 1959, Knox 1968, Craver and Knox 1971, Bojarski 1968, 1972, Kawski and Kaminski 1970, 1974, Rozman and Sichinava 1975, Bodunov 1976, 1977, Temenger and Pearlstein 1973): Bodunov gave the fullest treatment (Bodunov 1976, 1977). We shall illustrate the problem by the following simplified analysis.

Evidently, depolarization is maximal if excitation after leaving a donor never returns to it in the process of migration. In this case, the relative anisotropy is equal to the relative yield of the donor luminescence, that is, $r/r_0 = \eta/\eta_0$ where η is found from eq. (2.14). In fact, there is a certain "probability of return" of excitation. If the number of the initially excited molecules is n_1 and the number of molecules acquiring excitation from them is n_2 we can write down the following kinetic equations:

$$\dot{n}_1 = -n_1/\tau - wn_1 + \alpha wn_2, \qquad \dot{n}_2 = -n_2/\tau - wn_2 + \alpha wn_1, \tag{2.19}$$

Here $w = q_1(\tau t)^{-1/2}$ and the factor α takes into account the above probability of return. This factor, generally speaking, depends on time and concentration $(0 \leq \alpha \leq 1)$. Roughly approximating this factor by a constant we obtain

$$n_1(t) = n_0 \exp\left(-\frac{t}{\tau} - 2q_1\sqrt{\frac{t}{\tau}}\right)\cosh\left(2\alpha q_1\sqrt{\frac{t}{\tau}}\right), \tag{2.20}$$

and the anisotropy is

$$r = \frac{r_0}{\tau n_0}\int_0^\infty n_1(t)\,dt. \tag{2.21}$$

Note that for $\alpha = 1$, that is assuming only binary excitation (excitation always returns only to the same molecule), r/r_0 tends to $\frac{1}{2}$ with increasing concentration (Förster 1948, 1949, 1959). This resembles incoherent transfer between two identical molecules discussed in ch. 1.

Experimental studies of concentration depolarization of luminescence in high-viscosity solutions were carried out by many authors. The earlier studies were reviewed in the monographs of Feofilov (1959) and Sarzhevsky and Sevchenko (1971). Qualitatively, it is clear that depolarization is due to energy transfer. However, quantitative comparison of the theory and experimental data is complicated not only by the fact that the theory is

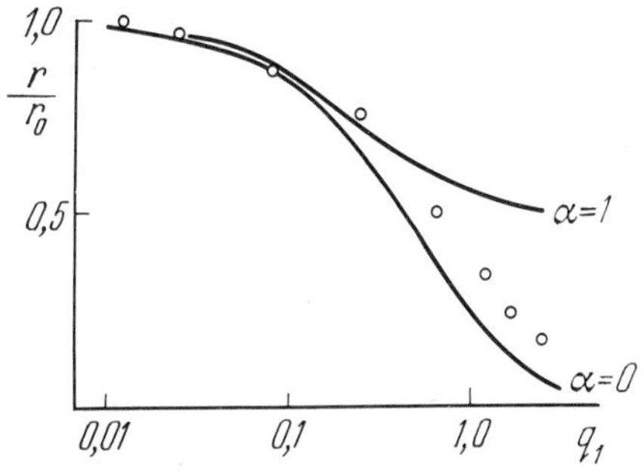

Fig. 2.5. The dependence of the relative anisotropy of luminescence on the concentration of rhodamine 6G solutions in glycerin; $q_1 = 3.14R_0^3c_A$ (see eq. (2.7)). Solid curves present the results calculated from eq. (2.20) with $\alpha = 0$ and $\alpha = 1$. The experimental points correspond to the data of Kawski (1963) for $R_0 = 52$ Å.

approximate but also by the difficulties presented by taking into account depolarization caused by secondary luminescence due to reabsorption whose degree of polarization is small (Galanin 1950). Many authors starting from Vavilov (1937, 1952) introduced corrections taking into account the effect of secondary luminescence (see, particularly, Budo and Ketskemety (1962)). Nevertheless, not all experimental results seem quite reliable. Another difficulty in comparing the theory with experimental data stems from the fact that concentration depolarization is usually accompanied with concentration quenching caused by resonance quenching by dimers. Owing to that effect the degree of polarization in some cases even starts to increase for high concentrations (Bojarski 1968, 1972). Therefore, the measurements of polarization to be compared with the theoretical predictions should be accompanied with measurements of lifetime or be done for those concentrations for which quenching is still low. Corrections should be made in the theory if the competing resonance transfer takes place.

This is illustrated by the experimental data presented in fig. 2.5 (Kawski 1963) for which no corrections for quenching have been made since the lifetime was not measured (according to the luminescence yield quenching is low). The theoretical curves for $\alpha = 0$ and $\alpha = 1$ are also given. We see that the experimental data are in a good agreement with the theory for low concentrations while for higher concentrations the experimental points lie above the curve for $\alpha = 0$ as it should be expected. This suggests that the "return effect" takes place.

1.5. "*Forbidden volume*" *and short-range order*

When we averaged the transfer probability we took $R = 0$ as the lower limit of integration for eq. (2.6). However, in reality the distance of closest approach for the donor and acceptor is $R = R_1$ where R_1 is the sum of the radii of the donor and acceptor (which determines the "forbidden volume"). We can modify calculations made to derive eq. (2.8) ignoring the possibility that at such small distances the dipole approximation can be invalid and other types of interaction (for instance, the multipole or exchange interactions) should be taken into account. In this way we can partially take into consideration the finite volume of molecules though the acceptors are still regarded as points (Sveshnikov and Shirokov 1962, Rikenglaz and Rozman 1974).

Assume for simplicity that the orientation factor is $\chi^2 = \frac{2}{3}$ (an analysis without this assumption is given by Rikenglaz and Rozman (1974). Then we obtain:

$$H(t) = \frac{2\pi}{3} R_1^3 \sqrt{y_1}\, c_A \int_0^{y_1} (1 - e^{-y}) y^{-3/2}\, dy$$

$$= \frac{4\pi R_1^3}{3} c_A \left(e^{-y_1} - 1 + 2\sqrt{y_1} \int_0^{\sqrt{(y_1)}} e^{-x^2}\, dx \right), \tag{2.22}$$

where $y_1 = (R_0/R_1)^6 (t/\tau_0)$.

The law of decay of the number of donors taking into account the forbidden volume differs from the Förster law (2.8) and only for $y_1 \gg 1$ it reduces to eq. (2.8). Even for $y_1 \geqslant 10^3$ the decay to 10% of the initial value is described by eq. (2.8) with the accuracy of better than 1% (Rikenglaz and Rozman 1974).

On the other hand, we can easily see that $y_1 \ll 1$ corresponds to the initial exponential part of the decay. Expanding eq. (2.22) in y_1 we obtain,

$$H(t) \approx wt, \tag{2.23}$$

where,

$$w = \frac{4\pi}{3} \frac{R_0^6}{\tau_0 R_1^3} c_A.$$

However, it should be borne in mind that the initial exponential part can be very short. For instance, for the dye molecules we have $R_0/R_1 \sim 5$ and the condition $y_1 \ll 1$ corresponds to $t \ll 10^{-4}\, \tau_0$ while the decay is described by eq. (2.8) starting from $t \sim 0.1\, \tau_0$.

We present below also the formulas for the yield and mean duration of donor luminescence which can be applied to solutions with low acceptor concentrations:

$$\eta = \eta_0\left[1 - w\tau_0 \arctan\left(\frac{R_0}{R_1}\right)^3\right],$$

$$\tau = \tau_0\left\{1 - \frac{w\tau_0}{2}\left[\arctan\left(\frac{R_0}{R_1}\right)^3 + \frac{R_0^3/R_1^3}{1 + R_0^6/R_1^6}\right]\right\}. \tag{2.24}$$

A comparison of eqs. (2.24) with eqs. (2.16) shows that the difference of the ratio between the initial slopes of the curves of τ and the yield from the 1:2 ratio can, in principle, be used to determine the ratio R_0/R_1 if it is not too large. A more detailed discussion of the effect of the "forbidden volume" was given by Rikenglaz and Rozman (1974).

In conclusion, we should note another factor determining the approximate nature of the above averaging of the transfer rate. Integration in eqs. (2.6) or (2.22) implies that the arrangement of acceptors with respect to donors is quite uniform statistically. In fact, even liquids, let alone crystals, have short-range order and the acceptor molecules are in more or less fixed positions relative to the donor. Moreover, the uniform distribution can be disrupted owing to other causes, for instance, if the acceptors and the donors are ions.

The short-range order should be especially important for small R_0. In this case integration should be replaced by a summation. The best approach is to take the sum over the coordination spheres around the donor. The resulting expression for the function $H(t)$ is:

$$H(t) = \frac{c_A}{N}\sum_i (1 - \exp[-W(R_i)t]\,\nu_i, \tag{2.25}$$

where ν_i is the number of sites in the ith coordination sphere, N is the total number of sites in the unit volume, and R_i is the radius of the ith coordination sphere. Since eq. (2.25) has been derived under the same assumptions as above it is valid only if $c_A/N \ll 1$, that is, if the number of sites occupied by acceptors in each coordination sphere is small compared with ν_i. This condition practically limits the acceptor concentration to less than 10% (see sect. 1.6).

Bodunov (1971, 1972, 1973) analyzed the effects of electrostatic interaction between ions. He assumed that the interaction between ions was described by a shielded Coulomb potential corresponding to the Debye potential (Bodunov 1971). Then the appropriate Boltzmann factor should be introduced into the spatial distribution of acceptors. Numerical calculations show that for the ions having the same charge the transfer rate in rigid solutions decreases by 10–20%. However, he demonstrated that in liquid solutions the effect of ion charges could significantly change the transfer efficiency (Bodunov 1971, 1972, 1973). It should be stressed that foreign ions in solutions can also exert a significant effect on energy transfer (Bodunov 1973, Shakhverdov and Bodunov 1973).

1.6. Averaging of the transfer probability for high acceptor concentrations

Golubov and Konobeev (1971, 1973, 1975) and Sakun (1972) extended the theory to the case of high concentrations of the acceptor molecules. They suggested the following approach to averaging the transfer probability. Consider a crystal lattice some of the sites in which are occupied by acceptors. Then the probability $\rho(t)$ of the absence of transfer, as defined above, can be expressed as

$$\rho(t) = \overline{\exp\left[-t \sum W(R_i)\right]}, \tag{2.26}$$

where the bar denotes averaging over various donors and the sum is taken over the acceptors in the vicinity of the given donor.

Now let us try to express this probability by taking a sum over all lattice sites rather than the sites occupied by acceptors. Now the probability of the absence of transfer can be expressed in terms of the probabilities to find a site occupied or unoccupied by an acceptor. Clearly, the former probability is c_A/N and the latter is $1 - c_A/N$. To find the probability of the absence of transfer we should multiply the above probability by $\rho_1(t) = \exp(-Wt)$ and 1, respectively, and add up the results so that the total probability is $1 - c_A/N + (c_A/N)\,\rho_1$. Taking the product of probabilities over all lattice sites we obtain the total probability of the absence of transfer:

$$\rho(t) = \prod_{i=1}^{N} \left[1 - \frac{c_A}{N} + \frac{c_A}{N}\rho_1(R_i)\right]. \tag{2.27}$$

Significantly, the number of factors in this product is N rather than N_A as in eq. (2.3). Therefore, the function $H(t)$ *introduced in eq.* (2.4) *can be written as*:

$$H(t) = -\sum_{i=1}^{N} \ln\left[1 - \frac{c_A}{N} + \frac{c_A}{N}\exp(-W_i t)\right]. \tag{2.28}$$

If $c_A/N \ll 1$ then we can expand the logarithm obtaining

$$H(t) \approx \frac{c_A}{N}\sum_{i=1}^{N} [1 - \exp(-W_i t)],$$

which is identical to eq. (2.25). In the other limit when $c_A/N \sim 1$ we obtain

$$H(t) \approx t \sum W_i, \qquad \text{or} \qquad \rho(t) = \exp\left(-t \sum W_i\right),$$

that is, exponential decay of the number of donors. In the case when

$c_A/N \sim 1$ it can be said, for instance, that each lattice site is an acceptor or each molecule, for instance solvent molecules, is an acceptor. The exponential dependence in this case should be expected since all the donors have practically identical surroundings and the cause for nonexponential decay ("depletion effect") disappears. However, in practical applications of the theory we should bear in mind that the Förster theory approximation can prove to be invalid for such small distances at which energy is transferred in this case. At high concentrations, even if we ignore the possible contributions of higher multipoles and exchange interaction, the positions of the donor and acceptor levels and, hence, the spectral overlap integral will be random functions related to the distribution of donors and acceptors. Generally speaking, this should be taken into account when averaging is carried out but we did not do that when deriving eq. (2.28).

The above qualitative analysis yielding eq. (2.28) follows a remark made by Golubov and Konobeev (1971, 1973); the relevant rigorous treatment is presented by Golubov and Konobeev (1971, 1973, 1975), Sakun (1972).

Recently Kalnin (1981) suggested the following simpler derivation of eq. (2.28) (see also Blumen and Mans 1979).

Consider an ensemble of macroscopically identical systems of volume V each containing NV lattice sites (N is the number of sites per unit volume) out of which at the moment $t = 0$, $N_D(0) = N_0$ sites are occupied by excited donors and N_A sites are occupied by acceptors. Define by $p(m_1 \ldots, m_r; n_1 \ldots, n_s, t)$ the probability that at the moment t the sites $\boldsymbol{m}_1 \ldots, \boldsymbol{m}_r$ of the system are occupied by excited donors and the sites $\boldsymbol{n}_1 \ldots, \boldsymbol{n}_s$ are occupied by acceptors. Variation of the state of the system with time can be described by the following chain of equations:

$$\frac{\mathrm{d}p(\boldsymbol{m}, t)}{\mathrm{d}t} = -\sum_{\boldsymbol{n}(\neq \boldsymbol{m})}^{NV} W(|\boldsymbol{m} - \boldsymbol{n}|)p(\boldsymbol{m}, \boldsymbol{n}, t),$$

$$\frac{\mathrm{d}p(\boldsymbol{m}, \boldsymbol{n}, t)}{\mathrm{d}t} = -W(|\boldsymbol{m} - \boldsymbol{n}|)p(\boldsymbol{m}, \boldsymbol{n}, t) - \sum_{\boldsymbol{n}(\neq \boldsymbol{m}, \boldsymbol{n})}^{NV} W(|\boldsymbol{m} - \boldsymbol{n}'|)p(\boldsymbol{m}, \boldsymbol{n}, \boldsymbol{n}', t),$$

and so on.

To terminate the chain we can use the Kirkwood superposition approximation (see also ch. 5 sect. 8):

$$p(\boldsymbol{m}, \boldsymbol{n}, \boldsymbol{n}', t) = \frac{p(\boldsymbol{m}, \boldsymbol{n}, t)p(\boldsymbol{m}, \boldsymbol{n}', t)p(\boldsymbol{n}, \boldsymbol{n}')}{p(\boldsymbol{m}, t)p(\boldsymbol{n})p(\boldsymbol{n}')}.$$

We shall neglect correlation in the distribution of acceptors on different sites, that is, we shall assume

$$p(\boldsymbol{n}, \boldsymbol{n}') = p(\boldsymbol{n})p(\boldsymbol{n}'); \qquad \boldsymbol{n} \neq \boldsymbol{n}' \qquad (p(\boldsymbol{n}, \boldsymbol{n}) = 0).$$

Assuming that the system is macroscopically homogeneous, that is,

$$p(\boldsymbol{m}_1, t) = p(\boldsymbol{m}_2, t) = \cdots = c_D(t)/N, \qquad p(\boldsymbol{n}_1) = p(\boldsymbol{n}_2) = c_A/N,$$

where $c_D(t)$ is the number of excited donors at the moment t and introducing the relative probabilities,

$$\eta(|\boldsymbol{l}|, t) = p(\boldsymbol{m}, \boldsymbol{n}, t)/p(\boldsymbol{m}, t)p(\boldsymbol{n}),$$

we obtain

$$\frac{dc_D(t)}{dt} = -c_D(t)\frac{c_A}{N}\sum_{\boldsymbol{l}} W(|\boldsymbol{l}|)\eta(|\boldsymbol{l}|, t),$$

$$\frac{d\eta(|\boldsymbol{l}|, t)}{dt} = -W(|\boldsymbol{l}|)\eta(|\boldsymbol{l}|, t) + \frac{c_A}{N} W(|\boldsymbol{l}|)\eta^2(|\boldsymbol{l}|, t).$$

The solution of the second of the above equations for the initial condition $\eta(|\boldsymbol{l}|, 0) = 1$ is,

$$\eta(|\boldsymbol{l}|, t) = \frac{\exp[-W(|\boldsymbol{l}|)t]}{1 - c_A/N + (c_A/N)\exp[-W(|\boldsymbol{l}|)t]}.$$

Since $c_D(t) = c_D(0)\rho(t)$ we obtain eq. (2.28).

1.7. Energy transfer in quasi-one-dimensional systems

The donor and acceptor molecules can be arranged along a quasi-one-dimensional chain. This is the case, for instance, when dyes are adsorbed by a DNA biopolymer (Weil and Calvin 1963, Borisova et al. 1968) and each excited molecule (energy donor) has only two closest neighbours (acceptors). In calculations of the transfer probability we can use, in first approximation, the assumptions made for the derivation of eq. (2.8). However, when we find the function $H(t)$ from eq. (2.4) we must take the integral along the one-dimensional chain rather than over volume. Assuming that the chain is a straight line we obtain:

$$H(t) = c_A \int_{-\infty}^{\infty} [1 - \exp(-\alpha x^{-6} t)]\, dx,$$

where $\alpha = \frac{3}{2}\chi^2 R_0/\tau_0$ and c_A is the linear density of the acceptor molecules (with dimension cm^{-1}). Integration yields:

$$H(t) = 2s(t/\tau_0)^{1/6},$$

where $s = \Gamma(\frac{5}{6})(\frac{3}{2}\chi^2)^{1/6} R_0 c_A$.

We see that this law of decay of the number of donors differs significantly from eq. (2.8). As a result, eqs. (2.15) can no longer be used for describing the yield and the mean lifetime of the donor luminescence as functions of the acceptor concentration. Taking the general equations (2.14) we can easily see that for low concentrations c_A, eqs. (2.16) are replaced by;

$$\eta/\eta_0 = 1 - 2s\Gamma(\tfrac{7}{6}); \qquad \tau/\tau_0 = 1 - 2s[\Gamma(\tfrac{13}{6}) - \Gamma(\tfrac{7}{6})] = 1 - \tfrac{1}{3}s\Gamma(\tfrac{7}{6}),$$

so that the ratio of the initial slopes of the $\eta(c_A)$ and $\tau(c_A)$ is 6 rather than 2.

The above calculations are just an illustration because when we compare experimental results and theoretical predictions we should take into account the discrete arrangement of the donors and acceptors along the polymer and its structure. To do this we can make use of eq. (2.25) and apply it to a specific model describing the arrangement and orientation of the donors and acceptors on the polymer. Such numerical calculations and respective experiments are reported by Borisova et al. (1968). The satisfactory agreement between theory and experimental results in this study supported the dipole–dipole mechanism of transfer and demonstrated the correctness of the theoretical models. Interestingly, the above-mentioned ratio of the initial slopes $[d\eta(c_A)/dc_A]/[d\tau(c_A)/dc_A]$ was calculated to be between 3 and 4 while in the three-dimensional case it was 2.

For further applications of the energy transfer theory to polymer studies see Trubitsin et al. (1971), Gurskiy et al. (1973).

2. *Dipole–dipole energy transfer in liquid solutions*

In low-viscosity liquid solutions molecules cannot be assumed stationary during the lifetime of the excited state.

The motion of molecules during energy transfer can have two-fold effects. First, it can cause additional mismatching of the wave function phases in the system of interacting molecules or, in other words, it can decrease the time of relaxation of the nondiagonal elements of the density matrix. Secondly, the motion of molecules can result in a change of the relative spatial arrangement of the donors and acceptors. For condensed medium and weak interaction the first factor is not significant since the spectral width is typically determined by intramolecular relaxation. Therefore, in those cases when we can apply the Förster theory and ignore the coherent effects ("the case of very weak coupling") we can also, in that respect, ignore the motion of molecules.

Thus, we shall assume that the motion of molecules only changes the spatial distribution of the donors and acceptors. The diffusion processes can serve as a good approximation to describe the motion of molecules in liquid solutions. In the limiting case of very high diffusion rates the depletion effect discussed above can be expected to disappear completely and the distribution of the acceptors relative to the donors will be uniform throughout the transfer process. Therefore, we shall start by treating models incorporating this assumption and then use a more rigorous theory to find the conditions of their applicability.

2.1. *Model of a continuous absorbing medium*

We can put forward a classical model for summation of the energy transfers from one donor to a large number of acceptors by analogy to the classical model of the elementary energy transfer event suggested by Galanin (1951, 1960). This model consists of a continuous absorbing medium with an absorption coefficient $\mu(\omega)$ determined by the uniform spatial distribution of acceptors. A classical model of the donor can be an electronic dipole oscillator with the electric moment varying with the frequency ω of the external field as

$$p = p_0\, \mathrm{e}^{\mathrm{i}\omega t}.$$

The oscillator is placed at the centre of a spherical cavity with the radius $R_1 \ll \lambda$ (where $\lambda = \lambda_0/n$ is the wavelength in the medium) cut out in the continuous absorbing medium (Galanin and Frank 1951). Clearly, we cannot consider the depletion effect and the relaxation processes in the framework of this model. In other words, in the respective microscopic treatment we should assume that the distribution of the acceptor molecules remains uniform throughout the transfer process and that relaxation is very fast.

To calculate the efficiency of energy transfer in this model we have just to solve the electrodynamic problem of the dipole radiation in the absorbing medium. The total energy flux from an oscillating dipole in the transparent medium with the index of refraction n is known to be

$$S_0 = \frac{1}{3}\frac{\omega^4 p_0^2 n}{c^3}. \tag{2.26a}$$

In the absorbing medium the energy flux decreases with distance as $S_0 \exp(-\mu R)$. Moreover, as a result of absorption an additional energy flux due to the quasistationary field of the dipole is found in the vicinity of the dipole at distances $R \ll \lambda$. The complex amplitudes of the components of the magnetic and electric fields produced by the dipole can be written as,

$$E_{0\vartheta} = -\frac{p_0}{\epsilon R^3}\sin\vartheta, \qquad H_{0\varphi} = \frac{\mathrm{i}\omega}{c}\frac{p_0}{R^2}\sin\vartheta, \tag{2.27a}$$

where the z-axis is parallel to the dipole axis, ϑ is the polar angle, φ is the azimuthal angle, $\epsilon = \epsilon' + \mathrm{i}\epsilon''$ is the complex dielectric constant, $\epsilon = (c^2/\omega^2)k^2$, $k = k' + \mathrm{i}k''$ (here $k' = \omega n/c$, $2k'' = \mu$), and the influence of the effective field is ignored (see, e.g., Mensel 1955). The respective energy flux is:

$$S(R, \vartheta) = \frac{c}{8\pi}(E_{0\vartheta}H^*_{0\vartheta} + E^*_{0\vartheta}H_{0\vartheta}) \approx \frac{\omega^3}{4\pi c^2}\frac{p_0^2}{R^5}\sin^2\vartheta\,\frac{k''}{k'^3}. \tag{2.28a}$$

Here we assumed that $k''/k' \ll 1$, or that absorption at the distance $\lambda\!\!\!^{-} = \lambda_0/2\pi_n$ is small ($\mu\lambda\!\!\!^{-} \ll 1$). This assumption is always valid, at least for solutions.

The total energy flux across the spherical surface can be found by averaging eq. (2.28a) over angles ($\overline{\sin^2\vartheta} = \frac{2}{3}$) and multiplying the result by the surface area ($4\pi R_1^2$):

$$S_{R_1} = \frac{1}{3}\frac{p_0^2}{R_1^3}\frac{c\mu}{n^3}. \tag{2.29}$$

The net energy flux is the sum of the flux S_0 due to the wave field and the additional flux given by eq. (2.29):

$$S = S_0 + S_{R_1} = \frac{1}{3}\frac{\omega^4 p_0^2 n}{c^3}\left[1 + \mu\lambda\!\!\!^{-}\left(\frac{\lambda\!\!\!^{-}}{R_1}\right)^3\right]. \tag{2.30}$$

Since the damping of the dipole oscillations is proportional to the energy drained from it we directly find that the damping time for the dipole oscillations decreases as

$$\frac{1}{\tau} = \frac{1}{\tau_0}\left[1 + \mu\lambda\!\!\!^{-}\left(\frac{\lambda\!\!\!^{-}}{R_1}\right)^3\right]. \tag{2.31}$$

Note that though $\mu\lambda\!\!\!^{-} \ll 1$ the second term in eq. (2.31) is not necessarily small since $R_1 \ll \lambda$.

The classical model of the dipole in the continuous absorbing medium treating energy transfer as the result of absorption in the near non-wave zone of the radiator evidently gives a correct qualitative description of the process. It is clear, however, that in most cases this model cannot lay claim to a quantitative description. Indeed, if the main contribution to energy transfer is made by the interaction with the nearest neighbours (for instance, when absorption is due to the solvent) the concept of a continuous medium becomes inapplicable and we have to turn to the microscopic theory (Agranovich and Dubovsky (1970) and ch. 4 sect. 9).

2.2. *"Complete mixing" model*

Let us now return to the molecular theory and consider the case when the molecules change their orientation and position many times during the lifetime of the excited state so that the depletion effect may be ignored. Below we shall estimate the diffusion coefficient for which this assumption is valid (see sect. 2.3).

In this case we should assume that the distribution of the acceptor molecules outside the forbidden volume (a sphere with the radius R_1) remains uniform throughout the transfer process and the transfer efficiency can be calculated just by finding the sum of the transfer probabilities over this uniform distribution.

The probability of transfer to the ith acceptor molecules is found from eq. (1.36):

$$W_i = \frac{3}{2}\chi^2 \frac{R_0^6}{\tau_0 R_i^6},$$

where

$$R_0^6 = \frac{3c^4}{4\pi n^4 c_A}\int F(\omega)\mu(\omega)\frac{d\omega}{\omega^4} = \frac{3\bar{\lambda}^4}{4\pi c_A}\int F(\omega)\mu(\omega)\left(\frac{\bar{\omega}}{\omega}\right)^4 d\omega.$$

The bar denotes the mean values in the spectral overlap region.

In accordance with the above assumptions we take $\chi^2 = \frac{2}{3}$ (see eq. (1.20)) and replace the sum $\Sigma\, W_i$ with the integral:

$$\sum \frac{1}{R^6} \to c_A \int_{R_1}^{\infty} \frac{4\pi R^2\, dR}{R^6} = \frac{4\pi c_A}{3R_1^3}.$$

Hence, the total transfer probability is

$$w = \sum W_i = \frac{4\pi}{3}\frac{R_0^6}{\tau_0 R_1^3} c_A = \frac{\bar{\lambda}^4}{\tau_0 R_1^3}\bar{\mu}, \tag{2.32}$$

where

$$\bar{\mu} = \int F(\omega)\mu(\omega)\left(\frac{\bar{\omega}}{\omega}\right)^4 d\omega. \tag{2.33}$$

Since the total transfer probability here is independent of time the mean time of donor decay is found from,

$$1/\pi = 1/\tau_0 + w. \tag{2.34}$$

A comparison of eqs. (2.34) and (2.31) shows that they are identical if we take $\mu = \bar{\mu}$ and $\lambda = \bar{\lambda}$ in the classical model.

It may be noted that w is just identical to the constant of the initial exponential part of the decay curve which is found for the stationary molecules when the forbidden volume is taken into account. This is understandable since the initial part of the decay curve corresponds to the period in which the initial uniform distribution does not have time to change significantly.

2.3. *Analysis of the dipole–dipole transfer taking into account the diffusion of molecules*

Let us now formulate the problem more exactly and take into consideration the diffusion motion of molecules. We shall employ the approach developed by Samson and Rozman and discussed in sect. 1 (see, Samson 1962, Agrest et al. 1969, Kilin et al. 1964) and the relevant discussion by Shirokov (1964), Samson (1964).

Now, when molecules are not stationary, the transfer probability is determined by the positions of the acceptor molecules at the moment prior to transfer rather than by their initial positions. Let us again consider one donor molecule and one acceptor molecule. The probability of the absence of transfer during the time t is now the sum of the respective probabilities while the acceptor travels from its initial position R_i to the position R (we assume that the donor is at the origin of the coordinates). Therefore, we shall now regard the function $\rho(R, t)$ defined in sect. 1 as a function of the initial position R_i of the acceptor and write it as follows:

$$\rho_1(R_i, t) = \int P(R_i, R, t)\,\mathrm{d}V, \tag{2.35}$$

Here $P(R_i, R, t)\,\mathrm{d}V$ is the probability that the acceptor is in the volume element $\mathrm{d}V$ at the distance R and no transfer has occurred in the period from 0 to t. The probability $P(R_i, R, t)$ varies owing to displacement of the molecules and transfer processes. Therefore, $P(R_i, R, t)$ as a function of R and t satisfies the following equation:

$$\frac{\partial P}{\partial t} = -\,W(R)P + D\nabla^2 P. \tag{2.36}$$

Here D is the sum of the coefficients of diffusion of the donor and acceptor.

The transformations resulting in eq. (2.4) are applicable to the function $\rho_1(R_i, t)$. Then the mean probability of the absence of transfer is,

$$\rho(t) = \exp\left\{-c_A \int [1 - \rho_1(R_i, t)]\,\mathrm{d}V_i\right\} = \exp[-H(t)]. \tag{2.37}$$

In contrast to the case of stationary molecules we cannot ignore here the forbidden volume and to obtain the convergent integral in eq. (2.37) we should take it over the whole volume with the exception of the sphere with the radius R_1.

Using the identity

$$1 - \rho_1(R_i, t) = -\int_0^t \frac{\partial \rho_1(R_i, t')}{\partial t'}\,\mathrm{d}t',$$

and remembering that $\rho_1(R_i, 0) = 1$ we obtain:

$$H(t) = -c_A \int \mathrm{d}V_i \int_0^t \frac{\partial \rho_1}{\partial t'} \mathrm{d}t' = -c_A \int \mathrm{d}V_i \int \mathrm{d}t' \int \frac{\partial P}{\partial t'} \mathrm{d}V. \tag{2.38}$$

Let us now introduce the function

$$\rho'(R, t) = \int P(R_i, R, t)\, \mathrm{d}V_i \tag{2.39}$$

(integration over the initial positions) which can be regarded as the normalized acceptor concentration in the vicinity of the donors in the "superposed" configuration at the moment prior to energy transfer, that is, as an analogue of the function $\rho_1(R, t)$ taking into account the diffusion. Clearly, the function $\rho'(R, t)$ satisfies the same equation as the function $P(R_i, R, t)$, that is, eq. (2.36). Substitution of $\partial P/\partial t$ from eq. (2.36) into eq. (2.38) yields:

$$H(t) = c_A \int_0^t \int W(R)\rho'(R, t')\, \mathrm{d}t'\, \mathrm{d}V - c_A D \int \frac{\partial \rho'}{\partial R} \mathrm{d}S. \tag{2.40}$$

The second term in eq. (2.40) was derived by transforming the integral of $\nabla^2 \rho' = \mathrm{div\,grad}\, \rho'$ over the volume into the integral over the surface of this volume; hence, this term depends on the boundary conditions. (For delta excitation the initial condition is, clearly, $\rho'(R, 0) = 1$.)

The boundary conditions are determined by the physical assumptions on the processes occurring at the boundary of the spherical forbidden volume with the radius R_1. The most general boundary condition is,

$$D \frac{\partial \rho'}{\partial R}\bigg|_{R=R_1} = v\rho'(R_1, t), \tag{2.41}$$

where v describes the relative rate of excitation transfer at the sphere boundary (the conditions for applicability of such approach are discussed in ch. 6 sect. 2). Hence, only when $v = 0$ do we assume that no additional processes, apart from energy transfer, take place (the so-called purely distant transfer). If $v \neq 0$ this implies that, apart from the long-range transfer, there occurs the so-called collision transfer caused by the process at the sphere boundary. The case of $v \to \infty$ (and, hence, the boundary condition $\rho'(R_1) = 0$) corresponds to the so-called black sphere in which energy transfer occurs immediately after an acceptor reaches the boundary.

Thus, the factor $v \neq 0$ can indirectly take into account the fact that the approximation of weak dipole–dipole interaction is not applicable at small distances.

If we have found the solution of the equation

$$\frac{\partial \rho'}{\partial t} = -W(R)\rho' + D\nabla^2\rho', \tag{2.42}$$

we can obtain $H(t)$ and the law of decay of the number of donors,

$$n(t) = n_0 \exp[-t/\tau_0 - H(t)].$$

Now we can find the luminescence yield and the mean lifetime of the donors from eqs. (2.14).

For the dipole–dipole interaction we have $W(R) = (1/\tau_0)(R_0/R)^6$ and eq. (2.42) can be solved only numerically. We shall discuss below the solutions found by Rozman and co-workers (Kilin et al. 1964, Agrest et al. 1969).

First, we shall consider the results of Tunitsky and Bagdasar'yan (1963) and Yokota and Tanimoto (1967) for the following specific case (Kurskiy and Selivanenko 1960, Vasil'ev et al. 1964, Gösele et al. 1975, 1976). Assume that a steady-state spatial distribution $\rho'(R, \infty)$ has been established. In contrast to eq. (2.4), eq. (2.42) allows such a steady-state distribution. Then the transfer rate is constant, $H(t) = wt$ and the number $n(t)$ of the excited donors decays as $\exp(-t/\tau_0 - wt)$. Taking $\partial\rho'/\partial t = 0$ and making the substitution $x = R_1/R$ we can transform eq. (2.42) into

$$\frac{\partial^2 \rho'}{\partial x^2} = b^2x^2\rho', \tag{2.43}$$

where

$$b = R_0^3/R_1^2\sqrt{D\tau_0}. \tag{2.44}$$

The solution of eq. (2.43) is known (Janke-Emde-Rosch 1960) to be

$$\rho'(x) = \sqrt{x}\,[C_1 I_{1/4}(bx^2/2) + C_2 I_{-1/4}(bx^2/2)], \tag{2.45}$$

where I_ν is the Bessel function with imaginary argument (the modified Bessel function). The constants C_1 and C_2 are determined by the boundary conditions. Let us consider two limiting cases: (1) the absence of the collision transfer, that is, $(d\rho'/dx)|_{x=1} = 0$; and (2) the "black" sphere, that is, $\rho' = 0$ for $x = 1$. In both cases we have $\rho'(x) \to 1$ for $x \to 0$. This last boundary condition yields

$$C_2 = 2^{-1/2}\Gamma(3/4)b^{1/8}c_A.$$

Using the rules of differentiation of the modified Bessel functions we obtain for the first limiting case,

$$C_1 = -C_2\frac{I_{3/4}(b/2)}{I_{-3/4}(b/2)}.$$

For the second case we have,

$$C_1 = -C_2 \frac{I_{-1/4}(b/2)}{I_{1/4}(b/2)}.$$

Analyzing these results we can find the conditions of applicability of the models of the uniform absorbing medium and "complete mixing". Assume that $b \ll 1$, that is, the diffusion coefficient is high. Then the argument of the Bessel functions in eq. (2.45) is small and we can approximate them by the first terms of their expansions:

$$I_\nu(z) \approx \frac{1}{\Gamma(\nu+1)}\left(\frac{z}{2}\right)^\nu.$$

When no collision transfer takes place this approximation yields

$$\rho'(x) \approx 1 - \tfrac{1}{3}b^2 x,$$

or

$$\rho'(R) \approx 1 - \tfrac{1}{3}b^2 R_1/R. \tag{2.46}$$

Here we have taken into account that

$$\frac{\Gamma(3/4)\Gamma(1/4)}{\Gamma(5/4)\Gamma(7/4)} = \frac{16}{3}.$$

When we substitute eq. (2.46) into eq. (2.40) for $H(t) = wt$ and integrate the result we can find the diffusion correction term for the transfer rate in the uniform medium model:

$$w = \frac{4\pi}{3}\frac{R_0^6}{\tau_0 R_1^3} c_A \left(1 - \frac{b^2}{4}\right). \tag{2.47}$$

Thus, the uniform medium model is applicable if the following condition is satisfied:

$$\frac{b^2}{4} = \frac{1}{4}\frac{R_0^6}{R_1^4 D \tau_0} \ll 1. \tag{2.48}$$

An essentially different picture is seen when collision transfer takes place. Then similar calculations for the black sphere yield:

$$\rho'(x) \approx 1 - x = 1 - R_1/R. \tag{2.49}$$

This is the distribution known in the theory of diffusion quenching of luminescence (Smoluchowski 1917, Sveshnikov 1935). When calculating the transfer rate we must take into account collision transfer whose rate is given by the well-known expression $4\pi D R_1 c_A$ (see also ch. 6 sect. 2).

Finally, we obtain

$$w = 4\pi D R_1 c_A (1 + \tfrac{1}{12} b^2). \tag{2.50}$$

We see that the major contribution in this case is made by collision transfer.

It should be remembered that these results are valid for the steady-state period of the transfer process since we have assumed at the beginning that the transfer rate is constant. However, if the diffusion coefficient is high the time needed to reach the steady state can be sufficiently short compared with τ_0 and the contribution of the non-steady-state period can be respectively small. In experiments this should be revealed by the exponential decay of the donor luminescence and by the similar dependences of the luminescence yield and the mean donor lifetime on the acceptor concentration.

When $b \gg 1$ the steady-state spatial distribution $\rho'(R)$ is identical in both limiting cases of the boundary conditions discussed above. When $z \gg 1$ we have $I_{1/4}(z) \approx I_{-1/4}(z)$ and $I_{3/4}(z) \approx I_{-3/4}(z)$. Therefore, we have $C_1 \approx -C_2$ irrespective of the boundary conditions. Calculating the function $H(t) = wt$ we obtain for distant transfer in this case,

$$w\tau_0 = 8\pi \frac{\Gamma(3/4)}{\Gamma(1/4)} (D\tau_0)^{3/4} R_0^{3/2} c_A \approx 4\pi \times 0.676 (D\tau_0)^{3/4} R_0^{3/2} c_A. \tag{2.51}$$

It can be readily seen that in this case we can ignore collision transfer since its rate is lower approximately by a factor of b.

In order to illustrate the problem we can express the transfer rate in terms of a certain effective radius. This can be done in two ways. We can find the effective radius $R_{\text{eff}}^{(1)}$ of the black sphere which would provide, by means of collision transfer only, the same transfer rate as the calculated distant transfer. A comparison of eqs. (2.51) and (2.50) yields:

$$R_{\text{eff}}^{(1)} = 2 \frac{\Gamma(3/4)}{\Gamma(1/4)} \frac{R_0^{3/2}}{(D\tau_0)^{1/4}} \approx 0.676 \sqrt{b}\, R_1. \tag{2.52}$$

On the other hand, we can find the radius $R_{\text{eff}}^{(2)}$ which would provide for the same rate of distant transfer for the uniform distribution $\rho'(R) = 1$ outside the sphere and $\rho'(R) = 0$ inside it. A comparison of eqs. (2.51) and (2.52) yields:

$$R_{\text{eff}}^{(2)} = \sqrt[3]{\frac{\Gamma(1/4)}{6\Gamma(3/4)} \frac{R_0^{3/2}}{(D\tau_0)^{1/4}}} \approx 0.79 \sqrt{b}\, R_1. \tag{2.53}$$

Thus, for $b \gg 1$ both effective radii are considerably larger than R_1 (and practically independent of R_1).

However, if $b \gg 1$ the results found for the steady-state case have only a limited application since the non-steady-state period is not necessarily

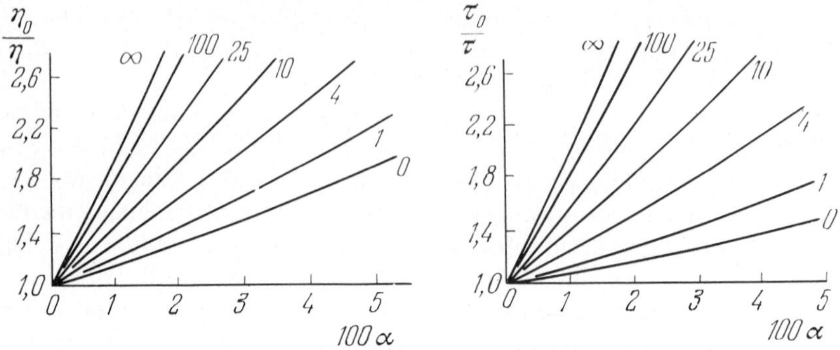

Fig. 2.6. Calculated quantum yield and lifetime for the dipole–dipole transfer involving diffusion (Agrest et al. 1936, 1969). The values of the parameter $\sigma = D\tau_0/R_1^2$ are given at the plots. The dimensionless concentration $\alpha = 4\pi R_1^3 c_A/3$.

short in comparison with τ_0. Then we must find the solution of the non-steady-state equation (2.42) which is determined not by one dimensionless parameter b but by two parameters; according to Agrest et al. (1969) these parameters may be $\sigma = D\tau_0/R_1^2$ and $\gamma = (R_0/R_1)^6$. Rozman et al. carried out numerical integration and found the relative luminescence yield and the mean donor lifetime as functions of the dimensionless acceptor concentration $\alpha = \frac{4}{3}\pi R_1^3 c_A$.

Figure 2.6 shows these dependences for $\gamma = 100$ (that is, for $R_0/R_1 \approx 2.15$) and various values of σ in the case of purely distant transfer ($v = 0$). Comparing the dependence for $\sigma = 0$ with dependences for high values of σ we can find how diffusion increases the transfer rate. At the same time, the differences between the concentration dependences of η_0/η and τ_0/τ for low values of σ show that the non-steady-state period plays a significant role. Agrest et al. (1969) gave detailed data obtained for various values of the parameters and various boundary conditions.

Now let us discuss experimental results for energy transfer in liquid solutions. An early discussion of the role of diffusion was given by Galanin (1951, 1960) where it was shown in experiments that the transfer efficiency increased with decreasing viscosity of the solution. As should be expected, this was accompanied by noticeable exponentialization of the donor decay dependence as evidenced by similarity of the luminescence yield and τ as functions of the acceptor concentration. Birks and co-workers (Birks 1968, Birks et al. 1970, Birks and Georghiau 1968) later observed such exponentialization directly from the curves of decay of the donor luminescence.

Bagdasar'yan and Muler (1965) attempted to compare some experimental results available at that time with the steady-state theory. Experimental data of Tomura et al. (1968) supported the theoretical result that the transfer rate was proportional to $D^{3/4}$ for low values of the diffusion coefficient D (that is, for $b > 1$).

Rozman and co-workers (Agrest et al. 1970) carried out the most thorough experimental study and comparison of the results with the theory (see also Elkana et al. 1968, Gösele et al. 1975, 1976).

It should be borne in mind that for liquid solutions a comparison of the experimental data with the theory is not such a straightforward procedure as for rigid solutions. The theory makes use of two parameters R_1 and D which can be estimated only with a limited accuracy. The third parameter – the rate v of energy transfer at the boundary of the forbidden volume sphere– should be measured in experiments.

It was shown by Agrest et al. (1970) that the experimental results in some cases cannot be approximated by the theoretical dependences for $v = 0$ and collision transfer plays a significant part for high diffusion coefficients and small values of R_0. In many cases the black sphere model is applicable and for certain relations between the parameters the transfer rate is practically fully determined by the processes occurring at the boundary of the forbidden volume. This is the most important physical result obtained in the studies of energy transfer in liquid solutions as it demonstrates the limits of applicability of the weak dipole–dipole interaction mechanism. The fact that $v \neq 0$ shows that when the intermolecular distance becomes R_1 other types of interaction start to play a decisive role.

In conclusion, let us estimate the parameter b according to eq. (2.44) for organic molecules of the dye type. As mentioned above, the typical values of the parameters in this case are $R_0 \sim 5$ nm, $R_1 \sim 1$ nm, and $\tau \sim 10^{-8}$ s. For liquid solvents we have, typically, $D \sim 10^{-5}$ cm^2 s. Hence, we obtain $b \sim 40$, that is, $b \gg 1$ and the model of uniform steady-state acceptor distrubution is inapplicable. For the above values the time of reaching the steady-state conditions is $t = R_{\text{eff}}^2/D \sim 10^{-8}$ s, that is, $t \sim \tau_0$ and the steady-state model is a poor approximation so that the role of diffusion cannot be described only by the parameter b.

2.4. *Migration of excitations over donor molecules*

The term "donor of energy" was applied above to the excited molecules. It was mentioned that under the conditions of weak excitation their concentration is typically low. When we discuss migration of energy over donors we imply energy transfer from the excited molecules to the unexcited molecules of the same species. Though this terminology is not consistent it is convenient since we usually consider two-component solutions and call the molecules of the second species the acceptors.

The donor concentration in that sense cannot be small and energy migration over the donors can play a significant role in energy transport to acceptors.

An example of such migration of energy over the donors is the transfer of energy from monomers to dimers analyzed by Rozman (1973). In this case the concentration of the donors (monomers) is much higher than the concentration of the acceptors (dimers). But energy migration over the donors plays a particularly important part in glasses and crystals doped with rare earth ions (Artamonova et al. 1972, Voron'ko et al. 1973, Basiev et al. 1974, 1975, Weber 1971, Zverev et al. 1975). In these systems migration over the donors often determines the efficiency of the concentration quenching since it is just migration over the donors that results in effective energy transport to the impurities.

The theory of the energy migration over the donors is close to that of the energy transfer by excitons which will be discussed in chs. 5 and 6. There we shall give a detailed discussion of the applicability of the diffusion approximation. If we assume that this approximation is applicable (that is, if the donor concentration is sufficiently high so that the lifetime of the excited donor state is much longer than the mean time of the donor–donor excitation transfer) and migration of excitation can be regarded as diffusion, then the migration plays the same role as the diffusion of the molecules themselves. Now we have to determine the diffusion coefficient and the boundary conditions in the vicinity of the impurity molecule. The boundary conditions depend on the system under consideration (see ch. 5). As for the diffusion coefficient, if the donor concentration is sufficiently high the simplest way is to take the elementary equation of gas kinetics

$$D = \tfrac{1}{3}lv,$$

where the mean free path l can be taken to be equal to the mean distance $\bar{R}$ between the donors ($\tfrac{4}{3}\bar{R}^3 = 1/c_D$ where c_D is the number of the donors per unit volume), and the "velocity" $v = \bar{R}/\tau$ where τ is the "mean" lifetime of the donor excitation defined by the relationship, $1/\tau = 1/\tau_0(R_{0D}/\bar{R})^6$. Here R_{0D} is the critical distance for the excited donor–unexcited donor energy transfer according to the Förster theory. Finally, we obtain:

$$D = \tfrac{1}{3}R_{0D}^6/\tau_0\bar{R}^4. \qquad (2.54)$$

In the more exact theory of Rozman (1973) the numerical factor is 0.336, rather than $\tfrac{1}{3}$, for $q_D = \tfrac{2}{3}\pi^{3/2}R_{0D}^3c_D \geqslant 3$ (Trlifai 1958).

For low donor concentrations when the condition $\tau \ll \tau_0$ is not satisfied, that is, if the number of hops of excitation during the lifetime of the excited state is small, the diffusion approximation is inapplicable.

Golubov and Konobeev (1971, 1973, 1975) analyzed this problem in detail using their accurate procedure of averaging. They found a numerical solution of the diffusion equation for the dipole–dipole transfer under the conditions of applicability of the diffusion approximation and demonstrated, in particular, that their numerical results can be described with an accuracy of 15% by the following analytical formula for decay of the number of the excited donors:

$$n(t) = n_0 \exp\left(-\frac{t}{\tau_0} - 2q\sqrt{\frac{t}{\tau_0}} - 0.68 \times 4\pi R_0^{3/2}(D\tau_0)^{3/4} c_A \frac{t}{\tau_0} \right). \tag{2.55}$$

Equation (2.55) graphically demonstrates the transition from the Förster decay equation (2.8) for short times to the steady-state diffusion equation (2.51) for the times $t > R_{\text{eff}}^2/D$. The low accuracy of eq. (2.55) is, probably, due to the fact that it fails to take into account the non-steady-state diffusion stage for short times $t < R_{\text{eff}}^2/D$.

If the donor concentration is low the effect of migration over donors cannot be reduced to diffusion and described by a diffusion coefficient.

The decay of the number of the excited donors for low concentrations of both donors and acceptors was described by the following function (Golubov and Konobeev 1975):

$$n(t) = n_0 \exp\left[-\frac{t}{\tau_0} - 2q\sqrt{\frac{t}{\tau_0}} - c_A c_D \Lambda\left(\frac{t}{\tau_0}\right) \right].$$

The function $\Lambda(t/\tau_0)$ was calculated by Golubov and Konobeev (1971, 1973). It can be approximated by the following analytical expressions:

$$\Lambda = 0.91 \times 2\pi^2 R_{0D}^3 R_0^3 t/\tau_0 \quad \text{for} \quad R_{0D} > 3R_0,$$
$$\Lambda = (2\pi/3)\, R_{0D}^5 R_0 (t/\tau_0) \quad \text{for} \quad R_{0D} < R_0.$$

For other approaches to analysis of energy migration over the donors see Shekhtman (1972), Burshtein (1972), Green (1974), Dubovskiy (1976), Blumen and Mans (1979).

CHAPTER 3

Effects of the Medium on the Energy Transfer Rate and the Spectrum of Molecular Crystals

1. *Introduction: the acting (local) field method*

It was stressed in ch. 1 that the resonance interaction $\hat{V}_{ab}$ between the impurity molecules a and b resulting in energy transfer can be determined, to a considerable extent, by the properties of the host lattice. Even qualitative arguments show that for sufficiently large distances R_{ab} between the impurity molecules (that is, for $R_{ab} \gg d$ where d is the host lattice parameter) the effect of the host lattice should be evidenced by the dependence of the resonance interaction on the dielectric constant tensor which is a macroscopic characteristic of the host lattice. This is, of course, self-evident if we recall the Coulomb law in a medium ($\hat{V}_{ab} \sim 1/\epsilon R_{ab}^3$ where ϵ is the dielectric constant of the medium).

In fact, the picture proves to be more complicated than it was shown in ch. 1 sect. 4 for two dipoles. This simplest model yielded the relationship $\hat{V}_{ab} \sim (1/\epsilon)[(\epsilon+2)/3]^2\, 1/R_{ab}^3$, that is, a considerably stronger dependence of the resonance interaction energy on the dielectric constant of the medium.

To what extent does this result remain valid if we take into more detailed consideration the structure of the crystal containing impurity molecules? What is the effect of crystal anisotropy on the results inferred from the Förster formula in ch. 1 sect. 4? To answer these questions we have to find the acting field in the crystal lattice. Thus, the problem of finding the effective (taking into account the effect of the host lattice) resonance interaction between the impurity molecules proves to be intimately related to the local field method which can also be used for obtaining the dielectric constant tensor and hence the optical properties of molecular crystals.

This chapter presents a consistent description of the local field method and some of its results. First, we should make a few brief notes.

It is well-known that the simplest approach to the study of the optical properties of condensed media is the macroscopic electrodynamics approach making use of the concept of the dielectric constant tensor $\epsilon_{ij}(\omega, k)$ where ω and k are the frequency and the wave vector of the light wave. Calculation of this tensor for a specific medium is, however, a problem of microscopic theory. For instance, the procedures for calculating the tensor $\epsilon_{ij}(\omega, k)$ for the exciton region of the spectrum in crystals are discussed by

Agranovich and Ginsburg (1983), Agranovich (1968). These procedures are based on using the different types of exciton states of the crystal (Coulomb or mechanical excitons, see ch. 4) which are treated as the zero-order states when we attempt to find a linear response to an external electromagnetic perturbation. These procedures are fairly general. However, this does not mean that we must know the exciton states of the crystal in order to calculate the dielectric constant tensor. We shall expound below on these points when we consider the spectra of molecular crystal (pure crystals, crystals containing impurities, and crystalline solutions) lowest singlet excited states where the mutual interaction between molecules does not result in charge transfer. The intermolecular interaction here is purely classical in character and is determined by the Van der Waals forces which only results in mixing of the molecular configurations (Agranovich 1968).

Numerous theoretical and experimental studies have been carried out in this field so that a whole branch of molecular optics – the optics of molecular crystals and molecular liquids – has been established. Even before Frenkel put forward his exciton concept the workers in this branch of optics had developed a variety of exact and approximate methods for the theoretical description of optical phenomena; many of these methods were also substantiated in experimental studies. However, after the discovery of excitons the use of these methods became increasingly rare and many of the results obtained with them have not been sufficiently understood in the framework of the exciton theory. Therefore, further development and generalization of these methods were impeded. On the other hand, since the results of pre-excitonic molecular optics were underestimated, the optical properties of crystals were treated in terms of only the exciton theory even in those cases when this could be done much easier by using the earlier, simpler and no less clear physical concepts. These circumstances could not, of course, fail to influence the development of the theory of the optics of crystals. The resulting situation is discussed by Agranovich (1974) and this chapter is largely based on the results presented therein. This paper was concerned mainly with the calculation of the dielectric constant tensor for crystals consisting of identical or different molecules, a discussion of their optical properties (dispersion and absorption of light), and a determination of the energy of the resonance dipole–dipole interaction between the impurity molecules resulting in excitation transfer as a function of the dielectric properties of the host crystal.

We can use the local field method to find the dielectric constant tensor for systems of this type. This method goes back to Lorentz who used it to derive the well-known formula for the optical refractive index in isotropic media (the Lorentz–Lorenz formula). Let us recall the derivation of this formula.

According to Lorentz, the electric field $\boldsymbol{E}'$ acting on a molecule in the isotropic medium and causing its polarization is not equal to the mean (macroscopic) field $\boldsymbol{E}$ which satisfies the phenomenological Maxwell equations, but is determined by,

$$\boldsymbol{E}' = \boldsymbol{E} + \frac{4\pi}{3}\boldsymbol{P} = \frac{\epsilon+2}{3}\boldsymbol{E}, \qquad \boldsymbol{P} = \frac{(\epsilon-1)\boldsymbol{E}}{4\pi},$$

where $\boldsymbol{P}$ is the polarization per unit volume, and ϵ is the dielectric constant of the medium. On the other hand, since the polarization per unit volume $\boldsymbol{P} = N_0 a \boldsymbol{E}'$ (where a is the polarizability of the molecule, and N_0 is the number of molecules per unit volume) the electric flux density vector $\boldsymbol{D}$ is given by the relationship,

$$\boldsymbol{D} = \boldsymbol{E} + 4\pi\boldsymbol{P} = \left[1 + \frac{4\pi}{3} N_0 a(\epsilon+2)\right]\boldsymbol{E} = \epsilon\boldsymbol{E},$$

which directly yields the Lorentz–Lorenz formula,

$$\frac{\epsilon-1}{\epsilon+2} = \frac{4\pi}{3} N_0 a.$$

However, this formula, which expresses the dielectric constant of the medium in terms of the polarizability of an individual molecule, is only a rough approximation even for cubic crystals with the Van der Waals forces acting between the isotropic molecules. For instance, the formula takes no account at all of spatial dispersion. Moreover, it does not take into account the contribution of the higher multipoles to the energy of intermolecular interaction which determines the difference between the optical properties of the crystal and those of molecules in low-density gases.

We shall show below (sects. 3 and 7) how these inaccuracies in the Lorentz–Lorenz formula can be eliminated. We shall employ the local field method to modify the formula and apply it to anisotropic molecular crystals of complex structure and we shall discuss a variety of their optical properties which were earlier analyzed in a less general form only in the framework of the exciton theory. Then we shall apply the local field method to find the dielectric constant tensor of the anisotropic molecular crystal taking into account spatial dispersion. Moreover, we shall discuss the effect of the medium on the energy of the resonance interaction between the impurity molecules.

Let the unit cell of the crystal contain σ identical molecules ($\alpha = 1, 2, \ldots, \sigma$) that are differently oriented with respect to crystallographic axes. According to Born and Huang (1954, sect. 30), when a plane electromagnetic wave of the amplitude $\boldsymbol{E}(\omega, \boldsymbol{k})$ propagates in the crystal the electric field $\boldsymbol{E}^{\alpha}$ acting on the molecule α is not equal to the mean field but is given by:

$$E_i^{\alpha} = E_i + \sum_{\beta j} Q_{ij}^{\alpha\beta}(\boldsymbol{k}) p_j^{\beta}, \tag{3.1}$$

where $\boldsymbol{p}^{\alpha}$ is the amplitude of the dipole moment induced in the molecules of the species α, and the coefficients $Q_{ij}^{\alpha\beta}(\boldsymbol{k})$ (the internal field coefficients) are determined by the lattice structure only. If $a_{ij}^{\alpha}(\omega)$ is the polarizability of the molecule with the orientation α (for simplicity, we assume here and below that the molecules have no static dipole moments) then we have,

$$p_i^{\alpha} = a_{ij}^{\alpha}(\omega) E_j^{\alpha}. \tag{3.2}$$

At the same time, of course, the molecules can have static moments of higher multipolarity both in the ground and in the excited state. Generally speaking, these moments are different in the ground and in the excited states. The molecules in these states have also different energies of interaction with the environment so that the frequencies of the intramolecular transitions are somewhat shifted with respect to the transition frequencies in vacuum. We shall assume below that the only difference between the tensor $a_{ij}(\omega)$ and the respective tensor for a molecule in vacuum is determined by this frequency shift.

Substitution of eq. (3.2) into eq. (3.1) yields:

$$E_i^{\alpha} = E_i + \sum_{\beta j_1} Q_{ij_1}^{\alpha\beta}(\boldsymbol{k}) a_{j_1 j}^{\beta}(\omega) E_j^{\beta}, \tag{3.3}$$

Now we can express the local fields $\boldsymbol{E}^{\alpha}$ in terms of $\boldsymbol{E}$:

$$E_i^{\alpha}(\omega, \boldsymbol{k}) = A_{ij}^{\alpha}(\omega, \boldsymbol{k}) E_j(\omega, \boldsymbol{k}). \tag{3.4}$$

If we know the tensor A_{ij} we can directly find the dielectric constant tensor for the crystal. Indeed, since the polarization per unit volume is,

$$P_i = \frac{1}{v} \sum_{\alpha} p_i^{\alpha}, \tag{3.5}$$

where v is the volume of the unit cell, then the dielectric constant tensor is

$$\epsilon_{ij}(\omega, \boldsymbol{k}) = \delta_{ij} + \frac{4\pi}{v} \sum_{\alpha j_1} a_{ij_1}^{\alpha} A_{j_1 j}^{\alpha}. \tag{3.6}$$

Equation (3.6) makes possible a very simple analysis of the problem of how the internal field correction affects the optical properties of crystals. A number of results for pure crystals in the dipole approximation are reported by Dunmur (1972), Cummins et al. (1973). For instance, Dunmur (1972) calculated the internal field coefficients for $\boldsymbol{k} = 0$ for the crystals of anthracene, naphthalene, benzene, phenanthrene, biphenyl and iodoform. Wünshe (1975a) analyzed the case of uniaxial crystals with one or two molecules in the unit cell and (Wünshe 1975b) the problem of the local field in the framework of nonlinear crystal optics (see also, Fulton 1974). It seems worthwhile to discuss this problem here since the literature in the field reveals considerable confusion. Many journal articles and even monographs have taken no account at all of the difference between the effects of the internal field correction in pure crystals and in crystals containing impurities.

Apart from the above-mentioned problems we shall discuss below the optical properties of mixed crystalline solutions, take into account some effects of spatial dispersion in the medium in the treatment of the optical properties of an impurity, and find the energy of the resonance interaction between the impurity molecules. Here we shall only note that the use of the tensor a_{ij}^{α} independent of $\boldsymbol{k}$ for describing the polarizability of an individual molecule means that we take into account only the dipole polarization of the molecule. This approximation is valid if we deal with the optical properties of a non-gyrotropic crystal in the region of resonances of the molecule corresponding to high oscillator strengths. However, this approximation is by no means essential for the following analysis. (This treatment using the tensor $a_{ij}(\omega, \boldsymbol{k})$ is no more phenomenological than the theory of small-radius excitons which assumes the wave functions and energies of the isolated molecule, rather than the tensor $a_{ij}(\omega, \boldsymbol{k})$, to be known (specified). The theory of point groups can be used to analyze the structure of this tensor in the regions of certain resonances of the molecule.)

Since we assume below that the polarizability of the molecule in vacuum is known we can use the more general expression $a_{ij}^{\alpha}(\omega, \boldsymbol{k})$ for it in eq. (3.2). Of course, the expression for the internal field should include the terms corresponding to the internal field in the lattice of quadrupoles, octupoles, etc., apart from the term given in eq. (3.1) and corresponding to the internal field in the lattice of dipoles. The electric field produced by these multipoles decreases with distances faster than the dipole field so that when we find the internal field corresponding to them we do not encounter the well-known difficulties with summing the series as were so adroitly overcome by Ewald for a lattice of dipoles (Born and Huang 1954). Thus, finding the internal field for the higher multipoles is a more or less routine procedure and we shall not discuss it below (these results in the most general form are presented by Khokhlov (1972) who gives also references to earlier studies; see also, Philpott and Mahan (1973), Cummings et al. (1974)).

As we emphasized above, the use of the tensor $a_{ij}^{\alpha}(\omega)$, rather than the tensor $a_{ij}^{\alpha}(\omega, \boldsymbol{k})$, in eq. (3.2) is quite unsuitable for the region of dipole-forbidden transitions. Therefore, when discussing the optical properties of non-gyrotropic pure and impure crystals, we shall consider at first only the spectral region of dipole-allowed transitions. However, a more general analysis can be carried out in a similar way (see sect. 7).

2. Dielectric constant of cubic crystals

As was shown by Born and Huang (1954) in cubic crystals with one molecule per unit cell the tensor $Q_{ij}^{\alpha\beta}$ is reduced to the scalar $Q_{ij} = (4\pi/3v)\delta_{ij}$ where v is the volume of the unit cell, if we ignore spatial dispersion. Moreover, $a_{ij} = a\delta_{ij}$ so that eq. (3.3) yields the tensor $A_{ij} = A\delta_{ij}$ where $A = [1-(4\pi a/3v)]^{-1}$. Substituting this expression into eq. (3.6) we obtain $\epsilon_{ij} = \epsilon\delta_{ij}$ where

$$\epsilon(\omega) = 1 + \frac{4\pi}{v}a\left(1 - \frac{4\pi}{3v}a\right)^{-1}, \tag{3.7}$$

so that $A = (\epsilon+2)/3$. Equation (3.7) directly yields the Lorentz–Lorenz formula:

$$\frac{\epsilon - 1}{\epsilon + 2} = \frac{4\pi}{3v}a. \tag{3.7a}$$

Let us now consider the dispersion of the quantity ϵ taking into account only one of the resonances of $a(\omega)$. In this approximation we obtain:

$$a(\omega) = \frac{F_1}{\omega_1^2 - \omega^2}, \qquad F_1 = \frac{2p_1^2\omega_1}{\hbar} = \frac{e^2}{m}f_1, \tag{3.8}$$

where f_1 is the oscillator strength (1.22), ω_1 is the frequency of the $0 \to 1$ transition in the isolated molecule, and p_1 is the corresponding dipole moment of the transition. Substitution of eq. (3.8) into eq. (3.7) yields:

$$\epsilon(\omega) = 1 + \frac{(4\pi/v)F_1}{\omega_1^2 - \omega^2 - (8\pi/3v\hbar)p_1^2\omega_1}. \tag{3.9}$$

Equation (3.9) shows that if we take into account the internal field correction, that is, the fact that $A \neq 1$, the oscillator strength for the transition is not changed and only the resonance frequency is shifted – the $\epsilon(\omega)$ resonance is shifted by $\Delta\omega \ll \omega_1$ towards lower frequencies from the frequency of the transition in the isolated molecule:

$$\Delta\omega = \frac{4\pi}{3v\hbar}p_1^2.$$

Decay is taken into account by adding the imaginary terms $-i\delta$ to the denominator of eq. (3.9). Since $\epsilon = (n + i\kappa)^2$ where n and κ are the refractive and absorption indexes, for $\delta \to +0$ we obtain,

$$\int_{-\infty}^{+\infty} 2n(\omega)\kappa(\omega)\,d\omega = 2\pi^2F_1/\omega_1 v. \tag{3.10}$$

In this case we have,

$$\kappa(\omega) = \frac{(2\pi/v)F_1\delta(\omega)}{(\omega_\perp^2-\omega^2)^2+\delta^2(\omega)}\,\frac{1}{n(\omega)},$$

where $\omega_\perp^2 = \omega_1^2 - (4\pi/3v)F_1$.

Let us now consider the crystal containing a certain number of substitutional impurity molecules (the treatment can be readily modified for the case of interstitial impurities). Now the internal field depends on the spatial distribution of the impurities. If we ignore fluctuations of this distribution and replace the internal field with its mean value we obtain,

$$\epsilon(\omega) = 1 + 4\pi N_0 a(\omega)\frac{\epsilon+2}{3} + 4\pi N_1[\tilde{a}(\omega) - a(\omega)]\frac{\epsilon+2}{3}, \tag{3.11}$$

where $\tilde{a}(\omega)$ is the polarizability of the impurity molecule, $N_0 = 1/v$, and N_1 is the concentration of impurity molecules (the accuracy of this approximation is discussed in sect. 5). If $\delta\epsilon$ is a small variation of ϵ due to the presence of impurities ($\delta\epsilon \sim N_1$ for $N_1 \ll N_0$) then eq. (3.11) yields

$$\delta\epsilon = \left(1 - \frac{4\pi N_0 a}{3}\right)^{-1} 4\pi N_1(\tilde{a} - a)\frac{\epsilon_0+2}{3},$$

where $\epsilon_0 = \epsilon$ for $N_1 = 0$. Hence (see eq. (3.7a)) we have

$$\epsilon(\omega) = \epsilon_0(\omega) + 4\pi N_1[\tilde{a}(\omega) - a(\omega)]\left(\frac{\epsilon_0+2}{3}\right)^2. \tag{3.12}$$

Using eq. (3.7a) we can rewrite eq. (3.12) as

$$\epsilon(\omega) = \epsilon_0 - \frac{c}{3}(\epsilon_0+2)(\epsilon_0-1) + 4\pi N_1\tilde{a}(\omega)\left(\frac{\epsilon_0+2}{3}\right)^2, \tag{3.12a}$$

where $c = N_1 v$. Assuming that for the impurity molecule we can write, instead of eq. (3.8), the relationship (taking into account decay)

$$\tilde{a}(\omega) = \frac{2\tilde{p}_1^2\tilde{\omega}_1/\hbar}{\tilde{\omega}_1^2 - \omega^2 - i\tilde{\delta}},$$

and that the frequency $\tilde{\omega}_1$ is in the region of transparency of the solvent, we obtain for the frequencies $\omega \sim \tilde{\omega}_1$:

$$2n\kappa = 8\pi N_1\left(\frac{\epsilon_0+2}{3}\right)^2 \tilde{p}_1^2 \frac{\tilde{\omega}_1\tilde{\delta}/\hbar}{[(\tilde{\omega}_1)^2 - \omega^2]^2 + \tilde{\delta}^2}.$$

Hence, integration in the region of the absorption band yields:

$$\int 2n\kappa\, d\omega = 4\pi^2 N_1\left[\frac{\epsilon_0(\tilde{\omega}_1)+2}{3}\right]^2 \frac{\tilde{p}_1^2}{\hbar}. \tag{3.13}$$

Thus, we have found that the coefficient of absorption of light by the impurity in a medium with the dielectric constant $\epsilon_0(\omega)$ and the integral on the left-hand side of eq. (3.13) known as the Kravets integral are proportional to the squared Lorentz factor. The variation of the oscillator strength formally means that under the effect of the medium the dipole moment $\tilde{p}_1$ of the transition in the impurity is replaced by the effective dipole moment $(\tilde{p}_1)_{\text{eff}} = \tilde{p}_1(\epsilon_0 + 2)/3$. (Evidently, the part of the impurity can be played by any crystal defect at which an excited state is localized. If we have $\epsilon_0(\tilde{\omega}_1) \gg 1$ for the transition frequency the intensity I of absorption at the centre increases as $(\tilde{\omega}_1 - \omega_0)^{-2}$ where ω_0 is the resonance frequency for $\epsilon_0(\omega)$. If we take into account the "blurring" effect we obtain $I \sim (\tilde{\omega}_1 - \omega_0)^{-3/2}$ as shown by Rashba (1972). A similar effect takes place for such "impurity" centres as biexcitons (in luminescence) and various other luminescence centres (for instance, exciton localized at the impurity, etc.). Since the experimentally measured quantity in eq. (3.13) is its left-hand side, a correct introduction of the internal field correction (the Lorentz factor in this case) makes it possible to find the oscillator strength for the isolated molecule from the data on dispersion and absorption for the molecules in solution. Of course, this can be done (a well-known fact) only if no chemical bonds arise between the molecules of the solute and the solvent, no aggregates of the impurity molecules are formed, and so on.

In the above discussion we have used a two-level model for the molecules of the impurity and the solvent. A natural question to consider now is how the existence of many resonances of the polarizability of the molecules affects the above results. When we derived eq. (3.12) we did not use the two-level model for describing the molecules of the solvent or the impurity. Therefore, this relation remains valid in the case of many resonances and only the right-hand side of eq. (3.13) is changed. Namely, when we have many levels $i = 1, 2, \ldots$ this term is transformed into a sum of contributions on the individual resonances:

$$\sum_i 4\pi^2 N_1 \left[\frac{\epsilon_0(\tilde{\omega}_i) + 2}{3}\right]^2 \frac{\tilde{p}_i^2}{\hbar}. \tag{3.13a}$$

Now let us consider the impurity-free crystal and assume that the isolated molecules of the crystal have the polarizability,

$$a(\omega) = a_0 + a_1(\omega), \tag{3.14}$$

where the function $a_1(\omega)$ is given by eq. (3.8) while a_0 is determined by the contributions of the far resonances to the polarizability of the molecule; it can be assumed to be a constant independent of ω in the frequency range $\omega \approx \omega_1$. Substituting eq. (3.14) into eq. (3.7) we obtain now,

$$\epsilon(\omega) = \epsilon_b + \frac{(4\pi/v)F_1[(\epsilon_b + 2)/3]^2}{\omega_\perp^2 - \omega^2}, \tag{3.15}$$

where

$$\epsilon_b = 1 + \frac{4\pi}{v} a_0 \left(1 - \frac{4\pi}{3v} a_0\right)^{-1} \tag{3.16}$$

is a quantity (see eq. (3.7)) corresponding to the dielectric constant of the crystal under the assumption that the polarizability of the crystal molecule lacks the resonance term for the frequency ω_1 (that is, $a_1(\omega) \equiv 0$). The resonance frequency $\omega_\perp$ is given by

$$\omega_\perp^2 = \omega_1^2 - \frac{4\pi}{3v} F_1 \left(\frac{\epsilon_b + 2}{3}\right).$$

If we introduce weak decay into eq. (3.15) then eq. (3.10) is replaced by

$$2 \int n(\omega)\kappa(\omega)\, d\omega = \frac{2\pi^2}{v\omega_\perp} F_1 \left(\frac{\epsilon_b + 2}{3}\right)^2. \tag{3.17}$$

We might call the quantity ϵ_b the background dielectric constant with respect to the resonance at the frequency $\omega_\perp$. Since $\epsilon_b \neq 1$ the existence of the background dielectric constant alters the oscillator strength of the transition. This is especially significant for the spectral range of low-intensity molecular transitions (see also Graig and Walmsley (1968)).

If the molecule has two or more close resonances in the given frequency range then we must also distinguish the terms $a_2(\omega)$, $a_3(\omega)$, etc., in eq. (3.14), apart from the terms indicated there. These terms can be readily taken into account though the resulting formulas for $\epsilon(\omega)$ become somewhat cumbersome.

Equations (3.15) and (3.17) are special cases of the more general formulas derived by Agranovich (1968); they demonstrate for the given crystal model how the mixing of molecular configurations due to intermolecular interaction in the crystal affects the oscillator strengths of the dipole transitions. We shall only stress here the significant difference between these equations and similar relationships (3.13) and (3.13a) for the impurity molecules in solutions. When we consider impurity molecules in a solution the Lorentz factor on the right-hand side of eq. (3.13) includes the dielectric constant of the solvent at the transition frequency of the impurity. In the case of pure substances we see from eq. (3.17) that this quantity in the Lorentz factor is replaced by the background dielectric constant which is not at all equal to the squared index of refraction of light in the crystal at the transition frequency. In general, in the two-level model we have $\epsilon_b = 1$ and then eq. (3.15) is transformed into eq. (3.9), and eq. (3.17) into eq. (3.10).

Finally, note that eq. (3.7a) can be used to obtain the resonances of the molecular polarizability from the data on dispersion and absorption of light in the crystal (Bakhshiev 1972). Since $\epsilon = (n + i\kappa)^2$ where n and κ are the index of refraction and absorption coefficient in the crystal eq. (3.7a) yields:

$$\operatorname{Im} a(\omega) = \frac{9n\kappa/2\pi N_0}{(n^2 - \kappa^2 + 2)^2 + 4n^2\kappa^2},$$

where N_0 is the number of molecules per unit volume. It has been shown by Bakhshiev (1972) for a number of examples that the maxima of the function $\operatorname{Im} a(\omega)$ which is proportional to the molecular absorption coefficient can be considerably shifted with respect to the maxima of the quantity $\kappa(\omega)$. It should be noted that this approach to finding the molecular frequencies can be justified only when in the vicinity of dipole transitions of sufficiently high intensity, and when we can neglect the contribution of higher multipoles (this is discussed in sect. 7).

3. *Acting field and dielectric constant of anisotropic crystals*

As an example of an anisotropic crystal, let us consider a molecular crystal with $\sigma \geqslant 1$ molecules per unit cell. If we are interested in the optical properties of the crystal in the frequency range $\omega \gtreqless \omega_1$ where ω_1 is the natural nondegenerate frequency of one of the dipole transitions of the isolated molecule, then its polarizability can be found from

$$a_{ij}(\omega) = \frac{F_1 l_i l_j}{\omega_1^2 - \omega^2}, \tag{3.18a}$$

where F_1 is a quantity proportional to the oscillator strength of the $0 \to 1$ transition and $\boldsymbol{l}$ is the unit vector parallel to the vector of the dipole moment of this transition. Since the different molecules in the unit cell have different orientations their polarizability tensor is

$$a_{ij}^{\alpha}(\omega) = \frac{F_1 l_i^{\alpha} l_j^{\alpha}}{\omega_1^2 - \omega^2}, \qquad \alpha = 1, 2, \ldots, \sigma. \tag{3.18}$$

When we substitute eq. (3.18) into eq. (3.3) and find the scalar product on the left with $\boldsymbol{l}^{\alpha}$ we obtain the following system of equations for the quantities $(\boldsymbol{E}^{\alpha}\boldsymbol{l}^{\alpha})$ where $\alpha = 1, 2, \ldots, \sigma$:

$$(\boldsymbol{E}^{\alpha}\boldsymbol{l}^{\alpha}) - \sum_{\kappa} \boldsymbol{M}_{\alpha\beta}(\omega, \boldsymbol{k})(\boldsymbol{E}^{\beta}\boldsymbol{l}^{\beta}) = (\boldsymbol{E}\boldsymbol{l}^{\alpha}). \tag{3.19}$$

Here,

$$M_{\alpha\beta}(\omega, \boldsymbol{k}) = \frac{F_1}{\omega_1^2 - \omega^2} \sum_{ij} Q_{ij}^{\alpha\beta}(\boldsymbol{k}) l_i^{\alpha} l_j^{\beta}. \tag{3.20}$$

For a crystal with one molecule per unit cell eq. (3.19) directly yields

$$(\boldsymbol{E}^1\boldsymbol{l}^1) = [1 - M_{11}(\omega, \boldsymbol{k})]^{-1}(\boldsymbol{E}\boldsymbol{l}^1),$$

so that the polarization per unit volume is

$$P_i = \frac{1}{v}\, a_{ij}^{(1)}(\omega) E_j^{(1)} = \frac{F_1 l_i^{(1)} l_j^{(1)}}{v(\omega_1^2 - \omega^2)} \frac{E_j}{1 - M_{11}(\omega, \boldsymbol{k})}.$$

Taking this relationship and eq. (3.20) we find the following expression for the dielectric constant tensor:

$$\epsilon_{ij}^0(\omega, \boldsymbol{k}) = \delta_{ij} + \frac{(4\pi F_1/v) l_i l_j}{\omega_1^2 - \omega^2 - F_1 \sum_{i_1 j_1} Q_{i_1 j_1}(\boldsymbol{k}) l_{i_1} l_{j_1}}. \tag{3.21}$$

The resonance of this expression occurs at $\omega = \Omega_1(\boldsymbol{k})$ where

$$\Omega_1^2(\boldsymbol{k}) = \omega_1^2 - F_1 \sum_{i_1 j_1} Q_{i_1 j_1}(\boldsymbol{k}) l_i l_j. \tag{3.22}$$

The tensor (3.21) is for a uniaxial crystal. If one of the coordinate axes, for instance, the x-axis, is taken parallel to $\boldsymbol{l}$ the tensor (3.21) is reduced to diagonal form with the following non-zero components:

$$\epsilon_2 = \epsilon_3 = 1, \qquad \epsilon_1 = 1 + \frac{4\pi F_1/v}{\Omega_1^2 - \omega^2}. \tag{3.21a}$$

Let us now consider crystals with two molecules per unit cell. The optical properties of the molecular crystals of this type are well-known now since this crystal group includes anthracene, naphthalene and other crystals of the aromatic series which were the objects of many experimental studies.

For crystals of the anthracene type there are symmetry operations that transform molecules with $\alpha = 1$ into molecules with $\alpha = 2$. Therefore, when $\boldsymbol{k} = 0$ or when vectors $\boldsymbol{k} \neq 0$ are parallel or perpendicular to the monoclinic axis the relationships $M_{11}(\omega, \boldsymbol{k}) = M_{22}(\omega, \boldsymbol{k})$ and $M_{12}(\omega, \boldsymbol{k}) = M_{21}(\omega, \boldsymbol{k})$ are satisfied. For such vectors $\boldsymbol{k}$ the solution of the system of eqs. (3.19) is less cumbersome. It can be easily seen that in this case we have

$$(\boldsymbol{E}^{(\alpha)}\boldsymbol{l}^{(\alpha)}) = \frac{1}{2}\left[\frac{L_j^{(1)}}{1 - M_{11}(\omega, \boldsymbol{k}) - M_{12}(\omega, \boldsymbol{k})} - \frac{(-1)^\alpha L_j^{(2)}}{1 - M_{11}(\omega, \boldsymbol{k}) + M_{12}(\omega, \boldsymbol{k})}\right] E_j, \qquad \alpha = 1, 2, \tag{3.23}$$

where

$$\boldsymbol{L}^{(1)} = \boldsymbol{l}^{(1)} + \boldsymbol{l}^{(2)}, \qquad \boldsymbol{L}^{(2)} = \boldsymbol{l}^{(1)} - \boldsymbol{l}^{(2)}. \tag{3.23a}$$

Substituting these relationships into eq. (3.5) we can find also the dielectric constant tensor (it was calculated in the framework of the Frenkel exciton theory by Agranovich and Ginsburg (1983), Khokhlov (1972):

$$\epsilon_{ij}(\omega, \boldsymbol{k}) = \delta_{ij} + \frac{2\pi}{v} F_1 \left[\frac{L_i^{(1)} L_j^{(1)}}{\Omega_1^2(\boldsymbol{k}) - \omega^2} + \frac{L_i^{(2)} L_j^{(2)}}{\Omega_2^2(\boldsymbol{k}) - \omega^2} \right], \tag{3.24}$$

Here,

$$\Omega_1^2(\boldsymbol{k}) = \omega_1^2 - F_1 \sum_{ij} Q_{ij}^{11}(\boldsymbol{k}) l_i^{(1)} l_j^{(1)} - F_1 \sum_{ij} Q_{ij}^{12}(\boldsymbol{k}) l_i^{(1)} l_j^{(2)},$$

$$\Omega_2(\boldsymbol{k}) = \omega_1^2 - F_1 \sum_{ij} Q_{ij}^{11}(\boldsymbol{k}) l_i^{(1)} l_j^{(1)} + F_1 \sum_{ij} Q_{ij}^{12}(\boldsymbol{k}) l_i^{(1)} l_j^{(2)}. \tag{3.25}$$

Since the quantities $Q_{ij}(\boldsymbol{k})$ are analytic functions of $\boldsymbol{k}$ the same is true for the frequencies $\Omega_{1,2}(\boldsymbol{k})$, as can be seen also from eq. (3.22). This is not surprising. As was emphasized by Agranovich and Ginsburg (1983) the resonances of the tensor $\epsilon_{ij}(\omega, \boldsymbol{k})$ occur at the so-called mechanical exciton frequencies which are analytic functions of $\boldsymbol{k}$, regardless of the model being used. The vectors $\boldsymbol{L}^{(1)}$ and $\boldsymbol{L}^{(2)}$ are orthogonal. Therefore, if the coordinate axes x and y are parallel to the vectors $\boldsymbol{L}^{(1)}$ and $\boldsymbol{L}^{(2)}$ the tensor ϵ_{ij} is transformed to the diagonal form with the non-zero components:

$$\epsilon_{xx}(\omega, \boldsymbol{k}) = 1 + \frac{2\pi}{v} \frac{F_{1,xx}}{\Omega_1^2(\boldsymbol{k}) - \omega^2},$$

$$\epsilon_{yy}(\omega, \boldsymbol{k}) = 1 + \frac{2\pi}{v} \frac{F_{1,yy}}{\Omega_2^2(\boldsymbol{k}) - \omega^2}, \tag{3.26}$$

$$\epsilon_{zz}(\omega, \boldsymbol{k}) = 1, \qquad F_{1,xx} = F_1 |\boldsymbol{L}^{(1)}|^2, \qquad F_{1,yy} = F_1 |\boldsymbol{L}^{(2)}|^2.$$

$$F_{1,xx}/F_{1,yy} = \cot^2(\theta/2), \qquad \cos\theta = l_1 l_2$$

These relationships show that absorption of light with the electric vector $\boldsymbol{E} \| \boldsymbol{L}^{(1)}$ propagating along the z-axis must occur at the frequency $\Omega_1(\boldsymbol{k})$. If the electric vector $\boldsymbol{E} \| \boldsymbol{L}^{(2)}$ light absorption occurs at $\omega = \Omega_2(\boldsymbol{k})$. Thus, though we assume that the oscillation in the isolated molecule at the frequency ω_1 is nondegenerate, the absorption spectrum of a crystal with two molecules per unit cell should have two differently polarized lines of light absorption. If we know the tensor ϵ_{ij} (see eqs. (3.24) and (3.26)) we can find the position of the absorption line for arbitrary polarizations and directions of propagation of the light and also take into account the spatial dispersion effects in crystals of arbitrary shape (the properties of monolayers are discussed by Agranovich and Dubovskiy (1966), Agranovich (1968).)

This phenomenon (Davydov splitting) has been studied in many systems. Interestingly, although it was found from the application of Frenkel's exciton theory to crystals having several molecules per unit cell, the theory of small-radius excitons was not needed to understand it. As was shown above, it is sufficient to generalize the Lorentz–Lorenz formula for the case of anisotropic crystals to describe this phenomenon. This cannot be said about the exciton spectra of semiconductors. The theory of large radius excitons had to be developed to describe them. This, of course, does not diminish the significance of development of the small radius exciton theory as stimulated by the studies of Frenkel, Peierls, Davydov and other authors. As we know, it is only this theory that makes it possible to account fully for such phenomena as transfer of electronic excitation energy in crystals (see the calculation of the diffusion coefficient of exciton in ch. 5), the optical properties of crystals at high excitation levels, the nonlinear optical effects, the fine details of the light absorption and luminescence spectra of crystals, and other matters (see chs. 4, 5 and 8).

Let us now continue the discussion of the optical properties of crystals in the region of impurity light absorption. In sect. 2 we treated only cubic crystals. Here we shall consider anisotropic crystals using the equations relating the local field and the mean field. It should be noted that, although the optics of impurity centres in crystals has a rich history, the interest in this branch of solid state spectroscopy is still high. One of the reasons is the use of crystals with impurities in new optical devices, other reasons are the discovery of the optical analogue of the Mössbauer effect, the studies of capture of excitons by impurities, energy transfer via the donors (see ch. 2 and ch. 5 sect. 9) and many other interesting optical phenomena occurring, for instance, in anisotropic matrixes.

In this connection, let us discuss how the dielectric properties of the matrix affect the optical properties of the impurity in the anisotropic medium.

4. *Dielectric constant of mixed crystalline solutions and polarization of impurity absorption bands*

Let us start with the simplest model of a crystal in which the polarizability of the molecules of the main substance in vacuum is given by eq. (3.18). For the sake of simplicity, we shall assume that the impurity is of the substitutional type for which the polarizability tensor $\tilde{a}^{\alpha}_{ij}(\omega)$ differs from eq. (3.18a) only in the resonance frequency and the oscillator strength so that we have,

$$\tilde{a}^{\alpha}_{ij}(\omega) = \frac{\tilde{F}_1 l_i^{(\alpha)} l_j^{(\alpha)}}{(\tilde{\omega}_1)^2 - \omega^2}, \quad \tilde{F}^1_1 \neq F_1, \quad \omega_1 \neq \tilde{\omega}_1. \tag{3.27}$$

If the impurity distribution is assumed to be uniform the local field, on the average, can be approximated (in the mean polarizability approximation) by

$$E_i^\alpha = E_i + \sum_{\beta j} Q_{ij}^{\alpha 3}(\boldsymbol{k})\bar{p}_j^\beta, \tag{3.28}$$

where $\bar{p}^\beta$ is the mean polarization of the site β,

$$\bar{p}_i^\beta = (1-c)a_{ij}^\beta E_j^\beta + c\tilde{a}_{ij}^\beta E_j^\beta, \tag{3.29}$$

and c is the ratio of the number of the impurity molecules to the total number of molecules in the crystal. Now, using eqs. (3.28), (3.29), (3.18) and (3.27) we can find an equation for the quantities $\boldsymbol{E}^\alpha \boldsymbol{l}^\alpha$, $\alpha = 1, 2, \ldots, \sigma$, similar to eq. (3.19). As can be readily seen, this equation has the form:

$$\boldsymbol{E}^\alpha \boldsymbol{l}^\alpha - \sum_\beta \tilde{M}_{\alpha\beta}(\omega, \boldsymbol{k})(\boldsymbol{E}^\beta \boldsymbol{l}^\beta) = (\boldsymbol{l}^\alpha \boldsymbol{E}^\alpha), \tag{3.30}$$

where

$$\tilde{M}_{\alpha\beta}(\omega, \boldsymbol{k}) = \sum_{ij} Q_{ij}^{\alpha\beta}(\boldsymbol{k}) l_i^\alpha l_j^\beta \left[\frac{(1-c)F_1}{\omega_1^2 - \omega^2} + \frac{c\tilde{F}_1}{\tilde{\omega}_1^2 - \omega^2}\right]. \tag{3.30a}$$

For crystals having one molecule per unit cell the polarization per unit volume is,

$$p_i = \frac{1}{v}[(1-c)a_{ij}^{(1)}(\omega)E_j^{(1)} + c\tilde{a}_{ij}^{(1)}(\omega)E_j^{(1)}],$$

so that we obtain:

$$\epsilon_{ij}(\omega, \boldsymbol{k}) = \delta_{ij} + \frac{4\pi}{v}\left[\frac{(1-c)F_1}{\omega_1^2 - \omega^2} + \frac{c\tilde{F}_1}{\tilde{\omega}_1^2 - \omega^2}\right][1 - \tilde{M}_{11}(\omega, \boldsymbol{k})]^{-1} l_i l_j. \tag{3.31}$$

When $c \to 0$ eq. (3.31) is transformed into eq. (3.21). Here we are interested in the case $c \ll 1$ when we can expand eq. (3.31) in powers of c and retain only the term linear in c along with ϵ_{ij}^0. It can easily be seen that in this case:

$$\epsilon_{ij}(\omega, \boldsymbol{k}) = \epsilon_{ij}^0(\omega, \boldsymbol{k}) + 4\pi N_1[\tilde{a}_{ij}(\omega) - a_{ij}(\omega)][1 + \xi(\epsilon_1^0 - 1)]^2, \tag{3.32}$$

where $\epsilon_{ij}^0(\omega, \boldsymbol{k})$ is the dielectric constant tensor for the pure crystal as defined by eq. (3.21),

$$\xi = \sum_{ij} vQ_{ij}(\boldsymbol{k}) l_i l_j, \qquad N_1 = c/v,$$

and ϵ_1^0 is given by eq. (3.21a). Thus, eq. (3.32) is similar to eq. (3.12a) derived above for cubic crystals and serves as its generalization for anisotropic crystals of the given type.

Before we proceed to the case of crystals with two molecules per unit cell we should note that the so-called mean polarizability approximation which makes it possible to idealize the crystal as an array of molecules having the mean polarizability given by eq. (3.29) is, actually, a very old approach. In former times (see, for instance, sect. 6 of Vol'kenstein (1951)), this approximation when applied to molecular systems with the Van der Waals interaction was referred to as the "additive refraction approximation". It should be stressed that, although eq. (3.31) provides just a very convenient extrapolation procedure, its accuracy increases with decreasing c so that the term linear in c in eq. (3.32) proves to be exact (but it does not take into account the effect of excitation delocalization discussed below).

Now let us consider crystals of the anthracene type having two molecules per unit cell. Since the system of equations (3.30) differs from eqs. (3.19) only in the replacement $M_{\alpha\beta}(\omega, \mathbf{k}) \to \tilde{M}_{\alpha\beta}(\omega, \mathbf{k})$ we can use eq. (3.23) for the quantities $(\mathbf{E}^\alpha \mathbf{l}^\alpha)$ satisfying the system of equations (3.30) and write directly,

$$(\mathbf{E}^\alpha \mathbf{l}^\alpha) = \frac{1}{2}\left[\frac{L_j^{(1)}}{1-\tilde{M}_{11}-\tilde{M}_{12}} - \frac{(-1)^\alpha L_j^{(2)}}{1-\tilde{M}_{11}+\tilde{M}_{12}}\right] E_j, \qquad \alpha = 1, 2. \tag{3.33}$$

Furthermore, since the dielectric constant tensor of the given medium is found from the relationship:

$$\epsilon_{ij}(\omega, \mathbf{k}) E_j = E_i + \frac{4\pi}{v}(1-c)\sum_\alpha a_{ij}^\alpha E_j^\alpha + \frac{4\pi}{v} c \sum_\alpha \tilde{a}_{ij}^\alpha E_j^\alpha, \tag{3.34}$$

we can substitute eq. (3.33) into eq. (3.34) and finally obtain:

$$\epsilon_{ij}(\omega, \mathbf{k}) = \delta_{ij} + \frac{2\pi}{v}\left[\frac{(1-c)F_1}{\omega_1^2-\omega^2} + \frac{c\tilde{F}_1}{\tilde{\omega}_1^2-\omega^2}\right]$$
$$\times\left[\frac{L_i^{(1)}L_j^{(1)}}{1-\tilde{M}_{11}(\omega, \mathbf{k})-\tilde{M}_{12}(\omega, \mathbf{k})} + \frac{L_i^{(2)}L_j^{(2)}}{1-\tilde{M}_{11}(\omega, \mathbf{k})+\tilde{M}_{12}(\omega, \mathbf{k})}\right]. \tag{3.35}$$

In this approximation (the additive refraction or mean polarizability approximation) eq. (3.35) with the addition of eq. (3.30a) fully determines the dependence of the dielectric constant tensor on the impurity concentration c. The optical properties of mixed crystals with large impurity concentrations are discussed in sect. 5. As we did above for crystals with one molecule per unit cell, we shall discuss here the case of small values of c when we can ignore the terms of the order of c^2, c^3, etc., in the expansion of the tensor (3.35) in powers of c. Then we obtain:

$$\epsilon_{ij}(\omega, \mathbf{k}) = \epsilon_{ij}^0(\omega, \mathbf{k}) + \frac{2\pi}{v} cF_1\left(\frac{1}{\tilde{\omega}_1^2-\omega^2} - \frac{1}{\omega_1^2-\omega^2}\right)$$
$$\times\left[\frac{L_i^{(1)}L_j^{(1)}}{(1-M_{11}-M_{12})^2} + \frac{L_i^{(2)}L_j^{(2)}}{(1-M_{11}+M_{12})^2}\right] \tag{3.36}$$

(for the sake of simplicity, here and below we assume $\tilde{F}_1 = F_1$ which is justified for an isotropic mixture). Using eq. (3.20) we can easily see that the term proportional to $c/(\omega_1^2 - \omega^2)$ in eq. (3.36) identically vanishes for $\omega \to \omega_1$. (A similar situation is found for cubic crystals and anisotropic crystals with one molecule per unit cell discussed above as can be seen from eqs. (3.12a) and (3.32), respectively.)

Thus, in the frequency range of impurity absorption, $\tilde{\omega}_1 \approx \omega$, to the only resonance term in eq. (3.36) has the form

$$\delta\epsilon_{ij}(\omega, \boldsymbol{k}) = \frac{2\pi}{v} c \frac{F_1}{\tilde{\omega}_1^2 - \omega^2} \left[\frac{L_i^{(1)} L_j^{(1)}}{(1 - M_{11} - M_{12})^2} + \frac{L_i^{(2)} L_j^{(2)}}{(1 - M_{11} + M_{12})^2} \right]. \tag{3.37}$$

However, if we recall what is said in sect. 2 about the role of the local field correction on impurity spectra we can write eq. (3.37) directly without expanding eq. (3.35) in powers of c.

Indeed, in the isolated molecule of an isotopic impurity having the orientation α the transition dipole moment is $\boldsymbol{p}^\alpha = p\boldsymbol{l}^\alpha$. The effective value corresponding to this dipole moment should be found from the condition

$$-\boldsymbol{p}^\alpha \boldsymbol{E}^\alpha = -\boldsymbol{p}_{\text{eff}}^\alpha \boldsymbol{E}. \tag{3.38}$$

Making use of eq. (3.23) we obtain

$$\boldsymbol{p}_{\text{eff}}^\alpha = \frac{p}{2} \left(\frac{\boldsymbol{L}^{(1)}}{1 - M_{11} - M_{12}} - \frac{(-1)^\alpha \boldsymbol{L}^{(2)}}{1 - M_{11} + M_{12}} \right). \tag{3.39}$$

On the other hand, we have by definition,

$$\delta\epsilon_{ij}(\omega, \boldsymbol{k}) = \frac{4\pi}{v} c \frac{F_1}{p^2} \sum_\alpha \frac{(\boldsymbol{p}_{\text{eff}}^\alpha)_i (\boldsymbol{p}_{\text{eff}}^\alpha)_j}{\tilde{\omega}_1^2 - \omega^2}, \tag{3.40}$$

and substitution of eq. (3.39) into eq. (3.40) directly yields eq. (3.37).

Equation (3.37) makes it possible to analyze the dependence of the intensity of impurity absorption on the polarization of the incident light. Rashba has treated this problem within the framework of the theory of small radius excitons (Agranovich 1968, Rashba 1972) for the isotopic substitutional impurities.

For the given crystal model the tensor $\epsilon_{ij}(\omega, \boldsymbol{k})$ is reduced to the diagonal form if the coordinate axes x and y are parallel to the vectors $\boldsymbol{L}^{(1)}$ and $\boldsymbol{L}^{(2)}$. Then we have:

$$\begin{aligned}
\epsilon_{11}(\omega, \boldsymbol{k}) &= \epsilon_{11}^0(\omega, \boldsymbol{k}) + \frac{2\pi}{v} c \frac{F_1}{\tilde{\omega}_1^2 - \omega^2 - i\eta} \frac{|\boldsymbol{L}^{(1)}|^2}{(1 - M_{11} - M_{12})^2}, \\
\epsilon_{22}(\omega, \boldsymbol{k}) &= \epsilon_{22}^0(\omega, \boldsymbol{k}) + \frac{2\pi}{v} c \frac{F_1}{\tilde{\omega}_1^2 - \omega^2 - i\eta} \frac{|\boldsymbol{L}^{(2)}|^2}{(1 - M_{11} + M_{12})^2}, \\
\epsilon_{33}(\omega, \boldsymbol{k}) &= 1,
\end{aligned} \tag{3.41}$$

where a possible weak decay has been taken into account ($\eta > 0$).

For instance, assume that the light is polarized along the x-axis (parallel to $\boldsymbol{L}^{(1)}$). Assuming that the matrix is transparent at the frequency $\omega \approx \tilde{\omega}_1$ we derive the following expression for the absorption coefficient of the impurity:

$$\kappa_{\mathrm{I}}(\omega) = \frac{\pi c}{2v\tilde{\omega}_1} \frac{1}{n_{\mathrm{I}}^0(\omega)} \frac{F_1|\boldsymbol{L}^{(1)}|^2 \delta(\omega - \tilde{\omega}_1)}{(1 - M_{11} - M_{22})^2}, \tag{3.42}$$

Here $n_{\mathrm{I}}^0(\omega) = \sqrt{\epsilon_{xx}^0(\omega, \boldsymbol{k})}$ is the refractive index of the matrix for light polarized along the x-axis. Similarly, we obtain for light polarized along the y-axis (parallel to $\boldsymbol{L}^{(2)}$):

$$\kappa_{\mathrm{II}}(\omega) = \frac{\pi c}{2v\tilde{\omega}_1} \frac{1}{n_{\mathrm{II}}^0(\omega)} \frac{F_1|\boldsymbol{L}^{(2)}|^2 \delta(\omega - \tilde{\omega}_1)}{(1 - M_{11} + M_{12})^2}, \tag{3.43}$$

where $n_{\mathrm{II}}^0(\omega) = \sqrt{\epsilon_{yy}^0(\omega, \boldsymbol{k})}$ can be found from eqs. (3.26). The ratio of the integral absorption intensities $\bar{\kappa}_{\mathrm{I,II}}$ (given by $\int \kappa(\omega)\,\mathrm{d}\omega$) corresponding to the absorption coefficients κ_{I} and κ_{II} is, evidently, equal to

$$\frac{\bar{\kappa}_{\mathrm{I}}(\tilde{\omega}_1)}{\bar{\kappa}_{\mathrm{II}}(\tilde{\omega}_1)} = \frac{n_{\mathrm{II}}^0(\tilde{\omega})}{n_{\mathrm{I}}^0(\tilde{\omega}_1)} \frac{|\boldsymbol{L}^{(1)}|^2 (1 - M_{11} + M_{12})^2}{|\boldsymbol{L}^{(2)}|^2 (1 - M_{11} - M_{12})^2}\bigg|_{\omega = \tilde{\omega}_1}. \tag{3.44}$$

In order to rewrite this relationship in a more explicit form let us turn back to the dielectric constant tensor (3.26) for the pure crystal. Using eqs. (3.20) and (3.25) we can rewrite eq. (3.44) in the following form:

$$\frac{\bar{\kappa}_{\mathrm{I}}(\tilde{\omega}_1)}{\bar{\kappa}_{\mathrm{II}}(\tilde{\omega}_1)} = \frac{n_{\mathrm{II}}^0(\tilde{\omega}_1)}{n_{\mathrm{I}}^0(\tilde{\omega}_1)} \frac{F_{1,xx}[\tilde{\omega}_1^2 - \Omega_2^2(\boldsymbol{k})]^2}{F_{1,yy}[\tilde{\omega}_1^2 - \Omega_1^2(\boldsymbol{k})]^2}. \tag{3.45}$$

If the frequency $\tilde{\omega}_1$ is close to the lattice absorption frequencies so that the difference $|\tilde{\omega}_1 - \Omega_{1,2}|$ is sufficiently small, eq. (3.45) can be also written as:

$$\frac{\bar{\kappa}_{\mathrm{I}}}{\bar{\kappa}_{\mathrm{II}}} = \frac{n_{\mathrm{II}}^0(\tilde{\omega}_1)}{n_{\mathrm{I}}^0(\tilde{\omega}_1)} \frac{F_{1,xx}(\tilde{\omega}_1 - \Omega_2(\boldsymbol{k}))^2}{F_{1,yy}(\tilde{\omega}_1 - \Omega_1(\boldsymbol{k}))^2}. \tag{3.46}$$

This relationship shows that when the frequency $\tilde{\omega}_1$ tends, for instance, to the frequency Ω_1 the quantity $\bar{\kappa}_{\mathrm{I}}$ can become anomalously high as compared with $\bar{\kappa}_{\mathrm{II}}$. Thus, absorption at the impurity frequency becomes sharply polarized even though sharp polarization of the impurity absorption may be absent far from the frequencies Ω_1 and Ω_2. Equation (3.46) for $\boldsymbol{k} = 0$ has been derived by Rashba (1972). The importance of taking into account spatial dispersion in eq. (3.46) will be discussed below. Now we shall make a few remarks on the nature of the effect described by eq. (3.46). As eq. (3.39) implies, this effect is due entirely to the internal field correction which increases when the frequency $\tilde{\omega}_1$ approaches the resonances of the dielectric constant tensor for the pure crystal. Of course, this mechanism determining the effect of the matrix occurs also in crystals of any structure and also for interstitial impurities. However, the structure and effect of the internal field can differ in different cases. For instance, for cubic crystals, as seen from eq. (3.12a), and for crystals with one molecule per unit cell, given the same dependence on the frequency $\tilde{\omega}_1$, the quantity $\bar{\kappa}$ proves to be proportional to the square of the transition oscillator strength for the matrix, rather than to its first power as in eq. (3.42).

Now let us continue the analysis of eq. (3.46). Assume that the frequency $\tilde{\omega}_1$ is in the region of a high-intensity dipole transition in the matrix and it is just this transition that determines dispersion of the refractive indexes $n^0_{\text{I,II}}(\omega)$ in the given frequency range. Then, if $\tilde{\omega}_1 < \Omega_1(0)$, $\Omega_2(0)$ and the differences $|\Omega_{1,2} - \tilde{\omega}_1|$ are small but yet great enough that we can neglect the spatial dispersion of the medium, we obtain:

$$n^0_{\text{I}}(\omega) \sim \sqrt{\frac{F_{1,xx}}{|\omega - \Omega_1(0)|}}, \qquad n^0_{\text{II}}(\omega) \sim \sqrt{\frac{F_{1,yy}}{|\omega - \Omega_2(0)|}}.$$

Under these conditions eq. (3.46) is transformed into:

$$\frac{\bar{\kappa}_{\text{I}}}{\bar{\kappa}_{\text{II}}} = \sqrt{\frac{F_{1,xx}}{F_{1,yy}}} \left| \frac{\tilde{\omega}_1 - \Omega_2(0)}{\tilde{\omega}_1 - \Omega_1(0)} \right|^{3/2}. \tag{3.47}$$

If the spatial dispersion must be taken into account then, since for small $\boldsymbol{k}$ we have:

$$\Omega_1(\boldsymbol{k}) = \Omega_1(0) + \mu_1(n^0_{\text{I}})^2, \qquad \Omega_2(\boldsymbol{k}) = \Omega_2(0) + \mu_2(n^0_{\text{II}})^2. \tag{3.48}$$

Equation (3.46) can be rewritten as,

$$\frac{\bar{\kappa}_{\text{I}}}{\bar{\kappa}_{\text{II}}} = \frac{F_{1,xx}}{F_{1,yy}} \left\{ \frac{\sqrt{n^0_{\text{II}}(\tilde{\omega}_1)}[\tilde{\omega}_1 - \Omega_2(0) - \mu_2(n^0_{\text{II}})^2]}{\sqrt{n^0_{\text{I}}(\tilde{\omega}_1)}[\tilde{\omega}_1 - \Omega_1(0) - \mu_1(n^0_{\text{I}})^2]} \right\}^2, \tag{3.49}$$

where $n^0_{\text{I,II}}$ satisfy the equations:

$$n^2_{\text{I}}(\tilde{\omega}_1) = \epsilon^0_{xx}(\tilde{\omega}_1, \boldsymbol{k}) = 1 + \frac{2\pi}{v} \frac{F_{1,xx}}{\Omega^2_1(0) + \mu_1 n^2_{\text{I}} - \tilde{\omega}^2_1},$$

$$n^2_{\text{II}}(\omega_1) = \epsilon^0_{yy}(\tilde{\omega}_1, \boldsymbol{k}) = 1 + \frac{2\pi}{v} \frac{F_{1,yy}}{\Omega^2_2(0) + \mu_2 n^2_{\text{II}} - \tilde{\omega}^2_1}. \tag{3.50}$$

The nature of the dependences $n_{\text{I}}(\omega)$ and $n_{\text{II}}(\omega)$ are known to be dependent on the signs of the coefficients μ_1 and μ_2, respectively (Agranovich and Ginsburg 1983). Therefore, the experimental results for the ratio (3.46) can, in principle, be used to identify the spatial dispersion effects of the crystal matrix. However, the decay of the excited states of both the impurity and the matrix makes this difficult. If the levels of these states are wide enough then it becomes practically impossible to "sneak up" on the frequency $\Omega_1(0)$ and to distinguish the impurity absorption from the matrix absorption. Nevertheless, when we compare the above relationships with the similar relationships derived within the framework of the exciton theory (Rashba 1972) we shall not take into consideration the broadening of the terms of the impurity and the crystal matrix since it is just in the immediate vicinity of the frequency $\Omega_1(0)$ that the above-mentioned spatial dispersion effects and the differences from the results of the exciton theory are manifested.

Let us examine these results in greater detail. It follows from a paper by Rashba (1972) that the integral absorption intensities $\bar{\kappa}_\mu(\mu = \text{I}, \text{II})$ are proportional to the quantities $F_\mu/(\tilde{\omega}_1 - \Omega_\mu)^2$, $F_\text{I} = F_{1,xx}$ and $F_\text{II} = F_{1,yy}$ that appear in eqs. (3.42) and (3.43) and also to the very same factor $|u(0)|^2$ which is given by the relationship (see ch. 4 sect. 7):

$$|u(0)|^{-2} = \frac{\tilde{\omega}_1 - \omega_1}{N_0} \sum_{k,\mu=1,2} \frac{|u_\mu(\boldsymbol{k})|^2}{[\omega_i - \Omega_\mu(\boldsymbol{k})]^2}, \tag{3.51}$$

where the coefficients $u_\mu(\boldsymbol{k})$ satisfy the normalization condition:

$$\sum_\mu |u_\mu(\boldsymbol{k})|^2 = 1, \tag{3.52}$$

and ω_i is the frequency of the transition in the impurity centre which starts to differ noticeably from $\tilde{\omega}_1$ as $\tilde{\omega}_1 \to \Omega_1(0)$. For the frequencies $\tilde{\omega}_1$ which are sufficiently far from $\Omega_1(0)$ the frequency $\omega_i \approx \tilde{\omega}_1$ and the value of $|u(0)|^2$ proves to be unity accurate to the small terms of the order of $\delta = |[\Omega_1(0) - \Omega_2(0)]/(\tilde{\omega}_1 - \Omega_1)|$. Therefore, in this frequency range the results derived above and the results of Rashba (1972) prove to coincide not only for the ratio $\bar{\kappa}_\text{I}/\bar{\kappa}_\text{II}$ but also for the quantities $\bar{\kappa}_\text{I,II}$ themselves. When $\tilde{\omega}_1$ approaches $\Omega_1(0)$ the value of $|u(0)|^2$ decreases and, for instance, if $\mu_1 > 0$ then $|u(0)|^2 \sim \sqrt{\Omega_1(0) - \tilde{\omega}_1}$. Under these conditions, as was shown by Rashba (1972), the excited state of the local centre already encompasses not only the impurity molecule but also a number of solvent molecules nearest to the impurity.

The quantity $|u(0)|^2$ has a quite clear physical meaning. Under these conditions it determines the probability that the impurity molecule itself is in the excited state and the decrease of this probability when $\tilde{\omega}_1 \to \Omega_1(0)$ is due to the "spreading" of the local excited state.

Typically, the papers dealing with impurity absorption interpret the sharp polarization of the impurity lines for $\tilde{\omega}_1 \to \Omega_1(0)$ precisely in terms of this "spreading". As follows from the above discussion, the main effect actually resulting in the sharp polarization is the effect of the internal field correction; the quantity $|u(0)|^2$ does not enter the ratio $\bar{\kappa}_\text{I}/\bar{\kappa}_\text{II}$ at all and in the expressions for $\bar{\kappa}_\text{I}$ and $\bar{\kappa}_\text{II}$ it only gives rise to additional, relatively weak root-law dependences on the frequency. (It becomes particularly clear that we are dealing with effects of different natures when we take into account the fact that the magnitude of the acting field at the site of the impurity is practically independent of the properties of the impurity and is not changed if, for instance, we replace the impurity with a vacancy.)

Interestingly, some experimental results actually demonstrate the relative weakness of the "spreading" effect. For instance, Broude et al. (1961) reports a variation of $\bar{\kappa}_{\mathrm{I}}$ by almost two orders of magnitude for deuterated naphthalene in naphthalene while, as estimated by Broude et al. (1961) themselves, even the minimum value of $|u(0)|^2$ is about 0.5 (according to Broude et al. (1961), the minimum value of $|u(0)|^2$ for deuterated benzene is as high as 0.8). Of course, it is not surprising that the internal field corrections remain unchanged when the excited state of the local centre is "spreading". This is due to the fact that, if the radius of this excited state is small compared with the wavelength of the light, the internal field correction is practically independent of this radius (the internal field corrections in ionic crystals are discussed by Smith and Dexter (1972)).

5. *Optical properties of mixed crystalline solutions*

Now we shall discuss some features of the electromagnetic wave spectrum in mixed crystalline solutions using the equations for the dielectric constant tensor of solution derived in the mean polarizability approximation. We know that the effects of concentration broadening of absorption spectra are lost in this approximation. However, Onodera and Toyozawa (1968), Dubovskiy and Konobeev (1970). Hoshen and Jortner (1970), Hong and Robinson (1970), who have actually studied the corrections to the mean polarizability approximation, have shown that this approximation is more or less suitable for treating such relatively coarse features of a spectrum as, for instance, the dependence of the position of the centre of an absorption band on the composition of the solution and some others.

Moreover, the mean polarizability approximation can yield highly accurate results for dispersion and optical anisotropy of crystalline solutions outside the absorption band. This is due to the fact that the concentration broadening in crystals of this type (with only Van der Waals interaction between the molecules) does not affect the integral oscillator strength of a transition. Furthermore, the mean polarizability approximation served as the basis for the procedure developed by Obreimov for the analysis of the composition of multi-component systems as applied to a wide variety of isotropic mixtures, both liquid and crystalline (for the details, see Obreimov 1945).

Clearly, the explanation of the high accuracy of the additive refraction approximation is that, though it yields an incorrect band structure, the error introduced into the index of refraction far from the absorption bands proves to be a small quantity of the order of $\delta/|\omega - \Omega| \ll 1$ where δ is the width of the resonance, Ω is the resonance frequency, and ω is the frequency in the region of transparency. Even for $\delta \approx 10^2\ \mathrm{cm}^{-1}$ and $|\omega - \Omega| = 10^4\ \mathrm{cm}^{-1}$ this error is of the order of 10^{-2} (therefore, the measurements of dispersion in the region of transparency can be used for analyzing the composition of a mixture the knowledge of which is needed for interpreting the results on concentration broadening and comparing them with theoretical predictions). We have not met in the literature any mention of the equivalence of the mean polarizability approximation and the additive refraction approximation, at least as applied to molecular crystals. More recently, the need for studying the properties of mixed crystalline solutions arose and workers in the field started to "discover" some methods developed long ago but, alas, forgotten.

Let us illustrate the mean polarizability approximation by applying it to the spectrum of a binary mixture of isotopically related molecules in a cubic crystal. In this case eq. (3.11) gives

$$\epsilon(\omega) = 1 + \frac{4\pi N_0 \bar{a}}{1 - (4\pi/3)N_0\bar{a}}, \qquad \bar{a} = (1-c)a + c\tilde{a}, \tag{3.53}$$

where c is the relative concentration of the impurity. Thus, the resonances of $\epsilon(\omega)$ correspond to the frequencies that depend on c and satisfy the equation

$$\frac{1}{A} = \frac{1-c}{\omega_1^2 - \omega^2} + \frac{c}{\tilde{\omega}_1^2 - \omega^2}, \tag{3.54}$$

where $A = (4\pi/3)N_0F_1$.

In deriving eq. (3.54) we have used the fact that the polarizabilities of the different isotopic molecules are given by eq. (3.8) and differ from one another only in the values of the resonance frequencies. Equation (3.54) enables us to determine the shift of the absorption lines of the mixture as a function of its composition in the given approximation. The dependence of the frequencies of the longitudinal waves on the composition of the mixture can be found from the equation $\epsilon(\omega) = 0$. Using eq. (3.53) we find that these frequencies satisfy the equation

$$-\frac{1}{2A} = \frac{1-c}{\omega_1^2 - \omega^2} + \frac{c}{\tilde{\omega}_1^2 - \omega^2}.$$

In a similar way we can find the dependence, for instance, of the frequency of the surface waves, on the composition of the mixture (from the condition $\epsilon(\omega) = -1$; the accuracy of determination of these frequencies is higher since they are farther from the resonance) and many other parameters of the mixture that are determined by its dielectric constant.

The treatment of anisotropic crystalline solutions can be also carried out very simply. For such solutions consisting of two types of isotopic molecules and having two molecules per unit cell the dielectric constant tensor has been derived above in the additive refraction approximation (see eq. (3.35)). This equation shows that in the polarization $\boldsymbol{L}^{(1)}$ the resonances of the tensor ϵ_{ij} correspond to the frequencies ω satisfying the equation

$$1-\tilde{M}_{11}(\omega, \boldsymbol{k})-\tilde{M}_{12}(\omega, \boldsymbol{k})=0, \tag{3.55}$$

and in the polarization $\boldsymbol{L}^{(2)}$ the respective frequencies satisfy the equation

$$1-\tilde{M}_{11}(\omega, \boldsymbol{k})+\tilde{M}_{12}(\omega, \boldsymbol{k})=0. \tag{3.56}$$

Using eq. (3.30a) we can transform eqs. (3.55) and (3.56) into

$$\frac{1}{A_{11}-A_{12}(-1)^{\rho}}=\frac{1-c}{\omega_1^2-\omega^2}+\frac{c}{\tilde{\omega}_1^2-\omega^2}, \qquad \rho=1,2, \tag{3.57}$$

where

$$A_{11}=F_1\sum_{ij} Q_{ij}^{11}l_i^1l_j^1, \qquad A_{12}=F_1\sum_{ij} Q_{ij}^{12}l_i^1l_j^2.$$

Equation (3.57) implies that a doublet of lines should correspond to each of the polarizations $\rho=1,2$ in the light absorption spectrum. If we denote the solutions of eq. (3.57) for each of the polarizations ρ by $\Omega_{\rho\lambda}^2$ ($\lambda=1,2$) then after expansion into elementary fractions we can write the tensor $\epsilon_{ij}(\omega, \boldsymbol{k})$ as

$$\epsilon_{ij}(\omega, \boldsymbol{k})=\delta_{ij}+\sum_{\rho,\lambda}\frac{\Pi_{\rho\lambda}}{\Omega_{\rho\lambda}^2-\omega^2}L_i^{(\rho)}L_j^{(\rho)}, \tag{3.58}$$

where

$$\Pi_{\rho\lambda}=\frac{2\pi}{v}\frac{F_1}{[A_{11}-(-1)^{\rho}A_{12}]^2}\left[\frac{1-c}{(\omega_1^2-\Omega_{\rho\lambda}^2)^2}+\frac{c}{(\tilde{\omega}_1^2-\Omega_{\rho\lambda}^2)^2}\right]^{-1}, \tag{3.59}$$

$$\rho=1,2, \quad \lambda=1,2.$$

The solutions of eq. (3.57) and eqs. (3.58) and (3.59) in this approximation fully determine the dependences on c of not only the intensities and positions of the light absorption bands for a crystal but also the dispersion of the refractive indexes. The above equations can be readily extended to the case of multi-component isotopic solutions. Then eqs. (3.30) and (3.58) retain their form but eqs. (3.30a), (3.57) and (3.59), respectively, are replaced by the more general equations:

$$\tilde{M}_{\alpha\beta}(\omega, \boldsymbol{k})=\sum_k\frac{F_1c_k}{\omega_k^2-\omega^2}\sum_{ij}Q_{ij}^{\alpha\beta}(\boldsymbol{k})l_i^{\alpha}l_j^{\beta},$$

(here c_k are the relative concentrations, $k = 1, 2, \ldots, s$, so that $\Sigma_k c_k = 1$)

$$\frac{1}{A_{11} - (-1)^\rho A_{12}} = \sum_k \frac{c_k}{\omega_k^2 - \omega^2}, \tag{3.57a}$$

$$\Pi_{\rho\lambda} = \frac{2\pi}{v} \frac{F_1}{[A_{11} - (-1)^\rho A_{12}]^2} \left[\sum_k \frac{c_k}{(\omega_k^2 - \Omega_{\rho\lambda}^2)^2} \right]^{-1}, \tag{3.59a}$$

where $\Omega_{\rho\lambda}^2$ are the solutions of eq. (3.57) for fixed $\rho = 1, 2$ and $\lambda = 1, 2, \ldots, s$.

It should be stressed that the results obtained here for the positions of the lines and the intensities of absorption in such solutions coincide with the results of Broude and Rashba (1961) which were derived in the framework of the theory of small radius excitons using an approximation identical with the mean polarizability approximation (the additive refraction approximation; the entire tensor $\epsilon_{ij}(\omega, \boldsymbol{k})$ was not found in this study).

Apart from the isotopic mixtures, the acting field method described here can be similarly applied to molecular mixtures of quite different molecules. Here we should only bear in mind the fact that in this more general case the resonance frequencies ω_k for individual molecules in solution become functions of the composition, in contrast to the case of isotopic molecules for which all the static multipoles can be considered independent of the isotopic composition to a high degree of accuracy. This dependence on composition can be easily understood since it is due to the shift of the resonance frequency of the individual molecule mentioned in sect. 1. As shown there, this shift is caused by the change upon excitation of the energy of interaction of the static multipoles of the molecule with the environment. This environment differs for different compositions of the mixture and this is the cause of the above dependence. We can take it into account by using the following notation: ω_k^0 is the resonance of the molecule in vacuum, and D_{kl} is the shift of this resonance in the environment consisting only of the molecules of the species l. Clearly, in the mean polarizability approximation we obtain for the mixture,

$$\omega_k \equiv \omega_k(c_1, c_2, \ldots) = \omega_k^0 + \sum_l D_{kl} c_l.$$

6. *Energy of the resonance interaction of the impurity molecules with one another*

In the preceeding sections of this chapter the local field method was applied mainly to some problems of the theory of optical properties of crystals and crystalline solutions. In this section we shall show how the above results can be used for determining the matrix elements of the operator for the energy of interaction of the impurity molecules with one another which corresponds to transfer of energy of intramolecular (electronic or vibrational) excitation from one impurity molecule to another. As shown in ch. 1 sect. 4 (see also sect. 1), this matrix element, generally speaking, can depend to a considerable extent on the properties of the matrix and of course, the probability of radiationless energy transfer between the impurity molecules which are sufficiently far from one another depend on it. In our analysis of the "mechanism" of the matrix effect we shall limit ourselves to the case when the main role is played by the dipole–dipole interaction as in the Förster mechanism. The operator of this energy (see also eq. (1.18)) in vacuum is

$$\hat{V}^0_{\mathrm{ab}} = \frac{R^2\hat{\boldsymbol{p}}_{\mathrm{a}}\hat{\boldsymbol{p}}_{\mathrm{b}} - 3(\hat{\boldsymbol{p}}_{\mathrm{a}}\boldsymbol{R})(\hat{\boldsymbol{p}}_{\mathrm{b}}\boldsymbol{R})}{R^5},$$

where $\hat{\boldsymbol{p}}_{\mathrm{a}}$ and $\hat{\boldsymbol{p}}_{\mathrm{b}}$ are the operators of the dipole moments of the impurity molecules a and b, and $\boldsymbol{R}$ is the vector between the centres of these molecules. If initially the molecule a was in the excited state f and the identical molecule b was in the ground state and in the final state the situation is reversed, the matrix element of the operator $\hat{V}^0_{\mathrm{ab}}$ corresponding to this transition would be:

$$M_{\mathrm{ab}} \equiv V^{f0,0f}_{\mathrm{ab}} = \frac{R^2\boldsymbol{p}^{f0}_{\mathrm{a}}\boldsymbol{p}^{0f}_{\mathrm{b}} - 3(\boldsymbol{p}^{f0}_{\mathrm{a}}\boldsymbol{R})(\boldsymbol{p}^{0f}_{\mathrm{b}}\boldsymbol{R})}{R^5}. \tag{3.60}$$

If the impurity molecules were in a polarizing medium, then the appropriate corrections should be made in the interaction energy operator and its matrix element. At first sight, it might seem that this can be done, for instance, for the isotropic medium and $c/\omega_{0f} \gg R \gg d$ (d is the crystal lattice parameter), just by dividing the matrix element M_{ab} by $\epsilon(\omega_{0f})$ where $\epsilon(\omega_{0f})$ is the dielectric constant of the solvent at the frequency ω_{0f}. However, this approach would not be quite correct even though it corresponds to the well-known result obtained within the framework of the phenomenological Maxwell equations. If we introduce a source dipole μ_{a} at a point a into these equations the electric field produced by this dipole is of the order of $\mu_{\mathrm{a}}/\epsilon R^3$ so that the resulting interaction energy is indeed of the order of

$$U \sim \mu_{\mathrm{a}}\mu_{\mathrm{b}}/R^3_{\mathrm{ab}}\epsilon(\omega_{0f}). \tag{3.61}$$

However, we should bear in mind that all the quantities entering into the phenomenological Maxwell equations should be regarded as phenomenological quantities which can be considerably different from the respective microscopic quantities. This fact is well known, for instance, for the dielectric constant $\epsilon(\omega)$. Here it is expressed as the difference between the polarizability of individual molecules and the polarizability of the crystal. The same is true for diagonal and off-diagonal elements of the dipole operator, that is, the phenomenological or effective value of this dipole in the medium differs considerably from the corresponding value in vacuum. In particular, as shown in ch. 1 sect. 4 for the model of two dipoles, eq. (3.60) is replaced by

$$(M_{\rm ab})_{\rm eff} \equiv (V_{\rm ab}^{f0,0f})_{\rm eff} = \frac{1}{\epsilon}\left(\frac{\epsilon+2}{3}\right)^2 V_{\rm ab}^{f0,0f}, \tag{3.62}$$

where ϵ is the dielectric constant of the medium at the transition frequency. This equation holds if we take into account the crystal structure of the matrix (for cubic crystals).

Equation (3.62) is derived within the framework of the microscopic theory in ch. 4 sect. 7 (Agranovich et al. 1969, Kamenogradskiy and Konobeev 1970) where anisotropic crystals are also treated and some microscopic corrections are taken into consideration (the shift of the impurity term and certain delocalization of excitation). (Mahan (1967), Mahan and Mazo (1968) derived eq. (3.62) for static dipoles ($\omega_{0f} = 0$) in cubic crystals. Nienhuis and Deutch (1972) reported calculations of the energy of interaction between charges and dipoles in a medium.) If the transition frequency is sufficiently far from the resonance of $\epsilon(\omega)$, as we shall assume below for the sake of simplicity, these corrections are not significant and we shall ignore them. Quite elementary and general analysis is sufficient to understand the new factors appearing in eq. (3.62) as compared with eq. (3.60) since these factors have a macroscopic nature.

Indeed, the effect of polarization of the medium on the dipole–dipole interaction energy is determined by the exchange of the virtual quanta of the electromagnetic field in the medium which is described by the phenomenological Maxwell equations. Therefore, we have to know the energy of interaction of the impurities with just this field. The dipole moment appearing in eq. (3.38) is not the dipole moment in vacuum but a certain effective dipole moment found by taking into account the internal field correction. In the isotropic medium this is $\mu_{\rm a} = [(\epsilon+2)/3]p_{\rm a}^{0f}$ which yields precisely eq. (3.62) if we take into account eq. (3.61). This conclusion is valid for the local centres of any nature in the nonconducting medium. The effect of polarization of the solvent on the resonance exchange energy in molecular media of any symmetry can be easily analyzed by taking into account the internal field corrections as discussed in sect. 3. (We mean here media in which the intermolecular interaction does not violate the neutrality of molecules. The specific effects found in ionic crystals are discussed by Smith and Dexter (1972)).

Proceeding from the above discussion, let us now find the general form of the energy of the dipole–dipole resonance radiationless interaction in anisotropic crystals (some more special results for crystals having one molecule per unit cell are reported by Agranovich et al. (1969), Kamenogradskiy and Konobeev (1970). To do this we must generalize somewhat the results of sect. 3. Assume that $a_\nu(\omega)$ are the principal values of the polarizability tensor of the individual molecule ($\nu = 1, 2, 3$) and $\boldsymbol{l}^\nu$ are the directions of its principal axes. Then we can write this tensor as

$$a_{ij}^{\alpha}(\omega) = \sum_{\nu} a_\nu(\omega) l_i^{\alpha\nu} l_j^{\alpha\nu}.$$

If we take into account only one of the excited states in the molecule this equation is replaced by eq. (3.18a). In the region of transparency many of the excited molecular states make comparable contributions to the polarizability of the molecule so that the approximation (3.18a) becomes insufficient. A similar situation occurs also in crystalline solutions. In this case, in the mean polarizability approximation (the additive refraction approximation) we should write, for instance, for the isotopic mixtures,

$$a_{ij}^{\alpha}(\omega) = \sum_{\nu} \bar{a}_\nu(\omega) l_i^{\alpha\nu} l_j^{\alpha\nu}, \qquad \bar{a}_\nu(\omega) = \sum_{\rho} c_\rho a_\nu^\rho(\omega),$$

where c_ρ is the relative concentration of isotopic component ρ, and $a_\nu^\rho(\omega)$ are the respective principal values of the polarizability tensor of the molecule ($\nu = 1, 2, 3$). Using eqs. (3.1) and (3.2) we find that the projections of the internal field $\boldsymbol{E}^\alpha$ on the directions $\boldsymbol{l}^{\alpha\nu}$, that is, the quantities $\boldsymbol{E}^\alpha \boldsymbol{l}^{\alpha\nu}$, satisfy a system of 3σ equations ($\alpha = 1, 2, \ldots, \sigma$; $\nu = 1, 2, 3$):

$$(\boldsymbol{E}^\alpha \boldsymbol{l}^{\alpha\nu}) = (\boldsymbol{E}\boldsymbol{l}^{\alpha\nu}) + \sum_{\beta\nu'} M_{\alpha\beta}^{\nu\nu'}(\omega, \boldsymbol{k})(\boldsymbol{E}^\beta \boldsymbol{l}^{\beta\nu'}),$$

where

$$M_{\alpha\beta}^{\nu\nu'}(\omega, \boldsymbol{k}) = a^\nu(\omega) \sum_{ij} Q_{ij}^{\alpha\beta}(\boldsymbol{k}) l_i^{\alpha\nu} l_j^{\beta\nu'}.$$

For fixed α the unit vectors $l^{\alpha\nu}$ make up a triad of mutually orthogonal vectors and, therefore, we can use the solution of the above system of inhomogeneous equations to find the form of the tensor $A_{ij}^{\alpha}(\omega, \boldsymbol{k})$ in eq. (3.4). Then, if $\boldsymbol{p}_{\mathrm{a}}^{0f,\alpha}$ is the value in vacuum for the matrix element of the dipole moment of the substitutional impurity at the site α, its effective value is

$$\mu_{\mathrm{a}i}^{0f,\alpha} = (\boldsymbol{p}_{\mathrm{a}}^{0f,\alpha})_j A_{ji}^{\alpha}(\omega, \boldsymbol{k}).$$

Therefore, if we ignore spatial dispersion the sought after energy of the dipole–dipole interaction is

$$V_{\mathrm{ab}}(\boldsymbol{R}) = -(\boldsymbol{p}_{\mathrm{a}}^{0f,\gamma})_{j_1} A_{j_1 i}^{\gamma}(\omega^{0f}) A_{j_2 j}^{\gamma'}(\omega^{0f})(\boldsymbol{p}_{\mathrm{b}}^{f0,\gamma'})_{j_2} \times \frac{\partial^2}{\partial x_i \partial x_j} \frac{1}{\sqrt{\epsilon_{\alpha\beta}^{-1} x_\alpha x_\beta}} (\det \hat{\epsilon}^{-1})^{1/2}, \tag{3.63}$$

where $\epsilon_{\alpha\beta}(\omega^{0f})$ is the dielectric constant tensor for the solvent at the frequency ω^{0f} (here γ and γ' denote the sites of the impurities a and b).

When we take into consideration spatial dispersion (Agranovich (1968) and ch. 4 sect. 7) we obtain in $V_{ab}(R)$ additional terms that decrease more quickly with increasing R. However, these terms are significant only for the frequencies ω^{0f} that are close enough to the intrinsic absorption frequencies of the medium. In the case of isotropic media eq. (3.63) is replaced by eq. (3.62). In the special case of a crystal having one anisotropic molecule per unit cell Kamenogradsky and Konobeev (1970) have found in the approximation (3.18a) a modification of eq. (3.63) using the theory of excitons of small radius as it was done by Agranovich et al. (1969).

Finally, note that both for pure crystals and for isotopically mixed crystals (for which $M^{\nu\nu'}_{\alpha\beta} \to \tilde{M}^{\nu\nu'}_{\alpha\beta}$, $a^{\nu}(\omega) \to \bar{a}^{\nu}(\omega)$) solution of the system of equations for the quantities $\boldsymbol{E}^{\alpha}\boldsymbol{l}^{\alpha\nu}$ makes it possible to generalize the above results and to take into account many resonances in the molecule (that is, to take into account the mixing of molecular configurations under the effect of the intermolecular interaction; these problems are discussed also in the excellent review by Mahan (1975)).

7. *Taking into account of higher multipoles in the local field method*

To illustrate how we can take into account the higher multipoles in the framework of the local field method let us consider, along with the dipole polarization, also the quadrupole q_{ij} and the octupole q_{ijl} polarizations of the molecule. In this approximation, the operator for the energy of interaction of the molecule with the external monochromatic electric field $E(\boldsymbol{r}, t)$ is:

$$\hat{V} = \sum_i \hat{p}_i E_i + \sum_{ij} \hat{q}_{ij} E_{ij} + \sum_{ijl} \hat{q}_{ijl} E_{ijl}, \tag{3.64}$$

where $\hat{p}_i$, $\hat{q}_{ij}$, and $\hat{q}_{ijl}$ are the operators for the dipole, quadrupole and octupole moments of the molecule, $E_{ij} = \partial E_i / \partial x_j$, and $E_{ijl} = \partial^2 E_i / \partial x_j \partial x_l$.

Using eq. (3.64) and the time-dependent perturbation theory (Landau and Lifshits (1965). sect. 40) we can find the multipoles induced by the external field. However, as noted in sect. 1, the local field serves as the perturbing field for the molecules in a crystal. Therefore, for the αth molecule ($\hat{q}^{\alpha}$ differ from the conventional tensors (see Landau and Lifshits 1962) in numerical factors) the multipoles can be written as:

$$p_i^{\alpha} = \sum_j a_{ij}^{\alpha} E_j^{\alpha} + \sum_{jl} a_{ijl}^{\alpha} E_{jl}^{\alpha} + \sum_{jlm} a_{ijlm}^{\alpha} E_{jlm}^{\alpha}, \tag{3.65a}$$

$$q^\alpha_{ij} = \sum_l b^\alpha_{ij,l} E^\alpha_l + \sum_{lm} b^\alpha_{ij,lm} E^\alpha_{lm} + \sum_{lmn} b^\alpha_{ij,lmn} E^\alpha_{lmn}, \tag{3.65b}$$

$$q^\alpha_{ijl} = \sum_m c^\alpha_{ijl,m} E^\alpha_m + \sum_{mn} c^\alpha_{ijl,mn} E^\alpha_{mn} + \sum_{mnp} c^\alpha_{ijl,mnp} E^\alpha_{mnp}. \tag{3.65c}$$

The tensors a^α, b^α, and c^α in eqs. (3.65), which depend on ω, have the form of the sum of resonance terms; each of the terms corresponds to a transition from the ground state to one of the excited states of the molecule. If, for the sake of simplicity, we take into consideration only one of the resonances corresponding to the transition from the ground state to the nondegenerate excited state of the molecule having the excitation energy $\hbar\omega_1$, the tensors a^α, b^α and c^α can be factorized so that the treatment becomes considerably simpler. In this approximation, which corresponds to ignoring the mixing of the molecular configurations, the tensor $b^\alpha_{ij,lmn}$, for instance, has the form

$$b^\alpha_{ij,lmn}(\omega) = \frac{2\omega_1}{\omega_1^2 - \omega^2} (q^\alpha_{ij})^{01} (q^\alpha_{lmn})^{10},$$

so that eqs. (3.65) can be rewritten as:

$$p^\alpha_i = \frac{2\omega_1/\hbar}{\omega_1^2 - \omega^2} (p^\alpha_i)^{01} V^{10}_\alpha,$$

$$q^\alpha_{ij} = \frac{2\omega_1/\hbar}{\omega_1^2 - \omega^2} (q^\alpha_{ij})^{01} V^{10}_\alpha, \tag{3.66}$$

$$q^\alpha_{ijl} = \frac{2\omega_1/\hbar}{\omega_1^2 - \omega^2} (q^\alpha_{ijl})^{01} V^{10}_\alpha,$$

where

$$V^{10}_\alpha = \sum_i (p^\alpha_i)^{10} E^\alpha_i + \sum_{ij} (q^\alpha_{ij})^{10} E^\alpha_{ij} + \sum_{ijl} (q^\alpha_{ijl})^{10} E^\alpha_{ijl}. \tag{3.66a}$$

Equations (3.65), which are a generalization of eq. (3.2), show that when we take into account higher multipoles within the framework of the local field method we must find not only the local field amplitude but also the amplitudes of its derivatives. Bearing in mind the above discussion and the results reported by Born and Huang (1954), Khokhlov (1972), we can write the local field and its derivatives in the following form:

$$E^\alpha_i = E_i + \sum_{\beta j} Q^{\alpha\beta}_{ij} p^\beta_j + \sum_{\beta jl} Q^{\alpha\beta}_{ijl} q^\beta_{jl} + \sum_{\beta jlm} Q^{\alpha\beta}_{ijlm} q^\beta_{jlm}, \tag{3.67a}$$

$$E^\alpha_{ij} = E_{ij} + \sum_{\beta l} \tilde{Q}^{\alpha\beta}_{ijl} p^\beta_l + \sum_{\beta lm} \tilde{Q}^{\alpha\beta}_{ijlm} q^\beta_{lm} + \sum_{\beta lmn} \tilde{Q}^{\alpha\beta}_{ijlmn} q^\beta_{lmn}, \tag{3.67b}$$

$$E^{\alpha}_{ijl} = E_{ijl} + \sum_{\beta m} \tilde{\tilde{Q}}^{\alpha\beta}_{ijlm} p^{\beta}_{m} + \sum_{\beta mn} \tilde{\tilde{Q}}^{\alpha\beta}_{ijlmn} q^{\beta}_{mn} + \sum_{\beta mnp} \tilde{\tilde{Q}}^{\alpha\beta}_{ijlmnp} q^{\beta}_{mnp}. \tag{3.67c}$$

The internal field coefficients $Q^{\alpha\beta}$, $\tilde{Q}^{\alpha\beta}$, and $\tilde{\tilde{Q}}^{\alpha\beta}$ depend only on the lattice structure; the explicit expressions for them are not given here, they can be found from the results of Born and Huang (1954), Khokhlov (1972). It is clear that eqs. (3.67) are a generalization of eq. (3.1) and are reduced to it for $q^{\alpha}_{ij} = q^{\alpha}_{ijl} = 0$.

Now, multiply eq. (3.67a) by $(p^{\alpha}_{i})^{10}$, eq. (3.67b) by $(q^{\alpha}_{ij})^{10}$, and eq. (3.67c) by $(q^{\alpha}_{ijl})^{10}$ and take a sum over the subscripts i, j and l. Using eqs. (3.66) and (3.66a) we can easily see that the quantities V^{10}_{α} (where $\alpha = 1, 2, \ldots, \sigma$) satisfy the following system of σ equations:

$$V^{10}_{\alpha} = K_{\alpha} + \sum_{\beta} M^{\text{eff}}_{\alpha\beta}(\boldsymbol{k}) V^{10}_{\beta}. \tag{3.68}$$

Here

$$\begin{aligned} M^{\text{eff}}_{\alpha\beta}(\boldsymbol{k}) = \frac{2\omega_1/\hbar}{\omega_1^2 - \omega^2} \Big\{ & \sum_{ij} Q^{\alpha\beta}_{ij}(\boldsymbol{k})(p^{\alpha}_{i})^{01}(p^{\beta}_{j})^{10} \\ & + \sum_{ijl} [Q^{\alpha\beta}_{ijl}(\boldsymbol{k})(p^{\alpha}_{i})^{01}(q^{\beta}_{jl})^{10} + \tilde{Q}^{\alpha\beta}_{ijl}(\boldsymbol{k})(q^{\alpha}_{ij})^{01}(p^{\beta}_{l})^{10}] \\ & + \sum_{ijlm} [Q^{\alpha\beta}_{ijlm}(\boldsymbol{k})(p^{\alpha}_{i})^{01}(q^{\beta}_{jlm})^{10} + \tilde{Q}^{\alpha\beta}_{ijlm}(\boldsymbol{k})(q^{\alpha}_{ij})^{01}(q^{\beta}_{lm})^{10} \\ & + \tilde{\tilde{Q}}^{\alpha\beta}_{ijlm}(\boldsymbol{k})(q^{\alpha}_{ijl})^{01}(p^{\beta}_{mn})^{10}] + \sum_{ijlmn} [\tilde{Q}^{\alpha\beta}_{ijlmn}(\boldsymbol{k})(q^{\alpha}_{ij})^{01}(q^{\alpha}_{lmn})^{10} \\ & + \tilde{\tilde{Q}}^{\alpha\beta}_{ijlmn}(\boldsymbol{k})(q^{\alpha}_{ijl})^{01}(q^{\beta}_{mn})^{10}] + \sum_{ijlmnp} \tilde{\tilde{Q}}^{\alpha\beta}_{ijlmnp}(\boldsymbol{k})(q^{\alpha}_{ijl})^{01}(q^{\beta}_{mnp})^{10} \Big\}, \end{aligned} \tag{3.69}$$

$$K_{\alpha} = \sum_{i} (p^{\alpha}_{i})^{10} E_i + \sum_{ij} (q^{\alpha}_{ij})^{10} E_{ij} + \sum_{ijl} (q^{\alpha}_{ijl})^{10} E_{ijl}. \tag{3.70}$$

In contrast to the tensor $M_{\alpha\beta}$ (3.20), the tensor $M^{\text{eff}}_{\alpha\beta}$ takes into account the dipole–quadrupole, dipole–octupole, quadrupole–quadrupole, octupole–octupole and quadrupole–octupole interactions, in addition to the dipole–dipole interaction. However, if we are dealing with crystals of the type of anthracene or naphthalene that consist of molecules having an inversion centre, then the matrix elements $(q^{\alpha}_{ij})^{01}$ of the quadrupole moment operator vanish for the dipole-allowed $0 \to 1$ transitions. In this case only the dipole–dipole, dipole–octupole and octupole–octupole interactions make contributions to eq. (3.69). Moreover, for the respective transitions we can omit in eq. (3.70) the second and third terms since for macroscopic fields $E_i(\boldsymbol{r}) = E_i \exp(i\boldsymbol{kr})$ their derivatives with respect to the coordinates are small (these terms are significant only when we analyze the gyrotropy effects in which case eq. (3.64) should take into account the interaction of the molecule with the magnetic field, and also in the frequency range of the dipole-forbidden transitions). Proceeding from the above discussion we come to the conclusion that the system of eqs. (3.68) is quite analogous to the system of eqs. (3.19) with the only difference that the matrix (3.69) gives a more accurate description of the intermolecular interaction. Therefore, in the region of the dipole-allowed transitions the taking into account of the higher multipoles (see eqs. (3.5) and (3.66)) only shifts somewhat the resonances of the dielectric constant tensor. All other results derived above remain unchanged. However, in the region of dipole-forbidden transitions the terms with E_{ij} and E_{ijl} should be retained and we have here

$$P_l = \frac{1}{v} \sum_{\alpha} (p_l^{\alpha} + \mathrm{i} q_{lj}^{\alpha} k_j - q_{ljm}^{\alpha} k_j k_m).$$

8. *Concluding remarks*

In this chapter we have applied the local field method to analyze the effect of the medium on the energy of the resonance interaction between the impurity molecules. Moreover, we have derived some equations which may be useful for treating a variety of optical effects in molecular and ionic crystals. Such a unified approach seems to be suitable since when we gradually analyze the transfer of electronic excitation energy in crystals (both pure crystals and crystals with impurities) we have to process a rather large amount of information on the optical properties of these crystals both for the intrinsic and impurity absorption regions. In the subsequent chapters of the book we shall continue the discussion of some of the above problems in the framework of the exciton theory. In this way we shall be able to obtain some results which cannot be obtained in the framework of the local field method used in this chapter. It should be stressed that we have used only a part of the possibilities of the local field method here. For instance, it would be very interesting to analyze gyrotropy of molecular crystals (both pure crystals and crystals with impurities) and, in particular, gyrotropy of crystalline solutions as a function of their composition. The mean polarizability approximation can, apparently, be used in the region of transparency for calculating the frequency dependence of the rotatory power of a crystalline solution. The accuracy of such calculations will not be poorer than that for the index of refraction in non-gyrotropic disordered media. It would be no less interesting to study the optical properties of molecular crystals as functions of magnetic fields (the Faraday effect) and static electric fields (the Stark effect). (The theory of the Stark effect for impurity molecules is developed by Dunmur and Munn (1975). These results are used for interpreting the experimental results for interstitial azulene molecules in naphthalene crystals reported by Hochstrasser and Noe (1969)). We mention here these effects not only because the information on them in molecular crystals is quite insufficient but also because their treatment in the framework of the local field method would, probably be, the most suitable and simplest approach available.

It should be recalled that the acting field method is based on eqs. (3.1) and (3.67) in which the internal field tensors are assumed to be known. Therefore, the simplicity and explicitness of this method are explained, evidently, by the possibility of efficiently using the earlier results of Born, Ewald and many other authors (Born and Huang 1954, Khokhlov 1972, see also Dunmur 1971) where the local field method was used for analyzing optical anisotropy of liquid (nematic) crystals.

A wide variety of optical properties and effects in crystals cannot be analyzed in the framework of the method described in this chapter. They will be discussed in the next two chapters of the book.

CHAPTER 4

Theory of Electronic Excitations in Molecular Crystals

1. *Introduction*

1.1. *Hamiltonian for a crystal in the representation of second quantization; processes of fusion and fission of molecular excitations*

The problems of the theory of electronic excitations in crystals have been discussed in a systematic way in a fairly large number of review papers and books (Agranovich and Ginzburg 1983, Agranovich 1968, Davydov 1971, Knox 1963, Dexter and Knox 1965, Craig and Walmsley 1968, Mahan 1975, Philpott 1973).

In particular, a fairly full treatment of the theory of small radius excitons is given in the monographs (Agranovich 1968, Craig and Walmsley 1968, Davydov 1971) and the review papers (Philpott 1973, Mahan 1975) in which the attention is focussed mainly on those problems of the exciton theory which are important for crystal optics. However, the problems of energy transfer by excitons have been treated in the review literature in a rather fragmentary way.

We shall attempt here a more systematic treatment of the theoretical aspects of energy transfer by excitons. We shall discuss also some results of the theory of electronic excitations in crystals which are relevant to the energy transfer problems; these results have been obtained in recent years and, therefore, have not been reflected in monographs. First, we shall discuss briefly (in sects. 1 and 2) the method of second quantization (as applied to the small radius excitons) which serves as a convenient tool for the analysis of a wide range of problems in the theory of excitons (exciton–phonon interaction, mechanisms of energy transfer, collective properties of excitons, etc.). For instance, using this method we shall derive the Hamiltonian for the electronic excitations of the crystal and find in it the so-called anharmonic terms which describe the processes of fusion and fission of molecular excitations in crystals. These terms play an important part in the processes of exciton–exciton annihilation, non-resonance energy transfer involving phonons, etc. (see ch. 5 sect. 8).

Apart from the above-mentioned problems we shall also deal with some questions of the theory of spectra of impurity molecules acting as donors and acceptors of the electronic excitation energy.

Application of the second quantization representation to molecular crystals is discussed in detail by Agranovich (1968). Therefore, we shall omit cumbersome calculations and formulas, in the following. However, we shall make all the necessary explanations on the formulas needed for subsequent treatment of transfer and transformations of the electronic excitation energy in crystals. We shall also make a no less detailed discussion of the starting assumptions or restrictions for subsequent calculations.

Now, starting the second quantization treatment, we shall assume that the wave functions and the corresponding energies for the isolated molecules are known. Let f determine a molecular term. If this term is r_0-fold degenerate we shall denote by (f, r) the corresponding states of the isolated molecule with the same energy $\epsilon_f (r = 1, 2, \ldots, r_0)$. If $\hat{H}_0$ is the Hamiltonian of the isolated molecule fixed at the origin of coordinates and $\varphi_0^{(f,r)}$ is its wave function in the state (f, r), then, by definition, we have

$$\hat{H}_0 \varphi_0^{(f,r)} = \epsilon_f \varphi_0^{(f,r)}. \tag{4.1}$$

(wave functions of isolated molecules will be assumed to be real if not specified otherwise).

Assume that the unit cell of the molecular crystal contains σ identical molecules whose orientations may be different (the molecules in the unit cell are denoted by $\alpha = 1, 2, \ldots, \sigma$). If $\boldsymbol{n}$ is the integral lattice vector,

$$\boldsymbol{n} = n_1 \boldsymbol{a}_1 + n_2 \boldsymbol{a}_2 + n_3 \boldsymbol{a}_3,$$

where n_i are integers $(i = 1, 2, 3)$ and $\boldsymbol{a}_i$ are the basis vectors, the Hamiltonian of the crystal is

$$\hat{H} = \sum_{n\alpha} \hat{H}_{n\alpha} + \tfrac{1}{2} \sum_{n\alpha m\beta}{}' \hat{V}_{n\alpha m\beta}, \tag{4.2}$$

where $\hat{V}_{n\alpha m\beta}$ is the operator of the Coulomb interaction between the molecules $\boldsymbol{n}\alpha$ and $\boldsymbol{m}\beta$ determined by the coordinates of their electrons and nuclei. The prime in eq. (4.2) means that the sum does not contain the term with $\boldsymbol{n}\alpha = \boldsymbol{m}\beta$. The operator (4.2) takes into account only the instantaneous Coulomb interaction between the charges comprising the crystal. Therefore, the excitons corresponding to the operator (4.2) will be called here the Coulomb excitons as it was done by Agranovich and Ginsburg (1983).

The operator $\hat{V}_{nm}$ of intermolecular interaction $(n \equiv \{\boldsymbol{n}\alpha\}, m \equiv \{\boldsymbol{m}\beta\})$ is fully equivalent to the matrix

$$\langle f_1 f_2 | \hat{V}_{nm} | f_1' f_2' \rangle \equiv \int \varphi_n^{*f_1} \varphi_m^{*f_2} \hat{V}_{nm} \varphi_n^{f_1'} \varphi_m^{f_2'} \, d\tau = \langle f_1' f_2' | \hat{V}_{nm} | f_1 f_2 \rangle^*, \tag{4.3}$$

which we can find if we know the wave functions of the nth and mth molecules. The above matrix element corresponds to the transition in the system of two molecules n and m which results in molecule n going from the state f_1' to state f_1, and molecule m going from state f_2' to state f_2.

In the two-level approximation (see ch. 1) in which f can have only two values – $f = 0$ (the ground state of the molecule) and $f = 1$ (the excited state of the molecule), the matrix corresponding to the operator $\hat{V}_{nm}$ has sixteen elements, only ten of which are independent, according to eq. (4.3). When we found the probability of transfer of the electronic excitation energy from the molecule a to the molecule b (see ch. 1) we took into consideration only one matrix element, namely, the element V_{12} which can be written in the notation of eq. (4.3) as

$$V_{12} \equiv \langle 0, 1|\hat{V}_{ab}|1, 0\rangle = \langle 1, 0|\hat{V}_{ab}|0, 1\rangle^* = V_{21}^*.$$

For two identical molecules a and b this matrix element relates the states 0, 1 and 1, 0 of the system which have the same energy; therefore, it is sometimes referred to as the resonance matrix element of excitation transfer. However, all the other matrix elements of intermolecular interaction can be sometimes significant, even in the two-level approximation.

The method of second quantization proves to be particularly convenient for analyzing various physical properties of the systems consisting of large numbers of identical interacting subsystems (atoms, molecules, etc.). When applied to the exciton spectra in molecular crystals (Agranovich 1959a, 1961, 1968, Davydov 1971) this method yields all the results which are conventionally obtained with the Heitler–London theory; the method also provides a very simple approach to analyzing higher approximations of the theory. To use the second quantization method we have to select a certain complete system of orthonormal functions describing the states of individual subsystems. In our model these functions can be the eigenfunctions φ_n^f of the operators $\hat{H}_n$ for the isolated molecules corresponding to the eigenvalues ϵ_f. (Ignoring the small overlap (for the ground and lowest excited states of the molecule) of the functions φ_n^f and $\varphi_m^{f'}$ where $n \neq m$ we shall also assume that $(\varphi_n^f, \varphi_m^{f'}) = \delta_{nm}\delta_{ff'}$.)

In the second quantization representation the isolated molecule n is described by the occupation number N_{nf} which is 1 if the molecule n is in the state f and zero if the molecule n is in any other state. Since the isolated molecule can be in only one of the stationary states f the occupation numbers N_{nf} satisfy the condition

$$\sum_f N_{nf} = 1. \tag{4.4}$$

The wave function of one of the possible states of the crystal in the second quantization representation is a function of the occupation numbers N_{nf} and is written down as

$$|\ldots N_{nf} \ldots\rangle. \tag{4.5}$$

All the operators in the second quantization method act on the functions of the occupation numbers. For instance, the operator $\hat{N}_{nf}$ of the occupation number is diagonal in the representation of occupation numbers

$$\hat{N}_{nf}|\dots N_{nf}\dots\rangle = N_{nf}|\dots N_{nf}\dots\rangle.$$

This operator can be expressed in terms of the two non-Hermitian operators b^+_{nf} and b_{nf} as $\hat{N}_{nf} = b^+_{nf}b_{nf}$. The latter operators act on the wave functions in the following way:

$$\begin{aligned} b^+_{nf}|\dots N_{nf}\dots\rangle &= (1-N_{nf})|\dots(N_{nf}+1)\dots\rangle,\\ b_{nf}|\dots N_{nf}\dots\rangle &= N_{nf}|\dots(1-N_{fn})\dots\rangle. \end{aligned} \tag{4.6}$$

The above relationships imply that we may call the operator b^+_{nf} the operator of creation of the state nf and the operator b_{nf} the operator of destruction of this state. According to eqs. (4.6), the operators b^+_{nf} and b_{nf} satisfy the commutation relations,

$$b^+_{nf}b_{nf} + b_{nf}b^+_{nf} = 1, \qquad b_{nf}b_{nf} = b^+_{nf}b^+_{nf} = 0. \tag{4.7}$$

Note also that the operators b_{nf} or b^+_{nf} relating to different values of n or f commute since they act on different variables of the wave function (4.5).

In our representation the Hamiltonian (4.2) can be written as:

$$\hat{H} = \sum \epsilon_f b^+_{nf}b_{nf} + \tfrac{1}{2}\sum{}' b^+_{nf'}b^+_{mg'}b_{mg}b_{nf}\langle f'g'|\hat{V}_{nm}|fg\rangle, \tag{4.8}$$

where f', f, g, g' are the quantum numbers describing various stationary states of the isolated molecule. In the first sum in eq. (4.8) summation is carried out over all values of n and f. In the second sum, summation is carried out over all n, m, f, f', g and g' for $n \neq m$. Since the states f, g, ... correspond to the neutral molecules the Hamiltonian (4.8) does not permit us to consider excitons with transport of charge (they are discussed below). Dealing only with a molecular excitation let us find the form of the operator (4.8) for the crystal whose molecules are described by the two-level model – 0 (the ground state) and f (the excited state of the molecule). Then, the first sum in eq. (4.8) can be written as

$$\sum_n (\epsilon_0 b^+_{n0}b_{n0} + \epsilon_f b^+_{nf}b_{nf}).$$

Since, according to eq. (4.4),

$$\hat{N}_{n0} = 1 - \hat{N}_{nf}, \tag{4.9}$$

we have

$$\sum_n (\epsilon_0 b^+_{n0}b_{n0} + \epsilon_f b^+_{nf}b_{nf}) = N\sigma\epsilon_0 + \Delta\epsilon_f \sum_n \hat{N}_{nf}, \tag{4.10}$$

where $\Delta\epsilon_f = \epsilon_f - \epsilon_0$ is the energy of excitation of the isolated molecule. The first term in the right-hand side of eq. (4.10) is equal to the energy of the ground state of the crystal if we do not take into account the interaction between molecules. The second term is the operator whose eigenvalues determine the energies needed to excite in the crystal various numbers of molecules which do not interact with each other.

Let us now consider the second sum in eq. (4.8). This sum is determined by the intermolecular interaction. Each term of this sum (for given n and m), in its turn, consists of many summands depending on the values of f, f', g, and g'. Let us start with the summand corresponding to $f = f' = g = g' = 0$. Using eq. (4.9) we can write it as

$$\tfrac{1}{2}\langle 00|\hat{V}_{nm}|00\rangle \hat{N}_{n0}\hat{N}_{m0} = \tfrac{1}{2}\langle 00|\hat{V}_{nm}|00\rangle(1 - \hat{N}_{nf} - \hat{N}_{mf} + N_{nf}N_{mf}). \quad (4.11)$$

Let us now consider the summands for which only one of the subscripts f, f', g, and g' differs from zero. It can be easily seen that these summands in eq. (4.8) have the following form:

$$\tfrac{1}{2}\langle f0|\hat{V}_{nm}|00\rangle(b^+_{nf}b_{n0} + b^+_{n0}b_{nf})(1 - N_{mf}) + \tfrac{1}{2}\langle 0f|\hat{V}_{nm}|00\rangle(b^+_{mf}b_{m0} + b^+_{m0}b_{mf})(1 - N_{nf}). \quad (4.12)$$

The summands for which two of the subscripts f, f', g, and g' differs from zero consist of three parts:

$$\begin{aligned} \mathrm{I} &= \tfrac{1}{2}\langle f0|\hat{V}_{nm}|f0\rangle\hat{N}_{nf}(1 - N_{mf}) + \tfrac{1}{2}\langle 0f|\hat{V}_{nm}|0f\rangle\hat{N}_{mf}(1 - \hat{N}_{nf}), \\ \mathrm{II} &= \tfrac{1}{2}\langle f0|\hat{V}_{nm}|0f\rangle b^+_{nf}b_{n0}b_{mf}b^+_{m0} + \tfrac{1}{2}\langle 0f|\hat{V}_{nm}|f0\rangle b^+_{n0}b_{nf}b^+_{mf}b_{m0}, \\ \mathrm{III} &= \tfrac{1}{2}\langle ff|\hat{V}_{nm}|00\rangle b^+_{nf}b^+_{mf}b_{n0}b_{m0} + \tfrac{1}{2}\langle 00|\hat{V}_{nm}|ff\rangle b^+_{n0}b^+_{m0}b_{nf}b_{mf}. \end{aligned} \quad (4.13)$$

Similarly, the summands for which only one of the subscripts f, f', g, and g' is zero have the form:

$$\tfrac{1}{2}\langle 0f|\hat{V}_{nm}|ff\rangle\hat{N}_{mf}(b^+_{nf}b_{n0} + b^+_{n0}b_{nf}) + \tfrac{1}{2}\langle f0|\hat{V}_{nm}|ff\rangle\hat{N}_{nf}(b^+_{mf}b_{m0} + b^+_{m0}b_{mf}), \quad (4.14)$$

and, finally, the summand for which all the above subscripts differ from zero has the form:

$$\tfrac{1}{2}\langle ff|\hat{V}_{nm}|ff\rangle\hat{N}_{nf}N_{mf}. \quad (4.15)$$

Equations (4.10)–(4.14) imply that the following new operators can be convenient:

$$P^+_{nf} = b^+_{nf}b_{n0}, \qquad P_{nf} = b^+_{n0}b_{nf}. \quad (4.16)$$

This definition indicates that, for instance, the operator P^+_{nf} acting on the molecule n in the ground state transforms it into the excited state and, hence, creates excitation f at the molecule n. On the contrary, the operator

P_{nf} destroys excitation f at the molecule n. Since $\hat{N}_{nf}\hat{N}_{n0}=0$ we can obtain from eq. (4.16) using eq. (4.7):

$$\begin{aligned} P^+_{nf}P_{nf} &= b^+_{nf}b_{n0}b^+_{n0}b_{nf} = \hat{N}_{nf}(1-\hat{N}_{n0}) = \hat{N}_{nf},\\ P_{nf}P^+_{nf} &= b^+_{n0}b_{nf}b^+_{nf}b_{n0} = \hat{N}_{n0}(1-\hat{N}_{nf}) = 1-\hat{N}_{nf}, \end{aligned} \tag{4.17}$$

and hence

$$P_{nf}P^+_{nf} + P^+_{nf}P_{nf} = 1. \tag{4.18}$$

The operators P_{nf}, P^+_{nf} and P_{mf}, P^+_{mf} are commutative for $m \neq n$ since they act on different variables of the wave function of the crystal. Thus, P_{nf} and P^+_{nf} are the Pauli operators. They satisfy the commutation relations which are a combination of the commutation relations for the Fermi operators (for $n = m$; see eq. (4.18)) and the Bose operators (for $m \neq n$).

The full Hamiltonian $\hat{H}(f)$ of the crystal in terms of the operators P and P^+ can be written as the sum:

$$\hat{H}(f) = \mathscr{E}_0 + \hat{H}_1 + \hat{H}_2 + \hat{H}_3 + \hat{H}_4, \tag{4.19}$$

where

$$\mathscr{E}_0 = N\sigma\epsilon_0 + \tfrac{1}{2}\sum_{n,m}{}' \langle 00|\hat{V}_{nm}|00\rangle \tag{4.20}$$

is a constant summand,

$$\hat{H}_1 = \sum_{n,m}{}' \langle f0|\hat{V}_{nm}|00\rangle(P^+_{nf} + P_{nf}), \tag{4.21}$$

and

$$\begin{aligned} \hat{H}_2 = &\sum_{n} (\Delta\epsilon_f + D_f)P^+_{nf}P_{nf} + \sum_{n,m} M^f_{nm}P^+_{mf}P_{nf}\\ &+ \tfrac{1}{2}\sum_{n,m}{}' M^f_{nm}\{P^+_{nf}P^+_{mf} + P_{nf}P_{mf}\}. \end{aligned} \tag{4.22}$$

Here,

$$D_f = \sum_{m} \{\langle 0f|\hat{V}_{nm}|0f\rangle - \langle 00|\hat{V}_{nm}|00\rangle\}, \tag{4.22a}$$

is the difference between the energies of interaction of the excited molecule and the unexcited molecule with all other molecules of the crystal, and

$$M^f_{nm} = \langle 0f|\hat{V}_{nm}|f0\rangle = \langle ff|\hat{V}_{nm}|00\rangle \tag{4.22b}$$

is the matrix element for energy transfer from the molecule n to the molecule m (the last equality here is satisfied if we do not take into account the exchange corrections; in molecular crystals the exchange corrections are really insignificant only for the lowest dipole allowed singlet excitations),

$$\hat{H}_3 = \sum_{n,m}{}' \{\langle 0f|\hat{V}_{nm}|ff\rangle - \langle f0|\hat{V}_{nm}|00\rangle\}(P^+_{nf} + P_{nf})P^+_{mf}P_{mf}, \tag{4.23}$$

$$\hat{H}_4 = \tfrac{1}{2}\sum_{n,m}{}' \{\langle ff|\hat{V}_{nm}|ff\rangle + \langle 00|\hat{V}_{nm}|00\rangle - 2\langle f0|\hat{V}_{nm}|f0\rangle\} P^+_{nf}P_{nf}P^+_{mf}P_{mf}. \tag{4.24}$$

In most cases in molecular crystals all the matrix elements for the intermolecular interaction are small compared with the excitation energy for the individual molecule. Therefore, when we calculate the energies of the excited states in the first approximation we can ignore in eq. (4.19) the operators $\hat{H}_1$ and $\hat{H}_3$ which do not conserve the number of the excited molecules in the crystal. (We can also ignore the third term in eq. (4.22). When we take into account all these operators we obtain corrections of the order of $|\langle fg|V_{nm}|f'g'\rangle|/\Delta\epsilon_f$ for the exciton energy which are, typically, much smaller than unity. The relevant questions are discussed in more detail by Agranovich (1968). At the same time, we should bear in mind that these operators can play a significant part in the analysis of nonlinear optical processes which accompany propagation of high-intensity light fluxes who's frequencies are in the range of the exciton absorption lines (Agranovich 1968)).

Interestingly, when we take into account only the dipole–dipole part of the intermolecular interaction these operators as well as the operator $\hat{H}_4$ and the quantity D_f vanish for the molecules with inversion symmetry (anthracene, naphthalene, etc.). In this approximation we have

$$\hat{V}_{n\alpha,m\beta} = \frac{(\hat{\boldsymbol{p}}_{n\alpha}\hat{\boldsymbol{p}}_{m\beta})|\boldsymbol{r}_{n\alpha,m\beta}|^2 - 3(\hat{\boldsymbol{p}}_{n\alpha}\boldsymbol{r}_{n\alpha,m\beta})(\hat{\boldsymbol{p}}_{m\beta}\boldsymbol{r}_{n\alpha,m\beta})}{|\boldsymbol{r}_{n\alpha,m\beta}|^5}, \tag{4.25}$$

where $\boldsymbol{r}_{n\alpha,m\beta}$ is the radius vector between the lattice sites $\boldsymbol{n}\alpha$ and $\boldsymbol{m}\beta$, and $\boldsymbol{p}_{n\alpha}$ is the dipole moment operator of the molecule $\boldsymbol{n}\alpha$. This follows from the fact that the mean values of the dipole moment of the molecule are zero both in the ground and the excited state:

$$\langle f|\hat{\boldsymbol{p}}_{n\alpha}|f\rangle = \langle 0|\hat{\boldsymbol{p}}_{n\alpha}|0\rangle = 0.$$

If in eq. (4.19) we take into consideration only the operators which conserve the number of excitations, the Hamiltonian of the crystal is commutative with the operator of the total number of the excited molecules

$$\hat{N} = \sum_n N_{nf} \equiv \sum_n P^+_{nf}P_{nf}. \tag{4.26}$$

This implies that the excited states of the crystal can be determined for a given number of excitations. The ground state of the crystal corresponds to the wave function $\Phi_0 = |00\ldots0\rangle$ which is an eigenfunction of the operator $\hat{H}$ in our approximation. Since $\hat{H}\Phi = \mathscr{E}_0\Phi$ we can conclude that the quantity $\mathscr{E}_0$ is equal to the energy of the ground state of the crystal, that is, the state of the crystal in which there are no excited molecules. In the crystal with one excited molecule the operator $\hat{N}_{nf}\hat{N}_{mf}$ is zero for $n \neq m$. This means that the operator $\hat{H}_4$ given by eq. (4.24) is also zero and the spectrum of the crystal with one excited molecule is fully determined by the operator

$$\hat{H}_2 = \sum_n (\Delta\epsilon_f + D_f)P^+_{nf}P_{nf} + \sum_{n,m}{}' M^f_{nm}P^+_{mf}P_{nf}. \tag{4.27}$$

If we consider the spectrum of a higher excited state of the crystal (that is, with a few excited molecules) we should take the operator $\hat{H} = \epsilon_0 + \hat{H}_2 + \hat{H}_4$ as the Hamiltonian.

Let us now drop the two-level approximation in treating the crystal molecules. It has been shown by Agranovich (1968) that in this case we still can express the Hamiltonian of the crystal as a power series in the operators (4.16). We shall limit ourselves to the third-order terms in P and P^+ and write:

$$\hat{H} = \sum_f \hat{H}(f) + \hat{H}'_2 + H''_2 + \hat{H}'_3, \tag{4.28}$$

where the operator $\hat{H}(f)$ is given by eq. (4.19),

$$\hat{H}'_2 = \tfrac{1}{2}\sum_{f\neq g}\sum_{n,m} \langle 0f|\hat{V}_{nm}|g0\rangle(P^+_{mg}P_{nf} + P_{mg}P^+_{nf}), \tag{4.29}$$

$$\hat{H}''_2 = \tfrac{1}{2}\sum_{n,f\neq g}(P^+_{ng}P_{nf} + P^+_{nf}P_{ng})\sum_m{}' \langle g0|\hat{V}_{nm}|f0\rangle,$$

and

$$\hat{H}'_3 = \sum_{f,f',g}\sum_{n,m} \langle f'0|\hat{V}_{nm}|fg\rangle(P^+_{nf'}P_{nf}P_{mg} + P^+_{nf}P_{nf'}P^+_{mg}). \tag{4.30}$$

The operators $\hat{H}'_2$ and $\hat{H}''_2$ are distinguished from the operator $\hat{H}_2$ given by eq. (4.22) by the following important feature. When these operators act on a state of the crystal with a molecule n in the excited state f they create a state of the crystal with the molecule m being in an excited state g. In those cases when the energies of these states are close or coincide the respective interaction results in noticeable mixing of the molecular configurations and renormalization of the exciton bands, (see Craig and Walmsley (1968) and Agranovich (1968) ch. II).

If the crystal contains substitutional impurities the interaction $\hat{H}'_2$ is just the one that results in the resonance energy transfer between them (see ch.

1 and sects. 7 and 10 of this chapter). Significantly, in both cases the number of excitations is not changed under the effect of $\hat{H}_2$; only the position of the excited molecule is changed.

Other virtual processes occur under the effect of the operator $\hat{H}'_3$ whose form implies that when it acts on a crystal state with two excited molecules it causes fusion of the excitations at one of the molecules. This operator causes also the reverse process, that is, fission of excitons. Only if these processes satisfy the law of conservation of energy they become real, rather than virtual, and their probability can be found even in first order by using the operator $\hat{H}'_3$ (see ch. 5 sect. 8; if we include in $\hat{H}$ the second-power terms, along with the third-power terms, in $\hat{P}$ we can analyze fusion of excitations at the molecule which was previously excited using second order perturbation). However, in many real situations (for instance, for crystals of the type of fluorite activated with ytterbium ions) precise resonance is lacking and in calculating the probability of the process we have to take into account the interaction between electronic excitations and phonons or photons in addition to the interaction $\hat{H}'_3$. Processes involving energy transfer between different impurity molecules or summation of the energy of their electronic excitations are typically referred to as the nonresonance processes (they are discussed in more detail in sect. 10 and in ch. 5 sect. 6). The luminescence which can occur owing to summation of excitations of two impurity molecules or ions (the so-called cooperative luminescence) can be regarded as sensitized luminescence (see chs. 1 and 2) in which the part of the acceptor is played by the previously singly excited centre or the centre which was earlier in its ground state.

In conclusion, note that for typical densities of electromagnetic radiation in crystals only a small fraction of molecules are excited, that is, the following inequality is satisfied for the mean values:

$$c = \langle P_s^+ P_s \rangle = \langle N_{nf} \rangle \ll 1. \tag{4.31}$$

Since we can rewrite eq. (4.18) in the identical form

$$P_{nf} P_{nf}^+ - P_{nf}^+ P_{nf} = 1 - 2\hat{N}_{nf}, \tag{4.32}$$

we can assume that the operators (4.16) approximately satisfy the conventional commutation relations for bosons, that is,

$$\begin{aligned} &P_{nf} \approx B_{nf}, \\ &P_{nf}^+ \approx B_{nf}^+, \qquad B_{n'f'} B_{nf}^+ - B_{nf}^+ B_{n'f'} = \delta_{nn'} \delta_{ff'}. \end{aligned} \tag{4.33}$$

This replacement of the Pauli operators with the Bose operators is the basis of the method of approximate second quantization, the fundamental concepts of which were developed by Bloch and further developed by Bogolyubov and Tyablikov (the method is described, for instance, by Tyablikov (1975) where the bibliography of early studies is given).

The occupation numbers for paulions can have the values 1 and 0 while the occupation numbers for bosons can have the values 0, 1, 2, . . . (that is, any positive integers). Therefore, the replacement of the operators P_n and P_n^+ with the Bose operators in the operator $\hat{H}$ (see eqs. (4.28)–(4.30)) introduces uncontrollable errors for all the states for which the number of bosons is greater than unity. These errors have been related to the "contribution of nonphysical states" (see, for example Tyablikov 1975). However, these errors can be eliminated if, after the above replacement, we add to the operator $\hat{H}(B)$ an additional operator which is usually referred to as the operator of kinematic interaction. In our two-level model of molecules we can find this operator in terms of Bose operators by using the exact formula for transition from paulions to bosons which was derived by Agranovich and Toshich (1967), see also Agranovich (1968). This approach was later developed by Lalovic et al. (1969), Marinkovic and Tosic (1975), by Marinkovic (1975), Tosic (1971), Tosic and Vujaklija (1971), and especially successfully by Goldhirsch et al. (1979). The expansions of Pauli operators in powers of normal products of Bose operators and their use in the quantum theory of magnetism are discussed by Rudoy and Tserkovnikov (1975) and by Goldhirsch et al. (1979).

As shown by Agranovich and Toshich (1967), for the operators P_n and P_n^+ satisfying eq. (4.18) this formula has the following form:

$$P_n = \left(\sum_{\nu=0}^{\infty} a_\nu B_n^{+\nu} B_n^{\nu}\right)^{1/2} B_n, \qquad P_n^+ = B_n^+ \left(\sum_{\nu=0}^{\infty} a_\nu B_n^{+\nu} B_n^{\nu}\right)^{1/2}. \tag{4.33a}$$

Here $a_\nu = (-2)^\nu/(1+\nu)!$ and B_n^+ and B_n are the Bose operators satisfying eqs. (4.33). As noted by Kompaneets in 1970, eqs (4.33a) can be rewritten in a more compact form (Goldhirsch et al. 1979):

$$P_n = \frac{1+(-1)^{\hat{N}_n}}{2\sqrt{1+\hat{N}_n}}\, B_n, \quad \text{and so on.}$$

However, this relationship, as eqs. (4.33a) gives a nonanalytic expression of the operators P and P^+ in terms of the operators B and B^+. Therefore, Agranovich (1968) also derived the expression:

$$P_n = \left[\sum_{\nu=0}^{\infty} b_\nu B_n^{+\nu} B_n^{\nu}\right] B_n, \qquad P_n^+ = B_n^+ \sum_{\nu=0}^{\infty} b_\nu B_n^{+\nu} B_n^{\nu},$$

which is equivalent to eqs. (4.33a) and specified the procedure for finding the coefficients $b_\nu (b_0 = 1,\ b_1 = -1,\ b_2 = \frac{1}{2}(1+\sqrt{3}/2)$ and so on). Substitution of these expressions into the Hamiltonian of the crystal written in terms of Pauli operators makes it possible to find the above-mentioned operator of the kinematic interaction. This operator expressed in terms of Bose operators contains terms of fourth and higher powers in B and B^+. Therefore, it is important to take into consideration this operator, for instance, in the

Note that for the dipole-forbidden $0 \to f$ molecular transitions or for the transitions corresponding to small oscillator strengths the matrix elements M^f_{nm} of the resonance interaction can be assumed to decrease faster than $|\boldsymbol{n}-\boldsymbol{m}|^{-3}$ with increasing $|\boldsymbol{n}-\boldsymbol{m}|$. Under such circumstances we can limit ourselves to the nearest neighbour approximation in calculations of sums of the type of eq. (4.37a) and rewrite this equation as,

$$L^f(\boldsymbol{k}) = 2M^1_f \cos \boldsymbol{ka}_1 + 2M^f_2 \cos \boldsymbol{ka}_2 + 2M^f_3 \cos \boldsymbol{ka}_3, \tag{4.38a}$$

where $\boldsymbol{a}_i$ are the basis vectors, and M^f_i are the matrix elements of the resonance interaction with the nearest neighbour ($i = 1, 2, 3$). In this case the exciton band width $\Delta E_e \leqslant 12M^f$ where $M^f = \max\{M^f_1, M^f_2, M^f_3\}$. Expanding the above expression, $L^f(\boldsymbol{k})$, in powers of $\boldsymbol{k}$ for small $\boldsymbol{k}$ we can readily obtain:

$$L^f(\boldsymbol{k}) = 2(M^f_1 + M^f_2 + M^f_3) + \sum_{i=1}^{3} \frac{\hbar^2 k_i^2}{2m_i^*},$$

where m_i^* is the effective mass of the exciton corresponding to the motion in the direction $\boldsymbol{a}_i$:

$$m_i^* = -\hbar^2/2M^f_i a_i^2.$$

By the order of magnitude we have $|m_i^*| \approx 2\hbar^2/a_i^2 \Delta E_e$; for instance, for $\Delta E_e \approx 200\ \mathrm{cm}^{-1}$ and $a_i \approx 6 \times 10^{-8}$ cm we obtain $|m_i^*| \approx 10\,\mathrm{m}$ where m is the mass of free electron.

The energy of the crystal with two excitons is given by the sum $E_f(\boldsymbol{k}) + E_f(\boldsymbol{k}')$ where $\boldsymbol{k}$ and $\boldsymbol{k}'$ are the wave vectors of the excitons. Using eqs. (4.16), (4.33) and (4.35) we can express the wave function of the crystal with one exciton, that is, the function $B_f^+(\boldsymbol{k})|0\rangle$ as,

$$B_f^+(\boldsymbol{k})|0\rangle = \frac{1}{\sqrt{N}} \sum_{\boldsymbol{n}} \mathrm{e}^{i\boldsymbol{kn}} b^+_{nf} b_{n0}|0\rangle.$$

Since the wave function of the ground state $|0\rangle = \Pi_{\boldsymbol{n}} \varphi^0_{\boldsymbol{n}}$ and, by definition, $b^+_{nf} b_{n0} \varphi^0_{\boldsymbol{n}} = \varphi^f_{\boldsymbol{n}}$ we find that the wave function of the crystal with one exciton $\boldsymbol{k}f$ in the coordinate representation has the form:

$$B_f^+(\boldsymbol{k})|0\rangle = \frac{1}{\sqrt{N}} \sum_{\boldsymbol{n}} \mathrm{e}^{i\boldsymbol{kn}} \chi^f_{\boldsymbol{n}},$$

where

$$\chi^f_{\boldsymbol{n}} = \varphi^f_{\boldsymbol{n}} \prod_{\boldsymbol{m} \neq \boldsymbol{n}} \varphi^0_{\boldsymbol{m}}.$$

This wave function of the crystal with one exciton, as well as the energy (4.38) are known to correspond to the Heitler–London approximation. A more general approach developed by Agranovich (1968), Mahan (1975) makes it possible to take into account the mixing of molecular configurations due to intermolecular interaction (see also sect. 3).

If the crystal contains $\sigma > 1$ molecules in the unit cell then $n \equiv (\boldsymbol{n}, \alpha)$, $m \equiv (\boldsymbol{m}, \beta)$ and the operator (4.34) can be diagonalized by means of the canonical transformation,

$$B_{n\alpha,f} = \frac{1}{\sqrt{N}} \sum_{k\mu} u_{\alpha\mu}(\boldsymbol{k})\, \mathrm{e}^{\mathrm{i}\boldsymbol{kn}}\, B_{f\mu}(\boldsymbol{k}),$$

$$\sum_{\alpha} u^*_{\alpha\mu}(\boldsymbol{k}) u_{\alpha\mu'}(\boldsymbol{k}) = \delta_{\mu\mu'}, \tag{4.39}$$

where the coefficients $u_{\alpha\mu}(\boldsymbol{k})$ are the components of the vector $u_{\mu}(\boldsymbol{k})$, $[u_{1\mu}(\boldsymbol{k}),\ u_{2\mu}(\boldsymbol{k}), \ldots, u_{\sigma\mu}(\boldsymbol{k})]$, which is the μth normalized vector of the matrix,

$$L_{\alpha\beta}(\boldsymbol{k}) = (\Delta\epsilon_f + D_f)\delta_{\alpha\beta} + \sum_{m} M^{f}_{n\alpha,m\beta} \exp[\mathrm{i}\boldsymbol{k}(\boldsymbol{r}_{n\alpha} - \boldsymbol{r}_{m\beta})], \tag{4.40}$$

These coefficients satisfy the system of equations:

$$\sum_{\beta} L_{\beta\alpha}(\boldsymbol{k})\, u_{\beta\mu}(\boldsymbol{k}) = E_{\mu}(\boldsymbol{k}) u_{\alpha\mu}(\boldsymbol{k}), \qquad \mu = 1, 2, \ldots, \sigma, \tag{4.41}$$

where $E_{\mu}(\boldsymbol{k})$ are the eigenvalues of the matrix $L_{\alpha\beta}(\boldsymbol{k})$. It can be readily seen that if we substitute eq. (4.39) into eq. (4.34) we obtain

$$\Delta\hat{H} = \sum_{\mu k} E_{\mu}(\boldsymbol{k}) B^{+}_{\mu}(\boldsymbol{k}) B_{\mu}(\boldsymbol{k}). \tag{4.42}$$

Equations (4.39) and (4.41) imply that

$$B_{f\mu}(\boldsymbol{k}) = \frac{1}{\sqrt{N}} \sum_{n\alpha} u^*_{\alpha\mu}(\boldsymbol{k})\, \mathrm{e}^{-\mathrm{i}\boldsymbol{kn}}\, B_{n\alpha,f}. \tag{4.43}$$

Using eq. (4.33) we obtain also

$$B_{f\mu}(\boldsymbol{k}) B^{+}_{f\nu}(\boldsymbol{k}') - B^{+}_{f\nu}(\boldsymbol{k}') B_{f\mu}(\boldsymbol{k}) = \delta_{\boldsymbol{kk}'}\delta_{\mu\nu}.$$

The form of the operator (4.42) and the above commutation relations indicate that the fth molecular term in the crystal corresponds to σ exciton states with the energies $E_{\mu}(\boldsymbol{k})$, $\mu = 1, 2, \ldots, \sigma$ governed by the Bose statistics (the Davydov splitting). According to eq. (4.43) the wave function of the crystal with one exciton $\mu\boldsymbol{k}$ in the coordinate representation has the form:

$$B^{+}_{f\mu}(\boldsymbol{k})|0\rangle = \frac{1}{\sqrt{N}} \sum_{n\alpha} \mathrm{e}^{\mathrm{i}\boldsymbol{kn}} u_{\alpha\mu}(\boldsymbol{k}) \chi^{f}_{n\alpha},$$

where,

$$\chi^f_{n\alpha} = B^+_{n\alpha,f}|0\rangle = \varphi^f_{n\alpha} \prod_{m\beta \neq n\alpha} \varphi^0_{m\beta}.$$

If we know the wave functions we can use the conventional procedures for determining the intensity and polarization of light absorption by excitons. Such results (Agranovich 1968) are identical to the results of the local field approach used in ch. 3 and, therefore, we shall not discuss them here. As for the structure of exciton bands, eqs. (4.38) and (4.41) show that their analysis necessitates calculation of sums of the type of eq. (4.37a). (The methods for calculating the energies of Coulomb excitons in molecular crystals of various shapes are discussed by Mahan (1975). The same results can be derived within the framework of macroscopic electrodynamics using the dielectric constant tensor as shown in ch. 3. It should be borne in mind that these energies correspond to the resonances of light absorption as in infinite media.) As discussed in ch. 3, for small values of the wave vector this analysis can be performed within the framework of macroscopic electrodynamics. We shall return to this question in relation to diffusion of excitons.

The wave functions of crystals with two or more noninteracting excitons can be found in a similar way. For instance, the wave function of the crystal with two excitons ($f\mu\boldsymbol{k}$ and $f\mu'\boldsymbol{k}'$) can be written as:

$$B^+_{f\mu}(\boldsymbol{k})\, B^+_{f\mu'}(\boldsymbol{k}')|0\rangle = \frac{1}{N} \sum_{\substack{n\alpha, m\beta \\ n\alpha \neq m\beta}} \exp(i\boldsymbol{k}\boldsymbol{n} + i\boldsymbol{k}'\boldsymbol{m})\, u_{\alpha\mu}(\boldsymbol{k})\, u_{\beta\mu'}(\boldsymbol{k}')\, \chi^f_{n\alpha,m\beta},$$

where,

$$\chi^f_{n\alpha,m\beta} = \varphi^f_{n\alpha}\varphi^f_{m\beta} \prod_{n'\gamma \neq n\alpha, m\beta} \varphi^0_{n'\gamma}.$$

The form of eqs. (4.7) directly indicates that the expression for $B^+B^+|0\rangle$ should not contain a term with $n\alpha = m\beta$. (The decay of the state with two excitons (in the anthracene crystal) giving rise to the electron–hole pair (auto-ionization of excitons) was discussed by Sang-il Choi and Rice (1963). The exciton–exciton interaction and, in particular, possible formation of biexciton (see sect. 6) was neglected in this study. The role played by these factors in exciton–exciton collisions is discussed in ch. 5 sect. 8.)

Let us now consider the part played by the exciton–phonon interaction in molecular crystals and the form of the operator for this interaction.

For weak exciton–phonon interaction the exciton states for the ideal lattice can be successfully used as the zero-approximation states; the wave vector proves to be a “good” quantum number for excitons in this case.

The energy of electronic excitation is transferred by wave packets with the wave vector indeterminacy $\Delta k \ll k < 1/a$ where a is the lattice parameter. Thus, the exciton wave packet covers a crystal region whose size $\Delta x \sim 1/\Delta k$ is much greater than the lattice parameter. It is precisely under the conditions of weak exciton–phonon interaction that the characteristic coherent properties of the exciton states are manifested resulting, for instance, in sharp polarization of the light absorption peaks for anisotropic crystals. A variety of relevant effects have been discussed in ch. 3 within the framework of the local field method. However, we have noted in ch. 3 that only the theory of excitons makes it possible to analyze the entire range of exciton properties and, in particular, the shapes of the exciton light absorption and luminescence spectra and exciton kinetics.

In order to determine the exciton–phonon interaction operator let us assume that the molecules of the crystal are displaced from their equilibrium positions and their orientations also differ from the equilibrium orientations. In this case the quantities D_f and M^f_{nm} in eq. (4.34) depend on the displacements from the equilibrium positions and orientations of the molecules. When we expand these quantities in powers of small displacements we find that the exciton–phonon interaction operator,

$$\hat{H}_{eL} \equiv \Delta\hat{H} - \Delta\hat{H}_0,$$

(where $\Delta\hat{H}_0$ is the operator (4.34) corresponding to the equilibrium positions of the molecules) can be written as:

$$\hat{H}_{eL} = \sum_n B^+_{nf} B_{nf} \sum_{qr} [U^*_{nr}(q)\, b_{qr} + U_{nr}(q) b^+_{qr}]\, \hbar\omega_{qr}$$

$$+ \sum_{m,n}{}' B^+_{mf} B_{nf} \sum_{qr} \hbar\omega_{qr} [I^*_r(\boldsymbol{q}, \boldsymbol{m}-\boldsymbol{n})\, U^*_{nr}(\boldsymbol{q}) b_{qr} + I_r(\boldsymbol{q}, \boldsymbol{m}-\boldsymbol{n})\, U_{nr}(\boldsymbol{q})\, b^+_{qr}], \tag{4.44}$$

where b^+_{qr} and b_{qr} are the operators of creation and annihilation of phonons, r is the number of the vibration branch, $U_{nr}(\boldsymbol{q}) = \gamma^*_{qr}(1/2N)^{1/2} \exp(-i\boldsymbol{q}\boldsymbol{n})$, N is the number of unit cells in the crystal, γ_{qr} are the coefficients determining the strength of the part of the interaction between the exciton and the rth phonon branch which is due to expansion of the quantity D_f in displacements, and $I_r(\boldsymbol{q}, \boldsymbol{m}-\boldsymbol{n})$ is a dimensionless quantity related to the dependence of the matrix element M^f_{mn} on the displacements. Equation (4.44) shows that $I_r(\boldsymbol{q}, \boldsymbol{m}-\boldsymbol{n})$ gives the ratio of the nondiagonal matrix element of the exciton–phonon interaction operator for the sites $\boldsymbol{n}$, $\boldsymbol{m}$ to the diagonal matrix element.

If we replace the operators B^+_{nf} and B_{nf} with the operators of creation (annihilation) of the excitons $\mu\boldsymbol{k}$ according to eqs. (4.39), then the operator $\hat{H}_{eL}$ can be written as:

$$\hat{H}_{eL} = \sum_{\mu,\mu',k} \sum_{qr} F(k+q, \mu; k, \mu'; qr) B_\mu^+(k+q) B_{\mu'}(k)(b_{qr} + b_{-qr}^+) + \dots . \tag{4.44a}$$

In eq. (4.44a), as in eq. (4.44), we have written only the linear terms in displacements of the molecules. The coefficients F in eq. (4.44a) satisfy the condition,

$$F(k+q, \mu; k, \mu'; qr) = F^*(k\mu'; k+q, \mu; -qr),$$

which follows from hermiticity of the operator $\hat{H}_{eL}$. The operator $\hat{H}_{eL}$ can be treated within the framework of the small perturbation theory only if the exciton–phonon interaction is weak as discussed above. However, as early as 1931 Frenkel (1931) discussed the other limiting case – that of strong exciton–phonon coupling. In this limiting case the time τ_1 of transition of the excitation from one molecule to another ($\tau_1 \approx \hbar/\Delta E_e$ where ΔE_e is the exciton band width) is large compared with the time τ during which the molecules are displaced to new equilibrium positions. This displacement is caused by the variation of the interaction forces which occurs when one of the molecules is excited and is determined by the dependence on displacements of the quantity D_f which enters into eq. (4.22) and is given by eq. (4.22a). The resulting local deformation (if E_B is the energy of local deformation then $\tau \approx \hbar/E_B$) travels together with the excitation along the crystal and in this case, as metaphorically described by Frenkel, the exciton, when travelling in the lattice, "drags with itself the entire load of atomic displacements" (similar effects can also take place for excimers (excimeric excitons, see Birks (1967), Chaiken and Kearns (1968), and for charge transfer excitons see, Agranovich and Zakhidov (1977), Haarer (1977)). For such excitons (that is, for $E_B > \Delta E_e$) the wave vector is not always a "good" quantum number. If, owing to strong exciton–phonon interaction, the exciton band width becomes so narrow that it is smaller than or of the order of $\hbar/\tau_e$ (τ_e is the mean free time of the excitons) then the band spectrum model ceases to be applicable for excitons. In this case the motion of exciton ("localized", according to Frenkel or incoherent exciton according to the current terminology) consists of random hops over the lattice sites and is largely similar to the motion of charge carriers with low mobility in semiconductors.

The mobility of excitons in molecular crystals is discussed in more detail in ch. 5. Here we shall only note that the theory of exciton mobility in molecular crystals and the theory of electron mobility in semiconductors encounter the same problem, namely, how to distinguish between the scattering effects due to the exciton–phonon interaction and the effect of renormalization of the exciton energy (Appel 1968). In the case of weak exciton–phonon interaction the excitons can be obtained without taking into account the exciton–phonon interaction and be described by the Boltzmann kinetic equation. If, however, the exciton–phonon interaction cannot be regarded as weak then the theoretical treatment of the exciton spectra and mobility we have to know, roughly speaking, what part of the operator $\hat{H}_{eL}$ changes the exciton spectrum and what part causes scattering of the new exciton states (determining their mean free path and so on). No general solution of the problem of the exciton–phonon interaction has been found yet. A number of models that have been suggested are formally analogous to those for electrons in semiconductors. We refer the reader to the detailed reviews by Appel (1968), Firsov (1968, 1975), Lang and Firsov (1962, 1963, 1967), Firsov and Kudinov (1964, 1965); see especially Klinger (1979) chapters III–VI. It should be noted that experimental studies of exciton mobility in molecular crystals are rather difficult and the fruitful work in this field has just started. Therefore, we shall discuss below only the theoretical predictions for two limiting cases – those of weak and strong exciton–phonon coupling. In the latter case a useful approach is the canonical transformation which is successfully used in the theory of small radius polaritons (see, Firsov 1975, Lang and Firsov 1962, 1963, 1967, Firsov and Kudinov 1964, 1965, Firsov 1968, Grover and Silbey 1971). This transformation allows us to take into account a significant part of the exciton–phonon interaction energy even in the zero approximation. We shall employ it to find the energy E_B of local deformation which accompanies creation of excitons and to determine the form of the operator of exciton–phonon interaction which causes the variation of the exciton states in the case under consideration.

The above-mentioned canonical transformation is performed by the unitary operators

$$\tilde{U} = e^{-\hat{S}},$$

where

$$\hat{S} = \sum_m \hat{S}_m B^+_{mf} B_{mf}, \qquad \hat{S}_m = \sum_{qr} [b^+_{qr} U_{mr}(q) - b_{qr} U^*_{mr}(q)].$$

If we take into consideration the exciton–phonon interaction the starting Hamiltonian $\hat{H}$ is the sum of the operators given by eqs. (4.34) and (4.44). The canonical transformation for the operator

$$\tilde{H} = e^{-\hat{S}} \hat{H}\, e^{\hat{S}},$$

yields,

$$\tilde{H} = \sum_n (\Delta\epsilon_f + D_f - E_B) B^+_{nf} B_{nf} + \sum_{n,m} M^f_{nm} B^+_{mf} B_{nf} \hat{\Phi}_{nm} + \Delta\hat{H},$$

where

$$E_{\mathrm{B}} = \frac{1}{2N} \sum_{qr} \hbar\omega_{qr} |\gamma_{qr}|^2,$$

is the energy of local deformation, and

$$\hat{\Phi}_{nm} = \exp(\hat{S}_m - \hat{S}_n) = \exp\left\{\sum_{qr} [b^+_{qr}\Delta_{qr}(n, m) - b_{qr}\Delta^*_{qr}(n, m)]\right\},$$

where

$$\Delta_{qr}(n, m) = U_{nr}(\boldsymbol{q}) - U_{mr}(\boldsymbol{q}).$$

The operator

$$\Delta\tilde{H} = -\sum_{n \neq m} B^+_{nf} B_{nf} B^+_{mf} B_{mf} \sum_{qr} \hbar\omega_{qr} \operatorname{Re}\{U^*_{nr}(\boldsymbol{q})\, U_{mr}(\boldsymbol{q})\}$$

leads to the exciton–exciton interaction due to exchange of phonons. We derived the above expression for $\hat{H}$ using the following rules of transformation of the exciton and phonon operators:

$$\tilde{B}_{nf} = B_{nf} \exp(-\hat{S}_n),$$
$$\tilde{B}^+_{nf} = B^+_{nf} \exp(\hat{S}_n),$$

$$\tilde{b}_{qr} = b_{qr} - \sum_n B^+_{nf} B_{nf} U_{nr}(\boldsymbol{q}),$$

$$\tilde{b}^+_{qr} = b^+_{qr} - \sum_n B^+_{nf} B_{nf} U_{nr}(\boldsymbol{q}).$$

If we omit the term with M^f_{nm} in the expression for $\hat{H}$ then the stationary states of $\hat{H}$ correspond to excitons localized on the lattice sites. In these states the phonon operators $\tilde{b}_{qr}$ differ from the respective operators in the absence of excitons only in the shift $\Sigma_n N_n U_{nr}(\boldsymbol{q})$ where $N_n = 1$ if there is an exciton on the nth site and $N_n = 0$ if the site is free. Thus, in this approximation in which we take only the terms linear in b^+ and b in the operator $\hat{H}_{eL}$ the local deformation is accompanied with displacements of the equilibrium positions of the lattice oscillators which depend on both the total number of excitons and their relative positions. Therefore, we can say that the operator $\tilde{B}^+_{nf}$ results in creation of exciton on the site n together with respective equilibrium deformation (we follow here the treatment of Lang and Firsov (1962, 1963, 1967), Firsov and Kudinov (1964, 1965), Firsov (1968, 1975) using the results for small radius polaritons); the terms with the matrix elements M^f_{nm} in the Hamiltonian $\tilde{H}$ lead to exciton motion in the lattice. What is the character of this motion? To give an answer to this highly important question let us add and subtract the term

$$\sum{}' M^f_{nm} B^+_{mf} B_{nf} \langle \hat{\Phi}_{nm} \rangle,$$

from the operator $\tilde{H}$ (here the angle brackets denote Gibbs averaging). Since we have,

$$\langle \hat{\Phi}_{nm} \rangle = \exp(-S_T|\boldsymbol{n}-\boldsymbol{m}|),$$

(see Lang and Firsov 1962, 1963, 1967, Firsov and Kudinov 1964, 1965, Firsov 1968, 1975) we assume here that the unit cell contains only one molecule), where T is the temperature of the crystal, and

$$S_T(|\boldsymbol{n}-\boldsymbol{m}|) = \frac{1}{2N} \sum_{qr} |\gamma_{qr}|^2 (1 - \cos \boldsymbol{q}(\boldsymbol{n}-\boldsymbol{m})) \coth\left(\frac{\hbar\omega_{qr}}{k_B T}\right),$$

the operator can be written as the sum

$$\tilde{H} = \tilde{H}_0 + \tilde{H}_{\text{int}} + \Delta \tilde{H},$$

where

$$\tilde{H}_0 = \sum_{k} E(\boldsymbol{k}) B_{\boldsymbol{k}}^{+} B_{\boldsymbol{k}},$$

$$B_{\boldsymbol{k}} = \frac{1}{\sqrt{N}} \sum_{n} \mathrm{e}^{\mathrm{i}\boldsymbol{k}\boldsymbol{n}} B_{nf},$$

$$E(\boldsymbol{k}) = \Delta\epsilon_f + D_f - E_B + \sum_{m} M_{nm}^{f} \exp[\mathrm{i}\boldsymbol{k}(\boldsymbol{m}-\boldsymbol{n})] \exp[-S_T(|\boldsymbol{n}-\boldsymbol{m}|)], \tag{4.44b}$$

$$\tilde{H}_{\text{int}} = \sum_{n,m}' M_{nm}^{f} B_{mf}^{+} B_{nf} \{\hat{\Phi}_{nm} - \langle \Phi_{nm} \rangle\}.$$

Since $\langle \Phi_{nm} \rangle < 1$ the new (restructured) exciton band $E(\boldsymbol{k})$ proves to have a smaller width than the starting exciton band. The operator $\tilde{H}_{\text{int}}$ leads to scattering of the "restructured" exciton states by phonons. If the width $\delta E(k)$ of the exciton level is much smaller than the width of the "restructured" exciton band then the exciton states found in this way are good states of the zero approximation and the exciton kinetics can be studied within the framework of the Boltzmann equation. During propagation of such excitons the phonon occupation numbers vary only under the effect of the operator $\tilde{H}_{\text{int}}$. During the mean free time of the exciton the energy of the lattice does not vary though the exciton motion is accompanied with rearrangement of the lattice equilibrium positions (it is in this model that exciton "drags with itself the entire load of atomic displacements"). In order to understand the exciton motion let us assume that a packet corresponding to localization of exciton on a site n is constructed of the coherent (band) exciton states. This packet spreads with time and the time of its spreading, for instance, up to the first coordination sphere is of the order of $z\tau_1 \approx \hbar z/\Delta E_e$ where ΔE_e is the width of the "restructured" exciton band and z is the number of the nearest neighbours.

However, the exciton can also hop to a neighbour under the effect of the operator $\tilde{H}_{\rm int}$. Evidently, when the hopping time τ_2 corresponding to $\tilde{H}_{\rm int}$ is much shorter than $z\tau_1$ there is not enough time for coherent spreading of the packet and the incoherent ("localized") exciton states should be used as the zero approximation states, rather than the band excitons. The hopping of the incoherent exciton from site to site is a result of all terms in the operator $\tilde{H}$ proportional to the matrix elements of the resonance interaction.

For temperatures T exceeding the Debye temperature we have,

$$S_T(|\boldsymbol{n}-\boldsymbol{m}|) = TS_0(|\boldsymbol{n}-\boldsymbol{m}|),$$

where

$$S_0(|\boldsymbol{n}-\boldsymbol{m}|) = \frac{1}{2N}\sum_{qr}\frac{|\gamma_{qr}|^2}{\hbar\omega_{qr}}[1-\cos \boldsymbol{q}(\boldsymbol{n}-\boldsymbol{m})],$$

so that the exciton band width $\Delta E_{\rm e}$ decreases exponentially with increasing T. As will be shown in ch. 5, the hopping probability $1/\tau_2$ increases as $\exp(-E_{\rm a}/k_{\rm B}T)$ (where $E_{\rm a}$ is the activation energy) under the same conditions. Therefore, if the temperature is sufficiently high the inequality $\tau_2 \ll z\tau_1$ is always satisfied and, hence, the incoherent exciton states must be realized. (The theory of exciton–phonon interaction in one-dimensional molecular crystals has been developed by Davydov and Kislukha (1973, 1976) where the references to earlier studies are also given.)

3. *Mechanical and Coulomb excitons and mixing of molecular configurations; some identities*

To compare the results of the theory of excitons with the results of the local field method derived in ch. 3 we shall need more general relationships for excitons in molecular crystals than those obtained above. We require the energies and wave functions of small radius excitons including the mixing with other molecular configurations (see sect. 7).

We can take this effect into account if in the Hamiltonian (4.8) we retain the matrix elements corresponding to transitions between various excited molecular states. In the second-power approximation in the Bose operators this Hamiltonian (see eqs. (4.22) and (4.29) for the operator $\hat{H}_2$) can be written as

$$\Delta\hat{H} = \hat{H} - \mathscr{E}_0 = \sum_{nf}\Delta_f B^+_{nf}B_{nf} + \tfrac{1}{2}\sum_{n,m,f,g} M^{fg}_{nm}(B^+_{mg}+B_{mg})(B^+_{nf}+B_{nf}),$$

$$\Delta_f = \Delta\epsilon_f + D_f. \tag{4.45}$$

(For details of the derivation procedure, see, Agranovich (1968)).

Diagonalization of the Hamiltonian $\Delta\hat{H}$ is performed by introducing new Bose operators B_ν^+ and B_ν defined by

$$B_{nf} = \sum_\nu [B_\nu u_\nu(nf) + B_\nu^+ v_\nu^*(nf)], \tag{4.45a}$$

where the coefficients u and v satisfy the normalization condition,

$$\sum_{nf} [|u_\nu(nf)^2| - |v_\nu(nf)|^2] = 1, \tag{4.45b}$$

and are determined by the system of equations:

$$(E - \Delta_f)u(nf) = -(E + \Delta_f)v(nf) = \sum_{mg} M_{nm}^{fg}\,\tilde{u}(mg),\ \tilde{u} = u + v.$$

The above equations yield:

$$\tilde{u}(nf) = \frac{2\Delta_f}{E + \Delta_f}\,u(nf).$$

Therefore, the system of equations for the new quantities $a(nf) \equiv \tilde{u}(nf)/\sqrt{\Delta_f}$ can be written as:

$$(E^2 - \Delta_f^2)\,a(nf) - \sum_{mg} \tilde{M}_{nm}^{fg}\,a(mg) = 0, \tag{4.46}$$

where $\tilde{M}_{nm}^{fg} = 2\sqrt{\Delta_f\Delta_g}\,M_{nm}^{fg}$. Owing to translational symmetry ($n \equiv \boldsymbol{n}, \alpha$) we have

$$a(nf) = \frac{1}{\sqrt{N}}\,\mathrm{e}^{\mathrm{i}\boldsymbol{kn}}\,a_\alpha(\boldsymbol{k}, f),$$

where $\boldsymbol{k}$ is the wave vector of the exciton. The system of equations:

$$(E_\mu^2 - \Delta_f^2)\,a_\alpha(\mu\boldsymbol{k}, f) - \sum_{g\beta} \tilde{L}_{\alpha\beta}^{fg}(\boldsymbol{k})a_\beta(\mu\boldsymbol{k}, g) = 0,$$

$$\tilde{L}_{\alpha\beta}^{fg}(\boldsymbol{k}) = \sum_{\boldsymbol{m}} \tilde{M}_{\boldsymbol{n}\alpha,\boldsymbol{m}\beta}^{fg}\exp[-\mathrm{i}\boldsymbol{k}(\boldsymbol{n} - \boldsymbol{m})], \tag{4.46a}$$

determining the coefficients $a(\boldsymbol{k}, f)$ is a generalization of the system of equations (4.41) and it is just the zeroes of the determinant of this system that give the possible exciton energies $E\mu(\boldsymbol{k})$, where μ is the number of the exciton band, taking into account the mixing of molecular configurations. (M. Tanaka and J. Tanaka (1973) put forward another method for taking into account the mixing of molecular configurations by using one-electron molecular orbitals instead of the molecular wave functions (see eq. (4.1)) and the random phase approximation (RPA) for the total Hamiltonian of the crystal.)

The normalization condition for the coefficients u and v yields

$$\sum_{nf} E_\mu |a(\mu \boldsymbol{k}, nf)|^2 = 1. \tag{4.46b}$$

Bearing in mind the condition (4.46b) we obtain the following expression for the Green function corresponding to eq. (4.46):

$$G_{nf,mg} = \frac{1}{N} \sum_{k\mu} \frac{a^{\mu}_{\alpha}(\boldsymbol{k}, f)\, a^{*\mu}_{\beta}(\boldsymbol{k}, g)\, E_\mu(\boldsymbol{k})}{E^2 - E^2_\mu(\boldsymbol{k})} \exp[i\boldsymbol{k}(\boldsymbol{n} - \boldsymbol{m})]. \tag{4.46c}$$

The wave function of the crystal with the exciton $\mu \boldsymbol{k}$ is

$$\Psi_{\mu k} = B^{+}_{\mu k}|0\rangle.$$

Let us find the matrix element for the dipole moment of the transition to the state with the exciton $\mu \boldsymbol{k}$ using the above wave function.

If $\boldsymbol{p}_n$ is the operator of the dipole moment of the molecule n then, going over to the representation of second quantization, we have the following expression for the Fourier component of the operator of the dipole moment of the crystal:

$$\hat{\mathscr{P}}(\boldsymbol{k}) = \sum_n \hat{\boldsymbol{p}}_n\, \mathrm{e}^{-i\boldsymbol{k}\boldsymbol{n}} = \sum_{\alpha,nf} \boldsymbol{p}^{0f}_{\alpha}\, \mathrm{e}^{-i\boldsymbol{k}\boldsymbol{n}}(B^{+}_{nf} + B_{nf}). \tag{4.47}$$

Therefore, the matrix element of the dipole moment of the crystal corresponding to the transition to the $\boldsymbol{q}\mu$ exciton state, that is, the quantity $\langle 0|\mathscr{P}(\boldsymbol{k})|\mu \boldsymbol{q}\rangle$, according to eq. (4.45a), is given by

$$\langle 0|\hat{\mathscr{P}}(\boldsymbol{k})|\mu \boldsymbol{q}\rangle = \delta_{\boldsymbol{k},\boldsymbol{q}} \sqrt{N}\, \boldsymbol{P}^{0\mu},$$

where

$$\boldsymbol{P}^{0\mu}(\boldsymbol{k}) = \sum_{f\alpha} \boldsymbol{p}^{0f}_{\alpha}\, \tilde{u}_\alpha(\mu \boldsymbol{k}, f) \equiv \sum_{\alpha} \boldsymbol{p}^{0\mu}_{\alpha}. \tag{4.48}$$

This relationship determines the polarization and oscillator strengths F_μ of the exciton transitions taking into account the mixing of molecular configurations. If we neglect spatial dispersion we obtain the following expression for the oscillator strength:

$$F_\mu = \frac{2m}{3\hbar^2 e^2} E_\mu(0)|\boldsymbol{P}^{0\mu}(0)|^2.$$

The above discussion dealt with the Coulomb excitons since we took into consideration all Coulomb interactions between charges. However, the dielectric constant tensor is easier to express in terms of so-called mechanical excitons (Agranovich and Ginzburg (1983), sect. 2), rather than the Coulomb excitons. The mechanical excitons are also determined by the system of equations (4.46a) where we must omit in the matrix $\tilde{L}^{fg}_{\alpha\beta}$

the contribution due to the longitudinal macroscopic field which is produced when exciton propagates in the crystal.

Now let us express the components of the dielectric constant tensor in terms of the energies and wave functions of the Coulomb excitons. Below, we shall need such expressions for a discussion of some aspects of the theory of energy transfer between impurity molecules and for analyzing the effect of the crystal matrix on the intensity of light absorption by impurity molecules. First of all, note that the matrix $\tilde{L}^{fg}_{\alpha\beta}$, as shown by Born and Huang (1954), can be written as the sum:

$$\tilde{L}^{fg}_{\alpha\beta}(\boldsymbol{k}) = \tilde{L}^{fg(1)}_{\alpha\beta}(\boldsymbol{k}) + \tilde{L}^{fg(2)}_{\alpha\beta}(\boldsymbol{k}),$$

where

$$\tilde{L}^{fg(2)}_{\alpha\beta}(\boldsymbol{k}) = \frac{8\pi}{v} \frac{(p^{0f}_{\alpha}\boldsymbol{k})(p^{0g}_{\beta}\boldsymbol{k})}{k^2} \sqrt{\Delta_f \Delta_g},$$

v is the volume of the unit cell of the crystal and for small $k \ll 1/a$, where a is the lattice parameter, $\tilde{L}^{fg(2)}_{\alpha\beta}(\boldsymbol{k})$ is a nonanalytic function of $\boldsymbol{k}$ while $L^{fg(1)}_{\alpha\beta}(\boldsymbol{k})$ is an analytic function of $\boldsymbol{k}$. The term $L^{(2)}(\boldsymbol{k})$ is determined by the macroscopic longitudinal electric field and, as mentioned above, it should be omitted if we seek the mechanical exciton states. Since the matrix $L^{(1)}(\boldsymbol{k})$ is an analytical function of $\boldsymbol{k}$ for small $\boldsymbol{k}$ the elementary excitations that corresopnd to it, that is, the mechanical excitons, have the energies $\tilde{E}_s(\boldsymbol{k})$ which are also analytical functions of $\boldsymbol{k}$, in contrast to the energies of the Coulomb excitons. It was shown by Agranovich and Ginzburg (1983) that owing to this fact the group-theoretical classification of the mechanical exciton states for $\boldsymbol{k} = 0$ is simpler than the respective classification of the Coulomb exciton states. For our treatment it is more important that (Agranovich and Ginzburg 1983) the dielectric constant tensor of the crystal which is due to the presence of exciton states has the form

$$\epsilon_{ij}(\omega, \boldsymbol{k}) = \delta_{ij} - \frac{8\pi}{v\hbar} \sum_s \frac{P^{0s}_i(\boldsymbol{k}) \overset{*}{P}{}^{0s}_j(\boldsymbol{k}) \tilde{\omega}_s(\boldsymbol{k})}{\omega^2 - \tilde{\omega}^2_s(\boldsymbol{k})}, \tag{4.49}$$

(here and below the sum over s means summation over mechanical exciton states). In eq. (4.49) $\tilde{\omega}_s(\boldsymbol{k}) = \tilde{E}_s(\boldsymbol{k})/\hbar$, $\tilde{E}_s(\boldsymbol{k})$ is the energy of the mechanical exciton s, and $\boldsymbol{P}^{0s}(\boldsymbol{k})$ is the matrix element of the dipole moment operator for the unit cell constructed with the wave functions of the ground state of the crystal and the state of the crystal with a mechanical exciton of the type $s\boldsymbol{k}$. The above discussion shows that the energies of mechanical excitons are found from the system of equations (compare with eqs. (4.46a)):

$$(\tilde{E}^2_s - \Delta^2_f) b_\alpha(s\boldsymbol{k}, f) - \sum_{g\beta} \tilde{L}^{fg(1)}_{\alpha\beta}(\boldsymbol{k}) b_\beta(s\boldsymbol{k}, g) = 0, \tag{4.50}$$

while

$$\boldsymbol{P}^{0s}(0)=\sum_{f\alpha}\boldsymbol{p}_{\alpha}^{0f}b_{\alpha}(s\boldsymbol{k},f)\sqrt{\Delta_f}. \tag{4.51}$$

The system of equations (4.50) for the vectors $\boldsymbol{b}$ ($b\equiv\{\dots b_\alpha(f)\dots\}$) can be formally written as

$$(E^2-\hat{L}^{(0)}-\hat{L}^{(1)})\boldsymbol{b}=0, \tag{4.52}$$

where $\hat{L}^{(0)}_{\alpha f,\beta g}=\Delta_f^2\delta_{\alpha\beta}\delta_{fg}$. The resolvent of eq. (4.52) can, evidently, be written as

$$(E^2-\hat{L}^{(0)}-\hat{L}^{(1)})^{-1}_{\rho\rho'}=\sum_s\frac{b_\rho(s\boldsymbol{k})\overset{*}{b}_{\rho'}(s\boldsymbol{k})\tilde{E}_s(\boldsymbol{k})}{E^2-\tilde{E}_s^2(\boldsymbol{k})}, \tag{4.53a}$$

where $\rho\equiv(\alpha f)$, so that eq. (4.51) can now be rewritten in the form

$$\boldsymbol{P}^{0s}(\boldsymbol{k})=\sum_\rho\boldsymbol{p}^{0\rho}b_\rho(s\boldsymbol{k})\sqrt{\Delta_\rho}. \tag{4.51a}$$

Using eq. (4.49) for ϵ_{ij} and $E\equiv\hbar\omega$ we obtain:

$$\begin{aligned}\epsilon_{ij}(E,\boldsymbol{k})k_ik_j&=k^2\left[1-\frac{8\pi}{v}\sum_s\frac{\tilde{E}_s(\boldsymbol{k})(\boldsymbol{k}\boldsymbol{P}^{0s})^2/k^2}{E^2-\tilde{E}_s^2(\boldsymbol{k})}\right]\\&\equiv k^2\left[1-\frac{8\pi}{v}\sum_{\rho\rho'}\left(\sum_s\frac{\tilde{E}_sb_\rho(s\boldsymbol{k})\dot{b}_{\rho'}(s\boldsymbol{k})}{E^2-E_s^2}\right)\frac{(\boldsymbol{k}\boldsymbol{P}^{0\rho})(\boldsymbol{k}\boldsymbol{P}^{0\rho'})}{k^2}\sqrt{\Delta_\rho\Delta_{\rho'}}\right]\\&\equiv k^2\{1-\mathrm{tr}[(E^2-\hat{L}^{(0)}-L^{(1)})^{-1}L^{(2)}]\}.\end{aligned}$$

The matrix $(E^2-\hat{L}^0-\hat{L}^{(1)})^{-1}L^{(2)}$ is a dyad and therefore we have,

$$1-\mathrm{tr}(E^2-\hat{L}^0-\hat{L}^{(1)})^{-1}L^{(2)}=\det[I-(E^2-\hat{L}^{(0)}-\hat{L}^{(1)})^{-1}L^{(2)}],$$

where I is the unit matrix. Hence, we obtain:

$$\begin{aligned}\frac{1}{\epsilon_{ij}k_ik_j}&=\frac{1}{k_2}\frac{\det(E^2-\hat{L}^{(0)}-\hat{L}^{(1)})}{\det(E^2-\hat{L}^{(0)}-\hat{L}^{(1)}-\hat{L}^{(2)})}\\&=\frac{1}{k^2}\det[I+(E^2-\hat{L}^{(0)}-\hat{L}^{(1)}-\hat{L}^{(2)})^{-1}\hat{L}^{(2)}].\end{aligned}$$

As above, we find that the matrix $(E^2-L^{(0)}-L)^{-1}$ containing the total interaction $\hat{L}=\hat{L}^{(1)}+\hat{L}^{(2)}$ can be expressed in terms of the Coulomb exciton states as

$$(E^2-\hat{L}^{(0)}-\hat{L}^{(1)}-\hat{L}^{(2)})^{-1}_{\rho\rho'}=\sum_\mu\frac{a_\mu(k\rho)\overset{*}{a}_\mu(k\rho')E_\mu(\boldsymbol{k})}{E^2-E_\mu^2(\boldsymbol{k})}. \tag{4.53b}$$

Hence, we have,

$$\det[I+(E^2-\hat{L}^{(0)}-\hat{L}^{(1)}-\hat{L}^{(2)})^{-1}\hat{L}^{(2)}]=1+\mathrm{tr}(E^2-\hat{L}^{(0)}-\hat{L})^{-1}\hat{L}^{(2)},$$

so that finally we obtain the relationship,

$$\frac{k^2}{\epsilon_{ij}k_ik_j} = 1 + \frac{8\pi}{v}\sum_{\mu}\frac{E_\mu(\mathbf{k})(\mathbf{P}^{0\mu}\mathbf{k})^2k^{-2}}{E^2 - E_\mu^2(\mathbf{k})}, \tag{4.54}$$

(Agranovich et al. 1969) which for isotropic medium reduces to the well-known expansion of the quantity $1/\epsilon$ in the poles corresponding to the longitudinal waves in the medium.

Let us now derive a more general relationship. Take the well-known formula for matrices,

$$(E^2 - \hat{L}^{(0)} - \hat{L}^{(1)} - \hat{L}^{(2)})^{-1}_{\rho\rho'} = (E^2 - \hat{L}^{(0)} - \hat{L}^{(1)})^{-1}_{\rho\rho'}$$
$$+ \sum_{\sigma,u}(E^2 - \hat{L}^{(0)} - \hat{L}^{(1)})^{-1}_{\rho\sigma}\hat{L}^{(2)}_{\sigma u}(E^2 - \hat{L}^{(0)} - \hat{L}^{(1)} - \hat{L}^{(2)})^{-1}_{u\rho'},$$

multiply both sides by $\sqrt{\Delta_\rho\Delta_{\rho'}}\,\mathbf{p}^{0\rho}\mathbf{p}^{0\rho'}$ and take the sum over all ρ and ρ'. Using eqs. (4.53a) and (4.53b) for the matrices $(E^2 - \hat{L}^{(0)} - \hat{L}^{(1)})^{-1}_{\rho\rho'}$ and $(E^2 - \hat{L}^{(0)} - \hat{L}^{(1)} - L^{(2)})^{-1}_{\rho\rho'}$ and eqs. (4.48) and (4.51) for $\mathbf{P}^{0\mu}$ and $\tilde{\mathbf{P}}^{0s}$ we obtain:

$$\sum_{\mu}\frac{E_\mu \mathbf{P}_i^{0\mu}(\mathbf{k})\mathbf{P}_j^{0\mu}(\mathbf{k})}{E^2 - E_\mu^2(\mathbf{k})} = \sum_s \frac{\tilde{E}_s\tilde{\mathbf{P}}_i^{0s}\tilde{\mathbf{P}}_j^{0s}}{E^2 - E_s^2}$$
$$+ \frac{8\pi}{v}\frac{1}{k^2}\left(\sum_s\frac{\tilde{E}_s(\mathbf{k})\tilde{\mathbf{P}}_i^{0s}(\tilde{\mathbf{P}}^{0s}\mathbf{k})}{E^2 - \tilde{E}_s^2(\mathbf{k})}\right)\left(\sum_\mu\frac{E_\mu(\mathbf{k})(\mathbf{P}^{0\mu}\mathbf{k})\mathbf{P}_j^{0\mu}}{E^2 - E_\mu^2(\mathbf{k})}\right).$$

Multiplying both sides of this expression by $\mathbf{k}_i$ and taking the sum over $i = 1, 2, 3$ and then using eq. (4.49) for the tensor ϵ_{ij} we come to,

$$\frac{8\pi}{v}\sum_\mu\frac{E_\mu(\mathbf{k})(\mathbf{k}\mathbf{P}^{0\mu})P_i^{0\mu}}{E^2 - E_\mu^2(\mathbf{k})} = \frac{k^2(\delta_{ij} - \epsilon_{ij}(E, \mathbf{k}))k_j}{\epsilon_{rt}(E, \mathbf{k})k_rk_t}. \tag{4.55}$$

Finally, we obtain the relationship:

$$\frac{8\pi}{v}\sum_\mu\frac{E_\mu(\mathbf{k})\mathbf{P}_i^{0\mu}(\mathbf{k})P_j^{0\mu}(\mathbf{k})}{E^2 - E_\mu^2(\mathbf{k})} = \delta_{ij} - \epsilon_{ij}(E, \mathbf{k})$$
$$+ \frac{(\delta_{i\gamma} - \epsilon_{i\gamma}(E, \mathbf{k}))k_\gamma(\delta_{j\sigma} - \epsilon_{j\sigma}(E, \mathbf{k}))k_\sigma}{\epsilon_{rt}(E, \mathbf{k})k_rk_t}. \tag{4.56}$$

Equations (4.54) and (4.55) follow from eq. (4.56). All these equations relate the Coulomb exciton states to the dielectric constant tensor of the crystal.

4. Vibrational excitons: biphonons

It has been stressed above that excitons in the spectrum of electronic excitations of molecule are only approximately characterized as bosons. We encounter a quite different picture when we are dealing with the elementary excitations of the crystal due to intramolecular vibrations of nuclei. In this case we naturally come to elementary excitations governed by the Bose statistics –phonons– when we consider the normal vibrations of the crystal lattice and subsequently describe them in quantum-mechanical terms. This is, of course, a well-known fact and the concept of phonons is usually described in text books and monographs on the solid state theory. Therefore, we shall make only a few remarks here.

In many molecular crystals the maximum frequency of the acoustic phonons is much lower than some frequencies of intramolecular vibrations. The interaction between such vibrations and the low-frequency lattice vibrations is relatively weak and, therefore, the elementary excitations of the crystal corresponding to them are largely similar to electronic excitons of small radius. In contrast to excitons of the electronic part of the spectrum, such elementary excitations are often referred to as vibrational excitons.

If B_n^+ and B_n are the Bose operators of creation and annihilation of intramolecular vibrational quantum with the energy $\hbar\Omega$ in the molecule n, then, by analogy with eq. (4.34), we can write in the harmonic approximation an expression for the Hamiltonian of the crystal. This enables us to analyze the effect of the intermolecular interaction on the spectrum of nondegenerate intramolecular vibrations in the frequency range of $\omega \sim \Omega$:

$$\Delta H_0 = \sum_n \hbar\Omega B_n^+ B_n + \sum{}' M_{nm} B_n^+ B_m. \tag{4.57}$$

Here M_{nm} is the matrix element of the energy of interaction between the molecules n and m corresponding to the transfer of the quantum from the molecule n to the molecule m. If the unit cell of the crystal contains σ molecules then $n \equiv (\boldsymbol{n}, \alpha)$, $\alpha = 1, 2, \ldots, \sigma$, and the spectrum corresponding to the Hamiltonian ΔH_0 has σ branches. If, moreover, the intramolecular vibration is degenerate (the degeneracy ν) the number of the exciton bands is equal to the product $\sigma\nu$ (the energy of vibrational excitons is $E_l(\boldsymbol{k})$, $l = 1, 2, \ldots, \sigma\nu$).

Though we have been discussing molecular crystals, our model, that is, the Hamiltonian (4.57) is applicable also to nonmolecular crystals whose optical phonon band width is much smaller than the phonon frequency. In this spectral region the atomic vibrations in the unit cell are similar to intramolecular vibrations in molecular crystals since the relatively narrow width of the phonon band evidences the weakness of interaction between atomic vibrations in different unit cells.

According to the treatment of Agranovich (1970), we shall use the Hamiltonian (4.57) to analyze excited states of the crystal in which two optical phonons are bound to each other and travel in the crystal as a unit (biphonons). Intensive theoretical and experimental studies of biphonons have been done in recent years (they are reviewed by Agranovich (1982); see also, Pitaevskiy (1976), Agranovich et al. (1976), Bogani (1978)). One of the reasons of the attention paid to biphonons is the fact that a biphonon is a simplest case of the bound state of two Bose quasiparticles (their possible role in the processes of transfer and transformation of the energy of vibrational excitons is discussed in sect. 5). These states have been found in the Raman and the optical absorption spectra of some crystals and the most convincing proof of their existence has been obtained in the studies of the Raman spectra for polaritons under the conditions of Fermi resonance (Polivanov 1978, 1979, Agranovich 1982).

If we ignore anharmonicity of vibrations, that is, use the Hamiltonian (4.57), the energy of the crystal with two vibrational excitons (the wave vectors $\boldsymbol{k}_1$ and $\boldsymbol{k}_2$) is

$$E_{l_1 l_2}(\boldsymbol{k}_1, \boldsymbol{k}_2) = E_{l_1}(\boldsymbol{k}_1) + E_{l_2}(\boldsymbol{k}_2),$$

where l_1 and l_2 are the numbers of the exciton bands. This state is described by two quasimomenta and, hence, is a two-particle state. Evidently, the width of the energy band of the two-particle state is a sum of the widths of the energy bands of the individual phonons.

In order to take into account anharmonicity we have to add the operator

$$\hat{H}_{\text{int}} = -A \sum_n (B_n^+)^2 B_n^2, \tag{4.58}$$

to the Hamiltonian (4.57) so that the total Hamiltonian is

$$\hat{H} = \Delta\hat{H}_0 + \hat{H}_{\text{int}}. \tag{4.59}$$

The factor A (the anharmonicity constant) in eq. (4.58) is $\frac{1}{2}(2\hbar\Omega - \epsilon_2)$ where ϵ_2 is the energy of excitation of the isolated molecule to the state with the vibrational quantum number $n = 2$. In the absence of intramolecular anharmonicity we have $\epsilon_2 = 2\hbar\Omega$ and $A = 0$. Therefore, it is clear that the operator (4.58) makes a major contribution to the anharmonicity only if the terms due to intermolecular interaction are relatively small (the reverse situation is discussed by Lalov (1974)). In isolated molecules, A is typically of the order of 1–3% of the quantum energy $\hbar\Omega$. For instance, for $\Omega \sim 500\ \text{cm}^{-1}$, the energy A is 5–15 cm^{-1}. At the same time in many crystals the energy of the intermolecular interaction which determines, in particular, the phonon band width in the given frequency range (as shown in some cases by the Davydov splitting measurements) is also of the order of 10 cm^{-1}. Therefore, in those cases when the phonon band width is of the order of A the spectrum of optical vibrations in the region of two-particle states can, generally, have a rather complicated structure.

The dependence of the spectrum in the frequency range $\omega \approx 2\Omega$ on the ratio of the quantities A and Δ can also be understood from the following simple considerations. If the crystal contains two quanta of molecular vibrations localized at different molecules and if the intermolecular interaction is ignored the crystal energy is $E = 2\hbar\Omega$. If both quanta are localized at the same molecule we have $E = 2\hbar\Omega - 2A$ owing to intramolecular anharmonicity. Speaking in terms of quasiparticles we can say that in this case intramolecular anharmonicity results in a decrease in the energy of the crystal when the distance between the quasiparticles decreases and, thus, it corresponds to their attraction. However, localization of both quasiparticles (intramolecular phonons) at one molecule results in increase of the kinetic energy of their relative motion. Since this energy is equal, by the order of magnitude, to the phonon band width the states of vibrational quanta bound to each other definitely appear if the anharmonicity energy A is large compared with the energy band width for vibrational exciton. In this case, for the overtone frequency range we obtain the energy band of two-particle states corresponding to independent motion of two vibrational excitons and, apart from that, the biphonon states which lie lower in the energy scale (for $A > 0$). In the reverse case of weak anharmonicity ($|A| \ll \Delta$) there are no biphonon states outside the band of two-particle states and the spectrum contains only a band (or bands) of two-particle states. (Pitaevsky (1976) has shown that the band of two-particle states always contains biphonons with low binding energy and wave vector $\mathbf{k} \neq 0$. These states have not been observed experimentally yet.)

Now we shall continue with a quantitative analysis of the problem. If we denote the wave function of the ground state by $|0\rangle$ the states with two vibrational quanta can be expressed as:

$$|2\rangle = \sum_{nm} \Psi(n, m) B_n^+ B_m^+ |0\rangle, \qquad \Psi(n, m) = \Psi(m, n), \tag{4.60}$$

where $\Psi(n, m)$ has the sense of the wave function of two phonons in the coordinate representation. This function should satisfy the Schrödinger equation,

$$\hat{H}|2\rangle = E_2|2\rangle, \tag{4.61}$$

where E_2 is the sought after energy of the crystal with two quanta. When we substitute eq. (4.60) into eq. (4.61) we find that $\Psi(n, m)$ satisfies the following equations:

$$(E_2 - 2\hbar\Omega + 2A\delta_{nm})\Psi(n, m) = \sum_l [M_{nl}\Psi(l, m) + M_{ml}\Psi(l, n)].$$

The solution of this system of equations is particularly simple for crystals with one molecule per unit cell (for details, see Agranovich (1970) where crystals with several molecules per unit cell are also treated). Here we shall only write the equation determining the biphonon energy E_2 and its dependence on the wave vector $\boldsymbol{K}$:

$$1 = -\frac{2A}{N}\sum_{\boldsymbol{k}} \frac{1}{E_2(\boldsymbol{K}) - 2\hbar\Omega - L(\boldsymbol{K}+\boldsymbol{k}) - L(\boldsymbol{K}-\boldsymbol{k})}, \tag{4.62}$$

where $L(\boldsymbol{K})$ is given by eq. (4.37a), and N is the number of unit cells in the crystal. Equation (4.62) is identical in form to the Lifshitz equation for the energy of the quantum of a local vibration in the neighbourhood of an isotopic impurity (see, Agranovich (1968) ch. 6 sect. 3 and Lifshits (1942, 1947, 1956)). In one-dimensional and two-dimensional crystals eq. (4.62) always has a solution E_2 which is outside the band $E(\boldsymbol{k}, \boldsymbol{k}')$ of the two-particle states,

$$E(\boldsymbol{k}, \boldsymbol{k}') = 2\hbar\Omega + L(\boldsymbol{k}) + L(\boldsymbol{k}'), \qquad \boldsymbol{k} + \boldsymbol{k}' = \boldsymbol{K}.$$

In three-dimensional crystals solutions of this kind for biphonon (outside the band of dissociated states) exist only for the values of A exceeding a certain value $A_{\min}$ which depends on the structure of the phonon bands. Since $A_{\min}$ is a function of $\boldsymbol{K}$ then, generally speaking, such situations are possible when the biphonon solution exists not for any $\boldsymbol{K}$ but only for the $\boldsymbol{K}$ for which we have $A \geqslant A_{\min}(\boldsymbol{K})$. However, if $A \gg |L(\boldsymbol{k})|$ the solution of eq. (4.62) for the biphonon exists always; with the accuracy of the terms $|L(\boldsymbol{k})|/A$ the solution is given by,

$$E_2(\boldsymbol{K}) = 2\hbar\Omega - 2A - \frac{1}{2AN}\sum_{\boldsymbol{k}} \{L(\boldsymbol{K}+\boldsymbol{k}) + L(\boldsymbol{K}-\boldsymbol{k})\}^2.$$

This expression shows that under the given conditions the width of the biphonon energy band is $\sim |L|^2/A$ and can be much smaller than the phonon band width. As it was stressed Agranovich (1970), under certain conditions in the presence of impurities or lattice defects this fact can result in formation of local and quasilocal states in the overtone spectrum range even when the states of this kind are absent in the fundamental tone region. The density of states in the spectral range under consideration was found within the framework of the coherent potential approximation (Agranovich et al. 1979) in connection with interpretation of experimental results for $N^{14}H_4Cl$ crystals with N^{15} as impurity.

Let us now discuss the spectrum of excited states of the crystal in the frequency region of the combination tone of two intramolecular vibrations. Take, for instance, the frequency range $\omega \approx \Omega_1 + \Omega_2$ where Ω_1 and Ω_2 are the frequencies of two different nondegenerate intramolecular vibrations. The model Hamiltonian in the given spectrum range can be written as

$$\hat{H} = \sum_n \hbar\Omega_1 B_n^{+(1)} B_n^{(1)} + \sum_{n,m}{}' M_{mn}^{(1)} B_n^{+(1)} B_m^{(1)} + \sum_n \hbar\Omega_2 B_n^{+(2)} B_n^{(2)}$$

$$+ \sum_{nm}{}' M_{nm}^{(2)} B_n^{+(2)} B_m^{(2)} - 2A \sum_n B_n^{+(1)} B_n^{+(2)} B_n^{(1)} B_n^{(2)}. \tag{4.63}$$

This Hamiltonian is commutative with respect to the operator $I_{nm} = B_n^{+(1)}B_m^{+(2)}B_n^{(2)}B_m^{(1)} + B_n^{+(2)}B_m^{+(1)}B_n^{(1)}B_m^{(2)}$ for transposition of the quanta $\hbar\Omega_1$ and $\hbar\Omega_2$ localized at the molecules n and m. This means that the wave functions of the crystal with two quanta of different species can be even or odd with respect to transposition of the coordinates n and m. Since in our case we have

$$|2\rangle = \sum_{nm} \Psi(n, m) B_n^{+(1)} B_m^{+(2)} |0\rangle,$$

then, generally speaking, we can obtain "even" states for which $\Psi(n, m) = \Psi(m, n)$ and "odd" states for which $\Psi(n, m) = -\Psi(m, n)$, as was stressed by Kozhushner (1971). For the "odd" states we have $\Psi(n, m) = 0$ and, therefore, our Hamiltonian (4.63) does not yield "odd" biphonon states. This directly follows from the fact that the anharmonicity operator in the Hamiltonian (4.63) acting on the "odd" wave functions for biphonon yields zero. If, apart from intramolecular anharmonicity proportional to A, the Hamiltonian (4.63) includes terms due to intermolecular interaction (for instance, the term $\Sigma_{n,m} K_{nm} I_{nm}$, where K_{nm} is the matrix element of the operator of intermolecular n, m interaction corresponding to transposition of the quanta), then the "odd" biphonon states can also "split off" from the band of two-particle states.

If we take, for instance, $M_{nm}^{(2)} = 0$ in eq. (4.63), that is, ignore possible motion of one of the quanta, we come to the case analyzed by Van Kranendonk (1959). Assuming that the quantum $\hbar\Omega_2$ is localized at the molecule $n = 0$ (in this state the eigenvalue of the operator $B_n^{+(2)}B_n^{(2)}$ is δ_{n0}) we can rewrite the Hamiltonian in the simpler form:

$$\hat{H} = \sum_n \hbar\Omega_1 B_n^{+(1)} B_n^{(1)} + \sum_{n,m}{}' M_{nm}^{(1)} B_n^{+(1)} B_m^{(1)} - 2A\delta_{n0} B_n^{+(1)} B_n^{(1)} + \hbar\Omega_2,$$

corresponding to the motion of the quantum $\hbar\Omega_1$ in a crystal with an "isotopic" substitutional impurity at the site $n = 0$ having a shifted vibrational frequency $\Omega_2' = \Omega_1 - 2A/\hbar$. Thus, in the Van Kranendonk model calculation of the biphonon energy reduces to the well-known problem of the energy of a vibration localized in the vicinity of a defect. In our case this energy is given by the equation:

$$1 = -\frac{2A}{N} \sum_k \frac{1}{E_2 - \hbar\Omega_1 - \hbar\Omega_2 - L(\boldsymbol{k})},$$

where $L(\boldsymbol{k})$ is determined by eq. (4.37a) for $M_{nm} \equiv M_{nm}^{(1)}$. Naturally, the odd states of biphonons do not appear in the Van Kranendonk approximation. If we take into account the motion of both quanta the energy $E_2(\boldsymbol{K})$ of the even biphonon is given in this model by the equation:

$$1 = -\frac{2A}{N}\sum_k \frac{1}{E_2(\boldsymbol{K}) - \hbar\Omega_1 - \hbar\Omega_2 - L_1(\boldsymbol{K}+\boldsymbol{k}) - L_2(\boldsymbol{K}-\boldsymbol{k})},$$

which is a simple generalization of eq. (4.52).

The Van Kranendonk model was developed for treatment of the IR rotational-vibrational spectrum of crystalline parahydrogen (Van Kranendonk 1959). However, this model proved to be useful for the analysis of spectra of low vibronic excitations in molecular crystals. These excitations correspond to a crystal state with an electronic exciton of a small radius and a quantum of intramolecular vibration. Since the width of the exciton band in molecular crystals is of the order of a hundred or hundreds of inverse centimetres and the band width of vibrational exciton is of the order of ten inverse centimetres the conditions of applicability of Van Kranendonk model are always satisfied. As reported by Sheka (1971), in some molecular crystals in the first vibronic excitation region the light absorption spectra contain the bands of two-particle states (electronic and vibrational exciton) and, apart from this band and outside it, the spectra exhibit sufficiently narrow peaks which are regarded as evidence for the formation of bound states of an electronic and a vibrational exciton. Attempts have been made recently (Lalov 1975) to analyze three-particle states (electronic exciton plus two vibrational quantums).

5. *Collective properties of a system of long-lived vibrational excitons*

Typically, it is precisely the anharmonicity of lattice vibrations resulting in decay and fusion of phonons that determines the lifetime of vibrational excitons. When such excitons were observed in crystals the widths of their levels were not less than about 0.1 cm^{-1} corresponding to lifetimes of the order of 10^{-10} s, even at very low temperatures. The small lifetimes of vibrational excitons prevent their luminescence since, as calculated by Bates and Poots (1953), for diatomic molecules having a dipole moment (CO, NO, HCl) the respective decay time $\tau \approx 1\text{–}10^{-2}$ s and for molecules lacking a dipole moment (O_2, N_2, H_2) $\tau \approx 10^{-6}\text{–}10^{-7}$ s. For more complex molecules the decay times for vibrational excitons are apparently, not shorter than for diatomics and the radiationless decay times should be even shorter owing to the large number of intramolecular vibrations. Therefore, observation of luminescence of vibrational excitons, generally speaking, is hardly feasible at present. As for obtaining sufficiently high concentrations of vibrational excitons, one promising approach is to use molecular crystals of diatomic molecules. Apparently, the longest radiationless decay times for vibrational excitons were obtained in molecular crystals with high frequencies Ω of intramolecular vibrations and low Debye temperatures Θ. In such cases ($\hbar\Omega \gg k_b\Theta$) radiationless decay of exciton involves creation of a large number N_p of phonons and must have a relatively lower probability. For instance, for the HCl crystal $\Omega \approx 2800\ \text{cm}^{-1}$ while $\Theta \approx 100$ K and we find approximately $N_p \sim 3 \times 10^3\ \text{cm}^{-1}/10^2\ \text{cm}^{-1}$ or about 30. Calculations of the radiationless decay time τ_0 of vibrational excitons for which $N_p \gg 1$ are not sufficiently reliable yet and too sensitive to the values of the anharmonicity constants. If this time were of the order of 10^{-4}–10^{-5} s this would create the conditions for achieving a vibrational exciton concentration of the order of 10^{17}–$10^{18}\ \text{cm}^{-3}$ even for $T \ll \hbar\Omega/k_b$.

It has been shown (Zel'dovich and Ovchinnikov 1971) that if the concentration of vibrational excitons is sufficiently high intramolecular anharmonicity can give rise to overpopulation of high excited vibrational levels even in a cold lattice. Inelastic collision of two excitons can lead to their fusion and formation of biexciton (biphonon). Collision of a biphonon and vibrational exciton results in formation of a complex of three quanta and so on. If the anharmonicity constant is sufficiently high in comparison with the energy of intermolecular interaction then all these processes are accompanied with energy transfer to the lattice and, thus, can occur at any low temperatures. The calculations of the distribution of molecules over the vibrational levels performed by Zel'dovich and Ovchinnikov (1971) assumed that the intermolecular interaction was weak and its effect on the excitation spectrum was ignored. In this approximation the spectrum of the excited states of the crystal is the spectrum of a system of noninteracting anharmonic oscillators. If E_r is the energy of such oscillator in the state r (where $r = 0, 1, \ldots, r_0$) then, for instance, for the Morse potential we have

$$E_r = r - \kappa r^2, \qquad r \leqslant r_0 = 1/2\kappa, \qquad \kappa = 1/4e_0, \tag{4.64}$$

where e_0 is the dissociation energy, r_0 is the number of the last vibrational level and we have used the system of units in which $\Omega = 1$, and $\hbar = 1$. If N_r is the number of molecules in the r-fold excited state we have

$$\sum_r N_r = N_0, \qquad \sum_r rN_r = N_q,$$

where N_0 is the total number of molecules in the crystal and N_q is the total number of quanta. If we replace N_r and N_q with $n_r = N_r/N_0$ and $N = N_q/N_0$ we obtain

$$\sum_r n_r = 1, \qquad \sum_r rn_r = N. \tag{4.65}$$

We see that n_r and N are the concentrations of rth excitations and the number of quanta (per one molecule).

If the total number of quanta is specified and their chemical potential $\mu \neq 0$, then at thermodynamic equilibrium we have

$$n_r = C \exp \beta(\mu r - E_r), \qquad \beta = 1/k_B T, \tag{4.66}$$

where T is the lattice temperature and the constants C and μ should be found from two normalization conditions (4.65). In the harmonic limit, that is, for $\kappa = 0$ we have $E_r = r$ and the resulting distribution,

$$n_r = (1-\gamma)\gamma^r, \qquad \gamma = N/(1+N), \tag{4.67}$$

is independent of temperature as should be expected since fusion and decay of excitations do not involve the lattice in this case. The distribution (4.67) monotonically decreases with increasing r and does not result in overpopulation of the highly excited vibrational states.

If we take anharmonicity into consideration the addition of a quantum to the $(r-1)$-fold excited molecule results in an energy decrease:

$$\Delta E_r = E_{r-1} + E_1 - (E_r + E_0) > 0 \qquad (r > 1). \tag{4.68}$$

Therefore, if the mobility of vibrational excitons is sufficiently high (or their concentration is sufficiently high) a practically irreversible process of coagulation of individual quanta ($r = 1$) occurs in the cold lattice resulting in formation of excited vibrational states with $r > 1$. The distribution function n_r can be found from the kinetic coagulation equation of Smoluchowski (see, Chandrasekar 1943). Here we can regard only the excited states with $r = 1$ as being capable of migrating in the lattice since in the presence of strong anharmonicity the energy band width decreases sharply for the excited states with $r > 1$. (Ovchinnikov 1969, Agranovich 1970). Under the given assumptions the kinetic equation for n_r can be written as:

$$\frac{dn_r}{dt} = W_r n_{r-1} n_1 - W_{r+1} n_r n_1 + \frac{1}{\tau(r+1)} n_{r+1} - \frac{1}{\tau(r)} n_r + i_0 \delta_{r,1}, \tag{4.69}$$

where $\tau(r)$ is the time of radiationless transition of the vibrational state r to the state $r-1$, and W_r is a kinetic coefficient determining the rate of fusion of vibrational exciton with a complex of $r-1$ quanta. As shown in ch. 6, we can write approximately:

$$W_r = 4\pi D R_r, \tag{4.70}$$

where D is the diffusion coefficient of the vibrational excitons, and R_r is the radius of capture determined by the character of interaction between exciton ($r = 1$) and a centre in the $(r-1)$th excited state ($r > 1$). For the fusion process to start from the stage $1 + 1 = 2$ a bound state of two quanta (biphonon) should be present. Otherwise $n_2 = 0$ and eq. (4.69) is suitable only for $r \geqslant 4$. As for the states with $r = 3$, their formation in this case requires three-particle collisions of vibrational excitons, rather than pair collisions, and this sharply decreases the probability of progressing to higher r values. Returning, however, to the case of sufficiently strong anharmonicity note that even in this case the process comes to a bottleneck due to a rapid decrease of W_r with increasing r. Ovchinnikov (1969) has shown that the increase of r leads to exponential decrease of the probability of fusion of the quantum ($r = 1$) with a neighbouring complex ($r - 1$) giving rise to the complex r. This means that eq. (4.70) can be rewritten as,

$$W_r \simeq A\, e^{-pr}, \tag{4.71}$$

where $p = 1 - 2$, according to Ovchinnikov (1969). In contrast to W_r, the time $\tau(r)$ depends on r comparatively weakly, particularly at not too high r values. If we ignore the dependence of τ on r completely and use eq. (4.71) we can find an exact solution of the system of equations (4.69) and obtain a power law decrease of n_r for $r \gtrsim 3$. The population of highly excited states in this case very strongly depends on the constants in eq. (4.69) which are largely unknown. Therefore, we shall not give here the detailed calculation procedure for n_r and shall make only a few remarks. If the values of the constants are favourable (long lifetimes of excitons, high mobility of vibrational excitons, etc.) the system can be in a strongly inverse state (with respect to the given lattice temperature) which can be used for the experimental observation of fusion of quanta. It is significant that inverse population can be found for the states with $r > 1$, that is, for frequencies different from the pumping frequency.

It should be borne in mind, however, that formation even of the states with $r = 2$ is possible only under the condition that $4\pi DR_1 n_1$ is greater or of the order of $1/\tau_1$ where τ_1 is the lifetime of vibrational exciton. Apparently, this condition is the strictest one since if it is not satisfied vibrational excitons cannot meet during their lifetime and even the states with $r = 2$ cannot be formed.

6. *Frenkel electronic biexcitons*

At present it is relatively easy to produce singlet exciton concentrations of the order of 10^{17}–$10^{18}\,\mathrm{cm}^{-3}$ in molecular crystals. For such exciton concentrations the time of exciton collision $\tau_{ee} = (8\pi DR_0N_e)^{-1}$ is of the order of 10^{-8} s, that is, of the order of the exciton lifetime (here D is the diffusion coefficient of excitons which is of the order of $10^{-4}\,\mathrm{cm}^2/\mathrm{s}$, $R_0 \approx 2\times 10^{-7}\,\mathrm{cm}^2$ is the collision "radius", and N_e is the exciton concentration). Therefore, we should analyze the various processes which can occur with exciton collisions, including formation of biexcitons. For molecular crystals we should consider the Frenkel biexcitons. It is clear that states of this type can manifest themselves in some way only in crystals for which they are actually present in the spectrum of excited states. The possible effect of electronic biexcitons on the course of exciton–exciton annihilation is discussed in ch. 5. Here we shall only consider the conditions of biexciton formation in the simple case of a cubic lattice having one molecule per unit cell (see, Efremov and Kaminskaya 1973).

Making use of eqs. (4.27) and (4.24) we can write the Hamiltonian in our case as:

$$\Delta\hat{H} = \hat{H}_2 + \hat{H}_4 = \sum_n \Delta P_n^+ P_n + \sum_{n,m}{}' M_{nm} P_n^+ P_m + \tfrac{1}{2} \sum_{n,m}{}' F_{n,m} P_n^+ P_m^+ P_n P_m, \tag{4.72}$$

where we have omitted the subscript f in the operators P and P^+, $\Delta \equiv \Delta\epsilon_f + D_f$, and

$$F_{n,m} = \langle ff|\hat{V}_{nm}|ff\rangle + \langle 00|V_{nm}|00\rangle - 2\langle f0|\hat{V}_{nm}|f0\rangle, \tag{4.73}$$

is the energy of interaction between the excitons at the lattice sites n and m. In contrast to the matrix element M_{nm} which for the dipole-active excitons contains dipole–dipole long-range terms decreasing in inverse proportion to the third power of the distance $|\boldsymbol{n} - \boldsymbol{m}|$, the function $F_{n,m}$ for crystals of the type of anthracene, naphthalene, etc. contains only terms which decrease faster with increasing $|\boldsymbol{n} - \boldsymbol{m}|$. In such crystals with inversion centres only high multipole–multipole interactions make a nonzero contribution to $F_{n,m}$ (as in D_f; see eq. (4.22a)) so that in a fairly good approximation we can take into account the exciton–exciton interaction only for nearest neighbours and describe it by a single constant F.

Determination of the biexciton states is largely similar to determination of the biphonon states (see sect. 4). Write the wave function of the crystal with two excitons in the form:

$$\Psi = \frac{1}{\sqrt{2}} \sum_{\boldsymbol{R},\boldsymbol{r}} f(\boldsymbol{r}, \boldsymbol{R}) P^+_{\boldsymbol{R}+\boldsymbol{r}/2} P^+_{\boldsymbol{R}-\boldsymbol{r}/2} |0\rangle,$$

where $f(\boldsymbol{r}, \boldsymbol{R})$ is the wave function of two excitons in the coordinate representation, $\boldsymbol{R}$ is the centre of mass of the biexciton ($\boldsymbol{R} = (\boldsymbol{n} + \boldsymbol{m})/2$), and $\boldsymbol{r}$ is the radius vector of relative motion ($\boldsymbol{r} = \boldsymbol{n} - \boldsymbol{m}$). Since the Hamiltonian (4.72) is invariant with respect to translation the biexciton states can be classified according to their total quasimomentum $\boldsymbol{K}$. If $\alpha_{\boldsymbol{K}}(\boldsymbol{r})$ is the wave function of relative motion of two excitons the total quasimomentum of which is $\boldsymbol{K}$, then in the coordinate representation the total wave function of two excitons with the total quasimomentum $\boldsymbol{K}$ is given by:

$$f(\boldsymbol{r}, \boldsymbol{R}) \equiv f_{\boldsymbol{K}}(\boldsymbol{r}, \boldsymbol{R}) = \frac{1}{\sqrt{N}} \alpha_{\boldsymbol{K}}(\boldsymbol{r})\, \mathrm{e}^{\mathrm{i}\boldsymbol{KR}}.$$

Substitute the function $\Psi \equiv \Psi_{\boldsymbol{K}}$ into the Schrödinger equation

$$\Delta \hat{H} \Psi_{\boldsymbol{K}} = E(\boldsymbol{K}) \Psi_{\boldsymbol{K}},$$

and multiply this equation from the left by $(1/\sqrt{2})\langle 0|P_1 P_s|$. Using the commutation relations for the Pauli operators we find that $\alpha_{\boldsymbol{K}}(\boldsymbol{r})$ satisfies the equation:

$$\sum_{\boldsymbol{r}'} H_{\boldsymbol{r}\boldsymbol{r}'}(\boldsymbol{K}) \alpha_{\boldsymbol{K}}(\boldsymbol{r}') = E(\boldsymbol{K}) \alpha_{\boldsymbol{K}}(\boldsymbol{r}), \tag{4.74}$$

where

$$\begin{aligned} H_{\boldsymbol{r}\boldsymbol{r}'}(\boldsymbol{K}) = {} & 2\Delta \delta_{\boldsymbol{r}\boldsymbol{r}'} + 2M_{\boldsymbol{r}\boldsymbol{r}'} \cos\tfrac{1}{2}\boldsymbol{K}(\boldsymbol{r}' - \boldsymbol{r}) + F_{\boldsymbol{r}} \delta_{\boldsymbol{r}\boldsymbol{r}'} \\ & - 2(\delta_{\boldsymbol{r}0} + \delta_{\boldsymbol{r}'0}) M_{\boldsymbol{r}'\boldsymbol{r}} \cos\tfrac{1}{2}\boldsymbol{K}(\boldsymbol{r}' - \boldsymbol{r}). \end{aligned}$$

Here $F_{\boldsymbol{r}} \equiv F_{\boldsymbol{n},\boldsymbol{m}}$. The last term in the above expression is the kinematic repulsion which does not allow two excitons to be on the same lattice site. This term appeared because we derived eq. (4.74) using the correct commutation relations for the Pauli operators. Since we deal here with states of identical excitons, the wave function of their relative motion is symmetric:

$$\alpha_{\boldsymbol{K}}(\boldsymbol{r}) = \alpha_{\boldsymbol{K}}(-\boldsymbol{r}).$$

Using this symmetry and introducing the Green function of two free excitons (without kinematic and dynamic interaction between them):

$$G_{\boldsymbol{r}\boldsymbol{r}'}(E) = \frac{1}{N} \sum_{\boldsymbol{p}} \frac{\cos \boldsymbol{p}\boldsymbol{r} \cos \boldsymbol{p}\boldsymbol{r}'}{E - E(\boldsymbol{K}, \boldsymbol{p})},$$

where $E(\boldsymbol{K}, \boldsymbol{p})$ is the energy of two free excitons with momentum, $\boldsymbol{K}/2 \pm \boldsymbol{p}$, we can write eq. (4.74) in the form,

$$\alpha_{\boldsymbol{K}}(\boldsymbol{r}) = \sum_{\boldsymbol{n},\boldsymbol{m}}^{1/2} G_{\boldsymbol{r}\boldsymbol{n}} \tilde{U}_{\boldsymbol{n}\boldsymbol{m}}(\boldsymbol{K}) \alpha_{\boldsymbol{K}}(\boldsymbol{m}),$$

where the summation over $\boldsymbol{n}$ and $\boldsymbol{m}$ is performed only over the half space,

$$\tilde{U}_{\boldsymbol{n}\boldsymbol{m}}(\boldsymbol{K}) = -4M_{\boldsymbol{m}} \cos\tfrac{1}{2}\boldsymbol{K}\boldsymbol{m} - 4M_{\boldsymbol{n}} \cos\tfrac{1}{2}\boldsymbol{K}\boldsymbol{n} + 2F_{\boldsymbol{n}} \delta_{\boldsymbol{n}\boldsymbol{m}},$$

and $M_{\boldsymbol{m}} \equiv M_{\boldsymbol{n},\boldsymbol{n}'}$ with $\boldsymbol{n} - \boldsymbol{n}' = \boldsymbol{m}$ ($M_{\boldsymbol{m}} \neq 0$ if $\boldsymbol{m} \neq 0$). Multiplying the equation for $\alpha_{\boldsymbol{K}}(\boldsymbol{r})$ from the left by $\tilde{U}_{\boldsymbol{n}\boldsymbol{r}}$ and taking a sum over $\boldsymbol{r}$ we obtain an equation for the new function $\varphi(\boldsymbol{K}, \boldsymbol{n}) = \Sigma_{\boldsymbol{r}}^{1/2} \tilde{U}_{\boldsymbol{n}\boldsymbol{r}} \alpha_{\boldsymbol{K}}(\boldsymbol{r})$:

$$\varphi(\boldsymbol{K}, \boldsymbol{n}) = \sum_{\boldsymbol{r},\boldsymbol{r}'}^{(1/2)} \tilde{U}_{\boldsymbol{n}\boldsymbol{r}}(\boldsymbol{K}) G_{\boldsymbol{r}\boldsymbol{r}'}(E, \boldsymbol{K}) \varphi(\boldsymbol{K}, \boldsymbol{r}').$$

For a given $\boldsymbol{K}$ the quantities $\varphi(\boldsymbol{K}, \boldsymbol{n})$ can be regarded as the components of the vector $\varphi(\boldsymbol{K})$; generally speaking, the number of nonzero components of this vector is infinite. However, if we take into account only the nearest-neighbour interaction for $M_{\boldsymbol{n}}$ (as for $F_{\boldsymbol{n}}$) then $\tilde{U}_{\boldsymbol{n}\boldsymbol{m}}$ is a fourth-order matrix (for the simple cubic lattice the nonzero values of $\tilde{U}_{nm}$ correspond to the vectors $\boldsymbol{n}$ or $\boldsymbol{m}$ which have only four values). In this approximation the equation for $\varphi(\boldsymbol{K})$ reduces to a system of four homogeneous equations. The condition of nontrivial solution for this system is

$$\det[\delta_{nm} - (\tilde{U}G)_{nm}] = 0.$$

The symmetry properties of the equation for $\varphi(\boldsymbol{K}, \boldsymbol{n})$ depend on the direction of $\boldsymbol{K}$. Therefore, its solutions can be conveniently classified in irreducible representations of the group of the wave vector $\boldsymbol{K}$. The matrix $(\tilde{U}G)_{nm}$ has the simplest form when the vector $\boldsymbol{K}$ is parallel to the [111] axis and the group of the wave vector is the subgroup Λ (Bouckaert et al. 1936). Factorization of the above determinant is performed by transition in the $\varphi(\boldsymbol{K})$ vector space to new unit vectors forming the basis of irreducible representations of the subgroup Λ. Then the equality of the determinant to zero is reduced to two equations. One equation determines the energy of the nondegenerate state corresponding to the irreducible representation Λ_1 (the s states) and has the form:

$$4MG_{00}(E, \boldsymbol{K}) \cos \frac{Ka}{2\sqrt{3}} = 2FG_{01}(E, \boldsymbol{K}),$$

The other equation determines the energy of the doubly degenerate state (the irreducible representation Δ_3 or d state) and has the form:

$$G_{11}(E) - G_{12}(E) = 1/2F.$$

In this case no p state is formed owing to the symmetry of the function $\alpha(\boldsymbol{r})$ noted above. When $F < 0$ the energies of the found biexciton states lie lower than the band of dissociated two-particle states and the energy of the s state is lower than the energy of the d state. The energy of the s state of biexciton found Efremov and Kaminskaya (1973)

$$E_s(K) = 2\Delta + F + 36\frac{M^2}{F}\cos^2\frac{Ka}{2\sqrt{3}}, \tag{4.75}$$

where a is the lattice parameter. The condition for the existence of a level below the bottom of the band of two-particle states proves to be very strict. This condition is $|F/M| \geqslant 6$. Figure 4.1 shows the calculated s and d biexciton energies for various ratios $|F/2M|$. Here δ is the band width and the biexciton energy is taken from the level 2Δ (see eq. (4.72)).

These results have been obtained for a very simplified model. Nevertheless, they point out the significant fact that strongly bound exciton states are formed only if the energy F of their interaction (attraction) is sufficiently large compared with the exciton band width. In contrast to M, the exciton interaction energy F is practically independent of the spin state

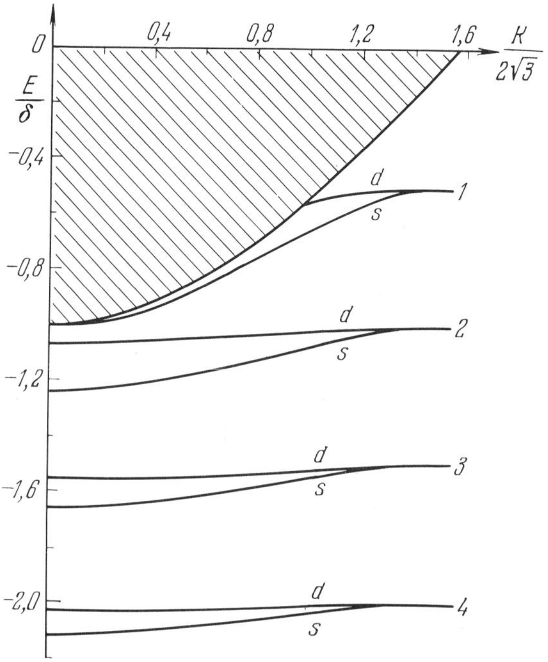

Fig. 4.1. Spectrum of small-radius biexcitons in a molecular crystal. The shaded energy region corresponds to the states of two free excitons. The curves (1), (2), (3), and (4) correspond to the $|F/2M|$ values of 3.1, 6, 9, and 12, respectively. For $F > 0$ the biexciton spectrum is above the region of the dissociated states and symmetric to the one shown here.

of excitons. Therefore, the probability of formation of biexcitons should be particularly high for the triplet excitons for which M and, hence, the exciton band width are much smaller than for singlet excitons. Equation (4.75) indicates the interesting fact (see also fig. 4.1) that the biexciton band width decreases sharply with increasing exciton interaction energy F and becomes smaller than the exciton band width. This fact should facilitate formation of biexciton states localized at lattice defects which is similar to the case with biphonons (see, Agranovich 1982, 1970).

The theory of biexcitons in molecular crystals has just started to develop and the interest in them will, undoubtedly, grow with improvement of the experimental techniques. It should be borne in mind that, apart from luminescence and absorption spectra, for sufficiently high excitation intensities biexcitons, generally speaking, can be manifested in spectral dependences of nonlinear susceptibilities giving rise to additional resonances in them (calculations of nonlinear susceptibilities including the bound states of quasiparticles are reported by Agranovich et al. (1971), Efremov and Kaminskaya (1972), and the first observations of such predicted new resonances in nonlinear processes are reported by Vu Duy Phach et al., (1970). It has been shown for excitons of large radius (Hanamura 1973, Flytzanis 1975) and for Frenkel excitons (Efremov 1975) that two-photon light absorption is an effective method of studying biexcitons.

7. *Theory of resonance interaction of impurity molecules; mechanism of virtual excitons and virtual phonons*

As we repeatedly emphasized above (see chs. 1, 2 and 3) a major problem in the theory of electronic excitation energy transfer between impurity molecules in crystals is the effect of the medium (matrix) on the intermolecular resonance interaction. We have discussed this problem within the framework of the acting field method in ch. 3 sect. 5. We shall use the exciton theory in this section to derive a more general expression for the effective intermolecular resonance interaction (Agranovich et al. 1969). This expression proves to be applicable for small distances between impurity molecules and it is more exact in taking into account the effect of the matrix on the states of impurity molecules that is the technique used in ch. 3.

Assume that two interstitial impurity molecules with the same orientation $\alpha = 1$ are at the points $\boldsymbol{p}$ and $\boldsymbol{q}$ of the crystal (the case of substitutional impurity molecules can be treated in a similar way, see Kamenogradskiy and Konobeev (1969). We shall employ the two-level model only for the impurity molecules but not for the matrix molecules ($\boldsymbol{p}^{01}$ is the transition dipole moment).

Since we are interested here in the resonance interaction between impurity molecules this approximation is quite justified. If, however, the excitation energy of the impurity is not at resonance with the excitation of the matrix the two-level model would be a poor approximation for the matrix molecules.

According to sects. 1 and 2, the Hamiltonian of the system consisting of the molecular crystal and two impurity molecules can be written in terms of the Bose creation and annihilation operators B^+_{ni} and B_{ni} for the ith excitation in the nth crystal molecule (for impurities $n = p, q$) in the form

$$\hat{H} = \hat{H}_0 + \hat{H}_1, \tag{4.76}$$

where $\hat{H}_0$ is the Hamiltonian of the ideal crystal given by eq. (4.45), and

$$\begin{aligned}\hat{H}_1 = {} & \Delta_1(B^+_{p1}B_{p1} + B^+_{q1}B_{q1}) \\ & + \sum_{mj} [M^{1j}_{pm}(B^+_{p1} + B_{p1})(B^+_{mj} + B_{mj}) + M^{1j}_{qm}(B^+_{q1} + B_{q1})(B^+_{mj} + B_{mj})] \\ & + M^{11}_{pq}(B^+_{p1} + B_{p1})(B^+_{q1} + B_{q1}).\end{aligned} \tag{4.77}$$

As was done in sect. 2, in order to find the energies of the elementary excitations corresponding to Hamiltonian (4.76) we shall introduce new Bose operators B^+_ν and B_ν defined by eqs. (4.45a). Now we have not only $n = (\boldsymbol{n}, \alpha)$ but also $n = (\boldsymbol{p}, 1)$ and $(\boldsymbol{q}, 1)$ and the factors u and v in eq. (4.45a) are given by the system of equations:

$$\begin{aligned}
(E-\Delta_i)u(\boldsymbol{n}i) &= -(E+\Delta_i)v(\boldsymbol{n}i) \\
&= \sum_{\boldsymbol{m}j} M^{ij}_{\boldsymbol{nm}}\tilde{u}(\boldsymbol{m}j) + M^{i1}_{\boldsymbol{np}}\tilde{u}(\boldsymbol{p}1) + M^{i1}_{\boldsymbol{nq}}\tilde{u}(\boldsymbol{q}1), \\
(E-\Delta_1)u(\boldsymbol{p}1) &= -(E+\Delta_1)v(\boldsymbol{p}1) \\
&= \sum_{\boldsymbol{m}j} M^{1j}_{\boldsymbol{pm}}\tilde{u}(\boldsymbol{m}j) + M^{11}_{\boldsymbol{pq}}\tilde{u}(\boldsymbol{q}1), \qquad (4.78) \\
(E-\Delta_1)u(\boldsymbol{q}1) &= -(E+\Delta_1)v(\boldsymbol{q}1) \\
&= \sum_{\boldsymbol{m}j} M^{1j}_{\boldsymbol{qm}}\tilde{u}(\boldsymbol{m}j) + M^{11}_{\boldsymbol{pq}}\tilde{u}(\boldsymbol{p}1),
\end{aligned}$$

where $\tilde{u} = u + v$. Equations (4.78) yield $\tilde{u}(\boldsymbol{n}i) = [2\Delta_i/[(E+\Delta_i)]u(\boldsymbol{n}i)$ and similar results for $\tilde{u}(\boldsymbol{p}1)$ and $\tilde{u}(\boldsymbol{q}1)$ so that for two molecules in vacuum (or for $M^{1j}_{\boldsymbol{pm}} = M^{1j}_{\boldsymbol{qm}} = 0$) we obtain the following system of equations for $\tilde{u}(\boldsymbol{p}1)$ and $\tilde{u}(\boldsymbol{q}1)$:

$$(E^2 - \Delta_1^2)\tilde{u}(\boldsymbol{p}1) - 2\Delta_1 M^{11}_{\boldsymbol{pq}}\tilde{u}(\boldsymbol{q}1) = 0,$$

$$(E^2 - \Delta_1^2)\tilde{u}(\boldsymbol{q}1) - 2\Delta_1 M^{11}_{\boldsymbol{pq}}\tilde{u}(\boldsymbol{p}1) = 0.$$

Equating the determinant of this system of equations to zero we find that if we take into account only the direct p, q resonance interaction we obtain two new terms $E_{1,2}$, instead of one term $E = \Delta_1$, and

$$E^2_{1,2} - \Delta_1^2 = \pm 2\Delta_1 |M^{11}_{\boldsymbol{pq}}|,$$

which for $|M^{11}_{\boldsymbol{pq}}| \ll \Delta_1$ yields

$$E_{1,2} = \Delta_1 \pm |M^{11}_{\boldsymbol{pq}}|.$$

We see that for two molecules in vacuum the splitting of the molecular term is $2|M^{11}_{\boldsymbol{pq}}|$.

To find the similar splitting of the impurity states in the crystal let us analyze the system of eqs. (4.78) in more detail. As in sect. 2, in order to consider the exciton states in the ideal crystal we introduce the new quantities $a(\boldsymbol{n}i) = \tilde{u}(\boldsymbol{n}i)/\sqrt{\Delta_i}$ for which the system of eqs. (4.78) is transformed into:

$$(E^2-\Delta_i^2)a(ni)-\sum_{mj}\tilde{M}^{ij}_{nm}a(mj)=\tilde{M}^{i1}_{np}a(p1)+\tilde{M}^{i1}_{nq}a(q1),$$

$$(E^2-\Delta_1^2)a(p1)=\sum_{mj}\tilde{M}^{1j}_{pm}a(mj)+\tilde{M}^{11}_{pq}a(q1),$$

$$(E^2-\Delta_1^2)a(q1)=\sum_{mj}\tilde{M}^{1j}_{qm}a(mj)+\tilde{M}^{11}_{pq}a(p1), \tag{4.78a}$$

where $\tilde{M}^{ij}_{nm}=2M^{ij}_{nm}\sqrt{\Delta_i\Delta_j}$. Using the Green function (4.46c) to obtain from the first of eqs. (4.78a),

$$a(ni)=\sum_{mj}G_{ni,mj}(E)(\tilde{M}^{j1}_{mp}a(p1)+\tilde{M}^{j1}_{mq}a(q1)), \tag{4.78b}$$

and substituting this expression into the second of eqs. (4.78a) we come to a system of two linear homogeneous equations for $a(p1)$ and $a(q1)$. Using the condition of nontrivial solution of this system of equations, that is, equating its determinant to zero we obtain the equation giving the energies of the local states which appear in the crystal in the presence of two impurity molecules. This equation has the form:

$$E^2-\Delta_1^2-\frac{1}{N}\sum_{k\mu}\frac{|A^p_\mu(k)|^2}{E^2-E^2_\mu(k)}=\pm\left[\tilde{M}^{11}_{pq}+\frac{1}{N}\sum_{k\mu}\frac{A^p_\mu(k)A^{*q}_\mu(k)}{E^2-E^2_\mu(k)}\right], \tag{4.79}$$

where N is the total number of unit cells in the crystal and,

$$A^p_\mu(k)=\sum_{mj}\tilde{M}^{j1}_{mp}\sqrt{E_\mu(k)}\,a^\mu(mj). \tag{4.80}$$

The expression for $A^q_\mu(k)$ has a similar form. The quantities $a^\mu(mj)$ entering into eqs. (4.46c) and (4.80) are found from an analysis of the exciton states in the ideal crystal; they have been discussed in sect. 2. When we derived eq. (4.79) we assumed, for simplicity of notation, that the identical molecules p and q occupy translationally equivalent positions though in different cells of the crystal (only then we have $|A^p_\mu(k)|=|A^q(k)|$). Clearly, analysis of the more general case will not encounter any fundamental difficulties.

Equation (4.79) can be used for the determination of the energies of local states for arbitrary distances between impurity molecules and arbitrary position of the impurity level Δ_1 relative to the exciton bands of the matrix. We shall be interested only in the case when the resonance splitting of the impurity term due to the p, q interaction is small in comparison with the width of the energy gap between the impurity term and the edge of the nearest exciton band (this condition is always satisfied if the distance between the impurity molecules is sufficiently large). In this case the magnitude of the right-hand side of eq. (4.79) is relatively small and we can include it as a small correction, writing, $E_{1,2} = E_0 \pm M_{pq}^{\text{eff}}$, where E_0 is given by,

$$E_0^2 - \Delta_1^2 - \frac{1}{N}\sum_{k\mu}\frac{|A_\mu^p(\mathbf{k})|^2}{E_0^2 - E_\mu^2(\mathbf{k})} = 0, \tag{4.81}$$

and

$$M_{pq}^{\text{eff}} = \frac{M_{pq}^{11} + \dfrac{1}{2\Delta_1 N}\displaystyle\sum_{k\mu}\frac{A_\mu^p(\mathbf{k})A_\mu^{*q}(\mathbf{k})}{E_0^2 - E_\mu^2(\mathbf{k})}}{\dfrac{E_0}{\Delta_1}\left[1 + \dfrac{1}{N}\displaystyle\sum_{k\mu}\frac{|A_\mu^p(\mathbf{k})|^2}{(E_0^2 - E_\mu^2(\mathbf{k}))^2}\right]}. \tag{4.82}$$

As follows from the above discussion, E_0 determines the impurity excitation energy for infinite dilution. The quantity M_{pq}^{eff} substantially depends on the distance between the molecules p and q and is not only determined by their direct interaction M_{pq}^{11} but also contains the contribution of the exciton states of the matrix to this interaction.

Let us now analyze eq. (4.82) more closely. First of all, note that, as follows from eqs. (4.78b) for $a(q1) = 0$, the coefficients $a(ni)$ for the local state are proportional to $a(p1)$, that is, we have

$$a(ni) = \frac{a(p1)}{N}\sum_{k\mu}\frac{a_\alpha^\mu(ki)\sqrt{E_\mu(\mathbf{k})}A_\mu^{*p}(\mathbf{k})}{E_0^2 - E_\mu^2(\mathbf{k})}\,\mathrm{e}^{i\mathbf{kn}}. \tag{4.83}$$

If now we make use of the normalization condition for the coefficients a corresponding to the local state, which is

$$E_0\Big(|a(p1)|^2 + \sum_{ni}|a(ni)|^2\Big) = 1,$$

(see also eq. (4.45b)) then taking eq. (4.83) we obtain

$$|a(p1)|^2 = \frac{1}{E_0}\left[1 + \frac{1}{N}\sum_{k\mu}\frac{|A_\mu^p(\mathbf{k})|^2}{(E_0^2 - E_\mu^2(\mathbf{k}))^2}\right]^{-1}. \tag{4.84}$$

Thus, eq. (4.82) can be rewritten in the following, more compact form:

$$M_{pq}^{\text{eff}} = \Delta_1|a(p1)|^2\left[M_{pq}^{11} + \frac{1}{2N\Delta_1}\sum_{k\mu}\frac{A_\mu^p(\mathbf{k})A_\mu^{*q}(\mathbf{k})}{E_0^2 - E_\mu^2(\mathbf{k})}\right]. \tag{4.82a}$$

Equations (4.81) and (4.84) indicate that in those cases when the impurity term is sufficiently far from the edge of the nearest exciton band we have $E_0 \to \Delta_1$ and $|a(p1)|^2 \to 1/\Delta_1$ so that we obtain $\Delta_1|a(p1)|^2 \to 1$ for the first factor in eq. (4.82a). Below we shall discuss some effects caused by the proximity of the impurity level to the edge of the exciton band of the matrix. Here we shall stress once again that the terms in the square brackets in eq. (4.82a) have essentially different natures. As noted above, the first term, that is, M^{11}_{pq}, corresponds to the energy of the resonance interaction between the impurity molecules p and q in vacuum. The second term is the contribution to the energy of the resonance interaction between the impurity molecules which is due to exchange of virtual excitons of the matrix. We can derive this term by using the operator (4.77) in the second order of perturbation theory in which the interaction between the impurity molecules and the matrix molecules is assumed to be weak and various virtual excitons (with any μ and $\boldsymbol{k}$) should be taken into account in the intermediate states. However, the result of this derivation will be less general than the term in eq. (4.28a) since the latter includes also the effect of delocalization of the impurity excitations. Its magnitude is related to the difference of the quantity $\Delta_1|a(p\,1)|^2$ from unity and this effect can be significant only if the impurity term approaches the bottom (or the top) of the exciton band providing that these extrema correspond to $\boldsymbol{k} \approx 0$.

It will be shown below that it is precisely the contribution of virtual excitons that gives rise to the dielectric constant tensor of the matrix in the expression for the resonance interaction energy discussed here; it also gives rise to the local field corrections discussed in ch. 3 sect. 5. We shall also show below for matrices with coherent excitons that inclusion of excitons gives rise to additional terms in the resonance interaction energy; these terms vanish when the exciton band widths tends to zero (or the exciton effective masses tend to infinity). If the distance between impurity molecules is large these terms decrease exponentially and they can be explained also within the framework of macroscopic electrodynamics taking into account spatial dispersion. (If the distance between impurity molecules is small, that is, of the order of the lattice parameter, then the expression for the resonance interaction energy derived from macroscopic electrodynamics is inapplicable, as was emphasized above. Under these conditions it is useless to find from eq. (4.82a) additional terms for the macroscopic expression.)

If the distance between impurity molecules is large the main contribution to the resonance interaction energy for the dipole transitions in impurity molecules is made by the dipole–dipole interaction between molecules. In this case the quantity $A^p_\mu(\boldsymbol{k})$ is equal to the product of a factor which will be given below and the energy of the interaction between the dipole of the transition in the impurity molecule p with the respective dipoles localized on the sites of the perfect lattice whose distribution is given by

$$\boldsymbol{p}^\mu_n \equiv \left(\sum_j \frac{a^\mu_\alpha(\boldsymbol{k}, j)\sqrt{\Delta_j}}{\sqrt{N}}\, \boldsymbol{p}^{0j}_\alpha \right) e^{i\boldsymbol{k}\boldsymbol{n}}. \tag{4.85}$$

Indeed, according to eq. (4.80) we have

$$A^p_\mu(\boldsymbol{k}) = \frac{2\sqrt{\Delta_1 E_\mu(\boldsymbol{k})}}{\sqrt{N}} \sum_{m,\beta,j} M^{j1}_{mp} a^\mu_\beta(\boldsymbol{k}, j)\, e^{i\boldsymbol{k}(\boldsymbol{m}-\boldsymbol{p})} \sqrt{\Delta_j}\, e^{i\boldsymbol{k}\boldsymbol{p}}, \tag{4.80a}$$

so that,

$$A^p_\mu(\boldsymbol{k}) = 2\mathrm{e}^{\mathrm{i}\boldsymbol{kp}}\sqrt{\Delta_1 E_\mu(\boldsymbol{k})}\,(-\boldsymbol{p}^{01}\mathscr{E}),$$

where $\mathscr{E}$ is the electric field produced by the dipoles with the distribution (4.85) at the point $(\boldsymbol{p}, \alpha)$. According to Ewald (see, Born and Huang (1954); we have used this procedure in ch. 3 sect. 1) this field can be written in the form:

$$\mathscr{E}_x = \mathscr{E}_{\mathrm{M}x} + \sum_{\gamma\beta} Q^{\alpha\beta}_{xy}(\boldsymbol{k})p^{0\mu}_{\beta y}(\boldsymbol{k}),$$

where the macroscopic part of the field is

$$\mathscr{E}_{\mathrm{M}} = -\frac{4\pi}{v}\frac{(\boldsymbol{P}^{0\mu}\boldsymbol{k})\boldsymbol{k}}{k^2},$$

and the vector $\boldsymbol{P}^{0\mu}(\boldsymbol{k})$ is determined by eq. (4.48) which is transformed into the following expression when we introduce the quantities a:

$$\boldsymbol{P}^{0\mu} = \sum_{j\alpha_1} \boldsymbol{p}^{0j}_{\alpha_1}\sqrt{\Delta_j}\, a^{\mu}_{\alpha_1}(\boldsymbol{k}, j) \equiv \sum_{\alpha_1} \boldsymbol{p}^{0\mu}_{\alpha_1}.$$

The internal field matrix $Q^{\alpha\beta}_{xy}(k)$ entering into the expression for $\mathscr{E}$ depends on the position of the impurity molecule $p(p \equiv \boldsymbol{p}, \alpha)$ and this fact is reflected in its form.

Thus, we come to

$$A^p_\mu(\boldsymbol{k}) = 2\sqrt{E_\mu(\boldsymbol{k})\Delta_1}\left[\frac{4\pi}{v}\frac{(\boldsymbol{p}^{01}\boldsymbol{k})(\boldsymbol{P}^{0\mu}\boldsymbol{k})}{k^2} - \sum_{xy\beta} p^{01}_x Q^{\alpha\beta}_{xy}(\boldsymbol{k})p^{0\mu}_{y\beta}(\boldsymbol{k})\right]\mathrm{e}^{\mathrm{i}\boldsymbol{kp}}, \qquad (4.86)$$

If we replace in eq. (4.86) p with q we obtain a similar expression for $A^q_\mu(\boldsymbol{k})$. Substituting these expressions into eq. (4.82a) and using eq. (4.56) and the expression

$$M^{11}_{pq} = \frac{4\pi}{vN}\sum_{\boldsymbol{k}}\frac{(\boldsymbol{kp}^{01})^2}{k^2}\mathrm{e}^{\mathrm{i}\boldsymbol{k}(\boldsymbol{p}-\boldsymbol{q})},$$

for the energy of the dipole–dipole interaction between identical dipoles $\boldsymbol{p}^{01}$ at the points $\boldsymbol{p}$ and $\boldsymbol{q}$ we find that if we ignore the dependence of the exciton energy $E_\mu(\boldsymbol{k})$ and the matrix elements $\boldsymbol{p}^{0\mu}(\boldsymbol{k})$ on $\boldsymbol{k}$ (we take these quantities at $\boldsymbol{k} = 0$) we can rewrite eq. (4.82a) as

$$M^{\mathrm{eff}}_{pq} = \Delta_1|a(p\,1)|^2\frac{4\pi}{vN}\sum_{kij}\frac{(k_\alpha R_{\alpha\beta}(E_0, 0)p^{01}_\beta)^2}{\epsilon_{tr}k_t k_r}\mathrm{e}^{\mathrm{i}\boldsymbol{k}(\boldsymbol{p}-\boldsymbol{q})}, \qquad (4.87)$$

where the tensor

$$R_{\alpha\beta}(E_0, 0) = \delta_{\alpha\beta} - (\delta_{\alpha\sigma} - \epsilon_{\sigma\sigma}(E_0, 0))\gamma_{\alpha\beta}, \qquad (4.87a)$$

and

$$\gamma_{x,y} = \frac{v}{4\pi}Q^{11}_{x,y}(0). \qquad (4.87b)$$

Equation (4.87) shows that the effective dipole moment of the transition determining the energy of the dipole–dipole resonance interaction in the medium is given by

$$(\boldsymbol{p}^{01}_{\text{eff}})_\alpha = |a(p\,1)|\sqrt{\Delta_1}\, R_{\alpha\beta}(E_0, 0) p^{01}_\beta,$$

where the tensor (4.87a) is expressed in terms of a macroscopic characteristic of the medium, namely, the dielectric constant tensor $\epsilon_{\alpha\beta}(E_0, 0)$, and the internal field tensor Q while the factor $\sqrt{\Delta_1}|a(p\,1)|$ has a purely microscopic nature and is determined by the character of interaction between the impurity molecule and its environment (see eq. (4.84)).

In cubic crystals we have $\epsilon_{\alpha\beta}(E_0, 0) = \epsilon(E)\delta_{\alpha\beta}$. If, moreover, the impurity molecules in a cubic crystal are at the points with maximum symmetry of the local group then we have $\gamma_{\alpha\beta} = \frac{1}{3}\,\delta_{\alpha\beta}$ (see, Born and Huang (1954) appendix VI). In this case eq. (4.87) yields

$$M^{\text{eff}}_{pq} = \Delta_1|a(p\,1)|^2 \frac{1}{\epsilon(E_0)}\left[\frac{\epsilon(E_0)+2}{3}\right]^2 M^{11}_{pq}. \tag{4.88}$$

This expression differs from that derived in ch. 3 sect. 6 only in that it contains the factor $\Delta_1|a(p\,1)|^2$ which, as noted above, is unity for transitions in the impurity molecules which lie in the region of transparency of the matrix. If the impurity excitation level E_0 approaches the edge of an exciton band then this factor can become smaller than unity.

Assume, for instance, that the energy E_0 is below the bottom of a $\tilde{\mu}$th exciton band and the difference $E_{\tilde{\mu}}(0)–E_0$ is small compared with the width of this band. Then in eq. (4.84) we can include only the sum

$$\frac{1}{N}\sum_{\boldsymbol{k}} \frac{|A^p_{\tilde{\mu}}(\boldsymbol{k})|^2}{(E_0^2 - E^2_{\tilde{\mu}}(\boldsymbol{k}))^2}.$$

As follows from the above arguments, the main contribution to this sum is made by small $\boldsymbol{k}$ so that we can employ the effective mass approximation for $E_{\tilde{\mu}}(\boldsymbol{k})$. If we ignore a possibility of $E_{\tilde{\mu}}$ being a nonanalytic function of $\boldsymbol{k}$ and assume that the effective exciton mass is isotropic we can readily find that the above sum is proportional to $(E_{\tilde{\mu}}(0)–E_0)^{-1/2}$. This implies that for $E_0 \to E_{\tilde{\mu}}(0)$ the quantity $|a(p\,1)|^2$ tends to zero according to $\Delta_1|a(p\,1)|^2 \approx [(E_{\tilde{\mu}}(0) - E_0)/M]^{1/2}$ where M is the exciton band width (this means that the excitation localized at the impurity is mainly not on the impurity molecules p but on the neighbouring matrix molecules). This expression is valid only for $(E_{\tilde{\mu}}(0) - E_0)/M \ll 1$ and it would be incorrect to substitute it into eqs. (4.87) or (4.88) since these equations contain the dielectric constant tensor $\epsilon_{\alpha\beta}\,(E_0, 0)$ ignoring spatial dispersion just because we have replaced $E_\mu(\boldsymbol{k})$ and $\boldsymbol{P}^{0\mu}(\boldsymbol{k})$ in eqs. (4.56) with $E_\mu(0)$ and $\boldsymbol{P}^{0\mu}(0)$, respectively.

What is the actual form of the effective resonance interaction when $E_0 \to E_{\tilde{\mu}}(0)$? To answer this question let us return to the starting equation

(4.82a). We see that for $E_0 \to E_{\tilde{\mu}}(0)$ the dependence on $\boldsymbol{k}$ may be significant only when we calculate the contribution to M^{eff} due just to the $\tilde{\mu}$th exciton band. We can find the contribution of the other, more distant bands as was done when we derived eq. (4.82a). Write M^{eff} as the sum

$$M^{\text{eff}} = M_1^{\text{eff}} + M_2^{\text{eff}}, \tag{4.89}$$

where M_1^{eff} is given by an expression of the type of eq. (4.87) or eq. (4.88) in which the tensor $\epsilon_{\alpha\beta}(E_0, 0)$ is replaced with a tensor $\epsilon'_{\alpha\beta}(E_0, 0)$ which differs from $\epsilon_{\alpha\beta}$ by the absence of the contribution of the $\tilde{\mu}$th band, and

$$M_2^{\text{eff}} = |a(p\,1)|^2 \frac{1}{2N} \sum_{\boldsymbol{k}} \frac{A_{\tilde{\mu}}^{p}(\boldsymbol{k}) A_{\tilde{\mu}}^{*q}(\boldsymbol{k})}{E_0^2 - E_{\tilde{\mu}}^2(\boldsymbol{k})}, \tag{4.90}$$

Thus, for $E_0 \to E_{\tilde{\mu}}(0)$ we, indeed, obtain $M_1^{\text{eff}} \sim [E_{\tilde{\mu}}(0) - E_0]^{1/2}$ while the dependence of M_2^{eff} on the position of the impurity level E_0 is more complicated.

The above discussion also indicates that M_1^{eff} decreases as the reciprocal of the third power of the distance $|\boldsymbol{p} - \boldsymbol{q}|$ (as M_{pq}^{eff} given by eq. (4.87) and M_{pq}^{11}). The quantity M_2^{eff} contains also terms which decrease exponentially with increasing distance between the molecules, apart from the terms which decrease as $|\boldsymbol{p} - \boldsymbol{q}|^{-3}$. (The exponentially decreasing terms are sometimes incorrectly identified as the contribution of virtual excitons, according to Haken (1958), Agranovich (1960). In fact, as noted above, virtual excitons make a significant contribution to M_1^{eff}, too. At the same time, as discussed below, M_2^{eff} vanishes if the matrix excitons are incoherent.)

Analysis of eq. (4.90) directly indicates the presence of terms exponentially decreasing with increasing distance $|\boldsymbol{p} - \boldsymbol{q}|$. They can also be easily explained within the framework of macroscopic treatment if we take into consideration that generally in a medium with spatial dispersion (in our case spatial dispersion is due to the contribution of the exciton states) the electric field produced by the dipole moment $\boldsymbol{p} = \boldsymbol{p}_0 \exp(\mathrm{i}\omega t)$ placed at the origin of coordinates is

$$\boldsymbol{E}(\boldsymbol{r}, t) = -\frac{1}{2\pi^2} \int \frac{\boldsymbol{k}(\boldsymbol{k}\boldsymbol{p}) \exp(-\mathrm{i}\boldsymbol{k}\boldsymbol{r})}{\epsilon_{ij}(\omega, \boldsymbol{k}) k_i k_j}\, \mathrm{d}\boldsymbol{k},$$

where $\epsilon_{ij}(\omega, \boldsymbol{k})$ is the dielectric constant tensor of the medium. If we neglect spatial dispersion, that is, tend the exciton band widths to zero and take $\epsilon_{ij} = \epsilon_{ij}(\omega, 0)$, then in the frequency range in which the principal values of the tensor ϵ_{ij} are positive (that is, in the region of transparency of the crystal) we obtain

$$E_i(\boldsymbol{r}, t) = p_k (\det \epsilon_{\alpha\beta}^{-1})^{1/2} \frac{\partial^2}{\partial x_i \partial x_k} \frac{1}{\sqrt{\epsilon_{rt}^{-1} x_r x_t}}, \tag{4.91}$$

so that for the isotropic medium with $\epsilon_{\alpha\beta}(\omega, 0) = \epsilon(\omega)\,\delta_{\alpha\beta}$ we have

$$E(r) = \frac{pr^2 - 3(pr)r}{\epsilon(\omega)r^5}. \tag{4.91a}$$

Hence, the energy of interaction between two dipoles p_1 and p_2 at the distance r is given by

$$M_{pq}^{12} = \frac{p_1 p_2 r^2 - 3(p_1 r)(p_2 r)}{\epsilon(\omega)r^5}.$$

If we include weak spatial dispersion then for the isotropic medium we can write $\epsilon_{ij}(\omega, k)k_i k_j \approx k^2[\epsilon(\omega) + \delta k^2]$ where δ depends only on ω. In this approximation eq. (4.91) is replaced with

$$E_l(r) = p_m \frac{\partial^2}{\partial x_l \partial x_m} \frac{1}{i\pi} \int_{-\infty}^{\infty} \frac{e^{ikr}}{\epsilon(\omega) + \delta k^2} \frac{dk}{k^2} = p_m \frac{\partial^2}{\partial x_l \partial x_m} \frac{1}{r\epsilon(\omega)} (1 - e^{ik_0 r}),$$

where $k_0 = k_0' + ik_0''$ is the solution of the equation $\epsilon(\omega) + \delta k^2 = 0$ in the upper complex half plane. If the frequency ω is close to the longitudinal exciton frequency $\omega_\parallel$ (for which $\epsilon(\omega_\parallel) = 0$) and is outside the spectral region of longitudinal waves, then $k_0' = 0$ and $k_0'' = [\epsilon(\omega)/\delta]^{1/2} > 0$ and its magnitude is small since in the region $\omega \approx \omega_\parallel$ the dielectric constant $\epsilon(\omega) = (\omega - \omega_\parallel)(d\epsilon/d\omega)_{\omega_\parallel}$ so that we have $k_0'' = (d\epsilon/d\omega)_{\omega_\parallel}^{1/2}(\omega - \omega_\parallel)/\delta)^{1/2}$, $(d\epsilon/d\omega)_{\omega_\parallel} > 0$. Thus, when we take into account spatial dispersion the expression for the electric field indeed includes exponentially decreasing terms of the order of $\exp(-rk_0'')$. The rate of their decrease depends on the position of the frequency $\omega = E_0/\hbar$; it declines as the frequency ω approaches the longitudinal wave frequency (in anisotropic crystals this effect is determined by the closeness of the frequency ω to the frequency of the Coulomb exciton). The estimates given by Dow (1968) show that typical values of $(k_0'')^{-1}$ are of the order of several lattice parameters and this supports the above conclusion that the contribution of the exponentially decreasing terms for low impurity concentration is insignificant.

As noted above, the exponentially decreasing terms can also be derived from eq. (4.90). Using eq. (4.80a) and omitting factors which are insignificant here we can rewrite eq. (4.90) as

$$M_2^{\text{eff}} \sim \int \frac{\varphi(k)\, e^{ik(p-q)}}{E_0^2 - E_{\bar{\mu}}^2(k)} dk,$$

where $\varphi(k)$ is a function independent of the distance between impurities. For large distances $|p - q|$ the main contribution to the above integral is made by small k values just as in the above derivation of eq. (4.88) from eq. (4.82a). The exponentially decreasing terms are due to the contribution of the pole of the integrand at $\boldsymbol{k} = \boldsymbol{k}_0$ satisfying the condition $E_0^2 - E_{\bar{\mu}}^2(\boldsymbol{k}_0) = 0$. Since the energy E_0, by assumption, does not lie in the exciton band, $\boldsymbol{k}_0 = \boldsymbol{k}_0' + i\boldsymbol{k}_0''$ is a complex-valued vector so that the residue at $\boldsymbol{k} = \boldsymbol{k}_0$ decreases as $\exp[-\boldsymbol{k}_0''(\boldsymbol{p} - \boldsymbol{q})]$ (we neglect here secondary effects due to anisotropy and the additional power-law decrease of the terms under consideration).

Similar calculations of the energy of resonance interaction between impurity molecules can be made for the impurity molecules in which the respective transitions are dipole forbidden. In such calculations the exchange interaction and higher multipoles should be taken into account. For instance, the quadrupole–quadrupole interaction was analyzed by Kamenogradskiy and Konobeev (1969). They have shown that at large distances between the impurity molecules the effective interaction (4.82a) in the isotropic medium (compare with eq. (4.88)) has the form:

$$M_{pq}^{\text{eff}} = \Delta_1 |a(p\,1)|^2 \frac{1}{\epsilon(E_0)} \left(\frac{2\epsilon(E_0) + 3}{5} \right)^2 M_{pq}^{11} \text{ (quadr–quadr)}$$

where M_{pq}^{11} (quadr–quadr) is the energy of the resonance quadrupole–quadrupole interaction in vacuum. Thus, the local field correction for the quadrupole $(2\epsilon + 3)/5$ differs from the respective correction for the dipole $(\epsilon + 2)/3$ so that, for instance, the energy of the quadrupole–dipole interaction includes the product $[(2\epsilon + 3)/5][(\epsilon + 2)/3]$ and so on. The above results and their generalization for anisotropic media can also be obtained within the framework of the local field method (see ch. 3). Note, however, that for short-range resonance interactions such as quadrupole–quadrupole and similar interactions the local field corrections may be practically useless since these interactions are weak at large distances between impurity molecules while at short distances general relationships of the type of eq. (4.82a) should be used.

Since crystals may contain not only excitons or plasmons, but other elementary excitations, such as photons, phonons or magnons in magnetic media, even a qualitative analysis shows that the exchange of such virtual excitations between impurity molecules should result in additional resonance interaction. Transverse photons give rise to retarded interaction which is effective only if the distance between impurity molecules is of the order of the wavelength of photons in resonance with impurities and also to reabsorption discussed in sect. 11 and ch. 6. (It should be recalled that in vacuum the exchange of virtual longitudinal and scalar photons forms the conventional Coulomb interaction between charges. The presence of a polarizing medium and, hence, the inclusion of the exchange of virtual excitons leads to the appearance of the dielectric constant, local field corrections and some other factors in the effective resonance interaction between impurities.)

In contrast to photons, phonons and magnons give rise to an additional resonance interaction which has a short-range character and their inclusion is significant only if the electronic excitation energy E_0 of the impurity is of the order of the phonon or magnon energy. This can occur, for instance, for pairs of paramagnetic ions (see, Orbach and Tachiki (1967) and references therein). In most cases of energy transfer between impurity molecules known to us the contribution of the exchange of virtual phonons for large E_0 is relatively small. Therefore, we shall only show how to derive a general expression for the transfer probability and discuss some of the results of its analysis (see also, Orbach and Tachiki 1967, Guari and Kozhushner 1970).

If the nuclei are shifted from their equilibrium positions the operator (4.21) can be expanded in powers of the displacement $\boldsymbol{u}_n$ of the nuclei. Since the matrix elements V_{nm} of the operator depend only on the differences between the displacements of the sites n and m (for the sake of simplicity, we ignore here the rotational degrees of freedom of the molecules) we can write in the first approximation,

$$\hat{H}_1 = \hat{H}_1^{(0)} + \hat{H}_1^{(1)},$$

where $\hat{H}_1^{(0)}$ is the operator (4.21) for the perfect lattice and

$$\hat{H}_1^{(1)} = \sum_{nm} \mathbf{A}_{nm}^{0f}(P_n + P_n^+)(\boldsymbol{u}_n - \boldsymbol{u}_m).$$

We shall not write down the expression for $\mathbf{A}_{nm}^{0f}$ here since the meaning of these quantities can be easily understood from the above discussion. Going over from the displacements $\boldsymbol{u}_n$ to the operators of creation and annihilation of phonons we obtain

$$\hat{H}_1^{(1)} = \sum_{n,\boldsymbol{q}r} K_n(\boldsymbol{q}r)(P_n + P_n^+)(b_{\boldsymbol{q}r} + b_{-\boldsymbol{q}r}^+)\,\mathrm{e}^{\mathrm{i}\boldsymbol{q}\boldsymbol{n}},$$

where $n \equiv \boldsymbol{n}, \alpha$; $m \equiv \boldsymbol{m}, \beta$; and

$$K_n(\boldsymbol{q}r) = \sum_m \mathbf{A}_{nm}^{0f}\left(\frac{\hbar}{2M\omega_{\boldsymbol{q}r}}\right)^{1/2}[\boldsymbol{e}_\alpha(\boldsymbol{q}r) - \boldsymbol{e}_\beta(\boldsymbol{q}r)\,\mathrm{e}^{\mathrm{i}\boldsymbol{q}(\boldsymbol{m}-\boldsymbol{n})}].$$

Here the vectors $\boldsymbol{e}_\alpha(\boldsymbol{q}r)$ are the unit vectors of displacements corresponding to the phonon $\boldsymbol{q}r$, and M is the mass of the crystal.

Now assume that two identical substitutional impurities are on sites n and m in the crystal. Regarding the operator $H_1^{(1)}$ as the cause of transfer of excitation from the impurity molecule n to the impurity molecule m we can find the additional resonance interaction $V_{nm}(P_n^+P_m + P_nP_m^+)$ to the second order of the perturbation theory where

$$V_{nm}^{\mathrm{ph}} = \sum_{\boldsymbol{q}r} \frac{2\hbar\omega_{\boldsymbol{q}r}|K_n(\boldsymbol{q}r)|^2}{E_0^2 - \hbar^2\omega_{\boldsymbol{q}r}^2}\,\mathrm{e}^{\mathrm{i}\boldsymbol{q}(\boldsymbol{n}-\boldsymbol{m})}.$$

If $E_0 \gg \hbar\omega_{qr}$ then for the nearest neighbours we have

$$V_{nm}^{\mathrm{ph}} \sim \left(\frac{\langle 0f|\hat{V}_{nm}|00\rangle}{E_0}\right)^2 \hbar\omega_{qr}^{\max}.$$

For large distances $|\boldsymbol{n}-\boldsymbol{m}|$ we obtain exponential decrease of V_{nm}^{ph} since for large E_0 the condition $E_0 = \hbar\omega_{qr}$ can be satisfied only for complex $\boldsymbol{q}$. According to Orbach and Tachiki (1967), V_{nm} decreases as $\exp[(-2|\boldsymbol{n}-\boldsymbol{m}|/a)\ln(2E_0/\hbar\omega_{\mathrm{D}})]$ where the frequency ω_{D} is of the order of the Debye frequency, and a is the lattice parameter. It has also been shown there that, for instance, for energy transfer between Cr^{3+} ions in ruby crystals (R_1 line) the exchange of virtual phonons is significant only for the ions within one lattice parameter. If the distance between ions is large the main contribution is that of the quadrupole–quadrupole interaction.

At the end of this discussion of the mechanisms of the resonance energy transfer due to contributions of virtual phonons and excitons we shall make two general remarks concerning calculations of the resonance interaction. The first of them concerns the asymptotic behaviour of this interaction at large distances between impurity molecules.

Since in all cases we have to find the asymptotic expression for the coefficient of the Fourier expansion of a function which is periodic and analytical for real $\boldsymbol{q}$ (see, for instance, the above expression for V_{nm}^{ph}), we should take care, when we replace the exciton energy $E_\mu(\boldsymbol{k})$ or the phonon energy $\hbar\omega_{qr}$ with an approximate expression, to prevent any additional discontinuities of the derivatives of $E_\mu(\boldsymbol{k})$ or ω_{qr} because these discontinuities lead to a power-law decrease of the energy of the resonance interaction, rather than the exponential decrease. For instance, the approximation of effective mass cannot be used for excitons, as noted in an early work (Agranovich 1960b) and the use of the Debye approximation for phonons leads to errors (Orbach and Tachiki 1967, Gurari and Kozhushner 1970).

The second remark concerns the processes of scattering of phonons and excitons which determine the finite path of quasiparticles. A general treatment of this problem has not been attempted. However, it is clear that for excitons these processes cannot be particularly significant provided they do not cause transformation of coherent excitons into incoherent excitons. Otherwise, the effect of these processes results in elimination of the exponentially decreasing terms in eq. (4.90) which are due to finite width of the exciton band. If the exciton–phonon interaction or scattering of excitons by impurities is weak (Burshtein and Kozhushner 1971), then the above relationships remain unchanged if we assume that the zero-approximation steady-state excitons states in these relationships take into account the scattering processes. Apparently, a similar situation is found for the phonon mechanisms of resonance interaction (Orbach and Tachiki 1967).

8. Dipole moment of the transition determining the intensity of light absorption by impurity

In the preceding section we have shown how we can find the excitation energies and the wave functions for impurity molecules in crystal matrix. These calculations took into account the total Coulomb interaction between charges and entirely ignored retarded interaction. As noted above, this approach corresponds to the so-called Coulomb gauge of potentials for which the vector potential $\mathbf{A}(\mathbf{r})$ of the electromagnetic field satisfies the condition div $\mathbf{A} = 0$ and the operator of the interaction between the charges and the electromagnetic field (the field of transverse photons) in the linear approximation in the field has the form

$$\hat{H}_{\text{int}} = -\frac{1}{c}\sum_n \frac{e}{m}\mathbf{A}(n)\hat{\boldsymbol{I}}_n,$$

where $\hat{\boldsymbol{I}}_n$ is the operator of momentum of the electrons of the molecule n. The vector potential in the crystal has the form

$$\mathbf{A}(\mathbf{r}) = \mathbf{A}(\mathbf{k}\rho)\,\mathrm{e}^{\mathrm{i}\mathbf{kr}},$$

where $\mathbf{A}(\mathbf{k}\rho)$ is the amplitude of the vector potential of the polariton with the energy $E_\rho(\mathbf{k})$. The matrix element of the operator $\hat{H}_{\text{int}}$ corresponding to the transition in the impurity centre from the ground state to the excited state with the energy E_0 given by eq. (4.81) and simultaneous absorption of a photon with the wave vector $\mathbf{k}$ can be written as

$$T(\mathbf{k}\rho) \equiv \langle 0|\hat{H}_{\text{int}}|E_0\rangle = -\frac{\mathrm{i}E_0}{c}\mathbf{A}(\mathbf{k}\rho)\Big\langle 0\Big|\sum_n \exp(\mathrm{i}\mathbf{k}\mathbf{r}_n)\hat{\mathbf{p}}_n|E_0\rangle,$$

where $\hat{\mathbf{p}}_n$ is the operator of the dipole moment of the molecule n.

In the second quantization representation we have

$$\hat{\mathbf{p}}_n = \sum_{ni} \mathbf{p}_n^{0i}(B_{ni} + B_{ni}^+).$$

Introduce new Bose operators B_ν and B_ν^+ (see eq. (4.45a)) for the crystal with one impurity molecule $p(p \equiv (\mathbf{p}, \alpha))$, as discussed in sect. 7, when the coefficients $a(ni)$ are given by eq. (4.83). The matrix element of the Fourier component of the dipole moment of the crystal $\mathbf{p}^{0\nu_0}(\mathbf{k}) \equiv \langle 0|\Sigma_s \mathbf{p}_s \exp(\mathrm{i}\mathbf{k}s)|E_0\rangle$ is given by the expression

$$\mathbf{p}^{0\nu_0}(\mathbf{k}) = \sum_{sj} \mathbf{p}_s^{0j}\sqrt{\Delta_j}\,a(sj)\exp(\mathrm{i}\mathbf{k}\mathbf{r}_s),$$

where summation over s is performed over all molecules including the impurity molecule. Using this expression and eqs. (4.83), (4.86) and (4.87b) we can write also:

$$p^{0\nu_0}(\boldsymbol{k}) = a(p\,1)\sqrt{\Delta_1}\left[\boldsymbol{p}^{01} + \frac{8\pi}{v}\sum_{\mu}\frac{E_\mu(\boldsymbol{k})(\boldsymbol{p}^{01}\boldsymbol{k})(\boldsymbol{P}^\mu(\boldsymbol{k})\boldsymbol{k})\boldsymbol{P}^\mu(\boldsymbol{k})}{k^2(E_0^2 - E_\mu^2(\boldsymbol{k}))}\right.$$
$$\left. - \frac{8\pi}{v}\gamma_{ij}\sum_{\mu}\frac{\boldsymbol{P}^\mu(\boldsymbol{k})p_i^{01}P_j^\mu(\boldsymbol{k})E_\mu(\boldsymbol{k})}{E_0^2 - E_\mu^2(\boldsymbol{k})}\right].$$

If we now take into account eq. (4.56) we finally find that for small $\boldsymbol{k}$ the matrix element of the dipole moment is given by the expression

$$p_\alpha^{0\nu}(0) = a(p\,1)\sqrt{\Delta_1}\, E_{\alpha\beta}(E_0)p_{1\beta}^{01},$$

where the tensor

$$E_{\alpha\beta} = \delta_{\alpha\beta} - (\delta_{\alpha\gamma} - \epsilon_{\alpha\gamma}(E_0))\gamma_{\gamma\beta}(p)$$
$$+ \frac{(\delta_{\alpha\gamma} - \epsilon_{\alpha\gamma}(E_0))k_\gamma}{\epsilon_{rt}(E_0)k_r k_t}[k_\beta - k_\gamma(\delta_{\gamma u} - \epsilon_{\gamma u}(E_0))\gamma_{u\beta}(p)]. \tag{4.92}$$

The oscillator strength of the transition $0 \to \nu$ in the impurity $F^{0\nu}$ is determined by the squared scalar product $\boldsymbol{e}\boldsymbol{p}^{0\nu}$ where $\boldsymbol{e} \perp \boldsymbol{k}$. Hence, we obtain

$$F^{0\nu} \sim \Delta_1|a(p\,1)|^2(e_\alpha E_{\alpha\beta}(E_0)p_\beta^{01})^2 E_0. \tag{4.93}$$

For the "deep traps" the product $\Delta_1|a(p\,1)|^2$ is close to unity (as mentioned in sect. 8). As for the tensor $E_{\alpha\beta}$ which determines the effective value of the dipole moment of the transition in the impurity, it differs from the tensor $R_{\alpha\beta}$ in eq. (4.87) in the third term which nonanalytically depends on $\boldsymbol{k}$ for $\boldsymbol{k} \to 0$ (it should be recalled that $\boldsymbol{k}$ is the wave vector of the incident light). It is true that in cubic crystals where the tensor $\epsilon_{\alpha\beta}(E_0)$ is diagonal this tensor makes no contribution to eq. (4.93) owing to the fact that $\boldsymbol{e}\boldsymbol{k} = 0$. However, in anisotropic media the contribution of this term does not vanish identically but is proportional to the "anisotropic" part of the tensor ϵ_{ij}. Note that, in contrast to the second term in eq. (4.92), the third term includes higher powers of the oscillator strengths of transitions in the matrix, rather than the first powers; the form of eq. (4.92) shows that this term becomes particularly significant when the energy E_0 approaches an exciton band (then the quantity $\epsilon_{ij}(E_0)k_ik_j$ becomes small and spatial dispersion also becomes significant under these conditions).

The fact that the tensor $E_{\alpha\beta}$ entering into eq. (4.93) differs from the tensor $R_{\alpha\beta}$ in an anisotropic medium implies a difference between the effective values of the matrix elements of the dipole moments manifesting themselves in the resonance interaction between impurities and in the oscillator strengths (in isotropic medium these values coincide: $\boldsymbol{p}_{\text{eff}}^{01} = \boldsymbol{p}^{01}(\epsilon + 2)/3$). If we recall the derivation of the Förster formula (see ch. 1) we can conclude that in an anisotropic medium eq. (1.38) can be violated, generally speaking. No experimental verification of this theoretical prediction has been reported.

9. Absorption and luminescence spectrum of an impurity centre

The expression for the dipole moment of the impurity transition can be used to derive the oscillator strength which determines the integrated intensity of light absorption of the impurity corresponding to this transition (see sect. 8). If we take into consideration the electron–phonon interaction for the impurity then the spectrum acquires a structure while the integrated intensity is conserved (we ignore here the mixing of electronic configurations in the impurity caused by phonons). The features of this structure are well-known now (Maradudin 1966, Rebane 1968, Perlin and Tsukerblat 1974). Since we shall need this knowledge here, for instance, for finding the integrals of the donor emission and acceptor absorption spectral overlap, we shall present the relevant results below without going into the details of analysis. However, we shall pay more attention to analyzing the mechanism of broadening of the absorption and luminescence lines of impurities which, as discussed in ch. 2, is due to the radiationless resonance transfer of the excitation energy from the impurity to the matrix, and to the effects of mixing of states with discrete and continuous spectra.

In ch. 2 we used the macroscopic approach when discussing the mechanism of broadening. (We calculated there the variation of the lifetime of the excited state of the impurity due to the probability of the resonance transfer of its energy to the environment. The additional broadening treated in this section is related by the expression $\Gamma \approx 2\pi\hbar(1/\tau - 1/\tau_0)$ to the lifetimes τ and τ_0 entering into eq. (2.31).) The macroscopic approach proves to be unsuitable for obtaining a quantitative estimate of the broadening in this case though its existence remains unquestioned.

If we ignore the resonance interaction between the impurity molecule and its environment then the following model Hamiltonian can be used for analyzing the impurity spectrum:

$$\hat{H} = \mathscr{E}_0 + BB^+\left[\Delta + \sum_\kappa \hbar\omega_\kappa (U_\kappa^* b_\kappa + U_\kappa b_\kappa^+ +)\ldots\right] + \sum_\kappa \hbar\omega_\kappa b_\kappa^+ b_\kappa. \tag{4.94}$$

Here B^+ and B are the operators of creation and annihilation of excitation in the impurity molecule (the level is assumed to be nondegenerate, for the sake of simplicity), b_κ^+ and b_κ are the operators of creation and annihilation of the phonon κ with the energy $\hbar\omega_\kappa$, $\mathscr{E}_0$ is the energy of the ground state of the system, Δ is the excitation energy of the impurity (the $0 \to 1$ transition) when the electron–phonon interaction is ignored, and U_κ are coefficients. Equation (4.94) explicitly displays only those terms of the electron–phonon interaction which are linear in b_κ and b_κ^+ and only these terms will be taken into consideration below.

The dipole moment operator for the impurity is

$$\hat{p} = \boldsymbol{p}^{01}(B^+ + B), \tag{4.95}$$

where $\boldsymbol{p}^{01}$ is the effective value of the matrix element corresponding to the $0 \to 1$ transition (see sect. 8). Take as a time-dependent perturbation the operator $\hat{H}' = -\mathscr{E}(t)\hat{p}$ where $\mathscr{E}(t)$ is the electric field in the light wave. Then, using a conventional procedure (that is, determining the probability of absorption in the first order in $\hat{H}'$ of the nonstationary perturbation theory, summing over the final states and averaging over the initial states) we find that the coefficient $\kappa(E)$ of light absorption by the impurity ($E = \hbar\omega$ where ω is the light frequency) is given by

$$\kappa(E) = C|\boldsymbol{p}^{01}|^2 E G^+(E), \tag{4.96}$$

where C is a constant,

$$G^+(E) = \frac{1}{2\pi\hbar} \int_{-\infty}^{\infty} \mathrm{e}^{\mathrm{i}Et/\hbar} \langle B(t)B^+(0)\rangle \,\mathrm{d}t, \tag{4.97}$$

and the angle brackets denote Gibbs averaging. We can find the spectral distribution of the luminescence light in a similar way. The total probability of emission of a photon with the energy E is calculated to be

$$A(E)\,\mathrm{d}E = C'E^3|\boldsymbol{p}^{01}|^2 G^-(E)$$

where C' is a constant and

$$G^-(E) = \frac{1}{2\pi\hbar} \int_{-\infty}^{\infty} \mathrm{e}^{\mathrm{i}Et/\hbar} \langle B^+(t)B(0)\rangle \,\mathrm{d}t.$$

It can be shown that

$$G^-(-E) = G^+(E)\,\mathrm{e}^{\beta(\mu - E)},$$

where μ is the chemical potential of excitations so that the above expressions for $\kappa(E)$ and $A(E)$ yield the well-known universal relation between the spectra of light absorption and light emission (see sect. 11).

If we ignore the electron–phonon interaction then we have $B(t) = \exp(-\mathrm{i}\Delta t/\hbar)B(0)$ and $\langle B(t)B^+(0)\rangle = \exp(-\mathrm{i}\Delta t/\hbar)$ so that $G^+(E) = \delta(E - \Delta)$ and the impurity spectrum is reduced to only one line at $E = \Delta$. If we include the electron–phonon interaction the dependence of the function

$\langle B(t)B^+(0)\rangle$ on t is more complex. If we include in eq. (4.94) only the terms linear in b_κ and b_κ^+ we obtain

$$G^+(E) = \frac{1}{2\pi\hbar}\int_{-\infty}^{\infty} \exp[i(E-\tilde{\Delta})t/\hbar + g(t)]\,dt, \tag{4.98}$$

where $\tilde{\Delta} = \Delta + A$, $A = \Sigma_\kappa |U_\kappa|^2 \hbar\omega_\kappa$,

$$g(t) = \sum_\kappa |U_\kappa|^2[(N_\kappa + 1)\,e^{-i\omega_\kappa t} + N_\kappa\,e^{i\omega_\kappa t} - (2N_\kappa + 1)], \tag{4.98a}$$

and $N_\kappa = [\exp(\hbar\omega_\kappa)/k_B T - 1]^{-1}$ is the mean number of phonons of the type κ at the temperature T.

In the frequency range where $E \approx \tilde{\Delta}$ the main contribution to eq. (4.98) is made by large t values and the terms proportional to the sums of $\exp(\pm i\omega_\kappa t)$ in eq. (4.98a) tend to zero (if local vibrations are absent; see, Maradudin (1966), Rebane (1968)) and can be ignored in the calculation of the function $G^+(E)$. Then we obtain

$$G^+(E) = e^{-2M}\delta(E - \tilde{\Delta}), \tag{4.99}$$

where $\exp(-2M)$ is the Debye–Waller factor in which

$$2M = \sum_\kappa |U_\kappa|^2(2N_\kappa + 1), \tag{4.100}$$

and which determines the probability that the state of the matrix has not changed in the process of light absorption. This is why the absorption line at the frequency $\tilde{\Delta}$ is usually referred to as the zero-phonon line. The quantity $2M$ increases with increasing temperature (when the temperature is higher than the Debye temperature, M increases linearly: $M = \Sigma_\kappa[|U_\kappa|^2/\hbar\omega_\kappa]k_B T]$ so that the intensity of the zero-phonon line drops. If the electron-phonon coupling is strong (the coefficients $|U_\kappa|^2$ are large) the impurity spectrum does not contain the zero-phonon line at any low temperature (and $\exp(-2M) \ll 1$).

Apart from the zero-phonon line, the impurity spectrum always contains the bands (the so-called phonon wings) corresponding to electronic transitions which are accompanied with creation or annihilation of a number of lattice phonons.

If the electron–phonon coupling is weak the shape of the phonon wing can be found from eq. (4.98) by expanding the integrand there in powers of $g(t)$. For instance, using eq. (4.98a) we find that the intensity of absorption accompanied with creation or annihilation of one phonon $\kappa(E = \tilde{\Delta} \pm \hbar\omega_\kappa)$ is determined by the function

$$G_1^+(E) = \frac{1}{\hbar}\overline{|U_\kappa|^2}\rho\left(\left|\frac{E-\tilde{\Delta}}{\hbar}\right|\right)\begin{cases} N_\kappa + 1, & \text{if} \quad E > \tilde{\Delta}, \\ N_\kappa, & \text{if} \quad E < \tilde{\Delta}, \end{cases}$$

where $\rho(\omega)$ is the density of phonons with the frequency ω and the bar above $|U_\kappa|^2$ denotes its averaging over all phonons with the frequency $\omega_\kappa = (1/\hbar)|E - \tilde{\Delta}|$. This relationship indicates that the measurement of impurity absorption in the region of the phonon wing can be used for finding the so-called reduced density of the phonon frequencies which is equal to $\overline{|U_\kappa|^2}\rho(\omega)$. Personov et al. (1971) made such measurements for organic crystals and studied various mechanisms of broadening of the zero-phonon lines.

Equation (4.99) indicates that for the above impurity model the width of the zero-phonon line is zero. The observed finite width of the zero-phonon lines ($\Gamma < 1\ \text{cm}^{-1}$) is due both to the finite lifetime of the impurity excitation (the radiation width, the radiationless processes in the impurity, etc.) and to the contribution of the processes determined by the second-power terms in b_κ^+ and b_κ of the electron–phonon interaction. When we take into account these widths (Maradudin 1966) we obtain the Lorentzian shape for the zero-phonon line,

$$\kappa_0(E) = C|p^{01}|^2 \tilde{\Delta}\, e^{-2M} \frac{(1/\pi)\Gamma}{(E - \tilde{\Delta})^2 + \Gamma^2},$$

where Γ is the half width at half maximum of the zero-phonon line (the theory of the width of the zero-phonon line is discussed also by Safaryan (1977)).

The energy $\tilde{\Delta}$ of the zero-phonon transition can differ for various impurity molecules owing to the differences in the crystalline field. This gives rise to the so-called inhomogeneous broadening resulting in a non-Lorentzian character for the observed zero-phonon line. The shape depends on the nature of the distribution of the excitation levels in the impurity molecules. If this distribution is Gaussian then the observed line shape is the convolution of Gaussian and Lorentzian. Personov and his co-workers (Al'shits et al. 1972) used this fact to determine the homogeneous and inhomogeneous widths of the zero-phonon lines thereby obtaining the temperature dependence of the homogeneous line width.

For crystals in which the inhomogeneous broadening is sufficiently significant overlapping of the phonon wings due to contributions of different centres produces wide and structureless spectra of luminescence or absorption. If under such conditions the exciting light is sufficiently monochromatic (for instance, the laser light) and its frequency lies in the frequency region of the zero-phonon transition lines of the impurity then absorption of light results in excitation of selected molecules (selective absorption); as shown by Personov et al. (1973), the inhomogeneous broadening disappears then and the luminescence spectra have a line form.

If there is a considerable inhomogeneous broadening the process of energy transfer via impurity molecules is distinguished by a number of interesting features, which will be discussed in ch. 5 sect. 9. Now we shall continue to treat the shapes of impurity absorption spectra.

When the electron–phonon coupling is strong the main contribution to the integral (4.98) is made by small t values (for E in the region of the absorption peak). We can find the function $G(E)$ by expanding $g(t)$ in powers of $t(t \ll \Omega_D^{-1}$, Ω_D^{-1} is the Debye frequency):

$$g(t) = \mathrm{i}At/\hbar - 4Bt^2/\hbar^2 + \dots, \tag{4.101}$$

Here A is defined above (see eq. (4.98)) and

$$B = 2\sum_{\kappa} |U_{\kappa}|^2(2N_{\kappa} + 1)\hbar^2\omega_{\kappa}^2. \tag{4.102}$$

If in eq. (4.101) we retain only the terms which are written here (this approximation is valid if $B \gg \hbar^2\omega_D^2$) and substitute it into eq. (4.98) we obtain after taking the integral,

$$G(E) = \frac{1}{\pi}\sqrt{\frac{\pi}{B}}\exp\left[-\frac{(E - \tilde{\Delta} - A)^2}{B}\right]. \tag{4.103}$$

Turning back to eq. (4.96) we can conclude that when the electron–phonon coupling is strong the absorption band has the Gaussian shape with the peak at $E_{max} = \tilde{\Delta} + A$ and the half-width $\Delta\omega = 2\sqrt{B \ln 2}/\hbar$. When temperature exceeds the Debye temperature B increases linearly with temperature as follows from eq. (4.102). Then the absorption band half-width $\Delta\omega \sim \sqrt{T}$.

The theoretical predictions and experimental results on light absorption by impurities were compared in a number of review papers and monographs (see, for instance, Maradudin (1966), Rebane (1968), Perlin and Tsukerblat (1974)). Therefore, we shall only discuss here how the resonance interaction between impurity and its environment which we ignored above, can affect the shape of the optical spectrum of the impurity.

Clearly, the broadening effect discussed here can be noticeable only if the electron–phonon interaction in the impurity itself is weak. At first, we shall assume that this interaction is entirely absent and find the broadening of the impurity line which is due only to the resonance interaction between the impurity and the matrix (this broadening mechanism for interstitial impurity molecules in a metal matrix was analyzed by Agranovich and Mekhtiev (1971), Mal'shukov (1974)). Let us consider interstitial impurities. As shown in sect. 7, the transition energy E_0 taking into account the interaction between the impurity and its environment in this case is given by eq. (4.81). However, this equation does not include the exciton–phonon interaction. Taking into account this interaction we can write the equation for the energy E_0 as

$$E^2 - \tilde{\Delta}^2 - \frac{1}{N}\sum_{k\mu} \frac{|A^p_\mu(\mathbf{k})|^2}{E^2 - E^2_\mu(\mathbf{k}) + 2E_\mu(\mathbf{k})M_\mu(\mathbf{k})} = 0, \tag{4.104}$$

where $M_\mu(\mathbf{k})$ is the mass operator which determines the variation of the exciton energy under the effect of the exciton–phonon interaction. Significantly, the mass operator is a complex quantity. Therefore, the solution of eq. (4.104) giving the excitation energy of the impurity is also a complex quantity $E_0 \to E_0 + \mathrm{i}\Gamma$ where for $\Gamma \ll E_0$ eq. (4.104) yields

$$\Gamma = \frac{1}{2E_0}\frac{1}{N}\sum_{k\mu} \frac{2E_\mu(\mathbf{k})|A^p_\mu(\mathbf{k})|^2 \operatorname{Im} M_\mu(\mathbf{k})}{[E_0^2 - E^2_\mu(\mathbf{k}) + 2E_\mu(\mathbf{k}) \operatorname{Re} M_\mu(\mathbf{k})]^2}. \tag{4.105}$$

For the sake of simplicity, we shall assume that the main contribution to eq. (4.105) is made by one of the bands μ but the distance of this band from the impurity level is much greater than the band width. Moreover, we shall assume that the shift of the exciton term due to the exciton–phonon interaction is much smaller than the difference $|E_0 - E_\mu|$. If we take into account that (see sect. 7)

$$|A^p_\mu(\mathbf{k})|^2 = 4E_\mu\tilde{\Delta}\left|\sum_n M^{1\mu}_{p,n}\, \mathrm{e}^{\mathrm{i}\mathbf{k}\mathbf{n}}\right|^2,$$

$$M^{1\mu}_{p,n\alpha} = \sum_j M^{1j}_{p,n\alpha}\, a^\mu_\alpha(\mathrm{j}, \mathbf{k}),$$

eq. (4.105) reduces to

$$\Gamma = \frac{1}{N}\sum_k \frac{\left|\sum_n M^{1\mu}_{p,n}\mathrm{e}^{\mathrm{i}\mathbf{k}\mathbf{n}}\right|^2}{(E_0 - E_\mu)^2} \operatorname{Im} M_\mu(E_0, \mathbf{k}). \tag{4.106}$$

Let us find the relationship between eq. (4.106) and the imaginary part of the dielectric constant. If we ignore anisotropy in the frequency range under consideration the imaginary part of the dielectric constant of the crystal is given by

$$\epsilon''(E, \mathbf{k}) = \frac{4\pi}{a^3}\frac{|\mathbf{P}^{0\mu}|^2}{(E - E_\mu)^2} \operatorname{Im} M_\mu(E, \mathbf{k}), \tag{4.107}$$

where a^3 is the volume of the unit cell (see, for instance, Agranovich and Ginzburg (1983) Chapter IV). Equation (4.107) is valid only for large wavelengths, that is, for $k \ll 1/a$, while the main contribution to eq. (4.106) is made by large $k \sim 1/a$. Therefore, strictly speaking, in a medium with spatial dispersion the quantity Γ given by eq. (4.106) cannot be expressed in terms of a macroscopic characteristic of the medium, such as the dielectric constant. However, if we assume that Im $M_\mu(E, \boldsymbol{k})$ in the spectral region under consideration is independent of $\boldsymbol{k}$, that is, for any $\boldsymbol{k}$ we have $\mathrm{Im} M_\mu(E, \boldsymbol{k}) \approx \mathrm{Im} M_\mu(E, 0)$ then we can write

$$\Gamma = \frac{\mathrm{Im} M_\mu(E, 0)}{(E_0 - E_\mu)^2} \frac{1}{N} \sum_{\boldsymbol{k}} \left| \sum_{\boldsymbol{n}} M_{p\boldsymbol{n}}^{1\mu} \, \mathrm{e}^{i\boldsymbol{kn}} \right|^2,$$

and

$$\frac{1}{N} \sum_{\boldsymbol{k}} \left| \sum_{\boldsymbol{n}} M_{p\boldsymbol{n}}^{1\mu} \, \mathrm{e}^{i\boldsymbol{kn}} \right|^2 = \sum_{\boldsymbol{n}} |M_{p\boldsymbol{n}}^{1\mu}|^2. \tag{4.107a}$$

Using eq. (4.107) we now obtain the sought relationship:

$$\Gamma = \frac{a^3}{4\pi} \frac{\sum_{\boldsymbol{n}} |M_{p\boldsymbol{n}}^{1\mu}|^2}{|\boldsymbol{P}^{0\mu}|^2} \epsilon''(E_0, 0).$$

Since $M_{p\boldsymbol{n}}^{1\mu}$ decreases as the reciprocal of the third power of the distance between the impurity molecule and the unit cell $\boldsymbol{n}$ the main contribution to eq. (4.107a) is made by the nearest neighbours of the impurity molecule. Therefore, we have

$$\sum_{\boldsymbol{n}} |M_{p\boldsymbol{n}}^{1\mu}|^2 \approx |\boldsymbol{p}^{01}|^2 |\boldsymbol{P}^{0\mu}|^2 z/a^6,$$

where z is the number of neighbours of the impurity molecule. Thus, we come to

$$\Gamma \approx \frac{z}{4\pi a^3} |\boldsymbol{p}^{01}|^2 \epsilon''(E_0, 0),$$

which coincides in order of magnitude with the result of the macroscopic treatment given in ch. 2. Here we shall note a possible strong temperature dependence of Γ. Indeed, if the impurity level lies in the region of the matrix spectrum in which Urbach's rule is satisfied, that is, in which the absorption index of the medium:

$$\kappa(\omega) = \kappa_0 \exp \frac{\sigma(\hbar\omega - E_\mu)}{k_B T}$$

(here σ and κ_0 are constants, T is the temperature of the medium, and k_B is the Boltzmann constant) we obtain an exponential dependence on temperature for Γ since $\epsilon'' = 2n\kappa$ (where n is the index of refraction):

$$\Gamma = \Gamma_0 \exp \frac{\sigma(E_0 - E_\mu)}{k_B T}.$$

(For the theory of Urbach's rule see Vinogradov (1970), Kogan (1975) where references to earlier studies can be found. Experiments are reviewed by Kurik (1971).)

Of course, a dependence of this type can be observed for the zero-phonon line only if the broadening mechanism under consideration is the main one.

Though we assume in the above discussion that the excited impurity level is outside the exciton bands of the matrix this broadening effect is actually possible only owing to the fact that the exciton–phonon interaction in the matrix gives rise to a "tail" in the density of the exciton states. Since the density of states $\rho(E)$ is proportional to the imaginary part of Greens function of the exciton which, in its turn, is proportional to the imaginary part Im $M_\mu(E, \boldsymbol{k})$ of the mass operator, indeed, we find (see, for instance, eq. (4.106)) that $\Gamma \sim \rho(E_0)$. At the same time, the existence of states with close or identical energies in the crystal requires, generally speaking, that we take into account their configuration interaction. If $|E_0 - E_\mu| \gg M_\mu$ this effect is small so that the matrix can be regarded as a dissipative subsystem the interaction with which results chiefly in broadening of the impurity absorption lines. However, the configuration interaction proves to be important for highly excited impurity states whose energies are either in an exciton band of the matrix or in the spectral region corresponding to excitation in the matrix, for instance, of a dissociated electron–hole pair. Fano (1961) has shown that for subsystems with discrete spectra (in our case the impurity; see, Shibatani and Toyozawa 1968) and continuous spectra (the matrix) this interaction affects the shape of the absorption bands giving rise to so-called antiresonances and, in some cases, significantly changing the character of dissipation of the impurity excitations. The results of Fano are useful also for treating the exciton absorption spectra in pure crystals (Phillips (1964) which deals with the exciton absorption which is "superposed" on the absorption due to the band–band transitions) and the mechanisms of exciton–exciton collisions (see ch. 6).

Following the Fano notation we shall denote by φ and ψ_E the approximate wave functions of the system with the Hamiltonian $\hat{H}$ which correspond to the discrete state and the continuous spectrum, respectively (for instance, the wave function φ may describe the state of the crystal with an excited impurity without taking into account its interaction with the crystal and so on). The matrix $\hat{H}$ with the wave functions φ and ψ_E is not diagonal and we have:

$$(\varphi\,|\,\varphi) = 1, \qquad (\varphi|\hat{H}|\varphi) \equiv E_\varphi,$$

$$(\psi_{E'}|\hat{H}|\varphi) \equiv V_{E'}, \qquad (\psi_{E''}|\hat{H}|\psi_{E'}) = E'\delta(E' - E'').$$

Diagonalization of the matrix $\hat{H}$ necessitates inclusion of the configuration interaction. This means that from the wave functions φ and ψ_E we have to go over to their linear combination

$$\Psi_E = a(E)\varphi + \int dE' b_{E'}(E)\psi_{E'},$$

where the coefficients $a(E)$ and $b_{E'}(E)$ are, of course, determined by the system of equations:

$$E_\varphi a + \int dE' V^*_{E'} b_{E'} = Ea, \tag{4.108a}$$

and

$$V_{E'} a + E' b_{E'} = E b_{E'}. \tag{4.108b}$$

The solution of eq. (4.108b) is

$$b_{E'} = \left[\frac{\mathcal{P}}{E - E'} + z(E)\delta(E - E')\right] V_{E'} a. \tag{4.108c}$$

Substitution of eq. (4.108c) into eq. (4.108a) yields

$$z(E) = \frac{E - E_\varphi - F(E)}{|V_E|^2},$$

where

$$F(E) = \fint dE' \frac{|V_{E'}|^2}{E - E'}.$$

The coefficient $a(E)$ is found from the normalization condition for the continuous spectrum. It can be shown (Fano 1961) that

$$a(E) = \frac{|V_E|^2}{(E - \bar{E}_\varphi)^2 + \pi^2 |V_E|^4}, \tag{4.108d}$$

where $\bar{E}_\varphi = E_\varphi + F(E)$. Equation (4.108d) shows that under the effect of the configuration interaction the state φ of the discrete spectrum spreads out in the continuous spectrum to the width $\Gamma = 2\pi |V_E|^2$. In other words, owing to the configuration interaction the discrete state φ acquires a finite lifetime $\tau \approx \hbar/2\pi |V_E|^2$ which is not, of course, related to any energy dissipation processes.

If we know the functions $a(E)$ and $b_{E'}(E)$ we can analyze the effect of mixing the states φ and $\psi_{E'}$ on the impurity absorption spectrum. Calculating the matrix elements of the operator of the electric dipole moment for the transition from the ground state of the crystal to the state Ψ_E (Fano 1961) we obtain finally

$$f(\epsilon, q) \equiv \frac{|(\Psi_E|\hat{p}|0)|^2}{|(\psi_E|\hat{p}|0)|^2} = \frac{(q + \epsilon)^2}{1 + \epsilon^2} = f(-\epsilon, -q), \tag{4.109}$$

where

$$\epsilon = 2[E - E_\varphi - F(E)]/\Gamma, \qquad q = \frac{(\Phi|\hat{p}|0)}{\pi V_E^*(\psi_E|\hat{p}|0)},$$

$$\Phi(E) = \varphi + \int \frac{V_{E'}\psi_{E'}}{E - E'}\,dE'. \tag{4.109a}$$

Equation (4.109) gives the ratio of the squared matrix elements of the dipole moment (that is, practically, the ratio of the respective oscillator strengths) found with inclusion of a discrete state (here in the presence of impurity) and without it. The parameter q entering into the ratio can be regarded as being approximately independent of E in the resonance region (for $|\epsilon| \sim 1$). Thus, if we take into account the configuration interaction between the discrete state and the continuous-spectrum states we obtain a dip of the absorption intensity at $\epsilon = -q$ when $f(\epsilon, q)$ vanishes (the antiresonance; see fig. 4.2). If $q > 0$ the effect occurs for $E < \bar{E}_\varphi$; if $q < 0$ the effect occurs for $E > \bar{E}_\varphi$.

The value of q^2 is determined by the ratio between the intensities of the optical transitions to the modified state Φ of the discrete spectrum and to the nonperturbed state $\psi_{E'}$ of the continuous spectrum. The sign of q can

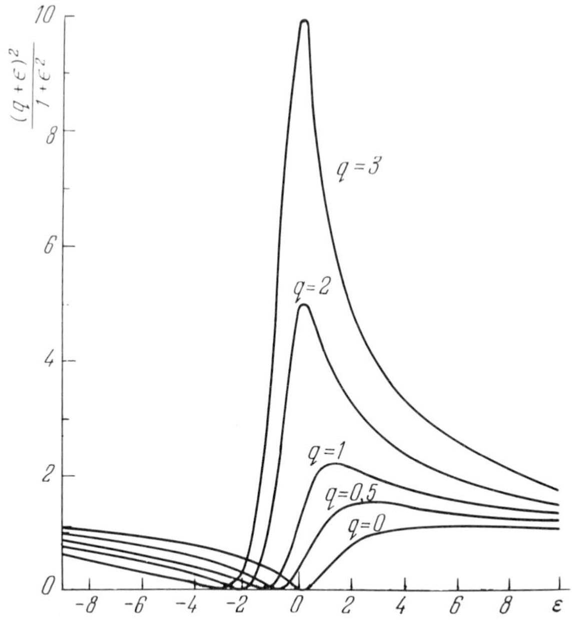

Fig. 4.2. The absorption intensity as a function of the photon energy in the antiresonance region for various values of the parameter q.

be found from the form of the spectrum. When we estimate q and Γ the following fact should be borne in mind. The dissipation processes that we ignored above (for instance, the exciton–phonon interaction), if their intensity is not too high, do not change the characteristic form of the absorption curves with antiresonance but prevent vanishing of the absorption intensity. An analysis of such situation is given by Shibatani and Toyozawa (1968), also by Fano (1961) where some generalizations of eq. (4.109) are also discussed.

10. *Probability of transfer of the electronic excitation energy from the donor to the acceptor and the role played by correlations of displacements of surrounding atoms; hot transfer*

In ch. 1 we took into account molecular vibrations when we derived the expression for the probability of transfer of the excitation energy from the donor to the acceptor. For this, we performed averaging over possible vibrational states of an impurity centre (donor or acceptor) for each given electronic molecular term. We assumed that the probabilities of distribution over these states were independent and therefore used the product of the probabilities. However, even purely qualitative considerations indicate that this approximation can be valid only for the impurity molecules the distance between which is sufficiently large. Otherwise, a strong correlation can exist between displacements of the nuclei of the impurity molecules, as is the case, generally speaking, for displacements of nuclei of one impurity centre. Therefore, we must use one distribution for describing the probabilities of occupation of vibrational states for two impurity molecules, rather than two independent distributions.

We shall follow Konobeev (1971), Soules and Duke (1971), Kozhushchner (1971), Golubov and Konobeev (1975), to discuss this problem in more detail, here, using the model Hamiltonian:

$$\begin{aligned}\hat{H} &= \epsilon_0 + \Delta_p B_p^+ B_p + \Delta_q B_q^+ B_q + M_{pq}^{\text{eff}}(B_p^+ B_q + B_p B_q^+) \\ &\quad + \sum_\kappa \hbar\omega_\kappa b_\kappa^+ b_\kappa + B_p^+ B_p \sum_\kappa \hbar\omega_\kappa (U_\kappa^*(p) b_\kappa + U_\kappa(p) b_\kappa^+) \\ &\quad + B_q^+ B_q \sum_\kappa \hbar\omega_\kappa (U_\kappa^*(q) b_\kappa + U_\kappa(q) b_\kappa^+). \qquad (4.110)\end{aligned}$$

This Hamiltonian includes the interactions of the impurity molecules p and q with each other and with the lattice vibrations (b_κ^+ and b_κ are the operators of creation and annihilation of phonon κ). If the excitation energies in the molecules p and q are not equal ($\Delta_p \neq \Delta_q$) the transfer process is known as nonresonance transfer. The effect of the matrix on the resonance interaction between the molecules p and q is described by M^{eff} (see, for instance, sect. 7). The linear approximation in displacements is used in eq. (4.110) for describing the interaction with phonons but the terms which are linear in b_κ and b_κ^+ and proportional to the operator $(B_p^+B_q + B_q^+B_p)(b_\kappa + b_\kappa^+)$ are, nevertheless, omitted. This omission can be justified only if the resonance interaction is sufficiently weak (large distances between the impurity molecules or small oscillator strengths). The quantities $U_\kappa(p)$ and $U_\kappa(q)$ in eq. (4.110) have a simple physical meaning; it can be readily seen that they are equal to a half of the displacement of the equilibrium position of the oscillator κ with normal vibrations caused by excitation of the molecule p or q and expressed in terms of the amplitudes of the zero-point vibrations.

We can write the Hamiltonian (4.110) in a convenient new representation by using the unitary transformation (see also sect. 2)

$$\tilde{H} = \mathrm{e}^{\hat{S}}\hat{H}\,\mathrm{e}^{-\hat{S}},$$

where

$$\hat{S} = B_p^+B_p\hat{S}_p + B_q^+B_q\hat{S}_q,$$

$$\hat{S}_p = \sum_\kappa [b_\kappa^+U_\kappa(p) - b_\kappa U_\kappa^*(p)].$$

The expression for $\hat{S}_q$ has a similar form. Now the transformed Hamiltonian can be written as

$$\tilde{H} = \hat{H}_0 + \hat{M}_{pq} - B_p^+B_pB_q^+B_q \sum_\kappa \hbar\omega_\kappa \operatorname{Re}(U_\kappa^*(p)U_\kappa(q)), \tag{4.111}$$

where

$$\hat{H}_0 = \Big(\Delta_p - \sum_\kappa \hbar\omega_\kappa |U_\kappa(p)|^2\Big)\tilde{B}_p^+\tilde{B}_p + \sum_\kappa \hbar\omega_\kappa b_\kappa^+ b_\kappa + \Big(\Delta_q - \sum_\kappa \hbar\omega_\kappa |U_\kappa(q)|^2\Big)\tilde{B}_q^+\tilde{B}_q, \tag{4.111a}$$

$$\hat{M}_{pq} = M_{pq}^{\text{eff}}\tilde{B}_p\tilde{B}_q + \text{h.c.}$$

$$\tilde{B}_p^+ = \exp(\hat{S}_p)B_p^+, \qquad \tilde{B}_p = B_p\exp(-\hat{S}_p). \tag{4.111b}$$

If we omit the operator $\hat{M}_{pq}$ in eq. (4.111) then the form of the operator $\hat{H}_0$ implies that in the new representation, for instance, the operator $\tilde{B}_p^+$ results in creation of a stationary excited state of the molecule p whose energy is not Δ_p (that is, not the "pure" excitation energy) but $\Delta_p - E_p$ where $E_p = \Sigma_\kappa \hbar\omega_\kappa |u_\kappa(p)|^2$ is the energy of deformation of the lattice due to transition of the molecule p to the excited state. Hence, in the new representation the operator $\tilde{B}_p^+$ gives rise not only to the electronic excitation but also to a local deformation of the lattice due to this excitation; thus, this operator is an analogue of the the operator of creation of exciton in the case of strong exciton–phonon coupling (see sect. 2). The third term in eq. (4.111) describes the interaction between the excitations of the impurity molecules p and q, if they exist simultaneously; this interaction is due to the phonon field with which these excitations interact according to eq. (4.110). However, this interaction is not manifested when we calculate the probability of transfer of the excitation from the molecule p to the unexcited molecule q. Regarding the operator $\hat{M}_{pq}$ in eq. (4.111) as a perturbation we find that in the first order of the nonstationary perturbation theory this probability is

$$W_{p\to q} = \frac{2\pi}{\hbar}\sum_{l,l'} \rho(E_l)\langle l|\hat{M}_{pq}|l'\rangle\langle l'|\hat{M}_{pq}|l\rangle\delta(E_l - E_{l'}), \tag{4.112}$$

where $\rho(E_l)$ is the diagonal density matrix, and $|l\rangle$ is the state vector corresponding to the Hamiltonian $\hat{H}_0$.

A convenient form of eq. (4.112) is

$$W_{p\to q} = \frac{1}{\hbar^2}\int_{-\infty}^{\infty} dt\langle \hat{M}_{pq}(t)\hat{M}_{qp}(0)\rangle,$$

where the angle brackets denote averaging with the density matrix $\rho(\hat{H}_0)$ and

$$\hat{M}_{pq}(t) = \exp(i\hat{H}_0 t/\hbar)\hat{M}_{pq}\exp(-i\hat{H}_0 t/\hbar).$$

Using eq. (4.111b) we find

$$W_{p\to q} = \frac{|M_{pq}^{\text{eff}}|^2}{\hbar^2}\int_{-\infty}^{\infty} dt\, e^{i\Delta t/\hbar}\langle e^{\hat{S}_p(t)}\, e^{-\hat{S}_q(t)}\, e^{\hat{S}_q(0)}\, e^{-\hat{S}_p(0)}\rangle, \tag{4.113}$$

where

$$\Delta = \Delta_p - \Delta_q + \sum_\kappa \hbar\omega_\kappa[|U_\kappa(q)|^2 - |U_\kappa(p)|^2]. \tag{4.113a}$$

is the variation of the energy of the system due to transition of the excitation from the molecule p to the molecule q. Integrals of the type of eq. (4.113) appear in the theory of mobility of small-radius polarons (see, Lang and Firsov 1962, 1963, 1967, Firsov and Kudinov 1964, 1965, Firsov 1968, 1975). We shall not go into details of their calculation and shall write here only the final result. As shown by Soules and Duke (1971), Kozhushchner (1971), Golubov and Konobeev (1975), eq. (4.113) can be written as

$$W_{p\to q} = \frac{|M_{pq}^{\mathrm{eff}}|^2}{\hbar^2} \int_{-\infty}^{\infty} \exp[\mathrm{i}\Delta t/\hbar + g_1(t)]\, \mathrm{d}t, \tag{4.114}$$

where

$$g_1(t) = \sum_{\kappa} |U_\kappa(p) - U_\kappa(q)|^2[(N_\kappa + 1)\,\mathrm{e}^{-\mathrm{i}\omega_\kappa t} + N_\kappa\,\mathrm{e}^{\mathrm{i}\omega_\kappa t} - 2N_\kappa - 1]. \tag{4.114a}$$

and $N_\kappa = [\exp(\hbar\omega_\kappa/kT) - 1]^{-1}$ is the equilibrium number of phonons of the species κ.

Expressions of the type of eq. (4.114) appear also in the theory of absorption of light and luminescence of a local centre (see sect. 9). In the absence of the molecule p the absorption spectrum $G^+(E)$ and the luminescence spectrum $G^-(E)$ for the molecule q are given by

$$G_q^{\pm}(E) = \frac{1}{2\pi\hbar} \int_{-\infty}^{\infty} \mathrm{d}t\, \exp\left[\frac{\mathrm{i}(E - \tilde{\Delta}_q)t}{\hbar} + g_q(\pm t)\right], \tag{4.115}$$

where

$$g_q(t) = \sum_{\kappa} |U_\kappa(q)|^2[(N_\kappa + 1)\,\mathrm{e}^{-\mathrm{i}\omega_\kappa t} + N_\kappa\,\mathrm{e}^{\mathrm{i}\omega_\kappa t} - (2N_\kappa + 1)].$$

Similar expressions can be written for the molecule p.

Therefore, the Förster integral for the overlap of emission and absorption spectra of the donor and acceptor molecules is given by

$$\int \mathrm{d}E G_q^+(E) G_p^-(E) = \frac{1}{2\pi\hbar} \int_{-\infty}^{\infty} \exp\left[\frac{\mathrm{i}(\tilde{\Delta}_p - \tilde{\Delta}_q)t}{\hbar} + g_p(t) + g_q(t)\right] \mathrm{d}t. \tag{4.116}$$

A comparison of eqs. (4.114) and (4.116) indicates that the probability (4.114) can be expressed in terms of the overlap integral as

$$W_{p\to q} = \frac{2\pi}{\hbar} |M_{pq}^{\mathrm{eff}}|^2 \int \mathrm{d}E G_q^+(E) G_p^-(E),$$

only if we ignore the term proportional to the product $U_\kappa(p)U_\kappa^*(q)$ in the expression for $g_1(t)$ which enters into eq. (4.114). To determine the role of this term which depends on the distance between the molecules p and q, let us calculate eq. (4.114) for the so-called strong coupling case when the main contribution to the integral (4.114) is made by small $t \ll \omega_{\mathrm{D}}^{-1}$ where ω_{D} is the Debye frequency (respective calculations for weak electron–phonon coupling are made in ch. 5 sect. 6; see eqs. (5.57) and (5.58)). As shown by Maradudin (1966), if the quantity B defined below is much greater than

$\hbar^2\omega_D^2$ we can approximate $g_1(t)$ by its expansion in powers of t retaining the terms up to the second power and obtain

$$W_{p\to q} = \frac{2}{\hbar}|M_{pq}^{\text{eff}}|^2 \sqrt{\frac{\pi}{B}} \exp\left[-\frac{(\Delta_p - \Delta_q - A)^2}{B}\right], \tag{4.117}$$

where

$$B = 2\sum_\kappa \hbar^2\omega_\kappa^2|U_\kappa(p) - U_\kappa(q)|^2(2N_\kappa + 1),$$

$$A = \sum_\kappa \hbar\omega_\kappa(|U_\kappa(p) - U_\kappa(q)|^2 - |U_\kappa(q)|^2 + |U_\kappa(p)|^2).$$

Within the framework of our model eq. (4.117) determines the effect of vibrations of the nuclei on the probability of energy transfer between the molecules p and q. The role played by correlations of displacements is described by contributions of the quantities,

$$\delta B = 4\sum_\kappa \hbar^2\omega_\kappa^2 U_\kappa(p)U_\kappa^*(q)(2N_\kappa + 1),$$

$$\delta A = 2\sum_\kappa \hbar\omega_\kappa U_\kappa(p)U_\kappa^*(q),$$

to eq. (4.117). Since $U_\kappa(p) = |U_\kappa(p)|\exp(i\kappa\boldsymbol{p})$ the quantities δA and δB decrease with increasing $|\boldsymbol{p} - \boldsymbol{q}|$ as the Fourier components of the respective functions. As in sect. 7, the analytic properties of the functions determine the character of the decrease which, generally speaking, follows a power law. The dependences of δA and δB on $|\boldsymbol{p} - \boldsymbol{q}|$ can be studied in detail only for a specific model of impurity molecules. However, even a purely qualitative analysis shows that the characteristic length entering into these dependences should be of the order of the lattice parameter a. Therefore, correlations of displacements discussed here are, apparently, insignificant at distances $R \equiv |\boldsymbol{p} - \boldsymbol{q}| \gg a$ as it was repeatedly found in studies of transfer of electronic excitation energy from the donor to the acceptor (according to Soules and Duke (1971), Kozhushchner (1971), Golubov and Konobeev (1975), δA and δB decrease as $1/R^2$ for $R \gg a$).

In conclusion, let us note another assumption typically made in the theory of resonance energy transfer between impurity molecules. When we derived the Förster formula we assumed that the initial state of the system of nuclei (the donor is in the excited state and the acceptor is in the ground state) was at thermodynamic equilibrium with the lattice. However, in some cases, generally speaking, a part of the energy transfer events can occur before vibrational relaxation in the donor has ended, that is, in the so-called "hot" state of the donor.

The characteristic time of vibrational relaxation is, typically, very short ($\tau_{\text{rel}} \sim 10^{-12}$–$10^{-10}$ s) so that the donor is in the equilibrium vibrational state most of the time (that is, for the time of the order of the luminescence decay time $\tau \sim 10^{-9}$ s or even longer). If, however, the electronic vibrational energy of the donor in this state is small compared with the energy of excitation of the acceptor then energy transfer ends practically after the end of vibrational relaxation. It is precisely under such conditions that a significant role can be played by transfer occurring before the end of vibrational relaxation, that is, the "hot" transfer the theory of which is treated by Dexter (1972), Khizhnyakov (1972), Tekhverand Khizhnyakov (1974). (Note, that, generally speaking, an excited electronic state can be described by a variety of quantum numbers (state, spin, etc.) and, hence, by a variety of relaxation times. The time of vibrational relaxation is the shortest among these times and it is just this process to which we can reasonably relate the concepts of "hot" transfer and "hot" luminescence; see below and Rebane et al. (1975), Toyozawa (1975, 1976).)

We shall not present the theory of "hot" transfer here; we shall only discuss the basic concepts and some results.

In order to analyze the role of energy transfer in the "hot" state we must study the development of the system donor–acceptor starting from the moment when a photon of the exciting light propagates in the system and the donor-acceptor pair is in the ground state. This can be done by using the density matrix approach (Toyozawa 1975, 1976) which makes it possible also to trace the development of the spectrum of secondary luminescence with time. If, however, we are not interested in the development of the spectrum then we can find the total probability that a photon with frequency ω' will be emitted after excitation by a photon with frequency ω by using the well-known formula of the second-order perturbation theory applied to the interaction of light and matter.

It has been shown (Dexter 1972, Khizhnyakov 1972, Tekhver and Khizhnyakov 1974) that the total probability of energy transfer from the donor to the acceptor contains the overlap integral of the absorption spectrum of the acceptor and the total spectrum of the non-equilibrium secondary emission of the donor, rather than the overlap integral of the equilibrium luminescence spectrum of the donor and the absorption spectrum of the acceptor which appears in the Förster theory. The non-equilibrium secondary emission of the donor includes, apart from the resonance Rayleigh and Raman scattering, the donor emission occurring during short times $t(\tau > t \gtrsim \tau_{rel})$ which is due to transitions from the excited vibrational states of the upper electronic state (that is, the so-called "hot" luminescence). The possibilities of distinguishing between the "hot" luminescence and the resonance scattering are discussed by Toyozawa (1975, 1976), Rebane et al. (1975). Similar problems appear also when we distinguish between the "hot" transfer and the direct processes of light absorption by the acceptor in the presence of the donor.

Experimental observations of the "hot" transfer are reported by Benderskiy et al. (1978) which describe energy transfer via higher vibrational states from the impurity molecules (tetracene) to the matrix molecules (anthracene).

11. *Polaritons, long-wavelength edge of the exciton absorption bands and the polariton mechanism of exciton luminescence*

When we analyzed the exciton spectrum in sects. 1–3 we made use of the Hamiltonian of electrons and nuclei of the crystal which, as we noted, ignored the so-called retarded interaction between the charges. This means that the interaction was described only in terms of the Coulomb forces. Since in reality the interaction between the charges is not instantaneous owing to the finite light velocity the theory which includes only the instantaneous Coulomb interaction (that is, assumes that the light velocity c is infinite) yields approximate results. From the viewpoint of quantum electrodynamics (Heitler 1954) the Coulomb interaction between the charges is due to the exchange of virtual scalar and longitudinal photons. As for the retarded interaction, it is obtained in quantum electrodynamics when we take into account the interaction of the charges with the transverse electromagnetic field and it is due to exchange of virtual transverse photons between the charges.

Under certain conditions the retarded interaction has a significant effect on the shape of the spectrum of the long-wavelength elementary Bose excitations (excitons, phonons, magnons). When the retarded interaction is taken into account the elementary excitations in the appropriate spectral regions are referred to as polaritons.

In order to understand the "mechanism" responsible for the appearance of polaritons let us ignore the interaction of the charges with the field of transverse photons for the moment. In this approximation the elementary excitations in the crystal are the Coulomb excitons (that is, the excitons that are obtained when only the Coulomb interaction is taken into account; we do not distinguish between excitons and optical phonons here) and the transverse photons.

For small values of the wave vector the exciton energy is

$$E_e(\boldsymbol{q}) = E_0 + \alpha_i q_i + \beta_{ik} q_i q_k + \dots, \qquad (4.118)$$

where E_0 is the limiting (that is, at $\boldsymbol{q} \to 0$) value of the exciton energy which, like the quantities α_i and β_{ik}, may depend on the direction of the vector $\boldsymbol{q}$.

Since the spectrum of transverse photons has the form

$$E_{\text{phot}}(\boldsymbol{q}) = \hbar c |\boldsymbol{q}|$$

the branches of the exciton energy and the transverse photon energy in the space of wave vectors always intersect at $q \approx q_0 = E_0/\hbar c$ (for the electronic excitons $q_0 \approx 10^5\,\text{cm}^{-1}$ and for phonons $q_0 \approx 10^3\,\text{cm}^{-1}$). At the intersection point the values of energy and momentum for the exciton and the photon are equal. Therefore, even a weak interaction between them (that is, the retarded interaction) gives rise to significant effects. Prior to analyzing these effects we shall make a remark on the role of interaction between electrons and transverse photons for an isolated atom in vacuum. It is well-known (see, for instance, Heitler 1954) that in this case such interaction determines the radiation width of the light emission line and the finite lifetime of the excited state of the atom so that the electromagnetic field of transverse photons acts chiefly as a dissipation subsystem. The renormalization of the electronic spectrum due to the retarded interaction (which makes contributions to the so called relativistic corrections for energy) for the valence electrons are, typically, negligible and they are taken into consideration only in special cases.

The situation for excitons and transverse photons is, generally speaking, quite different. Since conservation of quasimomentum must be satisfied during exciton–photon transformations in the ideal infinite crystal, the retarded interaction leads only to renormalization of the spectra of elementary excitations (see fig. 4.3). The decay of elementary excitations, that is, the decay of polaritons, is obtained only if we take into account the exciton–phonon interaction, the scattering of excitons by lattice defects or impurities, or exciton–exciton collisions and if we consider elementary excitations in a finite medium where the radiation width is restored to a certain extent (the same effect is found for one-dimensional and two-dimensional media; see, Agranovich (1968), Agranovich and Dubovskiy (1966), Freedhoff and Van-Kranendonk (1967), Philpott (1975)).

The spectrum of polaritons, that is, the dependence $\omega = \omega(\boldsymbol{q})$ was first derived in the framework of classical theory of light dispersion for the infrared spectral region in 1950 by Tolpygo (1950) and Huang (1951). These authors have shown that the conventional electrostatic approach to the calculation of vibration frequencies of the lattice of ionic crystals is not suitable if we deal with nonlongitudinal dipole vibrations with the phase velocity $v_p = E(\boldsymbol{q})/\hbar q$ which exceeds the light velocity in the medium. It has been found later that this is true for the exciton spectral region, too.

A consistent quantum theory of polaritons applicable both to optical phonons and excitons was developed by Fano (1956), Hopfield (1958) for isotropic media and by Agranovich (1959a) for crystals of arbitrary symmetry (the contribution of the Coulomb interaction was taken into account not quite accurately by Fano (1956) – in the random phase approximation). A detailed discussion of this theory is given, for instance, by Agranovich (1968), see also, Craig (1974). We shall note here only those

results of the theory which are significant for analyzing the long-wavelength edge of the exciton absorption bands, the polariton mechanism of exciton luminescence, and radiative energy transfer.

When we use the Coulomb gauge of the vector potential of the electromagnetic field the full Hamiltonian of the crystal including the energy of the electromagnetic field can be written as

$$H = \hat{H}_{\mathrm{coul}} + \hat{H}_{\perp} + \hat{H}_{\mathrm{int}}, \tag{4.119}$$

where $\hat{H}_{\mathrm{coul}}$ is the Hamiltonian of the crystal ignoring the retarded interaction, $\hat{H}_{\perp}$ is the Hamiltonian of the field of transverse photons, and $\hat{H}_{\mathrm{int}}$ is the operator of the interaction between charges and transverse photons. In the exciton spectral region the operator $\hat{H}_{\mathrm{coul}}$ can be expressed in terms of the Bose operators of creation and annihilation of the Coulomb excitons, $B_{\mu}^{+}(\boldsymbol{k})$ and $B_{\mu}(\boldsymbol{k})$, where $\boldsymbol{k}$ is the exciton wave vector, and the subscript μ denotes the exciton band (for the sake of simplicity we ignore here the vibrations of the lattice). If we neglect the exciton–exciton interaction the operator $\hat{H}_{\mathrm{coul}}$ can be assumed to be quadratic with respect to the operators B and B^{+}.

The operator $\hat{H}_{\perp}$ is also quadratic with respect to the operators of creation and annihilation of the transverse photons, $a_{\boldsymbol{k}j}^{+}$ and $a_{\boldsymbol{k}j}$, where $\boldsymbol{k}$ is the wave vector of photon, and $j = 1, 2$ denotes one of the two polarizations of photon. If we retain only those terms in the operator $\hat{H}_{\mathrm{int}}$ which are quadratic with respect to the operators B^{+}, B, a^{+} and a we can make use of the linear (u, v) transformation from the operators B^{+}, B, a^{+} and a to the new Bose operators $\xi_{\rho}^{+}(\boldsymbol{k})$ and $\xi_{\rho}(\boldsymbol{k})$ and obtain:

$$\begin{aligned}
\xi_{\rho}(\boldsymbol{k}) &= \sum_{\mu} [u_{\boldsymbol{k}\mu}^{*}(\rho) B_{\mu}(\boldsymbol{k}) - v_{-\boldsymbol{k}\mu}(\rho) B_{-\boldsymbol{k}\mu}^{+}] + \sum_{j} [u_{\boldsymbol{k}j}^{*}(\rho) a_{\boldsymbol{k}j} - v_{-\boldsymbol{k}j}(\rho) a_{-\boldsymbol{k}j}^{+}], \\
B_{\mu}(\boldsymbol{k}) &= \sum_{\rho=1,2} [\xi_{\rho}(\boldsymbol{k}) u_{\boldsymbol{k}\mu}(\rho) + \xi_{\rho}^{+}(-\boldsymbol{k}) v_{-\boldsymbol{k}\mu}^{*}(\rho)], \\
a_{\boldsymbol{k}j} &= \sum_{\rho=1,2} [\xi_{\rho}(\boldsymbol{k}) u_{\boldsymbol{k}j}(\rho) + \xi_{\rho}^{+}(-\boldsymbol{k}) v_{-\boldsymbol{k}j}^{*}(\rho)],
\end{aligned} \tag{4.120a}$$

where u and v are the transformation coefficients which are defined by Hopfield (1958), Agranovich (1959a, 1961), and the operator $\hat{H}$ given by eqs. (4.119) can be diagonalized as

$$\hat{H} = \sum_{\rho\boldsymbol{k}} \mathscr{E}_{\rho}(\boldsymbol{k}) \xi_{\rho}^{+}(\boldsymbol{k}) \xi_{\rho}(\boldsymbol{k}), \tag{4.120b}$$

where $\mathscr{E}_{\rho}(\boldsymbol{k}) \equiv \hbar\omega_{\rho}(\boldsymbol{k})$ are the energies of the new elementary excitations, that is, polaritons. Figure 4.3 shows the main features of the polariton spectrum in gyrotropic and nongyrotropic cubic crystals.

The above assumptions used for determining the polariton spectrum is fully equivalent to those that form the foundation of linear crystal optics based on the phenomenological Maxwell equations. We know that these equations yield the dispersion relation $\omega_\rho = \omega_\rho(\boldsymbol{k})$ only for normal electromagnetic waves with small wave vectors $\boldsymbol{k}$ (that is, for $ka \ll 1$ where a is the lattice parameter). However, the microscopic theory of polaritons applied to various exciton models enables us not only to obtain the results of the phenomenological theory but also to analyze effects which are outside its scope. The microscopic theory can treat the processes of interaction between polaritons and the short-wavelength or the long-wavelength excitations, take into account the non-Bose character of excitons and determine the anharmonicity terms in a system of polaritons which lead to scattering of polaritons by polaritons (see e.g., Agranovich 1968). The more general description of the polariton spectrum given by the microscopic polariton theory enables us to use it also for developing a theory of exciton luminescence (see below).

In the framework of the phenomenological Maxwell theory the properties of crystals are described by the dielectric constant tensor $\epsilon_{ij}(\omega, \boldsymbol{k})$ which depends on the light frequency ω and its wave vector. We know that the dielectric constant tensor determines the relationship between the electric flux density $\boldsymbol{D}$ and the electric field strength $\boldsymbol{E}$ in the electromagnetic plane wave:

$$D_i(\omega, \boldsymbol{k}) = \epsilon_{ij}(\omega, \boldsymbol{k})E_j(\omega, \boldsymbol{k}). \tag{4.121}$$

At the same time, the Maxwell equations yield the following relationship for the plane waves of the form $\boldsymbol{E}(\boldsymbol{r}, t) = E(\omega, \boldsymbol{k}) \exp i(\boldsymbol{kr} - \omega t)$ in the absence of external charges:

$$D = \frac{c^2}{\omega^2}[k^2\boldsymbol{E} - \boldsymbol{k}(\boldsymbol{kE})]. \tag{4.122}$$

When we substitute eq. (4.121) into eq. (4.122) we obtain three algebraic equations in the components $E_i(\omega, \boldsymbol{k})$ where $i = 1, 2, 3$. The condition of existence of a nontrivial solution of this system of equations yields

$$\det\left|\frac{\omega^2}{c^2}\epsilon_{ij}(\omega, \boldsymbol{k}) - k^2\delta_{ij} + k_ik_j\right| = 0, \tag{4.123}$$

that is, the equation which determines the dispersion relation $\omega_l = \omega_l(\boldsymbol{k})$, $l = 1, 2, 3, \ldots,$ for the long-wavelength polaritons of the normal electromagnetic waves in the medium. For instance, if we assume that the medium is isotropic and take

$$\epsilon_{ij}(\omega, \boldsymbol{k}) = \epsilon(\omega)\delta_{ij},$$

then eq. (4.123) yields

$$\epsilon(\omega) = k^2c^2/\omega^2 \tag{4.124a}$$

for the transverse waves and

$$\epsilon(\omega) = 0 \tag{4.124b}$$

for the longitudinal waves.

As mentioned above, it is precisely in the quantum description of the electromagnetic field that the normal waves correspond to elementary excitations – polaritons with the energy $\hbar\omega_l(\boldsymbol{k})$. For instance, if for an isotropic nongyrotropic medium in the resonance region $\omega \approx \Omega_\perp$ we take

$$\epsilon(\omega) = \epsilon_\infty + \frac{(\epsilon_0 - \epsilon_\infty)\Omega_\perp^2}{\Omega_\perp^2 - \omega^2}, \tag{4.125}$$

where ϵ_0 and ϵ_∞ are the low-frequency and the high-frequency values of the dielectric constant, then eq. (4.124a) for the transverse polaritons yields just the two branches of the spectrum ($\rho = 1, 2$) shown in fig. 4.3b. The same figure presents the branch of the spectrum corresponding to the longitudinal waves (longitudinal excitons). The velocity of light does not enter into eq. (4.124b). This fact implies that the retarded interaction does not affect the elementary excitations corresponding to the waves of the longitudinal electromagnetic field (that is, the longitudinal excitons and longitudinal optical phonons). A comparison of figs. 4.3a and 4.3b leads to the same conclusion. These figures show also that the situation is different for the quanta of the transverse field, and the effect of the retarded interaction is particularly strong in the region of small wave vectors, $q \lesssim q_0 = \Omega_\perp\sqrt{\epsilon_0}/c$. Equations (4.120a) shows that the transverse polaritons are a "mixture" (or superposition) of transverse excitons (optical phonons) and transverse photons and the ratio of their contributions to the polariton energy depends on the value of this energy (Agranovich 1959a, 1961, 1968, Hopfield 1958).

For large $q \gg q_0$ the upper and lower polariton branches are transformed, respectively, into the branch of transverse photons $\omega = cq/\sqrt{\epsilon_\infty}$ and the branch of transverse Coulomb excitons $\omega = \Omega_\perp$ (or transverse phonons if we deal with the infrared spectral region). For small $q \lesssim q_0$ the properties of transverse polaritons differ significantly from the properties of both transverse photons and transverse excitons.

The polariton structure can be quantitatively described by the function $S(\omega) = |u_{k\mu}(\rho)|^2$ $(\omega = \omega_\rho)$ which is equal to the exciton component of polariton as can be seen from eq. (4.120a). (Following Burstein (1969), Burstein and Mills (1969, 1970), the function $S(\omega)$ is usually referred to as the strength function. This function is naturally defined in the theory of scattering of polaritons by phonons, lattice defects, etc.) For instance, we obtain for the transverse polariton with $\omega = \omega_\rho(\boldsymbol{k})$, $\rho = 1, 2$, in the cubic crystal

$$S(\omega) = |u_k(\rho)|^2 = \frac{\beta^2 \omega_\rho}{\Omega_\perp[(\omega_\rho^2 - \Omega_\perp^2)^2 + \beta^2]},$$

$$\beta^2 = \frac{(\epsilon_0 - \epsilon_\infty)\Omega_\perp^4}{\epsilon_\infty}, \tag{4.126}$$

so that for the lower branch ($\rho = 1$) we have $S(\omega) \to 1$ for $\omega - \Omega_\perp$. If $\omega^2 \ll \Omega_\perp^2$ then we have

$$S(\omega) \approx \frac{\epsilon_0 - \epsilon_\infty}{\epsilon_0} \frac{\omega}{\Omega_\perp} \ll 1.$$

For the upper branch ($\rho = 2$) we have $S(\omega) \to \sqrt{\epsilon_\infty/\epsilon_0}$ for $\omega \to \Omega_\parallel$. If $\omega \gg \Omega_\parallel$ then we have

$$S(\omega) \approx \frac{\epsilon_0 - \epsilon_\infty}{\epsilon_\infty} \left(\frac{\Omega_\perp}{\omega}\right)^3 \ll 1.$$

The relationship (4.126) remains valid if we include the dependence of the exciton frequency $\Omega_\perp$ on the wave vector, that is, take into account the function $\Omega(\boldsymbol{k})$. The exciton component of polariton in anisotropic crystals can be calculated using the approach of Agranovich (1959a, 1961) (see also, Burstein 1969, Burstein and Mills 1969, 1970).

Since β^2 is proportional to the oscillator strength of the transition at the frequency $\Omega_\perp$ (see eq. (4.125)), eq. (4.126) indicates that the size of the polariton frequency region $\omega \ll \Omega_\perp$, for which the exciton component is still large, significantly depends on the oscillator strength of the transition and is given by

$$\Delta\omega \equiv |\omega - \Omega_\perp| \lessapprox \frac{\Omega_\perp}{2} \left(\frac{\epsilon_0 - \epsilon_\infty}{\epsilon_\infty}\right)^{1/2}. \tag{4.127}$$

In some cases (both for excitons and optical phonons) the size of this region can be as large as a few hundreds of cm^{-1}.

The fact that the exciton component can be fairly significant even for $\omega < \Omega_\perp$ results, as shown by Agranovich and Rukhadse (1959), in the long-wavelength edge of the exciton absorption bands whose dependence on the oscillator strength of the transition is nonlinear (see also, Agranovich and Konobeev 1961, Demidenko 1963, Tait and Weiher 1968, Agranovich 1968, Agranovich and Ginzburg 1983).

A monochromatic light wave incident on a plane crystal surface generates polaritons of the same frequency in the crystal; the normal component k_n of the quasimomentum of the polaritons is given by the condition

$$\hbar\omega = \mathscr{E}_\rho(\boldsymbol{k}_t, k_n),$$

where ω is the frequency of the incident light, $\boldsymbol{k}_t = \boldsymbol{Q}_t$, and $\boldsymbol{Q}_t$ is the tangential component of the wave vector of the incident light. Since energy in crystal is transported by polaritons the attenuation of light passing through a crystal is due to the interaction of polaritons, for example, with the lattice vibrations resulting in the Raman scattering of polaritons by phonons.

Let us explain the nature of this absorption effect analyzing, as above, the optical properties of a cubic crystal in the vicinity of an isolated resonance. When we take into account the attenuation of light we have to replace eq. (4.125) by the following more general relationship (for instance, for $\omega < \Omega_\perp$):

$$\epsilon(\omega) = \epsilon_\infty + \frac{(\epsilon_0 - \epsilon_\infty)\Omega_\perp^2}{\Omega_\perp^2 - \omega^2 - 2i\omega\gamma_1(\omega)},$$

Here $\gamma_1(\omega)$ is the frequency of collisions of a photon in the medium (in this case, a polariton of the lower branch, $\rho = 1$) with phonons resulting in variation of its frequency and wave vector. If we assume that the exciton–phonon interaction described by eq. (4.44a) is weak then we can treat it as a perturbation and obtain in the first order of the perturbation theory,

$$\begin{aligned}\gamma_1(\omega) = \frac{\pi}{\hbar}\sum_{q,r} |F(\boldsymbol{k}-\boldsymbol{q};\boldsymbol{k};\boldsymbol{q}r)|^2[S_1(\omega)S_1(\omega-\omega_{qr})\\ \times (N_{qr}+1)\delta(\hbar\omega - \mathscr{E}_1(\boldsymbol{k}-\boldsymbol{q}) - \hbar\omega_{qr})\\ + S_1(\omega)S_1(\omega+\omega_{-qr})N_{qr}\delta(\hbar\omega - \mathscr{E}_1(\boldsymbol{k}-\boldsymbol{q}) + \hbar\omega_{-qr})],\end{aligned} \quad (4.128)$$

where N_{qr} are the mean Planck occupation numbers for phonons (Agranovich (1968), Chapter IV). The contributions to eq. (4.128) are made by various processes of emission of phonon with the energy $\hbar\omega_{qr}$ which result in transformation of the given polariton with the energy $\mathscr{E}_1(\boldsymbol{k}) = \hbar\omega$ into a polariton with the energy $\mathscr{E}_1(\boldsymbol{k}-\boldsymbol{q})$ (inclusion of band-to-band transitions, see Tait and Weiher (1968)). Clearly, in the given spectral region we deal, actually, with the effects of the resonance Raman scattering ($\omega \lessapprox \Omega_\perp$) of the polariton on the rate of its decay in the crystal.

Note that if we ignore the polariton effects (or if they are small) the quantity β^2 in eq. (4.126) should be tended to zero. Then $|u_k(\rho)|^2$ also vanishes if $\omega < \Omega_\perp(0)$. This means that if we ignore the polariton effects we have $\gamma(\omega) \to 0$ in the region where $\omega < \Omega_\perp(0)$. If $\omega \geqslant \Omega_\perp(0)$ we have $\gamma(\omega) \neq 0$. In this case of low oscillator strengths (small β^2) the frequency of collisions is

$$\gamma_1(\omega) = \frac{\pi}{\hbar}\sum_{q,r} |F(\boldsymbol{k}-\boldsymbol{q};\boldsymbol{k};\boldsymbol{q}r)|^2\,\delta(\hbar\omega - E_\perp(\boldsymbol{k}-\boldsymbol{q}) - \hbar\omega_{qr}), \quad (4.128a)$$

so that $\omega = \Omega_\perp(0)$ is the long-wavelength edge of the exciton absorption band (for the sake of simplicity, we assume here that the crystal temperature T is zero).

The absorption index is related to $\gamma_1(\omega)$ by

$$\kappa(\omega) = c\gamma_1(\omega)/\omega v_{gr}(\omega), \tag{4.129}$$

where $v_{gr}(\omega)$ is the group velocity of light (that is, the group velocity of polariton) at the frequency ω (see, for instance, Agranovich (1968) Chapter IV). Thus, eq. (4.128), in contrast to eq. (4.128a), enables us to calculate the shape of the long-wavelength edge of the exciton absorption bands ($\omega \leqslant \Omega_\perp(0)$) with inclusion of retarded interaction (Agranovich and Rukhadze 1959, Agranovich and Konobeev 1961, Demidenko 1963); in a similar way, we can take into account the many-phonon processes resulting at $\omega \ll \Omega_\perp$ in absorption satisfying Urbach's rule (Agranovich 1968).

Let us now discuss the possible role of the above effects in the processes of formation of the exciton luminescence spectra of crystals. Usually, the term "exciton luminescence" is applied to the luminescence of a crystal caused by decay of excitons which previously were at thermodynamic equilibrium with the crystal lattice. Excitons can be at equilibrium with the lattice since the exciton lifetime which varies greatly in different crystals (from about 1 s for triplet excitons to about 10^{-9} s for singlet excitons) is always greater by many orders of magnitude than the characteristic times of exciton–phonon interaction ($t_0 \approx 1/\omega_D \approx 10^{-12}$ s).

If the exciton–phonon coupling is strong and excitons are incoherent their absorption and luminescence spectra are similar to the spectra of impurity molecules (see sect. 9). If excitons are coherent, renormalization of the spectrum due to inclusion of the retarded interaction can prove to be significant in certain cases. However, we shall, at first, neglect this renormalization.

If we ignore the effects of "mixing" of excitons and photons we have the Boltzmann distribution of excitons with the concentration N_0 in the band (typically, $N_0 \ll a^{-3}$ where a is the lattice parameter):

$$W(E)\,dE = \exp\left(\frac{\mu - E}{kT}\right)\rho(E)\,dE, \tag{4.130}$$

Here $\rho(E)$ is the density of states in the band, and μ is the chemical potential of excitons found from the normalization condition

$$N_0 = \int W(E)\,dE.$$

If $A(E, \omega)$ is the probability of annihilation of the exciton with the energy E accompanied with emission of the luminescence quantum $\hbar\omega$ then the exciton luminescence spectrum $L(\omega)$ can be written as

$$L(\omega) = \int A(E, \omega)\, W(E)\, \mathrm{d}E. \tag{4.131}$$

The function $A(E, \omega)$ is small at low temperatures if $\omega > E$ (that is, in the anti-Stokes region) and can reach far into the Stokes region ($\hbar\omega < E$) where the processes of exciton annihilation are accompanied with creation of one or several quanta of lattice vibrations. As for the zero-phonon processes of radiative decay of the exciton $\hbar\omega = E$, these processes can occur in the perfect infinite crystal (if we neglect the effects of "mixing" of excitons and photons) only for the long-wavelength excitons which correspond to the points of intersection with the branch of the transverse photon spectrum (see fig. 4.3a) and for which the exciton–photon transformation process conserves not only the energy $\hbar\omega = E$ but also the momentum.

If the bottom of the exciton band is in the region of wave vectors $k \approx k_0 \gg \Omega_\perp/c$ then only the part of the band with $k \approx k_0$ is populated at low temperatures so that the zero-phonon exciton luminescence is made difficult. If the bottom of the band is in the region of low $\boldsymbol{k}$ values then, according to our model, zero-phonon lines with the width of the order of $k_B T$ should be found in the spectrum of exciton luminescence (Segall and Mahan 1968). The experimental evidence, however, often does not support this conclusion (the line width is much greater than $k_B T$ (Heim and Weisner 1973) and the diagreement is the more noticable the greater the oscillator strength of the exciton transition. To understand such an effect we have to recall that the retarded interaction results in a significant renormalization of the exciton spectrum in the region of small $\boldsymbol{k}$ values and in formation of polaritons. At large $\boldsymbol{k}$ values ($k \gg \Omega_\perp/c$) polaritons have mainly the "exciton" energy (that is, the energy of electron motion). Therefore, these polaritons strongly interact with vibrations and defects of the lattice since it is just electrons, rather than the transverse electromagnetic field, that interact directly with the lattice, and it is for them that the processes of radiationless dissipation of the electronic ("excitonic") excitation energy to heat are most probable. On the contrary, for polaritons with small $\boldsymbol{k}$ values ($k < \Omega_\perp/c$) the exciton part of the energy is small and the dominating part is that of the energy of the transverse photon field. These polaritons weakly interacting with the lattice vibrations are capable of reaching the crystal surface and yield the luminescence light.

Therefore, the intraband transitions, that is, the transitions in which a polariton with a large quasimomentum $\boldsymbol{k}$ is converted into a polariton with a small $\boldsymbol{k}$, can result in transformation of the electronic excitation energy into observable light. Clearly, the reverse transition corresponds to the process of light reabsorption which is discussed in ch. 6.

The above qualitative model of the so-called polariton mechanism of luminescence was treated by Toyozawa (1959), Agranovich (1959b, 1960). The theoretical aspects of the polariton mechanism of luminescence were discussed by Hopfield (1966), Tait et al. (1967), Tait and Weiher (1969), Sumi (1975), Brodin et al. (1975), Bisti (1976). For instance, the analysis of exciton luminescence in case of high exciton concentration made by Bisti (1976) took into account the exciton–exciton collisions.

The direct calculations of Toyozawa (1959) have shown that, despite the existence of one (for instance, the lower) polariton branch, the time of "sliding" of polariton from the range of large $\boldsymbol{k}$ values range of small $\boldsymbol{k}$ values ($k \lesssim \Omega_\perp/c$) is always of the order of the "old" radiation lifetime, that is, it is greater than the mean free time and, hence, the time of thermalization for short-wavelength ($k \gg \Omega_\perp/c$) polaritons.

The region of the spectrum of lower-branch polariton where the dispersion curves of the Coulomb exciton and the transverse photon meet, is known as the "bottle neck" (see fig. 4.4). This region is narrow, indeed, since the mean (thermal) wave vector of excitons $k_T = (1/\hbar)\sqrt{3Mk_BT}$ (where M is the translational exciton mass, and T is the temperature) is usually much greater than the bottle neck size $k_0 \approx \Omega_\perp\sqrt{\epsilon_\infty}/c \approx 10^5\ \text{cm}^{-1}$ (for instance, at $T = 4$ K and M of the order of the free electron mass m_0 we have $k_T \approx 10^6\ \text{cm}^{-1}$; in molecular crystals $M \gg m_0$, for instance in anthracene $M \approx (5–7)m_0$, and k_T is respectively greater). Thus, the bottle neck corresponds to a relatively low density of states. However, another factor determining the long time needed for passing the bottle neck is the decrease of the "exciton" component of the polariton at $k < k_0$ accom-

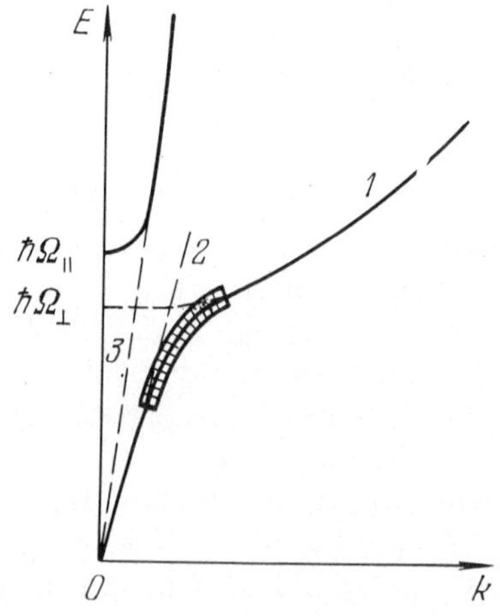

Fig. 4.4. Bottle neck region. (1) for $E = \hbar\Omega_\perp(0) + \hbar^2k^2/2M$. (2) for $E = \hbar ck/\sqrt{\epsilon_0}$. (3) for $E = \hbar ck/\sqrt{\epsilon_\infty}$. The dashed line denotes the polariton branches for low and high frequencies. M is the effective exciton mass.

panied with the declining capacity to be scattered by phonons or lattice defects.

Thererefore, polaritons in the region below the bottle neck rapidly leave the crystal so that their concentration becomes lower than that in the bottle neck region (rather than higher as would be the case under thermodynamic equilibrium). However, at low temperatures the polariton concentration in the region of wave vectors above the bottle neck also proves to be low. In this spectral region polaritons are strongly scattered by phonons, rapidly loose a part of their energy and get into the bottle neck region. Thus, at low temperatures a maximum of the polariton distribution is formed at the bottle neck region. In sufficiently thick crystals the width of this maximum can be of the order of $\Delta\omega$ (see eq. (4.127)). In terms of the above (exciton) treatment this maximum gives rise in the luminescence spectrum, that is, the spectrum of light passing out of the crystal, to a fairly broadened ($\delta \sim \Delta\omega \gg k_B T$) zero-phonon line of exciton luminescence which is displaced to lower energies from the exciton band bottom by ΔE which is also of the order of $\hbar\Delta\omega$.

The above qualitative ideas on the possible role played by the bottle neck in luminescence have been directly supported by the results of the experimental study reported by Heim and Weisner (1973). In this study, apart from observations of anomalous broadening and the shift of the zero-phonon line of exciton luminescence of CdS crystals (for A excitons, $\hbar\omega \approx 2.552$ eV), the authors measured also the decay time τ of luminescence as a function of the luminescence frequency in the same spectral region at 1.6 K. The maximum of the decay time ($\tau \approx 2.3 \times 10^{-9}$ s) is found in the bottle neck region and the width of the maximum is about the width of the emission line.

The role played by the defects of the crystal lattice was analyzed within the framework of the polariton mechanism of luminescence by Agranovich (1959b, 1960) (as well as the role of the impurities in the lattice; see, Konobeev (1962)). The lattice defects are known, generally speaking, to give rise to local electronic excitation levels. At low defect concentrations the contribution of these defects to the absorption spectrum is relatively weak. However, their contribution to luminescence can be significant owing to a possible transfer of the energy of excitons (or polaritons) to electrons of the defect and its subsequent radiation. This mechanism of sensitized luminescence is discussed in the other chapters of this book. Moreover, as shown by Agranovich (1959, 1960) defects can give rise to exciton luminescence without prior localization of the energy at the defect. This process can be readily treated in terms of polaritons since the only perturbation in this case is the perturbing effect of the defect. For any coupling strength the interaction between excitons and transverse photons is included even in the zero approximation and the above mechanism corresponds to the transition within the band from the region of large $\boldsymbol{k}$ values to the region of small $\boldsymbol{k}$ values at a constant energy. This transition is made possible precisely by the presence of defects which violate the translational symmetry of the crystal (additional bulk sources of reabsorption appear in this case; see, Agranovich et al. 1980).

If the exciton band minimum is at $\boldsymbol{k} = \boldsymbol{k}_0$ and $k_0 \gg \Omega_\perp/c$ (the point A_0 in fig. 4.3d), then at low temperatures when only a small part of the band with $\boldsymbol{k} \approx \boldsymbol{k}_0$ is populated the defects can give rise to a high-intensity narrow (the width $\delta \approx k_B T$) line $\hbar\omega = E_0 = E(k_0) < \hbar\Omega_\perp(0)$ in the luminescence spectrum. Naturally, this line should be present also in the long-wavelength absorption spectrum. However, its width should be greater and its intensity should be much lower than the absorption intensity at the frequency $\Omega_\perp(0)$. If the effective exciton mass is negative then in nongyrotropic crystals (see fig. 4.3d) the bottom of an exciton energy band is always not in the centre of the Brillouin zone. In gyrotropic crystals (see fig. 4.3f) the exciton band bottom is always in this position for the polaritons with the right-hand or left-hand polarization and this can lead to circular polarization of the long-wavelength lines of exciton luminescence (Agranovich 1959b, 1960).

We can conduct the general analysis of the exciton (polariton) luminescence spectrum by applying the principle of detailed balancing to the scattering of polaritons by the lattice defects and phonons. First, we assume that the crystal thickness is large compared with the mean free path of polaritons in the region we are interested in, that is, at the long-wavelength edge of exciton absorption. In this case we can ignore the boundary effects in the first approximation and assume that the medium is infinite.

If $h_{12}(\omega_1\boldsymbol{k}_1, \omega_2\boldsymbol{k}_2)\mathrm{d}\omega_2\,\mathrm{d}\theta_2$ is the probability that the quantum $\hbar\omega_1$ (per unit path) is scattered giving rise to a quantum in the frequency range $\mathrm{d}\omega_2$ propagating in the solid angle $\mathrm{d}\theta_2$ then, according to Landau and Lifshits (1966), we have

$$\exp\left(-\frac{\hbar\omega_1}{k_B T}\right)\epsilon(\omega_1)\omega_1^2 v_{\mathrm{gr}}^{-1}(\omega_1\boldsymbol{k}_1)h_{12}(\omega_1\boldsymbol{k}_1, \omega_2\boldsymbol{k}_2)$$

$$= \exp\left(-\frac{\hbar\omega_2}{k_B T}\right)\epsilon(\omega_2)\omega_2^2 v_{\mathrm{gr}}^{-1}(\omega_2\boldsymbol{k}_2)h_{21}(\omega_2\boldsymbol{k}_2, \omega_1\boldsymbol{k}_1), \tag{4.132}$$

where $v_{\mathrm{gr}}(\omega, \boldsymbol{k}) = \mathrm{d}\omega/\mathrm{d}k$. The extinction coefficient for the polariton $(\omega_1\boldsymbol{k}_1)$ is determined by the relationship

$$\mu(\omega_1\boldsymbol{k}_1) \equiv \frac{2\omega_1}{c}\kappa(\omega_1, \boldsymbol{k}_1) = \int h_{12}(\omega_1\boldsymbol{k}_1, \omega_2\boldsymbol{k}_2)\,\mathrm{d}\omega_2\,\mathrm{d}\theta_2.$$

At the same time, the number of polaritons $(\omega, \boldsymbol{k})$ emitted per 1 s, in 1 cm^3

per unit frequency $d\omega_1$ per unit solid angle $d\theta_1$ can be written as

$$L(\omega_1\mathbf{k}_1) = v_{gr}(\omega_1\mathbf{k}_1)\int h_{21}(\omega_2\mathbf{k}_2;\,\omega_1\mathbf{k}_1)\,N(\omega_2\mathbf{k}_2)\,\frac{k_2^2\,dk_2\,d\theta_2}{(2\pi)^3},$$

where $N(\omega, \mathbf{k})$ is the number of polaritons in one state. Using eq. (4.132) we now obtain

$$L(\omega_1\mathbf{k}_1) = \frac{\epsilon(\omega_1)\omega_1^2}{8\pi^3c^2}\int \exp\left[\frac{\hbar(\omega_2-\omega_1)}{k_BT}\right] h_{12}(\omega_1\mathbf{k}_1;\,\omega_2\mathbf{k}_2)N(\omega_2\mathbf{k}_2)\,d\omega_2\,d\theta_2. \tag{4.133}$$

If we ignore the polariton effects for the moment (that is, ignore "creeping" of excitons to the bottle neck region) and assume that the excitons in the band are at thermodynamic equilibrium with phonons, we have $N(\omega, \mathbf{k}) \neq 0$ only for $\hbar\omega \geqslant E(0)$ where $E(0)$ is the exciton band bottom. In this case $N(\omega, \mathbf{k}) = \exp[(\mu - \hbar\omega)/k_BT]$ where μ is the chemical potential of excitons given by the normalization condition

$$N_0 = \int N(\omega, \mathbf{k})\,\frac{k^2\,dk\,d\theta}{(2\pi)^3},$$

(N_0 is the exciton concentration). Using the above equations (Agranovich 1968) we can write now the universal relationship between the exciton luminescence spectrum and the index of absorption of light by excitons $\kappa(\omega, \mathbf{k})$:

$$L(\omega, \mathbf{k}) = \frac{\epsilon(\omega)\omega^3}{4\pi^3c^3}\,\kappa(\omega, \mathbf{k})\exp\left(\frac{\mu-\hbar\omega}{k_BT}\right). \tag{4.134}$$

If, however, the polariton effects are significant, as is the case for large $\Delta\omega$ (see eq. (4.127)), eq. (4.134) is only approximately satisfied for the frequencies within the bottle neck where the "exciton" component is still sufficiently large ($S(\omega) \lesssim 1$) providing that the form of the distribution $N(\omega, \mathbf{k})$ used for deriving eq. (4.134) is still valid. Below the bottle neck the relationship (4.134) is not applicable (Agranovich 1968). If $T(\omega)$ is the transmission coefficient for polariton in case of normal incidence on the crystal surface we find that the luminescence spectrum of a thick crystal in the direction normal to the crystal surface is the product $\Phi(\omega) = L(\omega)T(\omega)$ where $T(\omega) = 4n(\omega)/[n(\omega)+1]^2$.

In order to find the luminescence spectrum shape (4.134) we must know the function $\kappa(\omega)$ in the bottle neck region and above it. As the general treatment of this function is very cumbersome we shall analyze only a few special cases.

(a) The frequency range $\Delta\omega$ determined by eq. (4.127) is large compared with the frequencies of those long-wavelength lattice vibrations ω_{qr} which make the main contribution to the function $\gamma(\omega)$. In this approximation we have from eq. (4.128):

$$\gamma(\omega) \approx S^2(\omega)\Gamma_a$$

where $\Gamma_a = \text{const.}$

(b) After summation over $\boldsymbol{q}$, r eq. (4.128) yields approximately

$$\gamma(\omega) = S(\omega)\Gamma_b$$

where $\Gamma_b = \text{const.}$

(c) The main contribution to $\gamma(\omega)$ is made by one branch of optical vibrations of the crystal with the frequency $\omega_0(\boldsymbol{q})$. If we ignore dispersion of the phonon frequency we obtain

$$\gamma(\omega) \approx S(\omega)S(\omega - \omega_0)\Gamma_c$$

where $\Gamma_c = \text{const.}$

The positions of the peaks ω_{max} in the spectrum of long-wavelength luminescence are especially easy to find in the region of the exciton bands corresponding to low oscillator strengths. Then even for $k_B T > \Delta\omega$ (see eq. (4.127)) we have $S(\omega) \approx \beta^2/(\omega^2 - \Omega_\perp^2)^2$ in the given spectral region. Using this relation and ignoring dispersion of the index of refraction (this can be done here) we find that in the case (a) $\omega_{max} = \Omega_\perp(0) - 4k_B T$, in the case (b) $\omega_{max} \approx \Omega_\perp(0) - 2k_B T$, and in the case (c) the luminescence spectrum can have two peaks. For instance, for $\omega_0 \gg k_B T > \Delta\omega$ we have $\omega_{max}^{(1)} = \Omega(0) - 2k_B T$ and $\omega_{max}^{(2)} = \Omega_\perp(0) - \omega_0 - 2k_B T$. This situation is close to that discussed by Tait et al. (1967), Tait and Weiher (1969) in connection with the luminescence spectrum of the CdS crystal (in this crystal the frequency ω_0 corresponds to longitudinal optical vibrations of the lattice). The cases (a) and (b) can be realized in molecular crystals where a significant part is played by the scattering of excitons by acoustic vibrations.

Let us now discuss luminescence of small-size crystals, for instance, of a flat shape with the thickness $2L$ $(-L < z < L)$.

In this case excitons in the bottle neck and below it cannot be described by the distribution at thermodynamic equilibrium. In a greater part of this spectral region polaritons too rapidly escape from the thin crystal and there is not enough time for the equilibrium distribution to be established.

Take the z-axis perpendicular to the crystal. To find the distribution function $f(\boldsymbol{k}, z)$ for polaritons ($\boldsymbol{k}$ is the wave vector of polariton determining its energy $E = \hbar\omega(\boldsymbol{k})$) we can make use of the kinetic equation

$$-v_3(\boldsymbol{k})\frac{\partial f(\boldsymbol{k}, z)}{\partial z} + P_+(\boldsymbol{k}) - \frac{1}{T_2(\boldsymbol{k})}f(\boldsymbol{k}, z) - \frac{1}{T_1(\boldsymbol{k})}f(\boldsymbol{k}, z) = 0, \quad (4.135)$$

where

$$v_3(\boldsymbol{k}) = \left|\frac{d\omega(\boldsymbol{k})}{d\boldsymbol{k}}\right| \cos\theta$$

is the polariton velocity component along the z-axis (θ is the angle between $\boldsymbol{k}$ and the z-axis), $1/T_1(\boldsymbol{k})$ is the probability of polariton capture by impurities, $1/T_2(\boldsymbol{k})$ is the probability of polariton transition from the state $\boldsymbol{k}$ to other states owing to scattering by phonons or lattice defects, and $P_+(\boldsymbol{k})$ is the rate of creation of polaritons with the momentum $\boldsymbol{k}$ at the point z.

Following Bisti (1976), we shall assume that an external source maintains the Boltzmann exciton distribution in the region above the bottle neck. If the source power is independent of z (for instance, for two-photon pumping) then the exciton concentration above the bottle neck can be assumed independent of z; eq. (4.135) will be used only for polaritons in the region of the bottle neck and below it.

Note, however, that the use of eq. (4.135) is justified if the polariton wavelength λ in the given spectral region is small compared with the crystal thickness $2L$. Only then the conditions of applicability of the geometrical optics are satisfied and we can speak in terms of polariton coordinates.

The boundary conditions can take into account the fact that polaritons leave the crystal. Denote by $f^+(k, \theta, z)$ the distribution function for the polaritons travelling towards the upper boundary ($\theta < \pi/2$) and by $f^-(k, \theta, z)$ the distribution function for the polaritons travelling towards the lower boundary ($\theta > \pi/2$). Since the polariton either leaves the crystal or is regularly reflected at the boundary we have

$$f^+(k, \theta, L)[1 - T(k, \theta)] = f^-(k, \pi - \theta, L),$$

$$f^-(k, \theta, -L)[1 - T(k, \pi - \theta)] = f^+(k, \pi - \theta, -L),$$

where $T(k, \theta)$ is the polariton transmission coefficient at the crystal surface so that $1 - T = R$ is the reflection coefficient. (From the symmetry of the problem we have $f^+(k, \theta, z) = f^-(k, \pi - \theta, -z)$ so that it is sufficient to determine one of the functions f^+ or f^-.) The system of equations for $f(k, \theta, z)$ was solved by Bisti (1976) for the situation when the main cause of polariton scattering was the exciton–exciton collisions. Other situations can be treated in a similar way.

Sometimes it is enough to know the concentration of polaritons with the energy E averaged over the crystal thickness, that is, the function $f(E)$, to find the spectrum of long-wavelength luminescence. This function can be found (Sumi 1975) from the balance equation

$$I(E) + \int_0^\infty W(E', E) f(E')\, \mathrm{d}E' = f(E)[Q(E) + P(E) + R(E)], \tag{4.136}$$

where $W(E', E)$ is the probability of the $E' \to E$ scattering,

$$Q(E) = \int_0^\infty W(E, E')g(E')\, dE'$$

is the total probability of transition from the state E, $g(E')$ is the density of states, $R(E)$ is the rate of capture of polaritons by impurities, $P(E)$ is the probability of transformation of polaritons into vacuum photons at the boundary, and $I(E)$ is the pumping intensity (for monochromatic excitation we have $I(E) \sim \delta(E - E_0)$).

Using approximate values of the quantities entering into eq. (4.136) and solving this equation we can find the spectrum of polariton luminescence

$$\Phi(E) = \frac{4n(E)c}{(n(E)+1)^2} f(E),$$

where $n(E)$ is the index of refraction at the frequency $\omega = E/\hbar$, and c is the polariton velocity.

We can apply eq. (4.136) to thin films, too. Since such system can be regarded, to a large extent, as two-dimensional polaritons will again be described by the radiative decay time (Agranovich 1968). Interesting results can be obtained by taking into account the contribution of the surface polaritons into the exciton luminescence spectrum (Agranovich and Leshova 1976).

The frequencies of the surface polaritons lie in the region of longitudinal-transverse splitting of exciton. Their decay accompanied with luminescence can be due to the processes of both inelastic and elastic scattering by phonons, surface irregularities, etc. (Agranovich 1975, Bryskin et al. 1974) and, generally speaking, it should give rise to marked peaks of luminescence intensity in this spectral region. The first attempt at an experimental study of such effects is discussed by Brodin and Matsko (1979).

Experimental results on the polariton luminescence are reported by Heim and Weisner (1973), Brodin et al. (1975), Gross et al. (1971, 1972), Benoit a la Guillaume and Bonnot (1970), Sell et al. (1973), Saito and Shionoya (1974) (all for semiconductor crystals). In molecular crystals (Ferguson 1977) the polariton luminescence effects should be more pronounced owing to the relatively high oscillator strengths of the exciton transitions. Their theoretical treatment is reported by Brodin et al. (1975) where an attempt to interpret the experimental data of Brodin and Marisova is made. New experimental material and analysis of the part played by various phonons in relaxation of polaritons in anthracene are presented by Galanin et al. (1980).

12. A stochastic model of the exciton–phonon interaction in molecular crystals

The semi-phenomenological model suggested by Haken and Strobl (1967) is now frequently used for analyzing the dynamics of the small-radius triplet excitons.

This model assumes that the Hamiltonian of the crystal describing excitons plus phonons can be written as

$$\hat{H}_{\text{tot}} = \hat{H} + h(t), \tag{4.137}$$

where

$$\hat{H} = \sum_{n,m} \beta_{nm} B_n^+ B_m \tag{4.138}$$

is the time-independent Hamiltonian of excitons in the perfect crystal, and

$$h(t) = \sum_{n,m} h_{nm}(t) B_n^+ B_m \tag{4.139}$$

is the time-dependent part of the Hamiltonian. The quantities $h_{nm}(t)$ are assumed to be random for each moment t, the average values $\langle h_{nm}(t)\rangle$ of the matrix elements $h_{nm}(t)$ are zero but

$$\langle h_{nm}(t+\tau) h_{n'm'}(t)\rangle = [\delta_{nn'}\delta_{mm'} + \delta_{nm'}\delta_{mn'}(1-\delta_{nm})]\, 2\gamma_{m-n}\delta(\tau), \tag{4.140}$$

where γ_{m-n} is a function of the distance $|\boldsymbol{n}-\boldsymbol{m}|$ which determines the intensity of the so-called local (γ_0) exciton scattering for $n = m$ and the intensity of the nonlocal exciton scattering for $n \neq m$. Moreover, it is assumed that the averages of large numbers of the matrix elements $h_{nm}(t)$ can be expressed in terms of the averages (4.140), that is, the so-called cumulant averages (Kubo 1962) of the products of more than two matrix elements h_{nm} are zero (in other words, the quantities $h_{nm}(t)$ satisfy the normal distribution and correspond to a Markovian process; see, Haken and Strobl (1967)).

As shown by Haken and Strobl (1967), under the above assumptions the equation for the density matrix of the exciton subsystem $\hat{\rho}$ averaged over all possible fluctuations of $h(t)$ is

$$h\frac{\partial}{\partial t}\rho_{nm} = -\mathrm{i}[\hat{H}\hat{\rho}]_{nm} - 2\left(\sum_l \gamma_l\right)\rho_{nm} + 2\delta_{nm}\sum_l \gamma_{n-l}\rho_{ll} + 2(1-\delta_{nm})\gamma_{n-m}\rho_{mn}, \tag{4.141}$$

where

$$[\hat{H}\hat{\rho}]_{nm} = \sum_{m'} (\beta_{nm'}\rho_{m'm} - \rho_{nm'}\beta_{m'm}).$$

This is the master equation of this model. (Before averaging over fluctuations of the operator $h(t)$ the density matrix can be written as an infinite power series in $h(t)$. Equation (4.141) is obtained by averaging each term of the series and subsequent summation.) This model has been used for treating a wide variety of problems, for instance, diffusion of excitons, electron paramagnetic resonance (spin motion) and nuclear magnetic resonance (proton relaxation) for triplet excitons, shapes of exciton absorption and luminescence bands, etc.

The main advantage of the model determining its popularity is that it yields comparatively simple expression for a number of physical parameters without requiring any prior assumptions on the character of motion of excitons (coherent, incoherent, etc.). Various aspects of applicability of the model are discussed by Silbey (1976).

First we shall discuss some features of the model and its applicability.

The model is based on a semi-phenomenological approach since its parameters γ_{m-n} are assumed to be given, rather than calculated from the first principles, so to say. Moreover, no substantiation is given for the above-mentioned properties of the random quantities $h_{n,m}(t)$ postulated in the model. Therefore, the region of applicability of the model is, strictly speaking, unknown. Some studies, for instance (Ern et al. 1972), note that the model is applicable to excitons whose band width $\Delta \ll k_{\mathrm{B}}T$. This condition follows from eq. (4.141) and from the fact that the spectrum of fluctuations given by eq. (4.140) contains all frequencies with the same intensity.

It can be also easily seen that under the steady-state conditions, that is, when the density matrix is independent of time, the solution of eq. (4.141) has the form $\rho_{nm} = N^{-1}\delta_{nm}$ where N is the number of lattice sites (we are talking about the state of the crystal with one exciton). If excitons can be described by a band structure, that is, when excitons are coherent, the uniform distribution over the sites can be realized only if all the band states have the same population, that is, we come back to the above condition that the energy $k_{\mathrm{B}}T$ be larger than the exciton band width. Of course, for the incoherent excitons the inequality $k_{\mathrm{B}}T > \Delta$ is also satisfied. Therefore, we can consistently use eq. (4.141) if the exciton band width can be assumed to be sufficiently small. This condition is easily satisfied for triplet excitons and this is precisely the reason why the stochastic model is most often used for analyzing their properties (Haken and Reineker 1972, Haken and Strobl 1973, Reinaker 1974, Haken and Schwarzer 1974, Ovchinnikov and Erikhman 1974).

Note, however, that, apart from consistency, another requirement to the model is that it should yield the same results for the limiting cases of strong and weak exciton–phonon coupling as those obtained from analysis of the exciton–phonon interaction. If this requirement were satisfied we could justify to a smaller or greater extent the use of the stochastic model for describing the properties of excitons (for small Δ) in a wide temperature range corresponding to the transition from coherent excitons to incoherent excitons. Unfortunately, this is not the case; for instance, this requirement for the shapes of absorption and emission bands is satisfied only in the limit of weak exciton–phonon coupling. Indeed, in the stochastic model the normalized absorption and emission spectra are determined (Ern et al. 1972) by the spectral densities,

$$G^+(\omega, \boldsymbol{k}) = \frac{1}{2\pi}\int_{-\infty}^{\infty} \mathrm{d}t\, \mathrm{e}^{\mathrm{i}\omega\tau}\langle B_k(t)B_k^+(0)\rangle, \tag{4.142}$$

$$G^-(\omega, \boldsymbol{k}) = \frac{1}{2\pi}\int_{-\infty}^{\infty} \mathrm{d}t\, \mathrm{e}^{\mathrm{i}\omega t}\langle B_k^+(t)B_k(0)\rangle, \tag{4.142a}$$

where

$$B_k = N^{-1/2}\sum_n \mathrm{e}^{\mathrm{i}kn} B_n.$$

The angle brackets denote averaging over various matrix elements $h_{nm}(t)$ in the crystal without excitons in eq. (4.142) and in the crystal with one exciton in eq. (4.142a). Proceeding from the properties of the random quantities h_{nm} (Ern et al. 1972) we can obtain

$$\langle B_k(t)B_k^+(0)\rangle = \exp\left[-\frac{\mathrm{i}\Omega(\boldsymbol{k})t}{\hbar} - \Gamma|t|\right], \tag{4.143}$$

where $\Omega(\boldsymbol{k})$ is the frequency of the exciton with the wave vector $\boldsymbol{k}$ in the perfect crystal (that is, for $h_{nm}(t) = 0$), and $\Gamma = \Sigma_n \gamma_{|m-n|}$. Hence, we have

$$G^+(\omega, \boldsymbol{k}) = \frac{\Gamma}{\pi[(\omega - \Omega(\boldsymbol{k}))^2 + \Gamma^2]}, \tag{4.144}$$

and $G^-(\omega, \boldsymbol{k}) = G^+(\omega, \boldsymbol{k})$.

Thus, the model yields the Lorentzian shape for the absorption and emission bands irrespective of the intensity of fluctuations of the Hamiltonian $\hat{H}_{\mathrm{tot}}$. Since these bands have the Gaussian shape in the case of strong exciton–phonon coupling the stochastic model yields wrong results in this limiting case. Therefore, the model seems to be applicable only to the region of existence of coherent or weakly incoherent excitons, that is, the region of not very high temperatures. Adding this condition to the above condition $k_{\mathrm{B}}T > \Delta$ we come to the upper ($T < T_0$) and lower ($T > \Delta/k_{\mathrm{B}}$) temperature limits of applicability of the stochastic model (here $T > T_0$ is the region of existence of incoherent excitons; see sect. 2). Thus, the stochastic model is applicable in a sufficiently wide region only if

$T_0 > \Delta$. For singlet excitons, for instance in anthracene and naphthalene, we have $T_0 \leqslant \Delta$ (see ch. 5). Therefore, the stochastic model is not applicable to singlet excitons. (This model does not take into account the effect of excitons on the phonon subsystem. Therefore, it leaves out the possibility of exciton autolocalization (see sect. 2) and many other effects of strong exciton–phonon coupling.) However, for triplet excitons in anthracene and naphthalene the band width is of the order of $10\ \text{cm}^{-1}$. The coherent effects and, in particular, Davydov splitting, are observed in experiments at the temperatures up to $T \approx 370$ K. For instance, as reported by Ern et al. (1972), in anthracene Davydov splitting is $\Delta = 18\text{–}19\ \text{cm}^{-1}$ in the temperature range $118\ \text{K} < T < 371\ \text{K}$ while the width Γ (see eq. (4.144)) for the b and c Davydov components increases from 14 to $65\ \text{cm}^{-1}$ in this temperature range so that it is of the order of or even larger than the exciton band width. Since the quantity Γ in eq. (4.144) has the sense of the uncertainty of the energy of excitons due to their scattering by phonons the concept of the exciton band becomes meaningless and the exciton motion acquires a hopping-like character. Thus, according to the above discussion, the region of applicability of the stochastic model for anthracene corresponds, apparently, to the temperature range $20\ \text{K} < T < 150$ K.

In the next chapter we shall discuss the results of the stochastic model in connection with a exciton kinetics; in particular, we shall show how to use spectroscopic data for excitons to evaluate the diffusion coefficient of triplet excitons within the framework of the stochastic model.

CHAPTER 5

Theory of Kinetic Parameters Determining the Rate of Energy Transfer by Excitons

1. Introduction

The mobility of excitons is a major factor in electronic excitation energy transfer in crystals.

The capability of excitons to migrate in the crystal lattice can be described by the diffusion length $L = \sqrt{D\tau_0}$ where D is the coefficient of diffusion of excitons, and τ_0 is their lifetime. Experimental data show that for the large-radius exciton $D \approx 100\ \text{cm}^2/\text{s}$ and, for instance, for $\tau_0 \approx 10^{-6}$ s we obtain $L = 100\ \mu\text{m}$. For small-radius singlet excitons $D \approx 10^{-3}\ \text{cm}^2/\text{s}$, $\tau_0 \approx 10^{-8}$ s and $L \approx 100$ Å (at the room temperature in anthracene $L \approx$ 500 Å). But for triplet excitons the diffusion length is greater by more than two orders of magnitude owing to their long lifetimes (for instance, as reported by Ern et al. (1966) in anthracene $D \approx 2 \times 10^{-4}\ \text{cm}^2/\text{s}$, and $\tau_0 \approx 2 \times 10^{-3}$ s). We are discussing here diffusion of excitons corresponding to the lowest exciton energy bands. The contribution of high-energy excitons to energy transfer is, typically negligible owing to their short lifetimes. At the same time, these excitons can play a significant part in the exciton–exciton collisions (see sect. 8) and in the processes caused by fluxes of fast charged particles in crystals. In these cases considerable attention should be paid to analyzing various mechanisms of decay of high-energy excitons and plasmons; see Fano (1958), Klein and Voltz (1975).

Anselm and Firsov (1955, 1956) were the first to calculate the diffusion coefficient of large-radius excitons. The diffusion coefficient for small-radius excitons was calculated by Trlifai (1956, 1963) in the approximation of strong exciton–phonon coupling (that is, for incoherent or "localized" excitons) and by Agranovich and Konobeev (1959) in the approximation of weak exciton–phonon coupling (that is, for coherent or "free" excitons). A fundamental assumption of Ansel'm and Firsov (1955, 1956), Agranovich and Konobeev (1959) was that migration of excitons in the lattice can be described within the framework of the kinetic Boltzmann equation which determines the distribution of excitons over states found without taking into account the exciton–phonon interaction.

The well-known condition of applicability of the Boltzmann equation (Gurevich 1940, Greenwood 1958) is that the time of collision with a scattering centre should be short compared with the mean free time, that is,

$$\tau > \hbar/k_B T, \tag{5.1}$$

where T is the temperature, and k_B is the Boltzmann constant. For individual particles we can rewrite eq. (5.1) in the form $l > \hbar v/k_B T = \lambda$ where $l = v\tau$ is the mean free path, v is the mean velocity, $\lambda = h/p$ is the mean de Broglie wavelength, and $p = k_B T/v$ is of the order of the mean momentum (the wave packet size is of the order of λ). We shall return to this condition of applicability of the Boltzmann equation below and discuss in more detail the temperature range in which this equation can be applied to molecular crystals. Now we shall discuss Trlifai (1956, 1963) the range in which the diffusion coefficient was found for the case of strong exciton–phonon coupling.

In this limiting case the exciton wave vector is no longer a "good" quantum number and the exciton spectrum cannot be described in terms of a band structure. The exciton–phonon interaction causes localization of exciton at a lattice site and exciton travels by random hops from one lattice site to another. This character of exciton motion was, in fact, postulated by Trlifai (1956, 1963) on the basis of purely intuitive analysis. Later, this picture of exciton motion was consistently substantiated owing to the efforts of many theorists working on the formally similar problem of the mobility of small-radius polarons (these studies are reviewed by Lang and Firsov (1962, 1963, 1967), Firsov and Kudinov (1964, 1965), Firsov (1968, 1975), Klinger (1979)). Significantly, the theory discussed by Lang and Firsov (1962, 1963, 1967), Firsov and Kudinov (1964, 1965), Firsov (1968, 1975) can be used with a varying degree of accuracy for describing the exciton mobility in the case of arbitrary strength of the exciton–phonon interaction. The need for such theoretical predictions exists even now and it will grow with development of the experimental work in diffusion of both singlet and triplet excitons in molecular crystals. Referring the reader to Lang and Firsov (1962, 1963, 1967), Firsov and Kudinov (1964, 1965), Firsov (1968, 1975), Klinger (1979) discussing the results on the polaron mobility we should note here an important feature of excitons which polarons do not possess. We are talking here about the excitons which correspond to a nonzero dipole moment of the transition to the ground state of the crystal. In this case the exciton energy is a nonanalytical function of the wave vector $\boldsymbol{k}$ at small $\boldsymbol{k}$ values and this fact can significantly affect the value and anisotropy of the diffusion coefficient at low temperatures and its variation with temperature (Agranovich 1968). In this chapter we shall discuss the reported calculations of the kinetic parameters describing migration of excitons in molecular crystals. We shall start with the most important and general relationship – Kubo's formula for the diffusion coefficient (Kubo 1957). This formula has been effectively used in the theory of small-radius polarons (Lang and Firsov 1962, 1963, 1967,

Firsov and Kudinov 1964, 1965, Firsov 1968, 1975) and these results have been applied to the theory of excitons (Konobeev 1971, Grover and Silbey 1971).

2. General expression for the diffusion coefficient of small-radius excitons

According to Kubo (1957), the general quantum-mechanical expression for the tensor of the diffusion coefficient is

$$D_{ik} = \frac{1}{\beta} \lim_{s \to 0} \int_0^\infty e^{-st}\, dt \int_0^\beta d\lambda \, \langle \hat{V}_i(-i\hbar\lambda)\, \hat{V}_k(t) \rangle, \tag{5.2}$$

where $\beta = 1/k_B T$, $\hat{V}(t)$ is the velocity operator for a migrating particle in the Heisenberg representation, and the angle brackets denote the Gibbs averaging (it can be easily seen that for classical particles ($\hbar \to 0$) in the relaxation time approximation (that is, for $v_{0i}(t) = v_{0i} \exp(-t/\tau)$) eq. (5.2) leads to the well-known expression $D_{ij} = \frac{1}{3}\langle v^2 \tau(v) \rangle \delta_{ij}$ for an isotropic medium). If we assume that the crystal contains only one exciton then the radius-vector operator of its "centre of gravity" is

$$\hat{\boldsymbol{R}} = \sum_{n\alpha} \boldsymbol{R}_{n\alpha} B^+_{n\alpha} B_{n\alpha}, \tag{5.3}$$

where $\boldsymbol{R}_{n\alpha} = \boldsymbol{n} + \boldsymbol{r}_\alpha$, $\boldsymbol{n}$ is an integral lattice vector, and $\boldsymbol{r}_\alpha$ is the vector determining the position of the molecule in the cell ($\alpha = 1, 2, \ldots, \sigma$). If $\hat{H}$ is the Hamiltonian of the crystal then the velocity operator is

$$\hat{\boldsymbol{V}} = \frac{i}{\hbar} [\hat{H} \hat{\boldsymbol{R}}]. \tag{5.4}$$

Assume that the Hamiltonian has the form

$$\hat{H} = \sum_{n\alpha} \hat{H}_0(n\alpha) B^+_{n\alpha} B_{n\alpha} + \sum_{n\alpha,\, m\beta} \hat{M}_{n\alpha,\, m\beta} B^+_{n\alpha} B_{m\beta} + \sum_\kappa \hbar\omega_\kappa b^+_\kappa b_\kappa, \tag{5.5}$$

where the operators $\hat{H}_0(n\alpha)$ and $\hat{M}_{n\alpha,\, m\beta}$ are functions of the operators b^+_κ and b_κ, that is the operators of creation and annihilation of the phonon of the species $\kappa \equiv (\boldsymbol{q}r)$ with the energy $\hbar\omega_\kappa$. Substituting eq. (5.5) into eq. (5.4) we find that the operator of exciton velocity is

$$\hat{\boldsymbol{V}} = \frac{i}{\hbar} \sum_{n\alpha,\, m\beta} \boldsymbol{R}_{n\alpha,\, m\beta} \hat{M}_{n\alpha,\, m\beta} B^+_{n\alpha} B_{m\beta}, \tag{5.6}$$

where

$$\boldsymbol{R}_{n\alpha,\, m\beta} = \boldsymbol{R}_{n\alpha} - \boldsymbol{R}_{m\beta}.$$

We shall ignore the dependence of $\hat{M}_{n\alpha, m\beta}$ on the operators b_κ and b_κ^+ in eq. (5.6). If the exciton–phonon interaction is weak the inclusion of this dependence yields only small corrections to the expression for D_{ik}. Under these conditions it is sufficient to include the dependence on the operators b_κ and b_κ^+ only in the operator $\hat{H}_0(n\alpha)$. Thus, we shall replace the operator $\hat{M}_{n\alpha,m\beta}$ with the scalar $M_{n\alpha,m\beta}$ in eq. (5.6). Then, integration of eq. (5.2) over t and λ yields

$$D_{ik} = \frac{1}{2\hbar^2} \sum_{nm} \sum_{n'm'} R^i_{n,m} M_{n,m} R^k_{n',m'} M_{n',m'} \int_{-\infty}^{\infty} dt \langle B_n^+(t) B_m(t) B_{n'}^+(0) B_{m'}(0) \rangle, \quad (5.7)$$

where $n \equiv (n\alpha)$, $m \equiv (m\beta)$ and so on. We can rewrite eq. (5.7) in the form

$$D_{ik} = -\tfrac{1}{2} \sum_{k,k'} v^i_{\alpha\beta}(k) v^k_{\gamma\delta}(k') \int_{-\infty}^{\infty} dt \langle B_{k\alpha}^+(t) B_{k\beta}(t) B_{k'\gamma}^+(0) B_{k'\delta}(0) \rangle, \quad (5.7a)$$

where the Bose operators:

$$B_{k\alpha} = \frac{1}{\sqrt{N}} \sum_n e^{-ikn} B_{n\alpha}, \quad (5.7b)$$

and the vectors:

$$v_{\alpha\beta}(k) = \frac{1}{i\hbar} \sum_m e^{ik(n-m)} R_{n\alpha, m\beta} M_{n\alpha, m\beta}. \quad (5.7c)$$

In the crystals with one molecule per unit cell the vector

$$v(k) = \frac{1}{i\hbar} \sum_m e^{ik(n-m)} (n - m) M_{n,m} = \frac{1}{\hbar} \nabla_k E(k),$$

where

$$E(k) = \Delta + \sum_m e^{ik(m-n)} M_{n,m}$$

is the exciton energy. Thus, $v(k)$ is the group velocity of excitons with the wave vector k. In the crystals with several molecules per unit cell the quantities $v_{\alpha,\beta}(k)$ play only a secondary part.

As can be seen from eq. (5.7) the calculation of the tensor D_{ik} reduces to a calculation of the two-particle correlation functions. If the exciton–phonon coupling is sufficiently weak the solution of the equation for the correlation function $\langle B_{k\alpha}^+(t) B_{k\beta}^+(t) B_{k'\gamma}(0) B_{k'\delta}(0) \rangle$ is equivalent to the solution of the Boltzmann equation (the conditions of applicability of the Boltzmann equation were discussed above; derivation of the respective equation for the correlation function is given by Zubarev (1960)). If the exciton–phonon coupling is strong the condition (5.1) is not satisfied and the equation for the correlation function entering into eq. (5.7) is no longer equivalent to the Boltzmann equation.

We have noted in sect. 1 that a strong exciton–phonon coupling results in localization of the exciton at a lattice site so that exciton behaves as a classical particle and can hop from cell to cell. Using the results of the theory of small-radius polarons (Lang and Firsov 1962, 1963, 1967, Firsov and Kudinov 1964, 1965, Firsov 1968, 1975), we can show that in this limiting case the exciton motion is a series of uncorrelated hops (so that it is a Markovian process) and is described by the equation of random walk over the lattice sites.

If $P(m, t)$ is the probability that an exciton is at the lattice site m at the moment t then the random-walk equation for the function $P(m, t)$ has the form

$$\frac{\mathrm{d}P(m,t)}{\mathrm{d}t} = \sum_{m'} [W(\boldsymbol{m}', \boldsymbol{m})P(\boldsymbol{m}') - W(\boldsymbol{m}, \boldsymbol{m}')P(\boldsymbol{m})], \tag{5.8}$$

where $W(\boldsymbol{m}', \boldsymbol{m})$ is the probability of hopping from the site m' to the site m per unit time. However, even in the next approximation (with respect to the matrix elements of the resonance interaction) the hopping process is not Markovian (Lang and Firsov 1962, 1963, 1967, Firsov and Kudinov 1964, 1965, Firsov 1968, 1975). Then we have the following integro-differential equation for $P(m, t)$

$$\frac{\mathrm{d}P(m,t)}{\mathrm{d}t} = \sum_{m'} \int_0^t \tilde{W}(m', m, \tau)\,[P(m', t-\tau) - P(m, t-\tau)]\,\mathrm{d}\tau,$$

that is, the hopping process acquires a "memory". This problem concerning molecular crystals is discussed by Kenkre and Knox (1974) and Kenkre (1975). If we assume that variation of the function $P(m, t)$ at distances of the order of the lattice constant is small then eq. (5.8) reduces to the equation of diffusion. Indeed, if the vector $\boldsymbol{m}$ varies continuously we obtain

$$P(\boldsymbol{m}') = P(\boldsymbol{m}) + \sum_i (\boldsymbol{m}' - \boldsymbol{m})_i \frac{\mathrm{d}P(\boldsymbol{m})}{\mathrm{d}m_i} + \frac{1}{2}\sum_{ij} (\boldsymbol{m}' - \boldsymbol{m})_i (\boldsymbol{m}' - \boldsymbol{m})_j \frac{\mathrm{d}^2P(\boldsymbol{m})}{\mathrm{d}m_i\,\mathrm{d}m_j}.$$

Since $W(\boldsymbol{m}, \boldsymbol{m}') = W(\boldsymbol{m}', \boldsymbol{m})$, substitution of the above expression into eq. (5.8) yields

$$\frac{\partial P(\boldsymbol{r}, t)}{\partial t} = D_{ij} \frac{\partial^2 P(\boldsymbol{r}, t)}{\partial x_i \partial x_j},$$

where the diffusion coefficient tensor is

$$D_{ij} = \frac{1}{2}\sum_{m'} (\boldsymbol{m}' - \boldsymbol{m})_i (\boldsymbol{m}' - \boldsymbol{m})_j W(\boldsymbol{m}', \boldsymbol{m}). \tag{5.9}$$

In crystals with several molecules per unit cell, such as the anthracene crystal, which have symmetry operations exchanging molecules with different α values, the use of a similar procedure yields

$$D_{ij} = \frac{1}{2}\sum_{m\beta} (r_{n\sigma} - r_{m\beta})_i (r_{n\alpha} - r_{m\beta})_j W(n\alpha, m\beta). \tag{5.9a}$$

Thus, under the given conditions calculation of the diffusion coefficient reduces to determination of the probability $W(m, m')$ of hopping as it was done by Trlifai (1956, 1963).

The results of calculation of the tensor D_{ik} for strong exciton–phonon coupling are discussed in sect. 3. Some problems of exciton kinetics are treated in ch. 6 on the basis of random walk equations.

The lack of sufficiently detailed data on the structure of exciton bands and the constants of exciton–phonon interaction makes calculations of the diffusion coefficients of excitons in molecular crystals difficult for various exciton models. However, it will be shown below that the temperature dependences of the exciton diffusion coefficients significantly differ for coherent and incoherent excitons. Therefore, it is precisely the analysis of the temperature dependence of the exciton diffusion coefficient that is at present the main approach to determining the type of the exciton mechanism of energy transfer in crystals.

3. *Mobility of incoherent ("localized") excitons*

In sect. 2 we have noted that incoherent excitons correspond to relatively small matrix elements of the resonance interaction. Under such conditions we can take into consideration only the first (quadratic) terms of the expansion of eq. (5.7) in powers of $M_{n,m}$ and obtain

$$D_{ik} = -\frac{1}{\hbar^2}\sum_{nm} R^i_{n,m} R^j_{n,m} |M_{n,m}|^2 \int_{-\infty}^{\infty} dt \langle B^+_n(t) B_m(t) B^+_m(0) B_n(0)\rangle, \tag{5.10}$$

where the correlation function should be determined in terms of the Hamiltonian (5.5) at $\hat{M}_{n,m} = 0$. If we assume that the operator $\hat{H}_0(n\alpha)$ contains only the terms which are linear in b^+_κ and b_κ then we can use the unitary transformation described in ch. 4 sect. 2 for diagonalization of the operator (5.5) for $\hat{M}_{n,m} = 0$ with the states of incoherent ("localized") excitons (the operators of their creation and annihilation are $\tilde{B}^+_n$ and $\tilde{B}_n$). In this representation the expressions for the tensor D_{ik} have the same form but the operators B^+_n and B_n are replaced with $\tilde{B}^+_n$ and $\tilde{B}_n$.

Since the diffusion coefficient is independent of the exciton concentration we shall consider a crystal with one exciton. Using the translational symmetry of the crystal and eq. (5.9) we obtain

$$D_{ik} = -\frac{1}{\hbar^2}\sum_m R^i_{n,m}R^j_{n,m}|M_{n,m}|^2 \int_{-\infty}^{\infty} dt\langle \tilde{B}^+_n(t)\tilde{B}_m(t)\tilde{B}^+_m(0)\tilde{B}_n(0)\rangle,$$

or (see ch. 4 sects. 2 and 10)

$$D_{ik} = -\frac{1}{\hbar^2}\sum_m R^i_{n,m}R^j_{n,m}|M_{n,m}|^2 \int_{-\infty}^{\infty} dt \exp\left(\frac{\mathrm{i}\Delta_{mn}t}{\hbar}\right)\langle \mathrm{e}^{\hat{S}_n(t)}\mathrm{e}^{-\hat{S}_m(t)}\mathrm{e}^{\hat{S}_m(0)}\mathrm{e}^{-\hat{S}_n(0)}\rangle, \tag{5.11}$$

where Δ_{nm} is the difference between the energies of the crystal in the states with identical phonon occupation numbers which correspond to localization of the exciton on the lattice sites n and m. In the crystals of the type of the anthracene crystal in which the lattice symmetry operations exchange all the molecules of the unit cell the energy difference $\Delta_{nm} = 0$. In this case we should remember that the exciton states used here with the corresponding local deformation are stationary only because we ignore not only the terms with M_{nm} but also the quadratic terms in the phonon operators b^+ and b in eq. (5.5). If we take into account the finite lifetime of these states we obtain an additional factor $\exp(-2\Gamma|t|/\hbar)$ in the integrand of eq. (5.11) where Γ is the width of the level of the "localized" exciton. Therefore, in the crystals of the anthracene type we have

$$D_{ik} = -\frac{1}{\hbar^2}\sum_m R^i_{n,m}R^j_{n,m}|M_{n,m}|^2 \int_{-\infty}^{\infty} \exp\left(-\frac{2\Gamma|t|}{\hbar}\right)\langle \mathrm{e}^{S_n(t)}\mathrm{e}^{-S_m(t)}\mathrm{e}^{S_m(0)}\mathrm{e}^{-S_n(0)}\rangle\, dt,$$

or (see also ch. 4 sect. 10)

$$D_{ik} = \frac{1}{\hbar^2}\sum_m R^i_{n,m}R^j_{n,m}|M_{n,m}|^2 \int_{-\infty}^{\infty} \exp\left(-\frac{2\Gamma|t|}{\hbar} + g_{nm}(t)\right) dt, \tag{5.12}$$

where

$$g_{nm}(t) = \sum_{qr} |U_{nr}(q) - U_{mr}(q)|^2[(N_{qr}+1)\exp(-\mathrm{i}\omega_{qr}t) + N_{qr}\exp(\mathrm{i}\omega_{qr}t) - (2N_{qr}+1)]. \tag{5.12a}$$

The quantities $|M_{n,m}|^2$ vary at least as rapidly as $1/|\boldsymbol{n}-\boldsymbol{m}|^6$. Therefore, we need take into account only the nearest neighbours in the sum over m in eq. (5.12). We can ignore the differences between the functions $g_{nm}(t)$ for different neighbours since they are inessential for qualitative analysis. Then we obtain

$$D_{ik} = d_{ik}F(T), \tag{5.13}$$

where

$$d_{ik} = \frac{2\pi}{\hbar} \sum_m R^i_{n,m} R^j_{n,m} |M_{n,m}|^2,$$

$$F(T) = \frac{1}{2\pi\hbar} \int_{-\infty}^{\infty} \exp\left(-\frac{2\Gamma|t|}{\hbar} + g(t)\right) dt. \tag{5.13a}$$

The function $F(T)$ is identical to the function $G^+(E)$ (see eq. (4.98)) when $E = \tilde{\Delta}$, $\Gamma = 0$. Thus, we can use the results of ch. 4 sect. 2.9 for finding $F(T)$. The contribution of large t values ($t \gg \Omega_D^{-1}$) to the integral $F(T)$ is

$$F_1(T) = \frac{1}{2\pi\hbar} \int_{-\infty}^{\infty} \exp\left[-\frac{2\Gamma|t|}{\hbar} + g(\infty)\right] dt = \frac{\exp(-2\tilde{M})}{2\pi\Gamma}$$

$$2\tilde{M} = \sum_{qr} |\Delta U_{qr}|^2 (2N_{qr} + 1),$$

where ΔU_{qr} is the difference between the $U_{qr}(n)$ values for the neighbouring lattice sites. The contribution of small t values ($t \ll \Omega_D^{-1}$) to the integral $F(T)$ is given by

$$F_2(T) = \frac{1}{\pi}\sqrt{\frac{n}{\tilde{B}}} \exp\left(-\frac{\tilde{A}^2}{\tilde{B}}\right),$$

where

$$\tilde{A} = \sum_{qr} |\Delta U_{qr}|^2 \hbar\omega_{qr}, \qquad \tilde{B} = 2\sum_{qr} |\Delta U_{qr}|^2 (2N_{qr} + 1)\hbar^2\omega_{qr}^2.$$

Now we have the following approximation:

$$F(T) = F_1(T) + F_2(T) = \frac{1}{\pi}\left[\frac{\exp(-2\tilde{M})}{2\Gamma} + \sqrt{\frac{\pi}{B}} \exp(-\tilde{A}^2/\tilde{B})\right]. \tag{5.14}$$

The inclusion of only the second term in eq. (5.14) corresponds to the result of Trlifai (1956, 1963). This term increases with increasing temperature. In particular, when the Debye temperature is exceeded,

$$\tilde{B} = 4\tilde{A}k_B T,$$

the second term contains. $\exp[-E_a/k_B T)$ where $E_a = \tilde{A}/4$ is the hop activation energy. Since the first term in eq. (5.14) is exponentially small in the same temperature range we see that at sufficiently high temperatures T the diffusion coefficient of incoherent excitons has an exponential dependence on temperature:

$$D_{ik} \approx d_{ik} \exp(-E_a/k_B T), \qquad (E_a > 0) \tag{5.15}$$

The first term in eq. (5.14) corresponds to excitons hops during which the state of the phonon system is not changed. The factor $\exp(-2\tilde{M})$ in this term is analogous to the Debye–Waller factor which determines the intensity of zero-phonon lines in the impurity absorption spectrum (see ch. 4 sect. 9) and which appears in such situations (like the one discussed here) for which we can find a similarity to the Mössbauer effect.

Processes of this kind in the theory of the mobility of small-radius polarons were studied by Firsov and co-workers (Lang and Firsov 1962, 1963, 1967, Firsov and Kudinov 1964, 1965, Firsov 1968, 1975) and their possible contribution of exciton diffusion was noted by Konobeev (1971).

If in the calculation of the function $F(T)$ we ignore the effect of displacement correlation (see ch. 4 sect. 10) then the diffusion coefficient is proportional to the integral of overlapping of the exciton luminescence spectrum and the exciton absorption spectrum. The term F_1 is then replaced with an integral related to the overlapping of the zero-phonon lines and equal to $\exp(-4M)/2\pi\Gamma$. If we recall the definition of the quantity M (see ch. 4 sect. 9) we can obtain (n and m are neighbours)

$$4M - 2\tilde{M} = 2\,\mathrm{Re}\sum_{qr} U_{qr}(n)U^*_{qr}(m)(2N_{qr} + 1).$$

Generally speaking, this difference is of the order of $2\tilde{M}$ once more demonstrating that we cannot reduce the calculation of the exciton diffusion coefficient to the calculation of the overlap of the exciton emission and absorption spectra.

Since $\tilde{M}$ increases with increasing temperature the first term in eq. (5.14) can prove to be significant only at sufficiently low temperatures (we are talking about such crystals in which excitons are assumed to be incoherent even at low temperatures). The zero-phonon line should have a sufficiently high intensity here but there should be no coherent effects (for instance, the Davydov splitting), that is, the width of even the zero-phonon line should be large compared with the matrix elements of the resonance interaction.

In crystals in which excitons are coherent at low temperatures and become incoherent at higher temperatures the first term in eq. (5.14) can hardly be significant since it is relatively small.

Note that there are no reasons to believe that the constants of exciton–phonon interaction for singlet and triplet excitons are essentially different. Since the matrix elements $M_{n,m}$ of the resonance interaction for the triplet excitons are smaller (by magnitude) by more than an order of magnitude than the respective matrix elements for singlet excitons, the incoherent triplet excitons should often occur in molecular crystals. The coherent triplet excitons in a fairly wide temperature range can occur only in crystals with very weak exciton–phonon coupling (for instance, in anthracene in which the Davydov splitting of the triplet exciton lines is observed and the exciton band width $\Delta E_e \approx 10\,\text{cm}^{-1}$, see Ern et al. (1966, 1972)).

Finally, it should be remembered that the diffusion coefficient of incoherent excitons was calculated above, under the assumption that the main contribution was made by the exciton–phonon interaction operators linear with respect to the displacements of the nuclei. The inclusion of the quadratic terms leads to an expression of the type of eq. (5.12) for the tensor D_{ik}. However, in this expression $g_{nm}(t)$ contains, apart from the terms given in eq. (5.12a), additional terms which are due to the contribution of those terms in the operator of the exciton–phonon interaction which were neglected in the derivation of eq. (5.12). We shall not write down a more general expression for $g_{nm}(t)$ here because it is fairly cumbersome. Moreover, no calculations of the tensor D_{ik} in this general form have yet been reported. Instead, we shall discuss another limiting case in which we shall take into account only the quadratic terms with respect to displacements in the interaction of exciton with intramolecular vibrations (Munn and Siebrand 1970).

If B_n^+ and B_n are the operators of creation and annihilation of exciton at the site n then we can write for the exciton–phonon interaction in the model of Munn and Siebrand (1970)

$$\hat{H}' = -\tfrac{1}{2}m\omega_2^2 \sum_r x_n^2 B_n^+ B_n,$$

where x_n and m are the displacement and the mass of the intramolecular oscillator in the molecule n, and ω_2^2 is a constant of the exciton–phonon interaction which determines the variation of the frequency of the intramolecular vibration due to transition of the molecule to an excited (electronic) state. If ω_0 is the frequency of the intramolecular vibration in the ground state then the respective frequency after excitation of the molecule n is $\tilde{\omega}_0 = (\omega_0^2 - \omega_2^2)^{1/2}$. For the model of a one-dimensional molecular crystal discussed by Munn and Siebrand (1970) the potential energy of the displacements of the nuclei in the ground state (when no excitons are present) was taken in the form

$$V(x_1, x_2, \ldots, x_N) = \sum_{r=1}^{N} \tfrac{1}{2}m[\omega_0^2 x_r^2 + \omega_1^2 x_r(x_{r+1} + x_{r-1})].$$

The phonon frequency here is given by

$$\omega_k^2 = \omega_0^2 + \omega_1^2 \cos ka,$$

where $k = 2\pi p/Na$, a is the lattice parameter, $p = \pm 1, \pm 2, \ldots, \pm\tfrac{1}{2}N$, and the phonon band width $\Delta = \hbar\omega_1^2/\omega_0$.

If the crystal contains an exciton localized at one of the lattice sites then, apart from the above continuous vibrational spectrum, we also obtain the exciton-related local vibration with the frequency ω given by the relationship $\omega^2 = \omega_0^2 - (\omega_1^4 + \omega_2^4)^{1/2}$. Therefore, when the exciton hops to a neighbouring lattice site the normal vibrations of the lattice are rearranged.

The coupling between excitons and such intramolecular vibrations for which $\omega_2 \gg \omega_1$ was treated by Munn and Siebrand (1970). In this case the local vibration related to the exciton is, in fact, localized at one molecule. If we ignore the quantity ω_1 (that is, the phonon band width) when we calculate the matrix element for transition of exciton to the neighbouring site, then the wave function of the lattice with one exciton on the site n can be taken in the form of the product of the oscillator wave function

$$\chi_n = \varphi(v_n) \prod_{r \neq n} \varphi_0(v_r),$$

where v_r is the quantum number of the oscillator at the site r, φ is the wave function of the oscillator with the frequency $\omega = (\omega_0^2 - \omega_2^2)^{1/2}$, and φ_0 is the wave function of the oscillator with the frequency ω_0. If M is the matrix element of the resonance interaction corresponding to transition of exciton to a neighbouring site in the rigid lattice then we obtain for the matrix element for the $n \to n+1$ taking into account the exciton–phonon interaction:

$$|H_{n,\,n+1}| = M\langle \chi_{n+1}\{v'_r\} \mid \chi_n\{v_r\}\rangle.$$

Using the above expression for the functions χ_n we can write

$$|H_{n,n+1}| = M\langle \varphi(v_n)|\varphi_0(v'_n)\rangle\langle \varphi_0(v_{n+1})|\varphi(v'_{n+1})\rangle \prod_{r(r\neq n,\, n+1)} \langle \varphi_0(v_r)|\varphi_0(v'_r)\rangle$$

$$= MS(v_n, v'_n)S(v_{n+1}, v'_{n+1}) \prod_{r(r\neq n,\, n+1)} \delta_{v_r v'_r},$$

where $S(v_n, v'_n)$ is the overlap integral for the wave functions of the oscillators with the frequencies ω and ω_0 having the same equilibrium coordinate. The above expression for $|H_{n,n+1}|$ indicates that the $n \to n+1$ transition must satisfy the condition $v_r = v'_{r'}$ for $r \neq n$, $r \neq n+1$. Moreover, conservation of the total energy yields

$$\hbar\omega v_n + \hbar\omega_0 v_{n+1} = \hbar\omega_0 v'_n + \hbar\omega v'_{n+1},$$

and we obtain

$$(v_n - v'_{n+1})\omega = (v'_n - v_{n+1})\omega_0.$$

Since $\omega \neq \omega_0$ this condition is satisfied if both sides of the equation are equal to zero, that is, if $v_n = v'_{n+1}$ and $v'_n = v_{n+1}$ simultaneously. Thus, we have

$$|H_{n,n+1}| = MS^2(v_n, v_{n+1})\delta_{v_n, v'_{n+1}}\delta_{v_{n+1}, v'_n} \prod_{r(r\neq n,\, n+1)} \delta_{v_r v'_r}.$$

For a given initial phonon number distribution $\{v_r\}$ the expression for $|H_{n,n+1}|^2$ must be summed over all possible phonon occupation numbers $\{v'_r\}$. The result, $M^2S^4(v_n, v_{n+1})$, must then be averaged over the thermodynamic distributions for v_n and v_{n+1}. According to Munn and Siebrand (1970), we have

$$\langle S^4\rangle_T = (1-\sigma)/(1+\sigma) = \tanh(\beta\hbar\omega_0/2),$$

where

$$\sigma = \exp(-\beta\hbar\omega_0), \qquad \beta = 1/k_BT.$$

At finite temperatures the lattice contains a finite number of phonons $v_T = \sigma/(1-\sigma)$ and each of them has a continuous spectrum within the phonon band with the width $\Delta = \hbar\omega_1^2/\omega_0$. Therefore, the exciton transition $n \to n+1$ is realized in the system with the continuous spectrum when the transition probability

$$W(n, n+1) = \frac{2\pi}{\hbar}\,\overline{|H_{n,n+1}|^2}\,\rho(E),$$

where

$$\rho(E) = v_T/\Delta = v_T\omega_0/\omega_1^2,$$

is the density of the final states of the system. Hence, we obtain

$$W(n, n+1) = \pi\omega_0\left(\frac{M}{\hbar\omega_1}\right)^2\frac{\exp(-\theta/2T)}{\coth(\theta/2T)},$$

where $\theta = \hbar\omega_0/k_B$ is the characteristic temperature of the oscillator. The energy uncertainty for the initial state related to the probability of the transition $n \to n+1$ is determined by the quantity $\delta = 2\pi\hbar W(n, n+1)$. The above expression for $\rho(E)$ is valid if δ is small compared with the phonon band width Δ. Hence, we obtain the following condition of applicability of the above expression for $W(n, n+1)$:

$$\frac{\pi^2}{2}\xi^2\left(\frac{\sigma}{1+\sigma}\right) \ll 1,$$

Here $\xi = 2M/\Delta$. The diffusion coefficient for this system is $D = 2a^2W(n, n+1) = (\pi a^2M\xi/\hbar)[\sigma/(1+\sigma)]$. We see that in this model, as in the theory of Trlifai (1956, 1963) at high temperatures, the diffusion coefficient increases exponentially with increasing temperature. When $T \to 0$ the above expression for D tends to zero owing to a sharp decrease in the number of real phonons. It should be borne in mind, however, that excitons can become coherent at sufficiently low temperatures. In the three-dimensional crystals their mean free path determined by scattering by real phonons is $l \approx a^3/v_T\sigma_0$ where σ_0 is the cross section of scattering so that the diffusion coefficient $D \sim (1/\hbar)(\mathrm{d}E/\mathrm{d}k)l \sim \exp(\Theta/T)$ increases very rapidly with decreasing temperature (for detailed analysis see, Munn and Siebrand (1970)). Such a mechanism of exciton scattering by lattice defects of thermal origin (here a real optical phonon acts as a defect) is discussed in sect. 5 (Agranovich and Konobeev 1968). Now we shall only note that possible interference of different mechanisms of exciton–phonon interaction should be always borne in mind when theoretical predictions are compared with experimental results. Therefore, it is better to study any exciton diffusion mechanism in such temperature range where it is the main mechanism. In this connection, the mechanism treated by Munn and Siebrand (1970) seems to be suitable for analyzing the mobility of triplet excitons whose resonance interaction energy M is low. If we assume $\omega_0 \approx 300\ \mathrm{cm}^{-1}$, $a = 10$ Å, $M = 0.5\ \mathrm{cm}^{-1}$, and $\xi = 0.2$ for such excitons in anthracene (Munn and Siebrand 1970) we obtain the diffusion coefficient at the room temperature $D \approx 2 \times 10^{-4}\ \mathrm{cm}^2\ \mathrm{s}^{-1}$ which is in an order-of-magnitude agreement with the experimental data reported by Ern et al. (1966). (The diffusion coefficient $D \approx a^2\nu$ where ν is the frequency of excitons hops so that $1/\nu$ is the mean hop time. For the above D value we have $1/\nu \approx 2 \times 10^{-10}$ s.) However, the same experimental results indicate that for triplet excitons in anthracene D decreases with increasing temperature (in the temperature range from 100 to 400 K) so that the mechanism of exciton diffusion in this case remains unclear.

4. *Mean free path and diffusion coefficients of coherent excitons*

If the exciton–phonon interaction is weak it can be taken into account in the framework of perturbation theory. The variation of the shape and position of the exciton band due to this interaction is insignificant in this case and is, therefore, ignored. However, there always remains a part of the exciton–phonon interaction that makes coherent excitons nonstationary irrespective of the extent to which we can take into account the exciton–phonon interaction.

Generally speaking, scattering of excitons by phonons results in a variation of the wave vector and energy of excitons. Therefore, if $\delta E(\boldsymbol{k})$ is the width of the energy level of the exciton with the wave vector $\boldsymbol{k}$ determined by the exciton–phonon interaction then the respective uncertainty $\delta\boldsymbol{k}$ of the wave vector is related to $\delta E(\boldsymbol{k})$ by

$$\delta E(\boldsymbol{k}) = \hbar v(\boldsymbol{k})\delta\boldsymbol{k}, \tag{5.16}$$

where $v(\boldsymbol{k}) = (1/\hbar)(\mathrm{d}E/\mathrm{d}\boldsymbol{k})$ is the exciton group velocity.

The uncertainty of the wave vector of exciton indicates that, owing to the exciton–phonon interaction, the exciton state in the crystal is realized actually as a wave packet whose dimensions δx, δy, and δz can be estimated from the uncertainty relations $\delta x \delta k_x \sim 1$, etc., rather than as a plane wave. The motion of the exciton wave packet means the transport of energy so that in order to find the appropriate diffusion coefficient we must estimate the mean free path and the mean free time of the wave packet. Thus, the situation for the coherent excitons is quite similar to that of the phonon heat conductivity (Peierls 1955).

The diffusion coefficient for localized excitons is determined by the frequency of hops from one lattice site to other lattice sites; in contrast to that, the diffusion coefficient for coherent excitons as the mobility of electrons and holes in crystals (Ansel'm 1962) is determined by the relaxation time τ. According to Fröhlich (1937), we have

$$\frac{1}{\tau} = -\sum_{qr} \frac{\Delta k_z(\boldsymbol{q})}{k_z}[W_{\mathrm{a}}^{\boldsymbol{k}}(\boldsymbol{q}r) + W_{\mathrm{e}}^{\boldsymbol{k}}(\boldsymbol{q}r)], \tag{5.17}$$

where $\boldsymbol{k}$ is the exciton wave vector before the collision, $\Delta\boldsymbol{k}(\boldsymbol{q})$ is the variation of the exciton wave vector due to the collision with the phonon $\boldsymbol{q}r$, $W_{\mathrm{a}}^{\boldsymbol{k}}(\boldsymbol{q}r)$ is the probability of absorption of the phonon $\boldsymbol{q}r$, and $W_{\mathrm{e}}^{\boldsymbol{k}}(\boldsymbol{q}r)$ is the probability of emission of the phonon $\boldsymbol{q}r$ (both probabilities per unit time). The new exciton wave vectors, accurate to an integral reciprocal lattice vector, are $\boldsymbol{k}' = \boldsymbol{k} \pm \boldsymbol{q}$ for the cases of absorption and emission of phonon, respectively. The number of the phonon branch is r.

According to eq. (5.17) we can write for the relaxation time

$$1/\tau = 1/\tau^{\mathrm{ac}} + 1/\tau^{\mathrm{op}} \tag{5.17a}$$

where $1/\tau^{\mathrm{ac}}$ and $1/\tau^{\mathrm{op}}$ are the numbers of collisions between exciton and acoustic and optical phonons per 1 s. As is the case for electrons in semiconductors (Ansel'm 1962). τ is approximately equal to the mean free time of excitons with respect to collisions with phonons. Typically, a few collisions are sufficient for thermodynamic equilibrium between phonons and band excitons. Thus, if the exciton lifetime is considerably longer than the time τ we can assume, as was repeatedly noted above, that excitons are at thermodynamic equilibrium with the lattice prior to their decay due either to conversion into light or, for instance, capture by an impurity molecule.

The exciton diffusion coefficient is related to the relaxation time by

$$D = \tfrac{1}{3}\overline{\tau v^2} \approx \tfrac{1}{3}\bar{\tau}\overline{v^2} \tag{5.18}$$

where $\overline{v^2}$ is the mean squared group velocity of excitons, and $\bar{\tau}$ is the mean relaxation time.

Rigorously speaking, explicit expressions for wave packets should be used for calculations of the quantities in eq. (5.17). However, the resulting functions, that are dependent on the exciton wave vector, may vary only slightly in the interval of the order of the uncertainty of the exciton wave vector in the wave packet. Then the appropriate transition probabilities can be calculated by using the wave functions in the form of plane waves for excitons before and after scattering as can be done in all similar kinetical calculations (for electrons, holes, phonons, etc.). Moreover, in probability calculations we shall ignore the mixing of the states of excitons and transverse photons. The operator of the exciton–phonon interaction will be taken in the form given by eq. (4.44a) or eq. (4.44) if the effect of renormalization of the exciton band is significant. This approximation (the polariton effects are neglected) is valid if the oscillator strength corresponding to the exciton transition is small (for instance, if we deal with the dipole-forbidden singlet or any triplet excitons) or if, as noted in ch. 4 sect. 11, the wave vectors of most excitons in the band are much greater than the wave vector of the light wave in the given spectral range (not very low temperatures).

In this case we have

$$W_{\mathrm{e}}^{k}(qs) = \frac{2\pi}{N\hbar}|F(k-q;k;qs)|^2(\bar{N}_{qs}+1)\delta[E(k)-E(k-q)-\hbar\omega_s(q)]. \tag{5.19a}$$

$$W_{\mathrm{a}}^{k}(qs) = \frac{2\pi}{N\hbar}|F(k+q;k;qs)|^2\bar{N}_{qs}\delta[E(k)-E(k+q)+\hbar\omega_s(q)], \tag{5.19b}$$

At sufficiently low temperatures when the thermal motion energy $k_{\mathrm{B}}T$ is much smaller than the exciton band width most excitons at thermodynamic equilibrium are concentrated in the vicinity of the exciton band minimum in the wave vector space. If we denote the minimum by $\boldsymbol{k}_0$ and if

$$\boldsymbol{k} = \boldsymbol{k}_0 + \boldsymbol{\kappa}$$

then in the approximation of the isotropic effective mass the conservation of energy for absorption and emission of phonon by exciton can be written in the form

$$\frac{\hbar^2\kappa^2}{2m^*} \pm \hbar\omega_s(q) = \frac{\hbar^2(\kappa \pm q)^2}{2m^*}, \tag{5.20}$$

where m^* is the effective exciton mass. For the acoustic phonons we have $\omega_s(q) = v_0 q$ where q_0 is the sound velocity (for the sake of simplicity we ignore the dependence of v_0 on s and the direction of $\boldsymbol{q}$). Using eq. (5.20) we obtain

$$q = \mp 2\kappa\cos\theta \pm 2m^* v_0/\hbar, \tag{5.21}$$

where θ is the angle between the vectors $\boldsymbol{\kappa}$ and $\boldsymbol{q}$,

$$\frac{m^* v_0}{\hbar} \sim 10^5\left(\frac{m^*}{m}\right)\,\mathrm{cm}^{-1},$$

m is the electron mass in vacuum, and $v_0 \approx 10^5$ cm/s. Using the relationship

$$\frac{\hbar^2 \overline{\kappa^2}}{2m^*} = \tfrac{3}{2} k_B T$$

of the Boltzmann statistics we obtain for the mean value of κ at a given temperature T:

$$\bar{\kappa} \equiv (\overline{\kappa^2})^{1/2} = (1/\hbar)\sqrt{3m^* k_B T}. \tag{5.22}$$

Hence, we can ignore the second term in eq. (5.21) if

$$k_B T \gg \tfrac{1}{3} m^* v_0^2. \tag{5.23}$$

Using the above estimates we see that inequality (5.23) is satisfied if

$$T \gg \frac{0.1\,\mathrm{K}}{4}\left(\frac{m^*}{m}\right). \tag{5.24}$$

Below we shall assume that inequality (5.24) is satisfied and therefore we shall omit the respective term in the argument of the delta function in eqs. (5.19) and the second term in eq. (5.21). (The possibility of neglecting the last term in eq. (5.21) implies that exciton scattering is almost elastic. Only in this case we can use eq. (5.17) for calculation of the relaxation time. This is discussed in more detail by Ansel'm (1962), Chapter VII.) Now we can see from eq. (5.21) that in our model for a given value of κ the exciton can interact in the first approximation with the phonons whose wave vector is in the range from $q_{\min} = 0$ to $q_{\max} = 2\kappa$.

Since the exciton band width is assumed to be much larger than $k_B T$ the condition

$$\kappa a \ll 1, \tag{5.25}$$

where a is the lattice parameter, is satisfied for the overwhelming majority of excitons. Since we have $q < 2\kappa$ for the phonons which scatter exciton we can expand $F(\boldsymbol{k}+\boldsymbol{q}, \boldsymbol{k}; \boldsymbol{q}s)$ in powers of q and κ using eq. (5.25) and take only the first terms of this expansion. In this approximation the quantity $|F(\boldsymbol{k}+\boldsymbol{q}, \boldsymbol{k}; \boldsymbol{q}s)|^2$ is a linear function of q for the acoustic phonons and a constant for the optical phonons (see, Agranovich and Konobeev 1959, 1968). Moreover, if we ignore the dependence of this quantity on the directions of the vectors $\boldsymbol{q}$ and $\boldsymbol{\kappa}$ which is insignificant here and ignore the dependence of the phonon frequency on the direction of $\boldsymbol{q}$ and polarization we obtain approximately:

$$|F^{\mathrm{ac}}(\boldsymbol{k}+\boldsymbol{q}, \boldsymbol{k}; \boldsymbol{q}s)|^2 = |F_0^{\mathrm{ac}}|^2\, aq, \tag{5.26a}$$

$$|F^{\mathrm{op}}(\boldsymbol{k}+\boldsymbol{q}, \boldsymbol{k}; \boldsymbol{q}s)|^2 = |F_0^{\mathrm{op}}|^2, \tag{5.26b}$$

where F_0 are constants. Using eqs. (5.26) and the relationship

$$\int_0^{2\pi} \frac{\Delta k_z(\boldsymbol{q})}{k_z}\,\mathrm{d}\varphi = -2\pi \frac{q^2}{2(\boldsymbol{k\kappa})},$$

we find that the relaxation time due to exciton scattering by acoustic phonons is determined by

$$\frac{1}{\tau^{\mathrm{ac}}} = \frac{3|F_0^{\mathrm{ac}}|^2 a^4 m^*}{4\pi\hbar^3(\boldsymbol{k\kappa})\kappa}\left(\frac{k_\mathrm{B}T}{\hbar v_0}\right)^5 \int_0^{\xi} x^4 \frac{\mathrm{e}^{-x}-\mathrm{e}^{x}}{(\mathrm{e}^{x}-1)(\mathrm{e}^{-x}-1)}\,\mathrm{d}x, \tag{5.27}$$

where

$$\xi = 2\kappa\hbar v_0/k_\mathrm{B}T. \tag{5.28}$$

If $\xi < 1$, that is, if,

$$T > 12m^* v_0^2/k_\mathrm{B} \approx (m^*/m)\,\mathrm{K}, \tag{5.29}$$

(see eq. (5.22)) the integrand in eq. (5.27) can be replaced with its value for small x values. Replacing κ with $\bar{\kappa}$ (see eq. (5.22)) and performing integration of eq. (5.27) we now obtain

$$\frac{1}{\tau^{\mathrm{ac}}(\bar{\kappa})} = \frac{a^4 m^* |F_0^{\mathrm{ac}}|^2}{\overline{(\boldsymbol{k\kappa})}\bar{\kappa}\pi\hbar^3}\left(\frac{k_\mathrm{B}T}{hv_0}\right)\bar{\kappa}^4. \tag{5.30}$$

If the exciton band minimum is at the centre of the Brillouin zone, that is, if $\boldsymbol{k} = \boldsymbol{\kappa}$, then we can find from eqs. (5.22), (5.30) and (5.18) that in the temperature range in which inequality (5.29) is satisfied the diffusion coefficient is inversely proportional to the square root of temperature:

$$D \sim 1/\sqrt{T}. \tag{5.31}$$

This relationship was derived by Agranovich and Konobeev (1959), (see also Agranovich and Konobeev 1968); it can be seen from its derivation that it is independent of the model and, therefore, is valid also for large-radius excitons which are in the ground state.

This relationship can also be valid in the case when the exciton band minimum is not at the centre of the Brillouin zone, that is, if $\boldsymbol{k}_0 \neq 0$ but $k_0 a < 1$ and, apart from eq. (5.29) the condition,

$$k_\mathrm{B}T \gg E(0) - E(\boldsymbol{k}_0), \tag{5.32}$$

is satisfied.

In the region of very low temperatures when $\xi \gg 1$, that is, when the condition opposite to inequality (5.29) is satisfied but inequality (5.24) still holds, the main contribution to the integral (5.27) is made by large x values.

Replacing the integrand in eq. (5.27) with its asymptotic value at large x values and performing integration we obtain:

$$\frac{1}{\tau^{\mathrm{ac}}(\bar{\kappa})} = \frac{24a^4 m^*|F_0^{\mathrm{ac}}|^2}{5\pi\hbar^3\overline{(\boldsymbol{k}\kappa)}}\,\bar{\kappa}^4. \tag{5.33}$$

If the exciton band minimum corresponds to $\boldsymbol{k}_0 = 0$ then eqs. (5.22), (5.33) and (5.18) show that the diffusion coefficient due to scattering by acoustic phonons ceases to be temperature-dependent in the region of low temperatures. If $\boldsymbol{k}_0 \neq 0$, then, since $\overline{(\boldsymbol{k}\kappa)} = \overline{\kappa^2}$, the relationship

$$D = \tfrac{1}{3}\tau^{\mathrm{ac}}(\bar{\kappa})\bar{v}^2 \sim \mathrm{const} \tag{5.34}$$

holds for the overwhelming majority of excitons whether or not inequality (5.32) is satisfied. Thus, for $\xi \gg 1$, regardless of where in the Brillouin zone the exciton energy minimum is, the relationship $D \sim 1/\sqrt{T}$ for the exciton diffusion coefficient due to exciton scattering by acoustic phonons reduces to $D = \mathrm{const}$ with decreasing temperature.

Let us now analyze the effect of exciton scattering by phonons corresponding to an rth branch of optical crystal vibrations.

In the region of sufficiently high temperatures when $k_{\mathrm{B}}T > \hbar\omega_r$, that is, when the kinetic energy of most excitons is much higher than the variation of energy due to scattering, we can regard the scattering process as being approximately elastic and omit $\hbar\omega_r(\boldsymbol{q})$ in eq. (5.20) for exciton scattering by acoustic phonons. If we ignore the dispersion of optical phonons and make use of eq. (5.26b) we obtain

$$\frac{1}{\tau^{\mathrm{op}}(\bar{\kappa})} = \frac{2a^3 m^*|F_0^{\mathrm{op}}|^2}{\pi\hbar^3\overline{(\boldsymbol{k}\kappa)}\bar{\kappa}}\left(\frac{k_{\mathrm{B}}T}{\hbar\omega_r(0)}\right)\bar{\kappa}^4.$$

A comparison of this relationship with eq. (5.30) shows that the relaxation time determined by exciton scattering by optical phonons has the same temperature dependence for $k_{\mathrm{B}}T > \hbar\omega_r(0)$ as the relaxation time determined by exciton scattering by acoustic phonons in the temperature range given by the condition (5.29).

In the region of low temperatures where $k_{\mathrm{B}}T < \hbar\omega_s(0)$ the majority of excitons can only absorb optical phonons. Since the number of optical phonons is exponentially small, $\bar{N}_{qs} \sim \exp[-h\omega_s(0)/k_{\mathrm{B}}T]$, in this temperature region, the relaxation time τ^{op} proves to be much greater than the relaxation time τ^{ac} so that, according to eq. (5.17a), we have

$$\tau = \tau^{\mathrm{ac}}.$$

This means that, apart from the intermediate temperature range where $k_{\mathrm{B}}T \approx \hbar\omega_s(0)$, the temperature dependence of the diffusion coefficient of free excitons is determined as by scattering by acoustic phonons. As for the intermediate region it is just as if the constant of the exciton–phonon interaction varies there. In molecular crystals where the frequencies of optical intermolecular vibrations can be of the order of a few tens of cm^{-1} the above-mentioned anomalies due to the "cut-off" of the mechanism of exciton scattering by optical phonons with decreasing temperature can take place at temperatures of the order of a few tens of Kelvins. Then the anomalous increase of the exciton diffusion coefficient should be observed with decreasing temperature. The experimental results on the temperature dependence of the exciton diffusion coefficient are discussed in ch. 7.

Note in conclusion that applicability of the above $D(T)$ relationships is limited by the condition (5.1) of applicability of the Boltzmann equation. Using eq. (5.18) we see that eq. (5.31) implies the following dependence of the relaxation time on temperature: $\tau \sim 1/T\sqrt{T}$. Therefore, we should expect that the condition (5.1) is satisfied only for sufficiently low temperatures. Of course, the size of the respective temperature range depends on the exciton–phonon interaction constants and can be found at present only from experimental data.

The majority of the well-studied molecular crystals, such as anthracene, naphthalene, pyrene, etc., do not belong to the cubic crystal system but are very anisotropic. In this connection, it is natural to consider to what extent the above qualitative results on the temperature dependence of the exciton diffusion coefficient in cubic crystals are valid for anisotropic crystals (see also, Agranovich 1968).

It is clear, of course, that in anisotropic crystals the exciton diffusion coefficient can exhibit anisotropy owing, for instance, to anisotropy of the effective mass of excitons. However, the temperature dependence of the diffusion coefficient in this case has the same character ($D \sim 1/\sqrt{T}$) with the exclusion of the region of low temperatures ($T \leqslant T_0$), if only the effective masses of excitons in various directions do not differ by more than a few times (this seems to be just the case in the naphthalene crystal; see ch. 7).

A more interesting, though less well-studied case is that of such anisotropic crystals in which the lowest exciton band corresponds to high oscillator strengths. Under such conditions the exciton energy $E(\boldsymbol{k})$ is known to be a nonanalytic function of k for small $\boldsymbol{k}$ values (Agranovich and Ginsburg 1983). For uniaxial crystals (more general cases are discussed by Agranovich and Ginsburg (1983)) this function can be found from the equation of dispersion for extraordinary electromagnetic waves (polaritons)*

$$\frac{1}{n^2} = \frac{\cos^2\theta}{\epsilon_\perp(E)} + \frac{\sin^2\theta}{\epsilon_\parallel(E)},$$

*For ordinary (that is, transverse) polaritons $n^2 = \epsilon_\perp(E)$ so that the poles of $n^2(E)$ are independent of the direction of $\boldsymbol{k}$.

where θ is the angle between the wave vector $\boldsymbol{k}$ of the wave and the optical axis, and $\epsilon_{\perp}$ and $\epsilon_{\parallel}$ are the "transverse" and "longitudinal" components of the dielectric constant tensor. Since $n^2 = k^2c^2/\omega^2$ then for finite ω and k the equation for the frequencies of polarition (the Coulomb exciton) has the form

$$\epsilon_{\parallel}(\omega)\cos^2\theta + \epsilon_{\perp}(\omega)\sin^2\theta = 0, \tag{5.35}$$

if we ignore the retarded interaction, that is, take $c \to \infty$. If we deal with the vicinity of an exciton band which corresponds to transition dipole moments polarized perpendicular to the optical axis, then $\epsilon_{\perp} \approx \epsilon_0 + G/(E_0(\boldsymbol{k}) - E)$, $E = \hbar\omega$, $\epsilon_{\parallel} \approx \epsilon_0$, the mechanical exciton energy $E_0(\boldsymbol{k}) = E_0 + \hbar^2 k_i k_j/2M_{ij}$, and $1/M_{ij}$ is the inverse tensor of the effective exciton mass, so that eq. (5.35) yields

$$E(\boldsymbol{k}) = E_0 + \frac{G\sin^2\theta}{\epsilon_0} + \frac{\hbar^2 k_i k_j}{2M_{ij}}.$$

For the vicinity of an exciton band for which the transition dipole moment is polarized along the optical axis we have $\epsilon_{\perp} \approx \epsilon_0$, $\epsilon_{\parallel} \approx \epsilon_0 + G/(E_0(\boldsymbol{k}) - E)$ so that:

$$E(\boldsymbol{k}) = E_0 + \frac{G}{\epsilon_0}\cos^2\theta + \frac{\hbar^2 k_i k_j}{2M_{ij}}.$$

The above equations for $E(\boldsymbol{k})$ show that when $\boldsymbol{k} \to 0$ the value of $E(\boldsymbol{k})$ depends on the direction along which $\boldsymbol{k}$ tends to zero. Therefore, in the case under discussion here the limiting value of the exciton energy, that is, $E(0)$, spreads out into a band with the width G/ϵ_0.

The quantity G is proportional to the oscillator strength $F(G = 2\pi h^2 e^2 F/mvE_0$ where m is the free electron mass, and v is the lattice cell volume). Therefore, the effect under consideration can be particularly significant in the spectral range where the intensity of exciton transitions is sufficiently high. For instance, in anthracene the nonanalytic term in the exciton energy in the lowest band is of the order of 200 cm^{-1} so that in analysis of exciton diffusion its contribution is significant in the whole temperature range in which coherent excitons occur.

Katalnikov (1968, 1974) calculated the exciton diffusion coefficient in uniaxial crystals taking into account the nonanalytic energy term (see also, Katal'nikov and Rudenko 1977).

To do this Katalnikov used the method for solving the Boltzmann equation developed by Samoilovich et al. (1961) which is based on expanding the exciton distribution function in spherical harmonics. We shall ignore here the problems of substantiation of this procedure (for instance, convergence of the series must be analyzed) and note only the main results reported by Katal'nikov (1968, 1974).

Katalnikov has shown that, generally, in uniaxial crystals different components of the diffusion coefficient tensor have different temperature dependences. However, the temperature dependence for uniaxial crystals with two adjoining exciton bands is practically identical to that for cubic crystals. A similar result (Katal'nikov and Rudenko 1977) on the $D(T)$ dependence was obtained for not too low temperatures in the study of diffusion of small-radius excitons in anisotropic isolated band (anisotropy of D_{ik} can lead to one-dimensional or two-dimensional exciton motion (see, Agranovich 1968)).

5. *Scattering of coherent excitons by impurities and defects of crystal structure*

Real crystals can contain various chemical impurities in addition to defects in the crystal structure produced during its growth or due to thermal motion. Sometimes impurities are specially introduced into crystals (for instance, in experiments on energy transfer from host to impurity). Sometimes, it is just too difficult to get rid of certain impurity molecules during the process of crystal growth.

Impurities and crystal structure defects give rise to additional exciton scattering which we must take into account in an analysis of exciton diffusion just as we did with scattering by phonons. If τ_i and τ_d are the relaxation times corresponding to exciton scattering by impurities and defects then the total relaxation time τ is given by

$$\frac{1}{\tau}=\frac{1}{\tau^{ac}}+\frac{1}{\tau^{op}}+\frac{1}{\tau_i}+\frac{1}{\tau_d}. \tag{5.36}$$

If c is the concentration of the impurity molecules then the relaxation time τ_i is given by

$$\tau_i = 1/c\sigma_i\bar{v}, \tag{5.37}$$

where σ_i is the cross section of exciton scattering by impurity, and $\bar{v}$ is the mean exciton velocity. For molecular crystals the cross section of scattering of free exciton by impurity was found by Dubovsky and Konobeev (1965) who proceeded from a more general theory developed by Lifshits (1948).

Below, we shall follow the analysis of Dubovskiy and Konobeev (1965), Lifshits (1948) and consider the simplest case, namely, a cubic crystal with one molecule per unit cell in which one of the molecules (for instance, in the site $n = 0$) is replaced with an isotopic impurity.

The terms of the isotopic impurity are shifted with respect to the terms of the host material but the variation of the intermolecular interaction due to the presence of the isotopic impurity is small and can be ignored in the first approximation. As discussed in ch. 4 sect. 2, in this approximation we can add to eq. (4.34) the operator of perturbation of the excited crystal states

$$M^{(1)}_{n,m} = -\Delta\delta_{n0}\delta_{m0}, \tag{5.38}$$

where $\Delta = -\Delta'_{f1} + \Delta_f$ is the difference between the terms of the impurity molecule and the host molecule in the crystal (it should be remembered that $\Delta_f = \epsilon_f + D_f$; see ch. 4 sect 2). Then the equation for the coefficients u_n (here $v_n = 0$; see eq. (4.45a)) has the form

$$\sum_m M_{n,m}u(m) - \epsilon u(n) - \delta_{n0}u(n)\Delta = 0, \tag{5.39}$$

$$\epsilon \equiv E - \Delta_f,$$

This equation is identical (apart from notation) to the basic equation of Lifshits (1948) which treated scattering of elastic waves by isotopic defects in crystal lattice. We have to find the solution of eq. (5.39) which contains the incident wave $\exp(i\boldsymbol{kn})$ at infinity and a wave diverging from the origin of coordinates. The general solution of eq. (5.39) possessing the required asymptotic behaviour is

$$u(\boldsymbol{n}) = e^{i\boldsymbol{kn}} + \frac{\Delta u(0)}{N}\sum_{\boldsymbol{k}'}\frac{e^{i\boldsymbol{k}'\boldsymbol{n}}}{E(\boldsymbol{k}') - E(\boldsymbol{k}) - i\eta}, \tag{5.40}$$

$$u(0) = \left[1 - \frac{\Delta}{N}\sum_{\boldsymbol{k}'}\frac{1}{E(\boldsymbol{k}') - E(\boldsymbol{k}) - i\eta}\right]^{-1}, \tag{5.41}$$

where η is a positive infinitesimal determining the directions of circumventing the poles in eqs. (5.40) and (5.41). (In the case of isotopic impurities eq. (5.40) differs from eq. (4.83) in the solution of the homogeneous equation which should be taken into account if ϵ is in the exciton band. Equation for $u(0)$, that is, eq. (5.41), directly follows from eq. (5.40) if we take $n = 0$ in it.)

Below, we shall employ the approximation of the isotropic effective mass

$$E(\boldsymbol{k}) = E(0) + \frac{\hbar^2 k^2}{2m^*} \equiv E(0) + M\left(\frac{k}{k_m}\right)^2, \tag{5.42}$$

where $k_m = \pi/a$, and

$$M = \frac{\hbar^2}{2m^*}\left(\frac{\pi}{a}\right)^2, \tag{5.43}$$

is the exciton band width. When we replace the sum over $\boldsymbol{k}'$ in eqs. (5.40) and (5.41) with integration we obtain for long-wavelength excitons (with

$k \ll k_m$) in this approximation:

$$\frac{1}{u(0)} = 1 - \xi\left(1 - x^2 + \frac{\mathrm{i}\pi x}{2}\right), \tag{5.44}$$

where

$$\xi = \pi\Delta/2M, \qquad x = k/k_m. \tag{5.44a}$$

Moreover, if $|\boldsymbol{n}| \gg a$ we obtain

$$\frac{\Delta}{N}\sum_{\boldsymbol{k}'}\frac{\exp(\mathrm{i}\boldsymbol{k}'\boldsymbol{n})}{E(\boldsymbol{k}') - E(\boldsymbol{k}) - \mathrm{i}\eta} \approx \frac{a\xi}{2}\frac{\exp(\mathrm{i}k|\boldsymbol{n}|)}{|\boldsymbol{n}|}.$$

From the above equations we obtain the following expression for the amplitude A of elastic scattering of long-wave excitons by isotopic defects:

$$A(\boldsymbol{k}, \Delta) = \frac{a}{2}\frac{\xi}{1 - \xi + \xi x^2 - \mathrm{i}\pi x\xi/2}. \tag{5.45}$$

Hence, the total cross section of scattering of excitons by impurity is

$$\sigma_\mathrm{i} = 4\pi|A|^2 = \pi a^2 \frac{\xi^2}{(1 - \xi + \xi x^2)^2 + \pi^2 x^2 \xi^2/4}. \tag{5.46}$$

If $\xi \ll 1$, that is, if the displacement of the impurity term is much smaller than the exciton band width, the scattering cross section for excitons is small:

$$\sigma_\mathrm{i} \approx \pi a^2 \xi^2. \tag{5.47}$$

If $\xi \gg 1$, that is, if the displacement of the impurity term is much greater than the exciton band width, we obtain from eq. (5.46):

$$\sigma_\mathrm{i} \approx \pi a^2. \tag{5.48}$$

Equation (5.46) also indicates that the cross section becomes anomalously large when $\xi \to 1$ and it can be many times greater than the geometric cross section (5.48). It was shown by Dubovskiy and Konobeev (1964) that this effect corresponds to the situation when a local state is split off from the bottom of the exciton band; the energy of this state is close to the bottom of the exciton band and, therefore, it has a large radius.

Using the ratio $\rho = -\epsilon_\ell/M$ where ϵ_ℓ is the energy of the local state taken from the exciton band bottom we can write that the radius of this state is given by

$$r_\ell = a/\pi\sqrt{\rho}, \tag{5.49}$$

so that the cross section of scattering of excitons with extremely small wave vectors is

$$\sigma_\mathrm{i}(0) = 4\pi r_\ell^2, \tag{5.50}$$

that is, it is similar to the cross section of scattering of slow particles by a solid sphere with the radius r_ℓ.

Analyzing eq. (5.36) we can estimate the mean free path $l = \bar{v}\tau_i$ corresponding to exciton scattering by impurity. Assuming that the term displacement Δ is much greater than the exciton band width, that is, $\xi \gg 1$, and using eq. (5.48) we obtain

$$l = a/\pi\nu, \tag{5.51}$$

where $\nu = N_i/N$ is the ratio of the number of impurity molecules to the total number of molecules in the crystal. For instance, for the impurity concentrations of the order of 10^{-4} the mean free path (5.51) is greater than the lattice parameter by a factor of the order of 10^4 and considerably exceeds the mean free path of excitons corresponding to scattering by phonons.

In a variety of experiments the impurity molecules, apart from shifted terms, had the matrix elements of the operator of intermolecular interaction which were different from those for the host molecule. When the displacement of the impurity term is considerably greater than the variation of these matrix elements the cross section of exciton scattering by impurity depends mainly on the displacement of the term if only no exciton states are formed in the vicinity of the impurity molecule whose bonding energies are low and, hence, the radii are large. If the displacement Δ of the term is much greater than the band width then the properties of these states cease to be dependent on Δ (see Agranovich (1968), Chapter VII), and it is just these states that determine the cross section of scattering of excitons by impurity.

Let us now discuss scattering of excitons by the most important types of crystal defects which in molecular crystals include vacancies, disoriented molecules and, probably, dislocations. If we ignore the lattice deformation in the vicinity of the defect the cross section of exciton scattering by vacancy is approximately equal to the cross section of exciton scattering by isotopic defect with $\xi \gg 1$. In this case the cross section of exciton scattering by vacancy can be estimated from eq. (5.48). (When we take into account the lattice deformation in the vicinity of vacancy local states with large radii can appear. In this case the scattering cross section should be estimated from eq. (5.50) where r_ℓ is the radius of the state.)

In crystals with close packing, for instance, in crystals of benzene, anthracene, naphthalene, etc., interstitial molecules are extremely unfavourable energetically. Therefore, the concentrations of the defects of this type in the above molecular crystals are negligibly small and we neglect their contribution when we analyze exciton diffusion. However, in some crystals, even with close packing, increasing temperature can lead to possible formation of structural defects corresponding to changes of shapes or orientations of individual molecules in the crystal lattice. Each event of disorientation of a molecule necessitates overcoming of a potential barrier. The height of this barrier is, evidently, determined by the molecular structure and the structure of the crystal unit cell so that, in principle, we have a wide range of possibilities – from practically total absence of disorientation to almost free rotation. We shall ignore here the factors which contribute to lowering of the potential barriers preventing disorientation (for instance, the presence of axes of symmetry, low moments of inertia with respect to these axes, etc.). We shall note only that the concentration of the defects of this type can be estimated by

$$c = N \exp(-Q/k_B T), \tag{5.52}$$

where Q is the activation energy for disorientation, and N is the total number of molecules per unit volume. If σ is the cross section of exciton scattering by such defect the respective scattering length,

$$l = 1/c\sigma = \exp(Q/k_B T)/N\sigma, \tag{5.53}$$

drops exponentially with decreasing temperature. In those crystals where at sufficiently high temperatures scattering by disoriented molecules is the main mechanism of exciton scattering the diffusion coefficient of free excitons has an exponential dependence on temperature expressed as

$$D = D_0 \exp(Q/k_B T), \tag{5.54}$$

where the dependence of

$$D_0 = \bar{v}/N\sigma$$

on temperature is much weaker than exponential dependence.

The presence of the disoriented molecules can, apparently, be found not only in studies of exciton diffusion but also in studies of those physical properties of crystals which are sensitive to scattering of certain waves by defects. In this respect, it is interesting to analyze line broadening in absorption and Raman spectra, heat conductivity, electric conductivity, viscosity, and other properties of crystals which are fairly sensitive to structural defects of crystal lattice.

No experiments have been performed yet to show any significant effect of dislocations on exciton kinetics in molecular crystals. However, it is clear that at sufficiently low temperatures scattering of excitons by dislocations (as well as boundaries of microcrystals) can, generally speaking, have a significant effect on exciton diffusion as it is the case in heat conduction. Moreover, we have not seen any reports on calculations of cross sections for exciton scattering by dislocations in molecular crystals. As for exciton scattering by microcrystal boundaries the mean free path of excitons is approximately equal to the size of microcrystals in the temperature range where this scattering is significant. Then the diffusion coefficient is

$$D = \tfrac{1}{3} l\bar{v} \approx KL\sqrt{T}, \tag{5.55}$$

where K is a factor independent of temperature and the size of microcrystals, L is the microcrystal size, and T is the temperature.

Equation (5.55) is an analogue of Casimir's law in the theory of heat conductivity. It indicates that if exciton scattering by microcrystal boundaries becomes significant with decreasing temperature then the increase in the exciton diffusion coefficient due to scattering by phonons (for instance, as $D \sim 1/\sqrt{T}$; see sect. 4) must give way to its decrease according to the relation $D \sim \sqrt{T}$ where the proportionality coefficient must be proportional to the microcrystal size.

6. Capture of excitons by impurity molecules

As noted above, in many molecular crystals electronic excitation energy is transferred from the host material to impurities. If we deal with the exciton transfer mechanism we can conditionally identify two main stages of the process. The first one is diffusion of exciton towards impurity (if impurity concentration is high this stage can be absent; see ch. 6 sects. 3, 4 and 5) and the second is capture of the exciton by an impurity. The contributions of various mechanisms of exciton scattering to exciton diffusion have been discussed in the preceeding sections. Therefore, we shall present here only the calculated probabilities of exciton capture by impurities. At first, we shall assume that the limiting cases of exciton–phonon interaction are realized and excitons are fully incoherent (the hopping mechanism) or coherent (diffusion of wave packets). An attempt to analyze intermediate cases is reported by Hemenger et al. (1974) (for one-dimensional crystal model) where references to earlier studies in this direction are given.

6.1. Capture of incoherent excitons by impurity: "Nonresonance transfer"

As was done in calculations of the exciton diffusion coefficient in sect. 3, we can find the probability of capture of incoherent exciton by impurity from the probability of energy transfer from donor to acceptor calculated in ch. 4 sect. 10.

In our case the role of the donor can be played by any of the excited molecules which can be near or fairly far from the acceptor molecule. In the latter case we can, evidently, find the capture probability from the Förster formula. If the distance between donor and acceptor is short, that is, about one or two lattice parameters, we cannot write the effective matrix element of resonance transfer, for instance, for the dipole–dipole interaction in the form

$$M_{mn}^{\text{eff}} \sim \frac{p_1 p_2}{|\boldsymbol{n}-\boldsymbol{m}|^3} \frac{1}{\epsilon}\left(\frac{\epsilon+2}{3}\right)^2.$$

Owing to decrease in screening, we can use, instead of the dielectric constant ϵ of the crystal in rough estimates a certain "effective" quantity $\tilde{\epsilon}$ which is a function of the distance $|\boldsymbol{n}-\boldsymbol{m}|$ and tends to $\epsilon(\omega)$ when $|\boldsymbol{n}-\boldsymbol{m}| \to \infty$ (a similar procedure was employed by Haken (1958) in the theory of the Wannier–Mott excitons). Thus, at small distances $|\boldsymbol{n}-\boldsymbol{m}|$ the transfer probability W_{nm} increases with decreasing distance somewhat slower than $W_{nm} \sim |\boldsymbol{n}-\boldsymbol{m}|^{-6}$ and at small distances we also should take into account the correlation of displacements discussed in ch. 4 sect. 10.

The effective radius R_0 of exciton capture by a trap can be roughly estimated (in more detail see chs. 1 and 6) from the relationship

$$\tau W_{\boldsymbol{n}, \boldsymbol{n}+\boldsymbol{R}_0} \approx 1, \tag{5.56}$$

where τ is the exciton lifetime. Of course, if R_0 is much greater than the lattice parameter ($R_0 \gtrsim 10a$) the corrections to the Förster formula do not make any significant contribution.

When the crystal temperature is high or when the interaction between electronic excitations and the crystal vibrations is strong the quantity B in eq. (4.117) can be greater than $\hbar^2\omega_D^2$ where ω_D is the Debye frequency. Under such conditions the probability of capture of an exciton in the site p by an impurity in the site q is given by eq. (4.117). We can also derive this equation from the Förster formula if we ignore the correlation of displacements and assume that the spectra of exciton emission and donor absorption have the Gaussian shape.

When the interaction between electronic excitations and phonons in the crystal is weak the transfer probability (4.114) can be expanded into a power series in the constant of the electron–phonon interaction $|U_\kappa(\boldsymbol{n})-U_\kappa(\boldsymbol{m})|^2 = \bar{c}_\kappa(|\boldsymbol{n}-\boldsymbol{m}|)$. In the first approximation we obtain from eq. (4.114)

$$W_{nm} = \frac{|M_{nm}^{\text{eff}}|^2}{\hbar} \bar{c}(\boldsymbol{n}-\boldsymbol{m})[N(\omega_1)+1]\rho(\omega_1), \tag{5.57}$$

where $\rho(\omega)$ is the density of the frequencies of the lattice vibrations, $N(\omega)$ is the Planck mean number of phonons with the frequency ω, $\omega_1 = \Delta/\hbar$, and $\Delta = E_d - E_a$ is the difference between the energies of exciton (donor) and acceptor ($E_d > E_a$). If $E_d < E_a$ eq. (5.57) is replaced with the equation

$$W_{nm} = \frac{|M_{nm}^{\text{eff}}|^2}{h} \bar{c}(|\boldsymbol{n}-\boldsymbol{m}|)N(\omega_2)\rho(\omega_2), \tag{5.58}$$

where $\omega_2 = -\Delta/\hbar$. Equations (5.57) and (5.58) determine the probability of capture of incoherent exciton by acceptor accompanied with either emission or absorption of a lattice phonon. (Equations (5.57) and (5.58) indicate that for $k_B T \gg \hbar\omega_1$, $\hbar\omega_2$, the probability $W_{nm} \sim T$. Such dependence was observed in ruby (Selser et al. 1977) in the study of the rate of energy transfer over impurities with inhomogeneous broadening (R_1 line luminescence) in the temperature range $T < 50$ K.) Naturally, these probabilities are proportional to the frequency density for those phonons that are involved in the process. Using a similar procedure we can derive the probabilities of "nonresonance" processes with $E_d \neq E_a$ involving two phonons, etc. (a possible contribution of these processes is discussed in sect. 9).

6.2. *Capture of coherent exciton by a shallow trap*

Expressions of the type of eq. (4.114) are quite inapplicable to crystals in which energy transfer from the host material to impurities is affected by coherent excitons. Under such conditions in the initial state the electronic excitation energy belongs not to one molecule of the crystal but to a whole crystal region while the final state, as in the above-treated case corresponds to excitation of an impurity molecule (exciton trap).

We shall denote by Δ the trap depth, that is, the difference between the energy corresponding to the exciton band minimum and the electron excitation energy for the impurity molecule.

If the trap depth Δ is small in comparison with $k_B\Theta$ where Θ is the Debye temperature then one acoustic phonon is sufficient for capture of the exciton by impurity. Under such conditions the process has a noticeable probability even if the exciton–phonon interaction is weak. On the contrary, if the trap depth is large compared with $k_B\Theta$ then exciton capture by impurity is a multiphonon process. This process has a significant probability if exciton capture results in considerable displacement of the equilibrium positions of molecules.

Dubovsky and Konobeev (1964) calculated the probability of capture of free excitons by shallow traps in molecular crystals in the approximation of weak exciton–phonon interaction. They calculated in the first approximation of the perturbation theory the probability of capture of the exciton with the energy $E(\boldsymbol{k})$ to a local level formed by an impurity using the exciton–phonon interaction operator $\hat{H}_{eL}$ as the perturbation (only the lowest exciton band is considered below and the subscript μ denoting band number is, therefore omitted). It should be noted that it is difficult to estimate the capture cross sections owing to insufficient information on the constants of the exciton–phonon interaction and the shapes of exciton bands. Therefore, it would be interesting to derermine various qualitative characteristics of the capture cross section, for instance, its temperature dependence, which even now can be compared with the experimental data. This is way we shall discuss below mainly the qualitative features of the equation for the capture cross section.

In the presence of local excited states of electrons the operator $\hat{H}_{eL}$ is given by eq. (4.44a) with the addition of the term $\hat{H}'_{eL}$ corresponding to transformation of the exciton state into a local excited state accompanied with emission of an acoustic phonon:

$$\hat{H}'_{eL} = \frac{1}{N} \sum_{s,\, q,\, k} F_s(k, q, \ell) B^+_\ell B(k)(b_{qs} + b^+_{qs}) + \text{h.c.} \tag{5.59}$$

Here B^+_ℓ is the operator of creation of the local excited state ℓ, $B(k)$ is the operator of annihilation of exciton, and b^+_{qs} and b_{qs} are the operators of creation and annihilation of the phonon qs. Hence, the probability of capture of the exciton with the energy $E(k)$ to the local level E_ℓ is given by

$$W(k, \ell) = \frac{2\pi}{\hbar N^2} \sum_{s,\, q} |F_s(k, q, \ell)|^2 (1 + N_{sq}) \delta(E(k) - E_\ell - \hbar\omega_s(q)). \tag{5.60}$$

Since we assume that the trap is shallow we have the following condition for the energy difference at temperatures below the Debye temperature ($T \ll \Theta$) for most excitons:

$$E(k) - E_\ell \ll k_B \Theta. \tag{5.61}$$

When the inequality (5.61) is satisfied the main contribution to the sum in eq. (5.60) is made by the long-wavelength acoustic phonons for which $qa \ll 1$.

In the limit of infinitely long wavelengths of acoustic vibrations the exciton–phonon interaction must tend to zero since these vibrations do not alter the distances between molecules and their relative orientations. Therefore, as it can be readily shown for specific models, we have for the long-wavelength acoustic phonons,

$$|F_s(k, q, \ell)|^2 \approx |F_s^{(1)}(k, q/q)|^2 q. \tag{5.62}$$

Since $T \ll \Theta$ for most excitons k in eq. (5.62) can be replaced with $k = k_0$ where k_0 corersponds to the minimum of the exciton band. Moreover, if we ignore the dependence of $F_s^{(1)}$ and $\omega_s(q)$ on phonon polarization and the direction of q then eq. (5.60) readily yields

$$W(k, \ell) = \frac{3|F^{(1)}|^2 a^3}{\pi \hbar^2 v_0 N} \left[1 - \exp\left(-\frac{\hbar q_0 v_0}{k_B T}\right)\right]^{-1} q_0^3, \tag{5.63}$$

where v_0 is the velocity of sound and

$$q_0 = [E(k) - E_\ell]/\hbar v_0.$$

Using eq. (5.63) we find that the cross section of exciton capture is

$$\sigma_{\mathrm{i}} = \frac{3|F^{(1)}|^2 a^6}{\pi\hbar^2 v_0 \bar{v}} \left[1 - \exp\left(-\frac{\hbar q_0 v_0}{k_{\mathrm{B}}T}\right)\right]^{-1} q_0^3, \tag{5.64}$$

where $\bar{v}$ is the mean group velocity of excitons. For most excitons we have

$$E(\boldsymbol{k}) - E_\ell \approx \Delta + \tfrac{3}{2}k_{\mathrm{B}}T. \tag{5.65}$$

Hence, a low temperature when, apart from eq. (5.61), we have the condition

$$\Delta > \tfrac{3}{2}k_{\mathrm{B}}T, \tag{5.66}$$

the capture cross section is

$$\sigma_{\mathrm{i}} \approx \frac{3|F^{(1)}|^2 a^6}{\pi\hbar^2 v_0} \left(\frac{\Delta}{\hbar v_0}\right)^3 \left(\frac{m^*}{3k_{\mathrm{B}}T}\right)^{1/2}, \tag{5.67}$$

that is, according to the Bethe law (see, for instance, Landau and Lifshits (1965) sect. 140), the capture cross section increases with decreasing temperature as $T^{-1/2}$. It is also significant that the capture cross section is proportional to the third power of the trap depth. If we have

$$\Delta < \tfrac{3}{2}k_{\mathrm{B}}T, \tag{5.68}$$

then eq. (5.64) yields

$$\sigma_{\mathrm{i}} \approx \frac{3|F^{(1)}|^2 a^6}{\pi\hbar^2 v_0} \left(\frac{m^*}{3k_{\mathrm{B}}T}\right)^{1/2} \left(\frac{3}{2}\frac{k_{\mathrm{B}}T}{\hbar v_0}\right)^3, \tag{5.69}$$

that is, in this temperature range the capture cross section increases as $T^{5/2}$ with increasing temperature. Thus, the cross section of exciton capture by a shallow trap exhibits a minimum in the temperature range $T \approx \Delta/k_{\mathrm{B}}$.

The probability of capture differs from the capture cross section in the factor $\sqrt{T}$ and is a monotonic function of temperature. When $k_{\mathrm{B}}T < \Delta$ the probability is practically independent of temperature and it varies as T^3 with increasing T.

6.3. *Capture of coherent exciton by a trap accompanied with large displacements of the equilibrium configurations*

If the depth of the trap $\Delta \gg k_{\mathrm{B}}\Theta$ then, as noted above, exciton capture is a multiphonon radiationless process. The probability of capture can be relatively high only in those crystals where exciton capture is accompanied with considerable displacement of the equilibrium positions of molecules so that multiphonon transitions are allowed even in the first approximation.

Theoretical treatment of capture of coherent excitons by deep traps was reported by Agranovich (1958). As in the theory of capture of incoherent excitons, Agranovich (1958) employed adiabatic approximation but the contribution to the capture probability due to the nonadiabaticity operator was not taken into account. In this approximation capture of coherent exciton is due to the operator of interaction of the impurity molecule with other molecules of the crystal. Hence, the capture probability is given by

$$W = \frac{2\pi}{\hbar}\left|\int \psi_1^*(r, R_0^{(1)})\hat{H}_{ic}\psi_2(r, R_0^{(2)})\,d\tau\right|^2$$

$$\times \sum_{\dots N'_x \dots}\prod_x \left|\int \Phi_{N_x}(q_x - q_{x1})\Phi_{N'_x}(q_x - q_{x2})\,dq_x\right|^2$$

$$\times\, \delta\left[\Delta + \sum_x \hbar\omega_x(N_x - N'_x)\right].$$

where $\psi_1(r, R_0^{(1)})$ is the wave function of the electrons of the crystal corresponding to excitation of the impurity, $R_0^{(1)}$ are the equilibrium coordinates of the nuclei determining the positions and configurations of molecules in the crystal in this state, and $\psi_2(r, R_0^{(2)})$ is the wave function of the electrons of the crystal in the presence of an exciton. If $\boldsymbol{k}$ is the exciton wave vector then at large distances from the impurity the function $\psi_2(r, R_0^{(1)})$ corresponds to superposition of the incident wave $\exp(i\boldsymbol{kn})$ and diverging waves (see, for instance, eq. (5.40)). The trap depth

$$\Delta = E(\boldsymbol{k}) - E_\ell - \frac{\hbar}{2}\sum_\kappa (q_{\kappa 2}^2 - q_{\kappa 1}^2)\omega_\kappa \tag{5.70}$$

differs from the difference of energies of the electron subsystem with the configuration R_0 (which is the equilibrium configuration of the crystal in the ground state) by the energy of deformation of the lattice which occurs with exciton capture. Performing calculations similar to those in ch. 4 sect. 10 we obtain the probability

$$W = \frac{1}{\hbar}\left|\int \psi_1^*(r, R_0^{(1)})H_{ic}\psi_2(rR_0^{(2)})\,d\tau\right|^2 \exp\left[-\frac{(\Delta - A)^2}{B}\right], \tag{5.71}$$

where

$$B = 2\sum_\kappa \hbar^2\omega_\kappa^2(q_{\kappa 1} - q_{\kappa 2})^2(2\bar{N}_\kappa + 1) \tag{5.72}$$

and

$$A = \sum_\kappa \hbar\omega_\kappa(q_{\kappa 1} - q_{\kappa 2})^2. \tag{5.73}$$

It should be remembered that these calculations are valid only if the inequality

$$B \gg \hbar^2\omega_D^2, \tag{5.74}$$

where ω_D is the Debye frequency, is satisfied irrespective of the trap depth. We can readily estimate the value of B since $q_{\kappa 1}$ and $q_{\kappa 2}$ in eqs. (5.72) and (5.73) are the displacements of the equilibrium q_κ values from their values in the ground state of the crystal. Therefore, the fact that a coherent exciton is realized in the state ψ_2 implies that the quantities $q_{\kappa 2}$ are relatively small. If we ignore these quantities eq. (5.72) becomes equal to the squared half width of the impurity absorption line divided by 4 ln2 (see ch. 4 sect. 9). Hence, the width of the impurity absorption lines can serve as an indicator of the applicability of various methods for calculating the cross section of exciton capture. Indeed, if the impurity absorption line width is of the order of $k_B\Theta = \hbar\omega_D$ or smaller then the probability of capture by deep traps ($\Delta \gg \sqrt{B}$) is extremely low while capture by shallow traps ($\Delta \lessdot k_B\Theta$) is possible (see the above calculations of the cross section of exciton capture by shallow traps). If the impurity absorption line width is greater than $k_B\Theta$ the temperature dependence of the capture probability can be found from eq. (5.71).

If we retain only the dipole–dipole interaction in the operator $\hat{H}_{ic}$ the matrix element in eq. (5.71) can be found in the Born approximation. Indeed, the resonance interaction between the impurity molecule and the molecule n of the crystal at large distances from the impurity always satisfies the condition of applicability of the Born approximation (see, for instance, Landau and Lifshits (1965) sect. 130; for the sake of simplicity we assume that the crystal has one molecule per unit cell) since for large n values the energy of interaction between exciton and the impurity centre is

$$V_n \approx \frac{|\boldsymbol{p}^{0f}|\,|\boldsymbol{p}^{0f'}|}{|\boldsymbol{n}^3|} \ll \frac{\hbar^2}{m^*|\boldsymbol{n}|^2}, \tag{5.74a}$$

where $\boldsymbol{p}^{0f}$ and $\boldsymbol{p}^{0f'}$ are the matrix elements of the dipole moment of crystal molecules calculated with the wave functions of the ground and excited states of the crystal molecules and the impurity molecules, respectively, and m^* is the effective exciton mass.

If $|\boldsymbol{p}^{0f}|$ and $|\boldsymbol{p}^{0f'}|$ are not anomalously high the condition (5.74a) is valid also for the nearest neighbours, that is, for $|\boldsymbol{n}| = a$. In the Born approximation we have

$$\int \psi_1^* \hat{H}_{\pi,0}\psi_2 \, d\tau = \frac{1}{\sqrt{N}} \sum_n{}' e^{i\boldsymbol{k}\boldsymbol{n}} M_{n0}^{ff'}.$$

If the directions of the vectors $\boldsymbol{p}^{0f}$ and $\boldsymbol{p}^{0f'}$ coincide then

$$\left|\sum_{\boldsymbol{n}}{}' e^{i\boldsymbol{k}\boldsymbol{n}} M_{\boldsymbol{n}0}^{ff'}\right|^2 = \frac{F_1}{F}\,\epsilon^2(\boldsymbol{k}), \tag{5.75}$$

where F_1 is the oscillator strength for the transition in the impurity, F is

is satisfied for all $n \neq 0$. Therefore, eq. (5.77) can be solved by successive approximations. Assuming $n \neq 0$ and ignoring the dependence of $g_n(q)$ on time (that is, taking $t \gg 1/\Gamma$) we obtain

$$g_n(q) \approx -\tfrac{1}{2}\mathrm{i}\beta_n(\Gamma+\gamma_n)^{-1}(1-\mathrm{e}^{-\mathrm{i}qn})g_0(q). \tag{5.79}$$

Taking now $n = 0$ in eq. (5.77) and using eq. (5.79) for $n \neq 0$ we obtain the following equation for $g_0(q)$:

$$\frac{\partial}{\partial t}g_0(q) = -2[\Gamma-\gamma(q)]g_0(q) - g_0(q)\left\{\sum_m \beta_m^2[1-\cos(qm)]/(\Gamma+\gamma_m)\right\}, \tag{5.80}$$

Equation (5.80) for $q \to 0$ has the form

$$\frac{\partial}{\partial t}g_0(q) = -D_{ij}q_iq_jg_0(q), \tag{5.81}$$

where

$$D_{ij} = \sum_m m_im_j[\gamma_m + \tfrac{1}{2}\beta_m^2/(\Gamma+\gamma_m)]. \tag{5.82}$$

Since the function $g_0(q)$, according to eq. (5.76), is identical to the Fourier transform of the exciton concentration, when we go over from eq. (5.81) to the coordinate representation we obtain the conventional equation of diffusion for the exciton concentration $c(r, t) = \rho_{r,r}(t)$ with the diffusion coefficient D_{ij}:

$$\frac{\partial c(r,t)}{\partial t} = D_{ij}\frac{\partial^2 c(r,t)}{\partial x_i \partial x_j}. \tag{5.83}$$

If we assume that eq. (5.80) is satisfied not only for small q but for all q within the first Brillouin zone then we obtain the random walk equation by going over to the coordinate representation in eq. (5.80):

$$\frac{\partial}{\partial t}\rho_{nn}(t) = \sum_m \psi(n-m)\rho_{mm}(t) - \rho_{nn}(t)\sum_m \psi(m), \tag{5.84}$$

Here

$$\psi(m) = 2\gamma_m + \beta_m^2/(\Gamma+\gamma_m). \tag{5.85}$$

Let us now return to the equation of diffusion. The derivation of eq. (5.81) indicates that this equation is valid irrespective of the relationship between the parameters γ, Γ and β entering into eq. (5.82). At the same time, purely qualitative analysis shows that excitons are coherent, as noted above, only if the exciton level width Γ due to scattering (see ch. 4 sect. 12) is small compared with the exciton band width Δ. In this case, as shown by Haken and Strobl (1973) (and as follows from purely qualitative analysis), within the framework of this model the motion of excitons between collisions is, indeed, realized as the motion of wave packets (that is, as diffusion of wave packets). In the inverse limiting case the exciton motion is realized as random walk over lattice sites (the case when $\Gamma > \Delta$). The small term with β_m^2 can be omitted in eq. (5.82) in this case. Of course, intermediate situations are also possible. In all the above cases eq. (5.82) for D_{ij} is valid but the relative contributions of the terms in this equation differ. This constitutes the primary formal advantage of the model making it convenient for processing experimental data.

Now let us discuss some features of the temperature dependence of the diffusion coefficient. In this connection, it should be noted that the random quantities h_{nm}, strictly speaking, are functions of the coordinates of the nuclei and in the range of small displacements they include linear, quadratic and higher terms of the expansion in normal coordinates. Therefore, the quantities Γ and γ_m in the first approximation can be regarded as linear functions of the temperature T. Then eq. (5.82) yields that for coherent excitons ($\gamma \ll \beta^2/(\Gamma+\gamma)$) the diffusion coefficient $D_{ij} \sim 1/T$ and for incoherent excitons (the inequality is reversed) $D_{ij} \sim T$. The latter result (as the shape of the absorption band) contradicts the prediction of rigourous theory which gives $D_{ij} \sim \exp(-Q/k_B T)$ (see sect. 3). However, the result $D \sim 1/T$ is correct in the limit of coherent excitons; since for $k_B T > \Delta$ the entire band is populated, the squared group exciton velocity averaged over the band is independent of temperature, while the collision time $\tau \sim 1/T$, and we obtain $D \sim \overline{v^2}\tau \sim 1/T$.

To illustrate application of this model for interpreting experimental data let us again discuss the results for triplet excitons in anthracene reported by Ern et al. (1972). Table 5.1 presents the coefficient D_{aa} of exciton diffusion along the a-axis, the Davydov splitting and the half-widths Γ_b and Γ_c of the b and c component absorption lines. A sufficiently good approximation for triplet excitons is that which takes into account the resonance interaction only between the nearest neighbours. In this approximation we have

$$D_{aa} = a^2[\gamma_d + \tfrac{1}{2}\beta_d^2(\Gamma + \gamma_d)^{-1}],$$

where the subscript d corresponds to the nearest neighbours displaced by the vector $\frac{1}{2}(\boldsymbol{a} \pm \boldsymbol{b})$ (equal to the minimum distance between the molecules with $\sigma = 1$ and $\sigma = 2$ in the anthracene crystal). Since in this approximation the band term in the exciton energy is

$$\epsilon_k \approx 2\beta_b \cos(\boldsymbol{kb}) \pm 2\beta_d(\cos \tfrac{1}{2}\boldsymbol{k}(\boldsymbol{a}+\boldsymbol{b}) + \cos \tfrac{1}{2}\boldsymbol{k}(\boldsymbol{a}-\boldsymbol{b})],$$

the Davydov splitting $\Delta \approx 8\beta_d$. Therefore, we can use the experimental data

Table 5.1
Measured and calculated characteristics of triplet excitons in anthracene.

	Temperature (K)			
	118	160	298	371
D_{aa} (10^{-4} cm^2/s	4.0 ± 0.5	2.5 ± 0.3	1.5 ± 0.2	1.6 ± 0.3
Δ (cm^{-1})	18 ± 2	18 ± 6	17 ± 3	19 ± 2
Γ_b (cm^{-1})	14 ± 1	30 ± 2	51 ± 2	65 ± 2
Γ_c (cm^{-1})	14 ± 2	26 ± 6	54 ± 6	62 ± 2
γ_d (cm^{-1})	0.11 ± 0.08	0.10 ± 0.08	0.07 ± 0.04	0.07 ± 0.03
τ (10^{-3} s)	10–20		18–20	18–20

(that is, D_{aa}, Γ and $\beta_d = \Delta/8$) to find γ_d (see table 5.1 where the lifetime τ of triplet excitons is also given). Since Γ is much larger than γ_d in the above temperature range it is suggested by Ern et al (1972) that the contribution of nonlocal scattering is not significant ($\gamma_d \ll \gamma_0$). Interestingly, the measured Davydov splitting for triplet excitons in anthracene at low temperatures (4.2 K) is 22 cm^{-1} as reported by Clark and Hochstrasser (1967). The line width Γ in this case is small ($\Gamma \lesssim 1$ cm^{-1}) indicating the coherent nature of excitons. The absorption coefficient corresponding to the $S_0 \rightarrow T_1$ transition is very small ($\kappa \approx 10^{-4}$ cm^{-1}).

8. *Bimolecular annihilation of excitons*

In the last ten–fifteen years the demands of quantum electronics and nonlinear optics stimulate intensive studies of the physical properties of crystals with high concentrations of excitons.

The results of these studies demonstrate differences in behaviour between large-radius and small-radius excitons which lead to some very interesting consequences.

For semiconductors with large-radius excitons these studies have led to prediction, and later to experimental observation of biexcitons and/or electron–hole drops (see, for instance, Haken and Nikitine (1975) where theoretical and experimental results are reviewed). At sufficiently low temperatures the exciton system in such crystals can be thermalized within a short period (much shorter than the exciton lifetime) and goes over to the respective state with the minimal free energy.

For instance, in crystals in which this state corresponds to the gas of biexcitons biexciton luminescence is observed (for instance, in the CuCl crystals; Haken and Nikitine (1975)). This is accompanied with decay of biexcitons. The process yields excitons which are either in the ground state (the energy E_e^0) or in an excited state (the energy E_e^*). The energy of the emitted photons is given by

$$\hbar\omega = E_{be} - E_e^* = E_e^0 - u - (E_e^* - E_e^0),$$

where $E_{be} \equiv 2E_e^0 - u$ is the biexciton energy, and u is the bonding energy of excitons in the biexciton. Thus, decay of biexcitons gives rise to luminescence lines or bands which correspond to the wavelengths greater than the wavelength of exciton luminescence.

Long-wavelength luminescence of the same type occurs also in those cases (germanium and silicon crystals) when the exciton subsystem of a highly excited crystal forms a system of electron–hole drops. We cannot devote sufficient space here to a discussion of the interesting physical phenomena observed in the crystal containing electron–hole drops (ch. 6 sect. 5 and see Keldysh and Jeffries (1983). We shall only note here that in molecular crystals where excitons have small radius such phenomena have not been observed and this fact is clearly due to that the exciton–exciton interaction in semiconductors is, in a certain sense, much weaker than the exciton–exciton interaction in molecular crystals.

To explain this, let us analyze the state of the crystal with two excitons. Since the energy of excitation of such crystal is, typically, much greater than the forbidden gap width this state with two excitons can, generally speaking, go over to a state with the same energy corresponding to the crystal state with one electron–hole pair. Since such a state rapidly goes over to the state with one exciton the above transition, if it occurs, results in radiationless decay of at least one of the two excitons. To estimate the probability of this process we can use the theoretical results reported by Choi and Rice (1962) where the process (autoionization) was analyzed for the model of two excited hydrogen atoms at a distance R from each other. Assuming the existence of large-radius excitons and defining the effective charge $e^* = e/\sqrt{\epsilon}$ (where ϵ is the static dielectric constant) and the effective mass μ by $\mu^{-1} = (1/m_e + 1/m_h)$ where m_e and m_h are the masses of the electron and the hole we find that the probability (per 1 s) of the autoionization process is

$$W = \frac{1}{\pi}\left(\frac{a_B}{R}\right)^6 \frac{\mu e^4}{\epsilon^2\hbar^3} \approx 2\times 10^{16}\left(\frac{a_B}{R}\right)^6\left(\frac{\mu}{m}\right)\frac{1}{\epsilon^2}\,\text{сек}^{-1}, \tag{5.86}$$

where $a_B = \hbar^2\epsilon/\mu e^2$ is the Bohr radius of the exciton, and m is the electron mass in vacuum (Choi and Rice 1962) so that for $R = 2a_B$, $\epsilon = 30$ and $\mu = 0.1$ m we obtain $W \approx 10^{10}\ s^{-1}$.

If c is the exciton concentration and σ is the cross section of exciton–exciton collisions the mean free path determined by collisions is $l \approx 1/c\sigma$ and, for instance, for $c = 10^{15}\ cm^{-3}$ and $\sigma = 10^{-13}\ cm^2$ we have $l \approx 100\ \mu m$. This corresponds to the mean free time $t_0 = l/v$ where v is the exciton thermal velocity. At the temperature $T \approx 4$ K we have $v \approx 10^6$–10^7 cm/s and $t_0 \approx 10^{-8}$–10^{-9} s. The collision time is of the order of $a_B/v = 10^{-12}$–10^{-13} s so that autoionization cannot take place during this period.

Thus, if the concentration of large-radius excitons is not low the frequency of their collisions is high but they do not produce any noticeable variation of the total number of electrons and holes as noted above.

It should be noted also that owing to their large Bohr radius ($a_B \approx 10^{-6}$ cm) excitons in semiconductors fill up the total volume of the crystal even at such low concentrations as $c \approx c_0 \approx (a_B)^{-3} \approx 10^{18}\ cm^{-3}$. Therefore, excitons in semiconductors acquire new observable properties even at concentrations $c \approx 10^{-3}c_0$ or lower, that is, under the conditions when only a negligible fraction of the electrons in the valence band leave it. The behaviour of small-radius excitons has a number of special features since the above simplifications are not valid for them. As evidenced by eq. (5.86), the probability of autoionization for such excitons is higher than that for large-radius excitons owing to high values of μ and small values of ϵ. There are reasons to think (Northrop and Simpson 1958) that in some pure molecular crystals it is precisely autoionization that is the primary process of generation of charge carriers in the volume photoeffect. (This idea is confirmed by calculations of the probability of autoionization in the anthracene crystal reported by Choi and Rice (1963). In this study the calculated rate of generation of charge carriers was $3.7 \times 10^8\ cm^{-3}\ s^{-1}$ for the exciton concentration $c = 1.2 \times 10^{10}\ cm^{-3}$. This is in a good agreement with the measured rate of generation $\eta = 7 \times 10^8\ cm^{-3}\ s^{-1}$ which corresponds to $\gamma = \eta/c^2 \approx 5 \times 10^{-12}\ cm^3/s$, while similar calculations for incoherent excitons yield $\gamma = 3 \times 10^{-11}\ cm^3/s$ (Sharma 1967) where possible reasons for the poorer agreement with experimental data are also discussed.) Calculations of the autoionization rate are discussed also by Kearns (1963), Zgierski (1973).

However, the so-called bimolecular annihilation of excitons occurring in collisions of two Frenkel excitons proves to have a higher probability than autoionization. In contrast to autoionization, this process has basically a purely molecular nature and is not accompanied with formation of free electrons and holes.

If c is the exciton concentration then we can phenomenologically take into account the exciton–exciton collisions by using the equation

$$\frac{dc}{dt} = D\Delta c - \frac{c}{\tau} - \gamma c^2 + I, \tag{5.87}$$

where τ and D are the lifetime and the diffusion coefficient of excitons for low exciton concentrations, $I(t)$ is the rate of generation of excitons, γ is a phenomenological quantity determining the rate of exciton annihilation in exciton collisions.

In most studies, including those using picosecond exciton pumping pulses (Campillo et al. 1977), in which experimental results were interpreted on the basis of eq. (5.87) the quantity γ was assumed to be independent of time. Under such conditions the solution of eq. (5.87) with various boundary conditions makes it possible to determine the fluorescence intensity $\Phi(t) = 1/\tau \int c(r, t)\, dr$ as a function of the parameters D, τ and γ for various $I = I(t)$. The quantities D and τ can be found independently from the experiments using low exciton pumping intensities. Therefore, for many crystals γ has been determined from the data on the decay kinetics or the quantum yields of luminescence or phosphorescence of excitons when their concentrations are high. For instance, bimolecular annihilation of triplet excitons was first studied by Avakian et al. (1963). It has been found that triplet–triplet annihilation gives rise to singlet excitons which produce delayed fluorescence with the characteristic decay time equal to a half of the lifetime of triplet excitation.

Bimolecular annihilation of singlet excitons was first reported by Tolstoi and Abramov (1967) who found for anthracene that the linear dependence of the fluorescence intensity on the intensity I of pumping of excitations was replaced by a square-root dependence ($\sim\sqrt{I}$). (Equation (5.87) shows that for bulk excitation under the steady-state conditions when the diffusion term in this equation is not significant we have $c = \tau I$ for small I and $c \approx \sqrt{I/\gamma}$ for $I \gg c/\tau = (1/\tau)\sqrt{I/\gamma}$.) The available data (Bergman et al. 1967) indicate that the rate of bimolecular annihilation of singlet excitons in anthracene is determined by $\gamma_b \approx 10^{-8}\,\text{cm}^3\,\text{s}^{-1}$. The rate of autoionization found in the experiments on volume photoconductivity (Silver et al. 1963) corresponds to $\gamma_p = 10^{-12}\,\text{cm}^3\,\text{s}^{-1}$ which agrees with the theoretical estimates of Choi and Rice (1962); in eq. (5.87) we have $\gamma = \gamma_b + \gamma_p \approx \gamma_b$.

The initial theoretical estimates of γ were based on the following considerations.

As in the case of exciton capture by impurities, bimolecular reactions involving excitons occur in two stages – approach and collision. In the simplest approximation of the collision radius R we have

$$\gamma = 8\pi DR, \tag{5.88}$$

where D is the diffusion coefficient of excitons (eq. (5.88) differs from a similar equation determining the probability of exciton capture by impurity in the factor 2 which appears because we have to take into account diffusion of both excitons). Equation (5.88) is taken from the coagulation theory (Chandrasekar 1943); it is valid in the case of three-dimensional diffusion if the collision radius R is small in comparison with the exciton diffusion length $L = \sqrt{D\tau}$ and, at the same time, larger than the mean free path of excitons (or the hopping length for incoherent excitations; see ch. 6). If this condition is not satisfied the simple estimate given by Jortner et al. (1963) can be used.

Jortner et al. (1963) assumed that excitons migrating towards each other are incoherent, that the time needed for annihilation of two excitons is short compared with the time τ_h of exciton hopping to a neighbouring lattice site, and that annihilation occurs if excitons approach at a distance equal to the lattice parameter. The resulting estimate is

$$\gamma \approx Za^3/\tau_h, \tag{5.88a}$$

where a is the lattice parameter, and Z is the number of the nearest neighbours. Since in our model we have $R = a$ and $D = a^2/3\tau_h$ eq. (5.88) differs from eq. (5.89) only in a numerical factor. Note, however, that neither eq. (5.88) nor eq. (5.88a) is valid for quasi-one-dimensional or quasi-two-dimensional systems in which exciton motion is not three-dimensional owing to anisotropy of the resonance interactions. Under such conditions the quantity γ entering into eq. (5.87) is not constant but depends on the exciton concentration. This situation is very similar to that for the rate of exciton capture by impurities in one-dimensional and two-dimensional crystals (see ch. 6 sect. 3). The same is true, however, for three-dimensional crystals too, if only there is practically no exciton diffusion in these crystals. Under such conditions no stage of approach can be identified, too, and annihilation at any moment occurs at the mean distance $R \sim c^{-1/3}$. If annihilation is determined by the rate of dipole–dipole energy transfer the total number A of annihilation events per unit volume can be estimated as

$$A \sim (2/\tau)(R_0/R)^6 c = qc^3,$$

where $q \sim 2R_0^6/\tau$ which corresponds to $\gamma = \beta c$ (for a more rigourous treatment, see Agranovich et al. (1980)). This result reported by Benderskiy et al. (1976) is quite instructive. It demonstrates that even in three-dimensional crystals γ can be independent of time or exciton concentration only if the diffusion length for excitons is sufficiently large.

This important question has not been fully analyzed yet. To analyze the case of incoherent excitons we have to consider systems of coupled equations (Suna 1970) for one-particle, two-particle, etc. spatial distribution functions for excitons. Since we cannot discuss the relevant calculations in detail here we shall note only the main results and illustrate the methods being used.

First of all, note that the sought rate of annihilation of excitons per unit volume is given by

$$A(\boldsymbol{R}_0) = \sum_{\boldsymbol{R}} W(\boldsymbol{R})\rho_2(\boldsymbol{R}_0, \boldsymbol{R}_0 + \boldsymbol{R}; t), \tag{5.89}$$

where $W(\boldsymbol{R})$ is the probability of annihilation of excitons at the distance $\boldsymbol{R}$ which rapidly decreases with increasing R, and $\rho_2(\boldsymbol{R}_1, \boldsymbol{R}_2; t)$ is the two-particle distribution function for excitons which is equal to the probability that at the moment t the first exciton is on the lattice site $\boldsymbol{R}_1$ and the second exciton is on the lattice site $\boldsymbol{R}_2$. Formally, the function ρ_2 is typically written as

$$\rho_2(\boldsymbol{R}_1, \boldsymbol{R}_2; t) = \rho_1(\boldsymbol{R}_1, t)\rho_1(\boldsymbol{R}_2, t)N_{\boldsymbol{R}_1-\boldsymbol{R}_2}(t).$$

If the spatial distribution of excitons is uniform the one-particle distribution function $\rho_1 = vc$ where $v = 1/N_0$ and N_0 is the number of lattice sites. Therefore, if excitons travelled completely independently of each other the two-particle distribution function would be the product $\rho_1(R_1)\rho_1(R_2)$, that is, it would be equal to $(vc)^2$. In this approximation $N_{\boldsymbol{R}} = 1$. In reality, there is always some interaction between excitons so that $N_{\boldsymbol{R}} \neq 1$. Since such interaction decreases with increasing distance between excitons the difference of $N_{\boldsymbol{R}}$ from unity is especially large for small R. The kinematic part of the interaction between excitons does not allow the presence of two excitons on one lattice site so that $N_{\boldsymbol{R}} = 0$ for $R = 0$. At the same time, the dynamic interaction between excitons affects the probability of their hops and, moreover, results in annihilation of excitons. Therefore, for small R we have $N_{\boldsymbol{R}} < 1$ and this should be taken into account in the calculation of the annihilation rate A since it is precisely the summation over small R that makes the major contribution to A. Taking $A = \gamma c^2$ (see eq. (5.87)) and using eq. (5.89) we obtain

$$\gamma = v \sum_{\boldsymbol{R}} W(R)N_{\boldsymbol{R}}(t). \tag{5.89a}$$

Since $\rho_1 = vc$, eq. (5.87) is, in fact, an equation for the one-particle distribution function which is coupled to the distribution function $N_{\boldsymbol{R}}$ via eqs. (5.89) and (5.89a). The equation for this function including diffusion and annihilation of excitons (Suna 1970) can be written as

$$\frac{\mathrm{d}N_{\boldsymbol{R}}}{\mathrm{d}t} = -\frac{2}{\tau}N_{\boldsymbol{R}} - W(R)N_{\boldsymbol{R}} + \sum_{\boldsymbol{R}'} [\psi(\boldsymbol{R}', \boldsymbol{R})N_{\boldsymbol{R}'} - \psi(\boldsymbol{R}, \boldsymbol{R}')N_{\boldsymbol{R}}]$$

$$- cv \sum_{\boldsymbol{R}'} [W(|\boldsymbol{R}' - \boldsymbol{R}| + W(R')]N_{\boldsymbol{R},\boldsymbol{R}'} + I_2, \tag{5.89b}$$

where $\psi(\boldsymbol{R}, \boldsymbol{R}')$ is the probability per unit time that the radius-vector connecting excitons in the pair changes from $\boldsymbol{R}$ to $\boldsymbol{R}'$ owing to diffusion of the excitons, $I_2 = 2cv^2I$ is the rate of creation of the exciton pairs, and $N_{\boldsymbol{R},\boldsymbol{R}'}$ is a function determining the three-particle distribution function

$$\rho_3(\boldsymbol{R}_1, \boldsymbol{R}_1 + \boldsymbol{R}, \boldsymbol{R}_1 + \boldsymbol{R} + \boldsymbol{R}') \equiv (vc)^3 N_{\boldsymbol{R},\boldsymbol{R}'}.$$

The function $\psi(\boldsymbol{R}', \boldsymbol{R})$ depends only on the difference $|\boldsymbol{R} - \boldsymbol{R}'|$ and is equal to the probability of exciton hopping the distance $|\boldsymbol{R} - \boldsymbol{R}'|$.

In a similar way, we can derive the equation for the distribution function $N_{\boldsymbol{R},\boldsymbol{R}',\boldsymbol{R}''}$, and so on so that eqs. (5.87) and (5.89b) are, in fact, only the first two equations of the infinite hierarchic system of equations. The approximation

$$N_{\boldsymbol{R},\boldsymbol{R}'} = N_{\boldsymbol{R}} N_{\boldsymbol{R}'} N_{-\boldsymbol{R}-\boldsymbol{R}'}$$

typically used in the statistical theory of liquids (Kirkwood and Boggs 1942, Croxton 1975) was applied by Chabr and Williams (1978) to this system of equations. At first sight, this approximation should be sufficiently good only for describing such configurations of particles in gas in which not more than two particles are sufficiently close to each other. In fact, as shown by the results of computer simulation by Croxton (1975), the approximation of the three-particle distribution function suggested by Kirkwood and Boggs (1942) is sufficiently good not only for gases but even for liquids.

It has been shown by Chabr and Williams (1979) that under the steady-state conditions, provided the inequalities,

$$c \ll c_0 \quad \text{and} \quad \tau^{-1} + \gamma c \ll D/R_0^2,$$

are satisfied (here $c_0 \equiv \lambda/\gamma$,

$$\lambda \equiv \sum_{\boldsymbol{R}} W(R),$$

D is the exciton diffusion coefficient, and R_0 is the mean hop length) we obtain

$$\gamma = \gamma^\infty/(1 + \gamma^\infty/\lambda v),$$

where

$$\gamma^\infty = 4[D(\tau^{-1} + \gamma c)]^{1/2}$$

for one-dimensional excitons,

$$\gamma^\infty = -4\pi D/\{\ln[\tfrac{1}{2}(\tau^{-1} + \gamma c)^{1/2} R_0] + 0.577\}$$

for two-dimensional excitons, and

$$\gamma^\infty = 8\pi D R_0$$

for three-dimensional excitons ($\gamma = \gamma^\infty$ only for $\gamma^\infty \ll \lambda v$).

According to eq. (5.87) the quantity $\tau_{\text{eff}} = (\tau^{-1} + \gamma c)^{-1}$ is the exciton lifetime taking into account bimolecular annihilation. Therefore, the second of the above inequalities can be rewritten as

$$\sqrt{D\tau_{\text{eff}}} \gg R_0.$$

This implies that the above results for γ are valid if the diffusion length for excitons is much greater than the lattice parameter even if bimolecular annihilation takes place.

In conclusion note that a number of studies of bimolecular annihilation of incoherent excitons under nonstationary conditions have been recently reported. For instance, Agranovich et al. (1980) describes computer simulation of random walks of excitons and their bimolecular annihilation, while Ivanova (1980) reports the results of computer calculations for the system of equations for the first two excitons distribution functions using the Kirkwood method (Kirkwood and Boggs 1942, Croxton 1975). The results can be valuable for interpreting the data obtained in experiments with pulsed excitation.

Now we shall make a few remarks concerning calculations of the rate of bimolecular annihilation of coherent excitons. As noted above, in this case eq. (5.88) can be regarded as applicable if the effective radius R of the absolutely black sphere of capture is large in comparison with the mean free path l of excitons.

If, on the contrary, $l \gg R$, then we can calculate γ as it is done for the rate of capture of coherent excitons by impurities (see eq. (6.41) and Agranovich 1968)) using the formula of gas kinetics

$$\gamma = \sigma\bar{v},$$

where $\sigma = \pi R^2$ is the cross section of collision, $\bar{v}$ is the mean (thermal) relative velocity of excitons ($\bar{v} \sim \sqrt{T}$ in the approximation of isotropic effective mass) and T is the temperature of the crystal. Since the cross section σ should be regarded as being independent of temperature at sufficiently low temperatures we have $\gamma \sim \sqrt{T}$. Thus, both for coherent and incoherent excitons the phenomenological quantity γ entering into eq. (5.87) significantly depends on the relationship between the mean free path of excitons and the radius of the exciton–exciton interaction though this fact is not always fully appreciated (Ivanov and Koshushner 1980). Anisotropy of exciton diffusion may prove to be significant in the calculation of γ for coherent excitons. Further development of the theory is needed before it becomes clear to what extent the results of the theory of capture of coherent excitons by stationary traps (see ch. 6 sect. 2) are relevant for analysis of bimolecular annihilation. We shall not discuss the details here and shall focus our attention at the mechanism of the exciton–exciton collisions.

Note, first of all, that, apart from singlet–singlet and triplet–triplet collisions, there can occur singlet–triplet collisions of the type

$$S_1 + T_0 \rightarrow T_1 + S_0, \tag{5.90}$$
$$T_1 \rightarrow T_0, \tag{5.91}$$

where S_0 and S_1 denote the singlet ground state and the singlet excited state (singlet exciton) of the crystal and T_0 and T_1 are the triplet lowest state (triplet exciton) and the triplet first excited state of the crystal. The process described by eq. (5.90) corresponds to fusion of singlet exciton and triplet exciton giving rise to high-energy triplet exciton. It has been shown (Babenko et al. 1971) that in anthracene $\gamma_{ST} = (7 \pm 4 \times 10^{-9}\,\text{cm}^3\,\text{s}^{-1}$ for this process. Since the singlet exciton energy $E(S_1)$ is typically greater than the triplet exciton energy $E(T_0)$ steady-state excitation of singlet excitons in a crystal must always be accompanied with formation of triplet excitons: $S_1 \rightarrow T_0$.

Triplet excitons have longer lifetimes and in case of steady-state excitations their concentrations can be higher than the concentrations of singlet excitons. Under such conditions the fusion reaction (5.90) can have a strong effect on the kinetics of all processes involving singlet excitons as shown in a calculation of γ_{SS} by Babenko et al. (1971).

The excited triplet state T_1 produced in the process (5.90) can decay via one of the following reaction channels:

I. Transition to the ground triplet state: $T_1 \rightarrow T_0$.
II. Transition giving rise to a singlet exciton: $T_1 \rightarrow S_1$.
III. Transition to the ground singlet state: $T_1 \rightarrow S_0$.
IV. Autoionization: $T_1 \rightarrow$ electron + hole.

(The probability of the singlet–triplet and triplet–triplet autoionization was calculated by Kearns (1963) and the role played by the vibronic states (exciton + intramolecular phonons) in these processes was discussed by Zgierski (1973)).

Since the first reaction channel does not require a change of the spin and occurs at one centre it has the highest probability and this has been demonstrated in a number of experimental studies. For instance, in naphthalene the rate of the process I is higher by the factor of 10^6 than that of the process II (Keller 1969). The process III is similar to the process II. For the process IV it has been shown (Fourny et al. 1968) that γ for it is lower by approximately four orders of magnitude than γ_{ST}. This implies that the process IV is not the main channel of decay of the state T_1 produced in the $S_1 + T_0$ collision.

Similar various possibilities and channels of decay of the states produced in the exciton–exciton collisions can be found for the $S_1 + S_1$ and $T_0 + T_0$ reactions. We shall, however, continue to discuss the $S_1 + T_0$ reaction since sufficiently convincing estimates of the probability have been recently made precisely for this reaction (for the crystals of anthracene, chlorophyll-a and rhodamine 6G; Rahman and Knox (1973)).

Since in all these crystals excitons are incoherent at the room temperature (for instance, for anthracene this is shown by the measured temperature dependence of the coefficient of diffusion of the singlet excitons; see ch. 7) fusion of excitons can be treated within the framework of the Förster model (see eq. (1.36)). The probability of fusion, that is, of the resonance energy transfer from the singlet exciton to the triplet exciton is given by

$$W = \frac{1}{\tau_s}\left(\frac{R_0^{ST}}{R}\right)^6, \tag{5.92}$$

where τ_s is the lifetime of singlet excitons in the absence of any quenching centres, R is the distance between the triplet and the singlet excitons,

$$(R_0^{ST})^6 = \frac{9\chi^2 c^4}{128\pi^5}\int_0^\infty \frac{F_s(\omega)\sigma_{TT}(\omega)}{\omega^4 n^4(\omega)}\, d\omega, \tag{5.93}$$

χ^2 is an orientation factor which is taken to be equal to its mean value $\frac{2}{3}$, $F_s(\omega)$ is the normalized fluorescence spectrum of the singlet exciton, $\sigma_{TT}(\omega)$ is the cross section of the triplet–triplet absorption, and $n(\omega)$ is the index of refraction. The data on σ_{TT} used by Rahman and Knox (1973) for chlorophyll-a, anthracene and rhodamine 6G were taken, respectively, from Linschits and Sarkanen (1958), Porter and Windsor (1958), Berlman (1965), Kawski (1963, Buctner et al. (1969).

The following results for R_0^{ST} were obtained: 48 Å for chlorophyll-a, 39 Å for anthracene, and 46 Å for rhodamine 6G. These results show that when the distance between the excitons is shorter than approximately four lattice parameters the probability of fusion of the singlet exciton with the triplet exciton is higher than the probability of fluorescence. Apparently, the Förster radius R_0 must also have such large values in the S + S reactions and in the exciton–exciton collisions of other types and, most likely, this is a general property of most molecular crystals.

Typically, $D_T \ll D_S$ so that the motion of triplet excitons can be ignored and they can be treated as stationary traps of singlet excitons. Using the theory of exciton capture by traps (see chs. 2 and 6) and the values of R^{ST} and the diffusion coefficients D_S and D_T of the singlet and triplet excitons we can find γ_{ST} and compare it with the experimental results. For instance, in anthracene at room temperature $D_S = 3.6\times10^{-3}\,\text{cm}^2\,\text{s}^{-1}$ (Avakian and Merrifield 1964) and $D_T = 2\times10^{-4}\,\text{cm}^2\,\text{s}^{-1}$ and the value $\gamma_{ST} = 4\times10^{-9}\,\text{cm}^3\,\text{s}^{-1}$ found by Rahman and Knox (1973) is in good agreement with the measured value of $(7\pm4)\times10^{-9}\,\text{cm}^3\,\text{s}^{-1}$ (Babenko et al. 1971).

Note that the wave function of the triplet excited state (with the energy E) produced in the collision of the triplet and singlet excitons is a superposition of the state φ_{T_1} of the metastable triplet exciton and the continuous-spectrum states $\varphi_{E'}$ (electron in the conduction band and hole in the valence band) and φ_i (exciton in the state T_0 + highly excited vibrational continuum):

$$\Psi_E = a(E)\varphi_{T_1} + \sum_i b_i(E)\varphi_i + \int c_{E'}(E)\varphi_{E'}\,dE', \tag{5.94}$$

Here a, b_i and $c_{E'}$ are factors dependent on E. The states of the type of (5.94) which are produced by interaction between subsystems with discrete and continuous energy spectra and their manifestations in the absorption spectrum have been treated by Fano (1961); see also ch. 4 sect. 9. The states (5.94) in the case of exciton–exciton collisions have been analyzed by Choi (1967), Jortner (1968).

Since eqs. (5.92) and (5.93) include experimentally determined parameters (σ_{TT}, etc.) we can assume that the real function Ψ_E is automatically taken into account (in the appropriate matrix elements of transition). However, the details of this function also prove to be significant, for instance, for estimating the relative contribution of the fusion reactions $T_0 + S_1$ which yields electron–hole pairs (that is, for estimating the relative contribution of autoionization reactions). Furthermore, the quantity $|a(E)|^2$ determines the probability of formation of an excited molecular state due to fusion of excitons, and so on.

For instance, Jortner (1968) presents the following result (see also eq. (4.108d)):

$$|a(E)|^2 = \frac{|V_E|^2 + W_E}{(E - E_{T_1})^2 + \pi^2(|V_E|^2 + W_E)^2}, \tag{5.95}$$

Here $W_E = \langle\varphi_i|\hat{H}|\varphi_{T_1}\rangle^2\rho(E)$, $\rho(E)$ is the density of vibrational states, $V_E = \langle\varphi_E|\hat{H}|\varphi_{T_1}\rangle$, and $\hat{H}$ is the Hamiltonian of the crystal. Since the coupling of the molecular excited states with the "crystal" polar states is, typically, weak we have $|V_E|^2 \ll W_E$ and, hence,

$$|a(E)|^2 \approx \frac{W_E}{(E - E_{T_1})^2 + \pi^2 W_E^2}.$$

If at least one of the colliding excitons is coherent eq. (5.92) is not valid. However, since the width of the exciton band of triplet excitons is, typically, small in comparison with the band width for singlet excitons we can ignore the motion of triplet excitons, as noted above. This case is largely similar to the capture of coherent singlet excitons by stationary traps (see sect. 6). As in the interaction between coherent excitons and impurity molecules, capture can be preceeded by formation of the bound state of coherent exciton and trap, that is, of the biexciton in our case, if, of course, this is possible. Naturally, processes of such type are significant only at low temperatures $T < E_{be}/k_B$ where E_{be} is the binding energy of the biexciton. These processes have not been analyzed yet. Depending on the relationship between the biexciton radius R_b and the capture radius R_0 (see eq. (5.88)) formation of biexcitons can, apparently, intensify bimolecular quenching (when $R_0 > R_b$) or, on the contrary, inhibit quenching (when $R_b > R_0$).

The rate of direct capture of coherent excitons by triplet excitons must generally depend on the energy of the exciton–exciton interaction (this effect must occur for incoherent excitons, too, as shown by Suna (1970); quenching of localized excitations is discussed by Benderskiy et al. (1976). If this energy is sufficiently large (in comparison with the exciton band width) and the interaction is repulsive the capture is weak since the magnitude of the wave function of the coherent exciton in the region of the trap is small. In the case of strong attractive interaction the effect proves to be the same since a deep potential well behaves as a high barrier for elastic scattering owing to orthogonality of the states of the bound and free excitons (see, e.g., Landau and Lifshits 1965).

An attempt to calculate the probability of capture of the coherent exciton by the triplet excitation is reported by Onipko and Sugakov (1974). We shall not discuss it in detail since we have discussed the problem in sect. 6. We shall only note here that in all the processes of exciton fusion or decay the operator providing for the process is the operator of cubic anharmonicity of excitons (see ch. 4 sects. 1 and 2):

$$\hat{H}_3 = \sum_{n,m,p} (V_{nmp}^{f_1f_2f_3} B_n^{f_1} B_m^{f_2} B_p^{+f_3} + V_{nmp}^{*} B_n^{+f_1} B_m^{+f_2} B_p^{f_3}),$$

where $V_{nmp}^{f_1f_2f_3}$ are the matrix elements of the intermolecular interaction corresponding to the process in which the molecular excitations f_1 and f_2 disappear at the sites n and m and the excitation f_3 appears at the site p. In the case in question the main contribution is, clearly, made by the operator

$$\hat{H}_3' = \sum_n (V_{n00}^{S_1T_0T_1} B_n^{S_1} B_0^{T_0} B_0^{+T_1} + V_{n00}^{*S_1T_0T_1} B_n^{+S_1} B_0^{+T_0} B_0^{T_1}).$$

In conclusion, note that biexcitons have not yet been found in the crystals that have been studied, that is, those with small-radius excitons. However, such crystals can be expected to produce long-wavelength luminescence which is, to a certain extent, similar to biexciton luminescence. Indeed, if the potential energy of the interaction between two incoherent excitons is $F(R)$ (where R is the distance between the excitons; see eq. (4.73)) and corresponds to attraction (that is, $F(R) < 0$) then their energy is

$$E_2(R) = 2E_e + F(R) < 2E_e.$$

If radiative transition occurs in one of the excitons the energy of the luminescence quantum is $\hbar\omega = \hbar\omega_0 + F(R) < \hbar\omega_0$ where $\hbar\omega_0 = E_e$ is the energy of the luminescence quantum for the free exciton. Bimolecular quenching of excitons must result in a correlation between the quantum energy $\hbar\omega$ and the time of decay of luminescence at the frequency ω. Since bimolecular quenching dependent on R decreases the exciton lifetime the time of decay of luminescence must decrease with decreasing luminescence frequency ω (for $\omega < \omega_0$).

To analyze this correlation (Agranovich and Efremov 1980) assume that a short-duration pulse ($t \ll \tau_0$ where τ_0 is the lifetime of free excitons) has generated a sufficiently high concentration of excitons in a crystal. Let N_R be the number of exciton pairs per unit volume with the distance R between the excitons. If we ignore diffusion of excitons and triple correlations (that is, assume $N_{R,R'}(t) = 0$; see eq. (5.89b)) we obtain

$$N_R(t) = N_R(0) \exp\left[-\left(\frac{2}{\tau_0} + W(R)\right)t\right],$$

where $N_R(0)$ is the distribution of the exciton pairs at the starting moment (at the end of the exciting pulse). For crystals of the type of the anthracene crystal (we shall ignore anistropy) $F(R)$ is determined by the energy of the quadrupole–quadrupole interaction so that we have

$$F(R) = -D_0(a/R)^5,$$

where a is the lattice parameter, and the energy D_0 for the crystals of the type of the anthracene crystal is, apparently, of the order of a few hundreds of cm^{-1} (D_0 is approximately equal to the energy D_f given by eq. (4.22a) divided by the number of the nearest neighbours). Since the time of decay of the number N_R is $T(R) = [2/\tau_0 + W(R)]^{-1}$ we can use the above relationship between R and ω and the Förster formula

$$W(R) = \frac{2}{\tau_0}\left(\frac{R_0}{R}\right)^6,$$

and obtain

$$T(\omega) = \frac{\tau_0}{2}\left\{1 + \left[\frac{\hbar(\omega_0 - \omega)}{D_0}\right]^{6/5}\left(\frac{R_0}{a}\right)^6\right\}^{-1}.$$

The intensity of the resulting luminescence is proportional to the number of excitons rather than the number of exciton pairs. Therefore, the time of decay of excitons (those excitons which decay producing the luminescence quantum $h\omega$) is $\tau(\omega) = 2T(\omega)$, that is,

$$\tau(\omega) = \tau_0\left\{1 + \left[\frac{\hbar(\omega_0 - \omega)}{D_0}\right]^{6/5}\left(\frac{R_0}{a}\right)^6\right\}^{-1}.$$

If we include diffusion into the equation for N_R we obtain a nonexponential decay of the long-wavelength radiation. Additional features can be obtained by including anisotropy and triple correlations.

When excitation levels are high the kinetics and spectrum of fluorescence of molecular crystals can be also affected by formation of free or trapped charge carriers. Excitons can transfer their energy to charge carriers and this can result in quenching of both fast and delayed fluorescence and, simultaneously, in higher photoelectric current or photoelectric effect.

When we discuss the above phenomena we should bear in mind (Agranovich and Zakhidov 1979) the possibility of formation of the bound states exciton + charge. The electron polarizes the neighbouring molecules in the molecular crystal. A transition of one of these molecules to the excited states alters the energy of interaction between the electron and the neighbouring molecules and gives rise to the energy of the interaction between the exciton and the electron. For instance, in crystals consisting of molecules with an inversion centre (for a more general analysis see Agranovich and Zakhidov (1979)) when we take into account only the dipole polarization we have for the energy of the interaction between the exciton and the electron at the distance r:

$$V(r) = -\tfrac{1}{2}\Delta\alpha E^2(r),$$

where $\Delta\alpha = \alpha^f - \alpha^0$ is the variation of the static polarizability of the molecule due to excitation, $E(r) = e/\epsilon r^2$, e is the electron charge and ϵ is the dielectric constant. The variation $\Delta\alpha$ can be found from the Stark effect measurements (Hochstrasser 1972, 1973); for instance, for the lowest singlet excitation in anthracene $\Delta\alpha \approx 1.5 \times 10^{-23}\,\text{cm}^3$. For $r = 5$ Å when there is practically no static screening ($\epsilon \approx 1$) we have $V \approx 1500\,\text{cm}^{-1}$, for $r = 10$ Å we have $V \approx 200\,\text{cm}^{-1}$ and so on, that is, it is higher than or of the order of the widths of the lowest exciton bands and the lowest conduction band. Though the above expression for $V(r)$ gives a fairly rough approximation the estimates clearly indicate that the bound states exciton + charge can prove to be sufficiently stable in molecular crystals.

If this is so localized charges could serve as long-lived nuclei for condensation (clustering) of singlet or triplet excitons and, moreover, give rise to exciton fluorescence shifted towards lower energies by an amount of the order of the binding energy of the exciton and the charge.

If the exciton energy is transferred to the charge carrier (for instance, electron) a one-particle state is formed in the crystal (for instance, an electron in the continuous spectrum of the conduction band). This implies that the bound state exciton + charge is an autoionization state. The existence of the state must result in Fano antiresonances in the energy dependence of the cross section of electron scattering in the crystal which seem to have been found in experiments (Sanche (1979); theoretical analysis for two-dimensional systems is given by Golubkov et al. (1980)). This opens up exciting prospects in studies of various electron–exciton complexes in molecular crystals and, as noted above, can be significant for interpreting various optical effects occurring in molecular crystals at high excitation levels.

9. *Effects of inhomogeneous broadening and Anderson localization of excitations on the rate of energy transfer via impurity molecules; nonresonance transfer processes*

If the energy of the resonance interaction between the impurity molecules (at the average distance between them) in a crystal is considerably greater than inhomogeneous broadening of the impurity term then the impurity exciton band is formed. For some crystals we can trace formation of such bands in the studies of the absorption spectrum of the impurity molecules as a function of the impurity concentration. One of the first workers to carry out such studies were Broude and Sheka and their coworkers (Broude et al. 1961) who used the deuterium-substituted molecules of the host material as the impurity (similar IR results were obtained by Zhizhin (1975)).

The studies of transfer of electronic excitation energy via the impurity exciton bands have just been started; the most difficult theoretical question in these studies is how to take into account the cluster structure of the spectrum of excitations (Onodera and Toyozawa 1968, Dubovskiy and Konobeev 1970, Hoshen and Jortner 1970, Hong and Robinson 1970). The analysis of Lifshits (1964) shows that the random distribution of impurities leads to broadening of the impurity exciton band. If the average distance between the impurity molecules is sufficiently large the density of states has a minimum at the centre of the impurity band. This result can be qualitatively explained in the following way.

If E_0 is the excitation energy for the isolated impurity molecule then for a pair of molecules at the distance R the energy $E = E_0 \pm V(R)$ where $V(R)$ is the energy of the resonance interaction between the molecules. Only the pairs in which the distance between the molecules is much greater than the average distance make a contribution to the density of states at the centre of the impurity band. The number of such pairs is relatively small and this explains the above result. It is clear that the subdivision of the ensemble of impurity molecules into the pairs of interacting centres is valid only at sufficiently low impurity concentrations (that is, such concentrations for which the average distance between impurity molecules is not small in comparison with the effective radius of the resonance interaction).

Therefore, for instance, for the dipole–dipole resonance interaction the minimum of the density of states discussed above can be lacking.

Since the levels of different pairs of impurity molecules are shifted with respect to each other owing to the resonance interaction energy transfer between them, that is, actually, energy transfer via the impurity exciton band, can be considerably lower at sufficiently low temperatures when $k_B T \ll V(R_0)$, where R_0 is the average distance between impurity molecules, and when, moreover, the electron–phonon coupling is weak and the homogeneous width of impurity absorption band is small compared with $V(R_0)$. If one of these inequalities is reversed energy transfer via the impurity band can be considerable. The situation can be different in the case of inhomogeneous broadening $\Gamma \gg V(R_0)$, $k_B T$. The theoretical analysis of this situation is based on Anderson's study (Anderson 1958, 1970) of the conductivity of disordered medium (Mott and Davis 1971). It follows from Anderson's results that when the resonance interaction $V(R)$ decreases faster than $1/R^3$ with increasing R there is a critical impurity concentration c_{cr} below which any impurity excitations are predominantly localized and therefore cannot make any contribution to energy transfer. The Anderson localization actually implies that there is a negligibly small probability of the event when (for sufficiently low impurity concentrations) an unexcited impurity molecule whose excitation level permits resonance transfer is at a distance shorter than the effective length of the resonance interaction from a primarily excited molecule. Later Mott has shown (Mott and Davis 1971) that for impurity concentrations $c > c_{cr}$ not all the states are localized and the so-called mobility edge appears in the impurity spectrum. (As Anderson, Mott has treated electric conductivity of disordered medium (states, their spectrum and mobility in the framework of the one-electron problem with random potential), but formally this problem is equivalent to the calculation of the spectrum of impurity electronic excitations and their mobility, and we are making use of this.)

Let us illustrate the above discussion by fig. 5.1 which shows the absorption spectrum for an inhomogeneously broadened line with the centre at E_0 and the half-width Γ. The impurity states which are localized according to Anderson are shaded. This implies that for $c > c_{cr}$ only those impurity molecules whose states are in the central spectrum region can contribute to energy transfer; when c increases the quantity Δ increases and the region of localized states dimishes.

Recent experiments with crystals $CaWO_4$ with samarium impurity (Hsu and Powell 1975) and ruby crystals, that is Al_2O_3 crystals with chromium impurity, (Koo et al. 1975) have, apparently, demonstrated evidence of the Anderson localization of states in the studies of energy transfer via impurity molecules. These experiments have been performed with a frequency-tuned laser whose line width $\delta \ll \Gamma$. The use of such laser as the excitation source makes it possible to study energy transfer via impurity

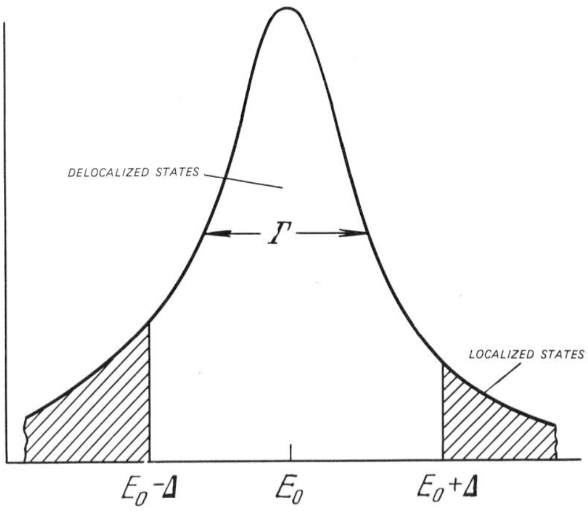

Fig. 5.1. Regions of localized and delocalized states (schematically) of an inhomogeneously broadened line for $c > c_{cr}$.

molecules to traps as a function of the excitation quantum energy $\hbar\omega_L$. The laser excites only those impurity molecules for which the excitation energy $E = \hbar\omega_L$ and if the energy E is the region of localized states no energy transfer to impurities must occur.

Figure 5.2 shows the results reported by Koo et al. (1975). The R_1 absorption line of the ion Cr^{3+} was excited in this study. The energy of such a single ion can be transferred to clusters of two neighbouring Cr^{3+} ions coupled by exchange interaction. The excitation level of such cluster (N_2 line) lies several hundreds cm^{-1} below the excitation level of isolated chromium ion and the clusters serve as traps. Figure 5.2 presents the ratio

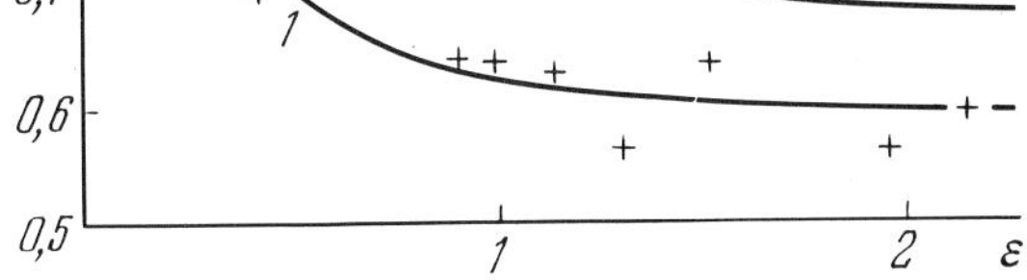

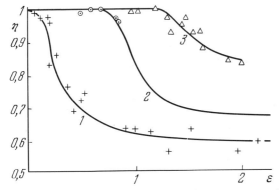

Fig. 5.2. The ratio $\eta = I_A/I_D$ as a function of excitation photon energy E (I_A is the intensity of acceptor luminescence for the N_2 line and I_D is the intensity of luminescence of the Cr^{3+} ions for the R_1 line). $\eta = I_A(E)I_D(E)I_A(E_0)$; $\epsilon = 2|E - E_0|/\Gamma$. (1) $\Gamma = 0.4\ cm^{-1}$, $c = 0.09\%$. (2) $\Gamma = 0.7\ cm^{-1}$, $c = 0.14\%$. (3) $\Gamma = 1.3\ cm^{-1}$, $c = 0.23\%$.

between the intensity of the luminescence of traps and the intensity of the luminescence of single chromium ions as a function of the excitation quantum energy for various trap concentrations c. The ratio of the intensities is normalized to unity at the centre of the line. Figure 5.2 shows that the energy transfer intensity fairly sharply decreases at certain values ϵ_{cr} which can be related to the above-mentioned mobility edge. According to Mott (Mott and Davis 1971), ϵ_{cr} decreases with decreasing impurity concentrations. However, for $\epsilon = \epsilon_{cr}$ the drop in the energy transfer intensity is not accompanied with vanishing of $\eta = I_A(E)I_D(E_0)/I_D(E)I_A(E_0)$.

This behaviour of $\eta = \eta(\epsilon)$ seems to be due to the fact that diffusion of excitations in the ruby is determined by the short-range exchange interaction and by the long-range electric dipole–dipole resonance interaction which is weaker (by an order of magnitude). If the latter interaction were absent we could expect vanishing of η beyond the mobility edge. However, the dipole–dipole transfer via the chromium ions results in a not too sharp drop at the mobility edge.

We should note also that localization of excitations should be hindered by the possibility of their nonresonance hopping accompanied with emission or absorption of lattice phonons. Such processes are especially significant, for instance, for admixtures of rare-earth elements in glasses where the inhomogeneous broadening is large ($\Gamma \sim 100\,\text{cm}^{-1}$). Since the frequencies of the short-wavelength phonons in this case are of the order of Γ they correspond to high densities of states so that the respective hop probability (see eqs. (5.57) and (5.58)) can be sufficiently high (a hop is realized if $W_D\tau_0 \gtrsim 1$ where τ_0 is the excitation lifetime in the absence of transfer and W_D depends on the distance between the impurity molecules). If Γ is small in comparison with the Debye frequency the contribution of the one-phonon processes should, generally speaking, be small. Under such conditions, as noted by Holstein et al. (1976), a more significant contribution at high temperatures can be made by the second-order processes corresponding to the transfer process accompanied with absorption and emission of short-wavelength phonons whose energy difference is of the order of Γ. (At low temperatures the rate of this process is low since it is proportional to $\exp(-\hbar\omega_D/k_BT)$. Apparently, it is precisely this fact that explains the linear dependence on temperature of the transfer rate found by Selser and Hamilton (1977) at sufficiently low temperatures; see also sect. 6.1, eqs. (5.57) and (5.58) for $\hbar\omega_1 \approx \Gamma \ll k_BT$.)

Analysis of energy transfer via impurities can serve as a new tool for studying the spectrum of disordered systems. Another factor which, undoubtedly, will stimulate studies in this field is a considerable practical interest to systems of the type of those treated by Hsu and Powell (1975), Koo et al. (1975), Holstein et al. (1976), Selser and Hamilton (1977) and used as laser materials.

CHAPTER 6

Phenomenological Theory of Exciton Migration

1. *Applicability of the diffusion approximation; quasi-one-dimensional and quasi-two-dimensional systems*

As noted above, during its lifetime an exciton undergoes a very large number of collisions with phonons. Therefore, in many cases we can describe exciton migration by the diffusion equation since under such conditions it serves as a sufficiently accurate approximation of the more general kinetic equation.

If $c(\boldsymbol{r}, t)$ is the exciton concentration at the point $\boldsymbol{r}$ of the crystal at the moment t (that is, the total number of excitons per unit volume) then the diffusion equation describing the variation of the function $c(\boldsymbol{r}, t)$ with time in space neglecting radiative transfer is

$$\frac{\partial c}{\partial t} = D\Delta c - Pc + I_0(t)k\,\mathrm{e}^{-kx}, \tag{6.1}$$

where D is the diffusion coefficient of excitons (anisotropy of the medium is ignored here), P is the probability of exciton decay per unit time, and k is the absorption coefficient for the exciting light. The last term on the right-hand side of eq. (6.1) gives the number of excitons generated by external radiation per unit volume per unit time. For the sake of simplicity, we shall assume that the crystal is a parallel-sided plate with the boundaries $x = 0$ and $x = d$ and that the incident light with the absorption coefficient k travels from the region $x < 0$ perpendicular to the plane $x = 0$.

Excitons can decay at the crystal surface. If v is the rate of surface annihilation of excitons then the boundary condition for $x = 0$ can be written as

$$D\left.\frac{\mathrm{d}c}{\mathrm{d}x}\right|_{x=0} = vc(0). \tag{6.1a}$$

If the crystal surface is contaminated or if the rate of capture to the surface exciton levels is high (see ch. 8) v becomes very high so that condition (6.1a) reduces to $c(0) = 0$. If, on the contrary, v is low, condition (6.1a) is transformed into $\mathrm{d}c/\mathrm{d}x = 0$ for $x = 0$.

Below we shall illustrate the use of eq. (6.1) for determination of various parameters which can be experimentally measured.

It should be borne in mind, however, that analysis of exciton kinetics sometimes necessitates going beyond the framework of the diffusion approximation. We shall consider such situations below in our discussion of exciton capture by impurities. They occur in the studies of energy transfer in thin crystal films, in case of metallic quenching of excitons (see ch. 8), etc. Generally speaking, the analysis of exciton migration in such situations is rather difficult owing to the complex form of the general kinetic equation (see ch. 4). However, as noted in ch. 4, this equation can be considerably simplified in the following two limiting cases: for coherent excitons and for incoherent (localized) excitons.

In the first limiting case the kinetic equation for excitons can be reduced to the Boltzmann equation.

It was noted in ch. 5 sect. 1 that the Boltzmann equation is applicable only in those cases when the mean free path of excitons is larger than the size of the wave packet (Peierls 1955) and that this condition can be satisfied only at sufficiently low temperatures. If the constants of the exciton–phonon interaction and the crystal temperature are such that the reverse inequality is satisfied the Boltzmann equation can be replaced with the kinetic equation describing random walk of excitons on the crystal sites.

At first, let us discuss the conditions under which the diffusion equation can be derived from the Boltzmann equation. In rigourous treatment if this equation is applicable it describes, apart from the exciton motion, the energy exchange between excitons and the lattice. This makes the analysis of this equation too complicated and, since the available experimental data on energy transfer at low temperatures are very limited, such analysis in most cases would be hardly justified. Therefore, we shall use the so-called one-velocity kinetic equation assuming that excitons have a constant (mean thermal) energy so that scattering of excitons results in variation of only the direction of the exciton velocity rather than its magnitude (this approach is substantiated in ch. 5 sect. 4).

One-velocity equations are widely used in the neutron physics and astrophysics and their derivation is discussed in many books (see, Galanin 1957). Therefore, we shall discuss here only those results which will be used below and omit inessential details.

Let $c(\boldsymbol{r}, \boldsymbol{n})\, \mathrm{d}V\, \mathrm{d}\Omega$ be the number of excitons in the volume $\mathrm{d}V$ at the point $\boldsymbol{r}$ whose velocity is in the solid angle $\mathrm{d}\Omega$ at the unit vector $\boldsymbol{n}$.

The balance equation of excitons in the volume $\mathrm{d}V\, \mathrm{d}\Omega$ yields the following equation for the function $c(\boldsymbol{r}, \boldsymbol{n})$:

$$\frac{\partial c(\boldsymbol{r}, \boldsymbol{n}, t)}{\partial t} = -v\nabla c(\boldsymbol{r}, \boldsymbol{n}, t) - \frac{v}{l}\, c(\boldsymbol{r}, \boldsymbol{n}, t) + \frac{v}{4\pi l_s} \times \int \mu(\Phi) c(\boldsymbol{r}, \boldsymbol{n}', t)\, \mathrm{d}\Omega' + \frac{1}{4\pi} I(\boldsymbol{r}, t), \tag{6.2}$$

Here v is the magnitude of the velocity ($\boldsymbol{v} = v\boldsymbol{n}$), and l is the total mean free path of excitons given by $1/l = 1/l_s + 1/l_a$ where l_s is the mean free path determined by scattering of excitons and l_a is the mean free path determined by absorption of excitons. The function $\mu(\Phi)/4\pi$ is the probability that scattering results in rotation of the velocity vector of exciton by the angle Φ. The function $I(\boldsymbol{r}, t)$ determined by exciton sources gives the number of excitons generated per unit time per unit volume at the point $\boldsymbol{r}$.

It can be shown (Galanin 1957) that we can go over from eq. (6.2) to eq. (6.1) for the function $c(\boldsymbol{r}, t) = \int c(\boldsymbol{r}, \boldsymbol{n}, t)\,\mathrm{d}\Omega$ only if the variation of this function along the length l is small, that is, if

$$l\left|\frac{\mathrm{d}c(\boldsymbol{r}, \boldsymbol{n}, t)}{\mathrm{d}x_i}\right| \ll c(\boldsymbol{r}, \boldsymbol{n}, t), \qquad i = 1, 2, 3, \tag{6.3}$$

if, furthermore, the dependence of $c(\boldsymbol{r}, \boldsymbol{n}, t)$ on $\boldsymbol{n}$ is sufficiently weak owing to the removal of the anisotropy of the excitons by collisions with phonons, and if, finally, $l \approx l_s$ so that $l_a \gg l_s$. Then we obtain the following relationship for the diffusion coefficient D entering into eq. (6.1):

$$D = \frac{1}{3}\frac{l_s v}{1 - \overline{\cos\Phi}}. \tag{6.4}$$

Here $\overline{\cos\Phi} = (1/4\pi)\int \mu(\Phi)\cos\Phi\,\mathrm{d}\Omega$. For isotropic scattering we have $\mu(\Phi) = 1$ and $\overline{\cos\Phi} = 0$ so that $D = \frac{1}{3}l_s v$. The quantity $l_t = l_s/(1 - \overline{\cos\Phi})$ is usually referred to as the transport length.

Generally speaking, the calculation of the quantity P in eq. (6.1) is more difficult. If the only processes resulting in exciton decay are the processes of spontaneous conversion of excitons into heat or luminescence photons we have $P = P_0 = 1/\tau$ where τ is the lifetime of excitons in the pure crystal. If, however, the crystal contains impurity molecules capable of capturing excitons the exciton distribution depends on the random distribution of impurities. (Below we assume that the impurity concentration is sufficiently low and its effect on the energy spectrum of the matrix can be neglected. The condition $l_a \gg l_s$ also requires that the impurity concentration be low.) Even if the mean impurity concentration is constant and the excitation is uniform (I = const) the exciton concentration can greatly decrease in the vicinity of traps so that the actual exciton distribution is extremely nonuniform. Fortunately, we do not need to know the real exciton distribution in order to find some important luminescent characteristics of crystals with impurities. In many cases we have to know only the volume distribution of the mean exciton concentration found by averaging the real concentration over a small volume which, however, contains a sufficient number of impurity molecules. It will be shown below (see sect. 2) that it is precisely this mean concentration that satisfies eq. (6.1) for crystals with impurities. Special efforts are needed here, however, to find the quantity P and sometimes this proves to be rather difficult.

Let us now discuss derivation of the diffusion equation in the model of random walk of excitons on the lattice. The random walk on the lattice was first analyzed by Polya (1921). Development of the random walk theory intensified in the last two decades in connection with analysis of some problems in solid state physics. A fairly full description of these results is given by Spitzer (1964), Montroll (1964, 1965, 1969). The studies of exciton energy transfer, in particular for high exciton concentrations, will, undoubtedly, stimulate further development of this theory. Since we shall repeatedly discuss exciton energy transfer to impurities within the framework of the random walk theory we shall consider here the main features of the relevant mathematical model in some detail (Polya 1921, Spitzer 1964, Montroll 1964, 1965, 1969). First of all, we shall assume that the three-dimensional discrete lattice of sites on which random walks occur has periodic boundary conditions as it is often assumed in the theory of crystals. This means that there is a very large integer N such that the site $s(s_1, s_2, s_3) \equiv s(s_1 + j_1N, s_2 + j_2N, s_3 + j_3N)$ where j_1, j_2 and j_3 are arbitrary integers. Let $P_n(s)$ be the probability that after the nth step the walker is at the point s. According to the periodic boundary conditions we obtain

$$P_n(s_1 + j_1N, s_2 + j_2N, s_3 + j_3N) = P_n(s).$$

Clearly, the probabilities $P_n(s)$ satisfy the recursion formula

$$P_{n+1}(s) = \sum_{s'} P(s - s')P_n(s'), \tag{6.5}$$

where $P(s)$ is the probability independent of n that a step results in a vector displacement s by a walker. Following the results of Montroll (1964, 1965), introduce the so-called structure function of the walk

$$\lambda(2\pi \boldsymbol{r}/N) = \sum_{s} P(s)\exp(2\pi i \boldsymbol{r}s/N). \tag{6.6}$$

In particular, if we consider only the jumps to the nearest neighbour and assume them to be equally probable then we obtain, for instance, for the simple cubic lattice $P(s) = \frac{1}{6}$ and

$$\lambda(\boldsymbol{\theta}) = \tfrac{1}{3}(\cos\theta_1 + \cos\theta_2 + \cos\theta_3), \tag{6.7}$$

where $\boldsymbol{\theta} = \boldsymbol{\theta}(\theta_1, \theta_2, \theta_3)$ and $\theta_i = 2\pi r_i/N$. A similar procedure can be used for finding the structure function of the walk for more complex lattices and for other assumptions on the possible length and symmetry of jumps. Note also that the assumption that the number of walkers is conserved yields the normalization condition for the probability

$$\sum_s P(s) = 1. \tag{6.8}$$

Equations (6.6) and (6.8) yield $\lambda(0) = 1$. To find $P_n(s)$ first introduce the random-walk generating function

$$P(s, z) = \sum_{n=0}^{\infty} z^n P_n(s), \qquad |z| \leq 1. \tag{6.9}$$

An equation for this function can be readily derived from eq. (6.5). Multiplying eq. (6.5) by z^n and taking a sum over n from $n = 0$ to $n = \infty$ we find that the function $P(s, z)$ satisfies an equation of the type of the equation for the Green's function

$$P(s, z) - z \sum_{s'} P(s - s')P(s', z) = \delta_{s0}, \tag{6.10}$$

if only we assume $P_0(s) = \delta_{s0}$. This assumption implies that, by definition, walks start from the origin, $s = 0$. The solution of eq. (6.10) has the form

$$P(s, z) = \frac{1}{(2\pi)^3} \iiint_{-\pi}^{\pi} \frac{\exp(-is\theta)\, d\theta}{1 - z\lambda(\theta)}, \tag{6.11}$$

from which, using eq. (6.9) we obtain

$$P_n(s) = \frac{1}{2\pi} \iiint_{-\pi}^{\pi} \lambda^n(\theta) \exp(-is\theta)\, d\theta. \tag{6.12}$$

Apart from the function $P_n(s)$, in our analysis of the rate of exciton capture by impurities we shall make use of the function $F_n(s)$ which, by definition, is equal to the probability that a walker starting from the point $s = 0$ visits the point s for the first time in the nth step. The generating function for this function is

$$F(s, z) = \sum_{n=1}^{\infty} z^n F_n(s). \tag{6.13}$$

The functions $F_n(s)$ and $P_n(s)$ are not independent. The function $P_n(s)$ is equal to the probability that after the nth step the walker is at the point s irrespective of whether he has visited this point before. Therefore, if the walker is at the point s for the first time only after the jth step $(0 < j \leq n)$ the probability that he visits this point again after the nth step is, evidently, equal to the product $F_j(s)P_{n-j}(0)$ where $P_{n-j}(0)$ is the probability of returning to the starting point in $n - j$ steps. Taking a sum over all possible j and using the starting condition $P_0(s) = \delta_{s0}$ we obtain

$$P_n(s) = \delta_{n0}\delta_{s0} + \sum_{j=1}^{n} F_j(s)P_{n-j}(0). \tag{6.14}$$

If we multiply both sides of eq. (6.14) by z^n and take a sum over all integers n from $n = 0$ to $n = \infty$ we obtain

$$P(s, z) = P(0, z)F(s, z) + \delta_{s0},$$

or

$$F(s, z) = [P(s, z) - \delta_{s0}]/P(0, z). \tag{6.15}$$

The probability that the walker visits the point $s \neq 0$ at least once is determined by the value of $F(s, 1)$ as can be seen from eq. (6.13). The functions $F_n(s)$ play a decisive role in the description of exciton capture by impurities within the framework of the random walk theory. It is clear that if the impurity on the site s captures an exciton on its very first visit then it is precisely the quantity $F_n(s)$ that determines the probability of capture as a function of the walk time $t = nT$ where T is the mean time of one jump. If the trapping properties of the impurity are such that on the average $r \geqslant 1$ visits are needed for exciton capture then we have to employ the functions $F_n^{(r)}(s)$ which determine the probability that the rth visit to the point s occurs at the nth step. These functions are also defined by Montroll (1965).

As noted in ch. 5 sect. 3, the passage time T, for instance, for singlet excitons, is of the order of 10^{-11}–10^{-13} s. Thus, in pure crystals the total number of steps during the exciton lifetime is $N \sim 10^3$–10^5. If the crystal contains exciton traps (impurities, etc.) the exciton lifetime and, hence, the total number of exciton hops decrease. It is significant, however, that in most real situations the mean number of exciton hops before capture is still so large that we can use asymptotic expressions for the functions $P_n(s)$ and $F_n(s)$ for $n \gg 1$.

This is, of course, true for triplet excitons, too, since the mean number of hops before capture depends mainly on the exciton lifetime, impurity concentration and, to a lesser extent, on the characteristics of the interaction between exciton and impurity. Proceeding from the above discussion let us now find the form of the function $P_n(s)$ for large n. The function $F_n(s)$ will be discussed in the following section in connection with calculations of the rate of exciton capture by impurities.

Equation (6.12) shows that for large n the main contribution to the integral is made by small θ for which $\lambda(\theta)$ is the highest (close to 1). For instance, for the simple cubic lattice (see eq. (6.7)) we have in this region θ, $\lambda(\theta) \approx 1 - \theta^2/6 \approx \exp(-\theta^2/6)$ so that for $n \gg 1$ we obtain

$$P_n(s) \approx \frac{1}{(2\pi)^3} \iiint_{-\infty}^{\infty} \exp\left(-n\frac{\theta^2}{6} - \mathrm{i}s\boldsymbol{\theta}\right) \mathrm{d}\boldsymbol{\theta}.$$

Integration yields

$$P_n(s) = c(r, t) = \frac{1}{(2\sqrt{\pi D t})^3} \exp\left(-\frac{r^2}{4Dt}\right), \tag{6.16}$$

where $r = as$, a is the lattice parameter, $t = nT$, $D = a^2/6T$, and the function (6.16) satisfies the equation of diffusion

$$\frac{\partial c}{\partial t} = D\Delta c + I_0(r, t). \tag{6.17}$$

This function is known to give the concentration of migrating material at the point r at the moment t in the presence of a delta source $I_0(r, t) = \delta(r)\delta(t)$. Thus, for sufficiently high n values the random-walk function $P_n(s)$, indeed, satisfies the equation of diffusion. However, the right-hand side of eq. (6.17) lacks the term Pc entering into eq. (6.1) and corresponding to possible spontaneous decay of excitons. This is due to the use of the conservation condition (6.8) in the random-walk model of Polya (1921), Spitzer (1964), Montroll (1964, 1965, 1969). If the probability of disappearance of the walker during the passage time T differs from zero eq. (6.8) can be replaced by

$$\sum_s P(s) = e^{-T/\tau}, \tag{6.18}$$

where $\tau \equiv P_0^{-1}$ is the walker's lifetime and $\exp(-T/\tau)$ is the probability that the walker survives by the end of the step. Using eq. (6.18) we readily obtain for the structure function of the walk

$$\lambda(\theta) = e^{-T/\tau}\lambda_0(\theta), \tag{6.19}$$

where $\lambda_0(\theta)$ is the structure function ignoring the possible disappearance of the walker (that is, for $\tau = \infty$). Hence (see eq. (6.12)) we obtain

$$P_n(s) = e^{-nT/\tau} P_n^0(s), \tag{6.19a}$$

where $P_n^0(s)$ is the random-walk function found for $\tau = \infty$. Thus, for $n \gg 1$ and $\tau \neq \infty$ eq. (6.16) is replaced by

$$P_n(s) \approx c(r, t) = \frac{1}{(2\sqrt{\pi D t})^3} \exp\left(-\frac{r^2}{4Dt} - \frac{t}{\tau}\right), \tag{6.20}$$

This function satisfies eq. (6.1) with $P = P_0 = 1/\tau$. In a quite similar way we can find asymptotic expressions for $P_n(s)$ for lattices of a more complex structure. For instance, for an arbitrary simple rectangular lattice and small θ values we have

$$\lambda(\theta) = 1 - \tfrac{1}{2}(\sigma_1\theta_1^2 + \sigma_2\theta_2^2 + \sigma_3\theta_3^2), \tag{6.21}$$

where

$$\sigma_i^2 = \sum_s s_i^2 P(s), \qquad i = 1, 2, 3, \tag{6.22}$$

so that eq. (6.20) is replaced by

$$P_n(s) = c(r, t) = \frac{\exp(-t/\tau)}{(2\sqrt{\pi t})^3 \sqrt{D_1 D_2 D_3}} \exp\left[-\left(\frac{x^2}{4D_1 t} + \frac{y^2}{4D_2 t} + \frac{z^2}{4D_3 t}\right)\right],$$

where

$$D_i = a_i^2 \sigma_i^2, \qquad i = 1, 2, 3. \tag{6.23}$$

Equation (6.22) makes it possible, for instance to estimate for various models of resonance interaction how hops of lengths greater than the lattice parameter affect the coefficient of diffusion. It is clear, however, that when we take into consideration long-range hops but ignore spatial dispersion we have in case of the dipole–dipole interaction $P(s) \sim 1/s^6$ and the contribution of long-range hops accounts for just a few tenths of the value of the diffusion coefficient. As for shorter-range transfer mechanisms (for instance, the exchange interaction) the contribution of the long-range hops for them is even less significant.

In the above discussion we have always assumed that the exciton motion is three-dimensional. In fact, as noted above (see ch. 4 and 5 and also Agranovich (1968)), in a variety of crystals the resonance interaction between molecules in some spectral regions is sharply anisotropic. This can lead to considerable anisotropy of the exciton diffusion coefficient providing for two-dimensional or one-dimensional character of the exciton motion in the limiting cases. Hochstrasser et al. (1975) were the first to carry out experimental studies of quasi-one-dimensional triplet excitons, and in particular, energy transfer involving these excitons.

For instance, for such quasi-two-dimensional system in which the diffusion coefficient is isotropic in a plane we can replace eq. (6.1) with the system of equations

$$\frac{\partial c_n}{\partial t} = D\Delta^{(2)} c_n - Pc_n + \alpha(c_{n+1} + c_{n-1}) - 2\alpha c_n + I_n(\rho, t), \tag{6.24}$$

where $c_n(\rho, t)$ is the exciton concentration at the point $\rho \equiv (x, y)$ of the nth plane at the moment t, $\alpha \equiv \frac{1}{2}T^{-1}$ is the probability per unit time of exciton hopping to the neighbouring plane, and $I_n(\rho, t)$ is the intensity of generation of excitons in the plane n. If the conditions of excitation are such that $|c_n - c_{n-1}| \ll c_n$ then we obtain

$$\alpha(c_{n-1} + c_{n+1}) - 2\alpha c_n \approx D_\perp \frac{\mathrm{d}^2 c(\rho, z)}{\mathrm{d}z^2},$$

where the new coordinate $z = nd$, d is the distance between the nearest planes, and $D_\perp = \alpha d^2$ is the "transverse" coefficient of exciton diffusion. Clearly, transverse diffusion is not significant at all if $D_\perp \ll D$ and, moreover, if the exciton generation intensity I_n varies slowly when $n \to n+1$. The two-dimensional equation of diffusion can, naturally, be derived from the two-dimensional Boltzmann equation or from the equations describing the random walks in a plane. In the former approach eq. (6.4) is replaced by the following equation for the coefficient of diffusion in a plane:

$$D = \frac{\frac{1}{2} v l_s}{1 - \overline{\cos \Phi}}, \tag{6.25}$$

Here $\overline{\cos \Phi} = (1/2\pi) \int_0^{2\pi} \mu(\Phi) \cos \Phi \, d\Phi$, and $\mu(\Phi)/2\pi$ is the probability that scattering results in rotation of the exciton velocity by the angle Φ. Applying the random-walk theory, for instance, to the simple square lattice we obtain $D = a^2/4T$ where a is the lattice parameter, and T is the mean hop time.

2. *Theory of exciton capture by impurities; the case of low impurity concentrations*

It has been noted above that at not too high impurity concentrations the effects of the impurity are especially significant if it acts as the exciton trap. This leads to a decrease in the exciton lifetime in the crystal which in the framework of the equation for the mean concentration corresponds to an increase in the coefficient P of the term linear in the concentration (see eq. (6.1)). Usually, P is written as

$$P = P_0 + P_1 n, \tag{6.26}$$

where P_0 is the probability of exciton decay in the absence of impurity, n is the impurity concentration, and P_1 is a quantity independent of n for sufficiently low n. Thus, the expression (6.26) is based on the assumption that the dependence $P(n)$ is analytic for small n. This assumption is, indeed, valid for three-dimensional systems but not for two-dimensional and one-dimensional systems which will be discussed below. It should be noted here that P_1 can be determined only within a specific model of the motion of excitons and their interaction with impurities. Such a model is needed also for determination of the range of small impurity (acceptor) concentrations for which we can employ the expression for P which is linear in the impurity concentration.

Assume, at first, that excitons are incoherent and that the capture of an exciton which is at the point $\boldsymbol{r}$ by an impurity which is at the point $\boldsymbol{R}_i$ can be described by a function $W(|\boldsymbol{r}-\boldsymbol{R}_i|)$ equal, by definition, to the capture probability per unit time. Then we have a parametric dependence of the exciton concentration $\tilde{c} \equiv c(\boldsymbol{r}, t; \boldsymbol{R}_1, \boldsymbol{R}_2, \ldots)$ on the trap coordinates $(R_1, R_2, \ldots)$. In the diffusion approximation this function satisfies the equation

$$\frac{\partial \tilde{c}}{\partial t} = D\Delta\tilde{c} - P_0\tilde{c} - \sum_i W(\boldsymbol{r}-\boldsymbol{R}_i)\tilde{c} + I(\boldsymbol{r}, t), \tag{6.27}$$

where I is the intensity of the exciton source.

In most real situations the mean distance between the traps is much smaller than the characteristic lengths at which the variation of the intensity of exciton generation is noticeable. Therefore, we can always take such a volume V_0 of the crystal in which the source intensity is practically constant but the number of traps is sufficiently large. If the point $\boldsymbol{r}$ is inside the volume V_0 the function $\tilde{c}(\boldsymbol{r}, t; \boldsymbol{R}_1, \ldots, \boldsymbol{R}_N)$ depends essentially on the positions $(\boldsymbol{R}_1, \ldots, \boldsymbol{R}_N)$ of the traps in the volume V_0.

The mean concentration of excitons $c(\boldsymbol{r}, t)$ at the point $\boldsymbol{r}$ is, clearly, given by

$$c(\boldsymbol{r}, t) = \frac{1}{V_0}\int_{V_0} \tilde{c}(\boldsymbol{r}+\rho, t, \boldsymbol{R}_1, \ldots, \boldsymbol{R}_N)\,\mathrm{d}\boldsymbol{\rho}, \tag{6.28}$$

where $\boldsymbol{\rho}$ is the radius vector between an arbitrary point in the volume V_0 and the point $\boldsymbol{r}$. Applying eq. (6.28) to eq. (6.27) we obtain eq. (6.1) for the mean concentration where the total number of captures per unit volume is, by definition,

$$P_1 n c(\boldsymbol{r}, t) = \frac{1}{V_0}\int_{V_0} \sum_{i=1} W(\boldsymbol{r}+\boldsymbol{\rho}-\boldsymbol{R}_i)\tilde{c}(\boldsymbol{r}+\boldsymbol{\rho}, t, \boldsymbol{R}_1, \ldots, \boldsymbol{R}_N)\,\mathrm{d}\boldsymbol{\rho}, \tag{6.29}$$

where n is the trap concentration. Equations (6.28) and (6.29) are used as a basis for calculating the quantity P_1 entering into eq. (6.1) for the mean exciton concentration $c(\boldsymbol{r}, t)$. However, in order to employ these equations we must know, at least approximately, the function $\tilde{c}$ corresponding to the diffusion equation (6.27) in the presence of an enormous number of arbitrarily distributed sinks.

Below we shall ignore the fluctuations of the mean distance between the impurities and use a method for determination of the function $\tilde{c}$ and then the quantity P_1 which is similar to the Wigner–Seitz cell method in the theory of electronic states in crystals (these fluctuations make a significant contribution for very low impurity concentrations when $n \lesssim (1/\sqrt{D\tau})^3$; the relevant effects are discussed in sect. 4). Employing this method we shall assume that the traps are at centres of the spheres of the radius L (so that $4\pi L^3/3 = V_0/N = v_0$ where N is the total number of the traps in the volume V_0) and that these spheres are closely packed in the volume V_0. Then the function $\tilde{c}$ becomes a periodic function of $\boldsymbol{r}$ and eqs. (6.28) and (6.29) are replaced with

$$c(t) = \frac{1}{v_0} \int_{v_0} \tilde{c}(\boldsymbol{r}, t)\, \mathrm{d}\boldsymbol{r}, \tag{6.28a}$$

$$P_1 = \frac{1}{n} \sum_{i=1} \int_{v_0} W(\boldsymbol{r} - \boldsymbol{R}_i)\tilde{c}(\boldsymbol{r}, t)\, \mathrm{d}\boldsymbol{r} \Big/ \int_{v_0} \tilde{c}(\boldsymbol{r}, t)\, \mathrm{d}\boldsymbol{r}, \tag{6.29a}$$

where integrations are performed inside only one cell.

In most real cases the characteristic length R_0 for variation of the function $W(\boldsymbol{r})$ is small in comparison with the mean distance between the impurities. Therefore, in eq. (6.29a) we can take into account only the term which corresponds to the trap at the centre of the sphere v_0 for which we perform integration in eqs. (6.18a) and (6.29a). For the same reason, the function $\tilde{c}(\boldsymbol{r}, t)$ inside the cell can be found not from eq. (6.27) but from the equation

$$\frac{\partial \tilde{c}}{\partial t} = D\Delta\tilde{c} - P_0\tilde{c} - W(\boldsymbol{r})\tilde{c} + I(t). \tag{6.27a}$$

In experiments excitons are often generated under steady-state conditions (the excitation duration $T_0 \gg \tau$). Under such conditions the intensity I in eq. (6.27a) is independent of time and therefore the function $\tilde{c}$ is also independent of time. Hence, according to eqs. (6.28a) and (6.29a), under steady-state generation of excitons the quantity P_1 is independent of time and thus is equal to its steady-state value.

In the other limiting case, namely, that of a pulsed exciton source with $I = I_0\delta(t)$ (so that $T_0 \ll \tau$) we have for $t > 0$ the intensity $I = 0$ in eq. (6.27a) but the function $\tilde{c}$ is a function of time corresponding to the initial condition $\tilde{c}(\boldsymbol{r}, t \to 0) = c_0$. The quantity P_1 also depends on time and tends to its steady-state value $P_1(\infty)$ only for $t \to \infty$.

Since the variation of the intensity of the exciton source at distances of the order of L is assumed to be small we can ignore exciton flow between neighbouring cells when we calculate P_1. This implies that at the cell surface the exciton concentration satisfies the following boundary condition:

$$\left.\frac{\partial \tilde{c}}{\partial r}\right|_{r=L} = 0. \tag{6.30}$$

It follows from the above discussion that the boundary condition (6.30) significantly depends on the selected cell shape. However, for sufficiently low impurity concentrations ($L \gg R_0$) if we ignore their fluctuations we can assume that the cell shape in the case of close packing cannot have any significant effect on the calculated values of P_1. It should be also noted that for sufficiently low impurity concentrations the solution of eq. (6.27a) with the boundary condition (6.30) gives an especially accurate description of the behaviour of the function $\tilde{c}(r, t)$ precisely in the trap vicinity. As can be seen from eq. (6.28a), in this case the mean exciton concentration, to within small terms of the order of n, coincides with the asymptotic $\tilde{c}(r, t)$ which is obtained here even when $r > R_0$ but, at the same time, $r < L$ (R_0 is the radius of action of the trap and L is the cell radius). To find $\tilde{c}(r, t)$ here we can tend L to infinity and replace the boundary condition (6.30) with the condition $\tilde{c}(r, t) \to c(t)$ for $r \to \infty$ where $c(t)$ is the mean exciton concentration. In the case of pulsed excitation when $\tilde{c}(r, t) = c_0$ for $t = 0$ the mean concentration for $L \to \infty$ is $c_0 \exp(-P_0 t)$ so that we have $\tilde{c}(r, t) \to c_0 \exp(-P_0 t)$ for $r \to \infty$. Now, a convenient new function to use in eq. (6.27a) is $\psi(r, t) \equiv \tilde{c}(r, t) \exp(P_0 t)/c_0$ which satisfies the equation

$$\frac{\partial \psi}{\partial t} = D\Delta\psi - W(r)\psi, \tag{6.31}$$

Clearly, $\psi(r, t) = 1$ for $t = 0$. The solution of eq. (6.31) can be written as the sum (Shekhtman 1972):

$$\psi(r, t) = \psi^{st}(r) + \tilde{\psi}(r, t),$$

where $\psi^{st}(r)$ satisfies the stationary diffusion equation

$$D\Delta\psi^{st} - W(r)\psi^{st} = 0 \tag{6.32}$$

with the asymptotic condition $\psi^{st}(r) \to 1$ for $r \to \infty$ while $\tilde{\psi}$ satisfies the initial nonstationary equation for ψ with the starting condition $\tilde{\psi}(r, 0) = 1 - \psi^{st}(r)$.

According to eq. (6.29) we have

$$P_1 = \int W(r)\psi(r, t)\,\mathrm{d}r = P_1(\infty) + \Delta P_1(t),$$

where

$$P_1(\infty) = \int W(r)\psi^{st}(r)\,\mathrm{d}r \tag{6.33}$$

and

$$\Delta P_1(t) = \int W(r)\tilde{\psi}(r, t)\,\mathrm{d}r. \tag{6.34}$$

Since $\tilde{\psi}(r, t) \to 0$ for $t \to \infty$ the quantity $\Delta P_1(t)$ also tends to zero for $t \to \infty$.

We shall use the above results for calculations of P_1 for various trap models assuming that the trap concentration is low.

2.1. *Model of absolutely black sphere of capture*

This case is similar to that treated by the theory of coagulation (Chandrasekar 1943). As shown by Galanin and Chishikova (1956), in the case of exciton capture by impurities we can assume

$$P_1 = 4\pi DR, \tag{6.35}$$

where D is the exciton diffusion coefficient, and R is the radius of the sphere of capture around the impurity. Expression (6.35) corresponds to the assumption that all the excitons which have reached the surface of the sphere are captured by the impurity so that the sphere can be regarded as being absolutely black for them.

Despite the clear physical meaning of eq. (6.35), it is, in fact, a rather rough approximation; this expression is not general and it expresses the phenomenological parameter P_1 in terms of two parameters D and R which are not generally independent. It will be shown below that in a more rigourous treatment of exciton capture by impurities the calculations of P_1 formally based on eq. (6.35) often yield expressions for R which are dependent on the exciton diffusion coefficient D and some other parameters describing the system. To explain this and to derive expressions more accurate than eq. (6.35) let us discuss how eq. (6.35) is obtained in the black sphere model (see also ch. 2 sect. 3). Indeed, let an impurity located at the origin be represented by the absorbing sphere of the radius R and let the exciton concentration in the infinite volume of the crystal at the initial moment $t = 0$ be constant and equal to c_0. For $t > 0$ the exciton concentration should decrease owing to exciton capture by the impurity. To find the rate of capture we have to obtain the nonstationary solution of the exciton diffusion equation

$$\frac{\partial c}{\partial t} = D\Delta c - \frac{1}{\tau} c, \tag{6.36}$$

with the boundary condition $c(R, t) = 0$. Introducing a new function $\psi(r, t) = c(r, t) \exp(t/\tau)$ we replace eq. (6.36) with

$$\frac{\partial \psi}{\partial t} = D\Delta\psi, \tag{6.37}$$

while the initial and boundary conditions remain unchanged. The solution of this equation is

$$\psi(r, t) = c_0\left[1 - \frac{R}{r} + \frac{2R}{r\sqrt{\pi}} \int_0^{\frac{r-R}{2\sqrt{Dt}}} \exp(-x^2)\,dx\right], \tag{6.38}$$

so that the number of exciton captures per 1 s, equal to the total fiux of excitons at the surface of the capture sphere, is

$$4\pi R^2(-D\nabla c)_{r=R_0} = 4\pi DR_0\left[1+\frac{R_0}{(\pi Dt)^{1/2}}\right]c_0\,e^{-t/\tau}. \qquad (6.39)$$

In this case of one trap the exciton concentration averaged over the volume for any finite time t, that is, the quantity

$$\bar{c}(r,t) \equiv \frac{1}{V}\int c(r,t)\,dv$$

is, clearly, equal to $c_0\exp(-t/\tau)$ to a great accuracy. This implies that the number of exciton captures is proportional to the mean exciton concentration and the proportionality factor is

$$P_1(t) = 4\pi DR\left[1+\frac{R}{(\pi Dt)^{1/2}}\right]. \qquad (6.40)$$

Thus, eq. (6.35) determines only $P_1(\infty)$, that is, the asymptotic value of $P_1(t)$ for $t \gg R^2/\pi D$. It can be easily seen, however, that eqs. (6.35) and (6.40) are not general in character, even in the black sphere model.

Indeed, since we derived eq. (6.40) using the solution of the exciton diffusion equation with the appropriate boundary condition at the surface of the sphere we thus assumed that the radius of the sphere was large in comparison with the exciton mean free path l. (This discussion is also useful for determining the applicability of the results given in ch. 2 sect. 2. In particular, the "collision" radius defined there should be much greater than the mean hop length for the migrating molecules.) If, on the contrary, $l \gg R$ as is the case for excitons at low temperatures in some crystals, then the number of captures by one impurity per 1 s is no longer dependent on the exciton mean free path l and is equal to the value given by gas kinetics, that is, $Svc(r,t)/4$ where $S = 4\pi R^2$ is the surface area of the sphere of capture, v is the mean exciton velocity, and $c(r,t)$ is the exciton concentration so that we have

$$P_1 = \pi R^2 v. \qquad (6.41)$$

Equations (6.40) and (6.41) determine P_1 only in the above limiting cases so that a natural question is how to calculate P_1 for $l \approx R$. To do this we must go outside the framework of the diffusion equation. For free (coherent) excitons whose distribution function satisfies the Boltzmann equation the appropriate approximate solution can be taken from the theory of nuclear reactors. Indeed, if we ignore the anisotropy of the exciton masses and the exciton–phonon interactions and assume that all excitons have the same velocity $v = v_T$ (where v_T is the mean thermal velocity) and the same mean free path l the kinetic equation for excitons is identical to the one-velocity equation derived by Peierls. Using the solutions of this equation (see, e.g., Galanin 1957) we can show that within the framework of the diffusion equation we can obtain the correct results for P_1 in the black sphere model for the intermediate values of the ratio l/R if we replace the condition $c(R, t) = 0$ at the surface of the sphere of the radius R with the so-called effective boundary condition

$$c = \gamma l \left.\frac{\mathrm{d}c}{\mathrm{d}r}\right|_{r=R},$$

where γ is a function of the ratio l/R. The plot of this function is shown in fig. 6.1 where it can be seen that $\gamma \to 0$ for $R/l \gg 1$ and $\gamma = \frac{4}{3}$ for $R/l \ll 1$.

Using this effective boundary condition we obtain

$$P_1(t) = \frac{4\pi DR^2}{\gamma l}\left\{1 - \left(1 + \gamma\frac{l}{R}\right)^{-1}[1 - \mathrm{e}^{\alpha^2 t}\,\mathrm{erfc}\,(\alpha\sqrt{t})]\right\}, \tag{6.42}$$

where

$$\alpha = \frac{\sqrt{D}}{\gamma l}\left(1 + \gamma\frac{l}{R}\right), \qquad \mathrm{erfc}\,(x) = \frac{2}{\sqrt{\pi}}\int_x^\infty \mathrm{e}^{-t^2}\,\mathrm{d}t.$$

It should be borne in mind that the term $\exp(\alpha^2 t)\,\mathrm{erfc}\,(\alpha\sqrt{t}) \to 0$ for $\alpha \to \infty$. The same is true for $t \to \infty$ so that we have

$$P_1(\infty) = 4\pi DR \Big/ \left(1 + \gamma\frac{l}{R}\right). \tag{6.42a}$$

When $l/R \ll 1$ eq. (6.42) reduces to eq. (6.40). In the other limiting case

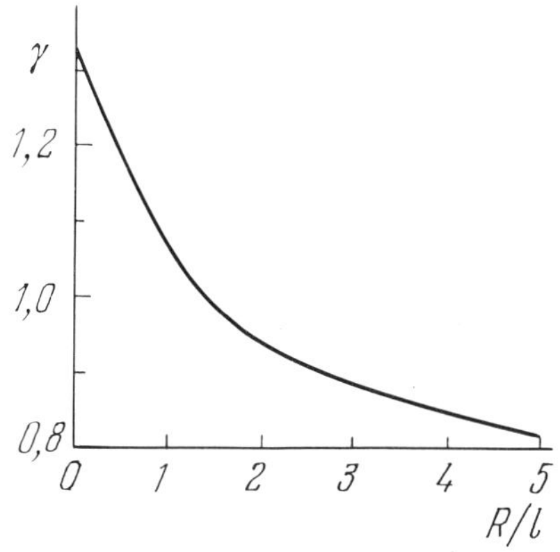

Fig. 6.1. The dependence of γ on the ratio R/l. R is the radius of the sphere of capture, l is the mean free path $\gamma(0) = \frac{4}{3}$.

when $l/R \gg 1$ the quantity $P_1(t)$ is practically independent of time and given by eq. (6.41) since $\gamma(l/R) \to \frac{4}{3}$ for $l/R \gg 1$ (see fig. 6.1).

We have used the model of coherent (free) excitons to obtain eq. (6.42) for the rate of capture of excitons by the impurity with radius R. However, at high temperatures excitons are strongly scattered by the lattice vibrations and excitons are better described by the model of random walks (hops) as shown in ch. 4. The description of electronic excitation energy transfer in pure (defect-free) crystals within the framework of the theory of random walks is discussed in sect. 2. For long migration times this theory yields asymptotically the diffusion equation. Therefore, eq. (6.40) where D is the diffusion coefficient (6.23) of incoherent excitons remains valid in the model of the black sphere under the condition $R \gg a$ where a is the parameter of the crystal lattice (in the random walks model the mean hop length which is an analogue of the mean free path is approximately equal to a; see sect. 1). If $R \leqslant a$ eq. (6.40) is no longer valid and of course, eq. (6.42) derived for coherent excitons is not applicable, too.

2.2. *Exciton capture in the random walk model; the trap model of Powell and Soos*

At first, let us analyze the simplest trap model. We shall assume that an impurity molecule capable of trapping excitons is on a lattice site and that an exciton is trapped if after the step n of random walk it gets to the site $s = 0$. In full accordance with the starting conditions for eq. (6.40), we shall assume that at the moment $t = 0$ the exciton has equal probabilities $1/N$ to be on any of the sites $s \neq 0$ where $N + 1$ is the total number of molecules in the crystal ($N \gg 1$). Then the probability that the exciton is trapped after the nth step is

$$G_n = \Delta_n/N, \tag{6.43}$$

where

$$\Delta_n \equiv \sum_{s\neq 0} F_n(s).$$

The generating function corresponding to Δ_n,

$$\Delta(z) = \sum_{n=1}^{\infty} \Delta_n z^n, \tag{6.44}$$

according to eq. (6.13) is equal to

$$\Delta(z) = \sum_{s=0}^{\infty} F(s, z).$$

Using eq. (6.15) we can rewrite this as

$$\Delta(z) = \left[\sum_{s=0}^{\infty} P(s, z)/P(0, z) - 1\right]. \tag{6.45}$$

On the other hand, eqs. (6.9) and (6.19) yield

$$\sum_{s=0}^{\infty} P(s, z) = \frac{1}{1 - \alpha z}, \qquad \alpha = e^{-T/\tau},$$

so that we obtain

$$\Delta(z) = -1 + [(1 - \alpha z)P(0, z)]^{-1}. \tag{6.46}$$

Generally, the expression for the function $\Delta(z)$ has a complicated form and cannot be expanded in powers of z. But since we are interested in Δ_n values for $n \gg 1$ we can replace the function $\Delta(z)$ with its asymptotic expression for $z \to 1$. The function $P(0, z)$ entering into eq. (6.45) can be found from eq. (6.11) and has the form

$$P(0, z) = \frac{1}{8\pi^3} \iiint_{-\pi}^{\pi} \frac{d\theta}{1 - z\lambda(\theta)}. \tag{6.47}$$

It remains finite for $z \to 1$. Therefore, for these values of z we obtain

$$\Delta(z) \approx -1 + \frac{1}{P(0, 1)} \frac{1}{1 - \alpha z}, \tag{6.48}$$

or

$$\Delta(z) \approx -1 + \frac{1}{P(0, 1)} \sum_{n=0}^{\infty} z^n e^{-nT/\tau},$$

so that for $n \gg 1$ we have

$$\Delta_n = \frac{1}{P(0, 1)} e^{-nT/\tau}. \tag{6.49}$$

Substituting eq. (6.49) into eq. (6.43) we find

$$G_n \approx \frac{v}{P(0, 1)} c(n),$$

where v is the crystal volume per one molecule, $c(n) = \exp(-nT/\tau)V^{-1}$ is the mean exciton concentration at the moment $t = nT$, and $V = (N + 1)v$ is the crystal volume.

If we multiply G_n by the number of traps n per unit volume and divide the result by the hop time T we obtain the number of trapped excitons per 1 s per unit volume at the moment t. Comparing the resulting expression with the product $P_1 c(\mathbf{r}, t)$ we see that in this model,

$$P_1 = v/TP(0, 1). \tag{6.50}$$

The quantity $P(0, 1)$ significantly depends on the crystal structure; for instance, for the simple cubic lattice $P(0, 1) \approx 1.52$, for the body-centred lattice $P(0, 1) \approx 1.40$, and for the face-centred lattice $P(0, 1) \approx 1.34$ (Watson 1939). It is clear, however, that eq. (6.50) can always be written as

$$P_1 = 4\pi DR,$$

where D is the exciton diffusion coefficient in the model of random walks, and R is of the order of the lattice parameter (for the simple cubic lattice $R \approx \gamma a$ where a is the lattice parameter and $\gamma = 4\pi/6P(0, 1) \approx 4\pi/9$). Here R can be regarded as the effective radius of capture.

Thus, eq. (6.50) is a generalization of eq. (6.35) for the case when the effective radius of capture is of the order of the lattice parameter. The fact that here P_1 is independent of t is explained by the approximation for $\Delta(z)$ being insufficiently accurate as $z \to 1$. Being back to eq. (6.46) we can express $P(0, z)$ as the sum

$$P(0, z) = P(0, 1) - \delta, \tag{6.51}$$

where

$$\delta = \frac{1 - z}{(2\pi)^3} \iiint_{-\pi}^{\pi} \frac{\lambda(\theta)\, d\theta}{[1 - \lambda(\theta)][1 - z\lambda(\theta)]}. \tag{6.52}$$

For $z \to 1$ the main contribution to the integral (6.52) is made by small θ values. Therefore, we can employ the approximation (6.21) for $\lambda(\theta)$ and show that (Montroll 1965):

$$\delta \approx \frac{[\frac{1}{2}(1 - z)]^{1/2}}{\pi \sigma_1 \sigma_2 \sigma_3}. \tag{6.53}$$

As shown by Montroll (1965), the use of this approximation for $P(0, z)$ in eq. (6.46) leads to

$$\Delta_n = E\left[1 + \frac{2E}{(2\pi)^{3/2}\sigma_1\sigma_2\sigma_3} \frac{1}{\sqrt{n}} + \cdots\right], \tag{6.54}$$

where the terms of the order of $1/n$ are omitted and $E = 1/P(0, 1)$ is sometimes referred to as the power of the trap.

Interestingly, eq. (6.54) has the same form even for traps of more complicated structure when an exciton is captured not only on the site $s = 0$ but also on A arbitrary neighbouring sites (the trap model of Soos and Powell discussed by Soos and Powell (1972)). Neither is the form, eq. (6.54), changed when we take into consideration possible trapping of

Table 6.1
Values of E for simple cubic lattice (Maradudin et al. 1960, Vineyard 1963).

Trap characteristics		Random walks	
Volume $A+1$	Shape	Isotropic ($\rho = 1$)	Anisotropic ($\rho = 8$)
1	Point	0.659	0.459
	Spherical	1.936	1.337
7	Linear (x or y)	2.263	1.749
	Linear (z)	2.263	1.030
	Spherical	3.156	2.067
27	Linear (x or y)	6.213	4.942
	Linear (z)	6.213	2.452

excitons, for instance, in the presence of long-range resonance interaction. In all these cases only the trap power E is changed (Soos and Powell 1972). Table 6.1 presents the results of the calculations reported by Vineyard (1963) Maradudin et al. (1960), which demonstrate how the trap power depends on the shape and dimensions of the trap and the character of the exciton-trap interaction. Table 6.2 presents the results of the calculations of Soos and Powell (1972) of the trap power for the simple cubic lattice under the conditions when excitons can hop only to the nearest neighbours, but the hop probability depends on the hop direction so that the structure function (compare with eq. (6.7)) is

$$\lambda(\theta) = \frac{1}{2+\rho}(\cos\theta_1 + \cos\theta_2 + \rho\cos\theta_3).$$

If the probability of hops of the length r varies as r^{-6} then for the isotropic

Table 6.2
Dependence of E on the maximum step length (pa) and trap size in the simple cubic lattice.

			E	
p	$\beta(p)$	$3\sigma^2(p)$	$A = 0$	$A = 26$
1	1.000	1.000	0.659	3.156
$\sqrt{6}$	1.365	1.451	0.763	4.381
$\sqrt{14}$	1.389	1.600	0.784	4.963
$\sqrt{27}$	1.396	1.690	0.791	5.161
$\sqrt{29}$	1.400	1.782	0.795	5.297

model we obtain

$$P(s)=P(|s|)=(6\beta s^6)^{-1}, \qquad \beta=1+\tfrac{1}{6}\sum s^{-6}, \qquad s>1,$$

$$\sigma^2=\frac{1}{18\beta}\sum_s s^{-4}, \qquad D=\tfrac{1}{2}\sigma^2a^2/T.$$

and this leads to some increase in E.

If we go over from Δ_n to G_n and find P_1 as discussed above we can obtain from eq. (6.54)

$$P_1(t)=\frac{vE}{T}\left[1+\frac{2E\sqrt{T}}{(2\pi)^{3/2}\sigma_1\sigma_2\sigma_3}\frac{1}{\sqrt{t}}+\cdots\right],$$

or

$$P_1(t)=4\pi DR\left[1+\frac{R}{(\pi Dt)^{1/2}}+\cdots\right], \tag{6.55}$$

where

$$R=\frac{E(A)a}{2\pi\sigma^4}, \qquad D=\frac{\sigma^2a^2}{2T}=(D_1D_2D_3)^{1/3},$$

$$\sigma=\sqrt[3]{\sigma_1\sigma_2\sigma_3}, \qquad a=v^{1/3}, \qquad D_i=\frac{\sigma_i^2a_i^2}{2T}, \qquad i=1,2,3. \tag{6.56}$$

The first two terms of the expansion of $P_1(t)$ in powers of $1/\sqrt{t}$ have the same form as those obtained from the diffusion equation (see eq. (6.40)). Therefore, we can regard eq. (6.55) as a generalization of eq. (6.40) for localized excitons in the case of small effective trapping radius R and for the case of anisotropic traps of arbitrary shape. At the same time, the derivation of eq. (6.55) implies that eqs. (6.40) and (6.55) are not valid for small times t when $R(DT)^{-1/2}\sim 1$ or larger. We shall estimate the region of applicability of eqs. (6.40) and (6.55) below in connection with the analysis of the experimental data on the kinetics of luminescence of the mixed crystals. Here we shall only note that the probability of hops to impurity molecules can differ significantly from the probabilities of hops between the host molecules. This fact has not been taken into account by Soos and Powell (1972) so that it is difficult to apply these results for analysis of the dependence of the effective trapping radius R (see eq. (6.56)) on the diffusion coefficient of excitations in the matrix. However, it would be interesting to compare the results of such analysis with the experimental data on the temperature dependence of the effective radius of exciton trapping by impurities (see ch. 7) and the data on the energy transfer from the donor impurity molecules to the acceptor impurity molecules. In the latter case the mean length of exciton hops via donors is determined by their concentration so that the trapping radius depends on the concentration of donors.

2.3. *Trapping of incoherent excitons in the case of multipole exciton–acceptor interaction**

For multipole interaction we have $W(r) = P_0(R_0/r)^{2m}$ where $m = 3, 4, 5$ for the dipole–dipole, dipole–quadrupole and quadrupole–quadrupole interaction, respectively. The first to find the effective trapping radius for the case of the dipole–dipole interaction ($m = 3$) were Yokota and Tanimoto (1967) who employed the scattering length method. We shall also consider the case of $m > 3$ (Galanin 1977).

According to eq. (6.32), the function $\psi^{st}(r)$ in this case satisfies the equation

$$\frac{d^2y}{dr^2} - \frac{1}{\lambda^2}\left(\frac{R_0}{r}\right)^{2m} y = 0, \tag{6.57}$$

where $\lambda = \sqrt{D/P_0}$ is the exciton diffusion length in the crystal without traps, and $y(r) = r\psi^{st}(r)$. Equation (6.57) reduces to the Bessel equation (Kamke 1959). This equation with the conditions $\psi^{st}(r) \to 1$ for $r \to \infty$ and $\psi^{st}(r) \to 0$ for $r \to 0$ has the solution

$$\psi^{st}(r) = \frac{a^\alpha}{\pi} \sin \alpha\pi \Gamma(1-\alpha) \sqrt{\frac{R_D}{r}} \mathcal{K}_\alpha \left[2\alpha \left(\frac{R_D}{r}\right)^{1/2\alpha} \right], \tag{6.58}$$

where $\alpha = \frac{1}{2}(m-1)$, $\Gamma(x)$ is the gamma function, and

$$\mathcal{K}_\alpha(z) = \frac{\pi}{2 \sin \alpha\pi} [I_{-\alpha}(z) - I_\alpha(z)],$$

is the MacDonald function in which $I_\alpha(z)$ is the Bessel function of the imaginary argument, and the radius

$$R_D = \left(\frac{P_0 R_0^{2m}}{D}\right)^{1/2(m-1)}. \tag{6.59}$$

Using eqs. (6.58) and (6.33) we obtain

$$P_1(\infty) = 4\pi \int_0^\infty W(r)\psi^{st}(r) r^2 \, dr = 4\pi D R_D A(\alpha), \tag{6.60}$$

where the factor

$$A(\alpha) = \alpha^{2\alpha-1}\Gamma(1-\alpha)/\Gamma(\alpha) \tag{6.61}$$

*See also ch. 2 sect. 2 for the results in the case of the dipole–dipole interaction.

is equal to 0.676, 0.669 and 0.688 for the dipole–dipole, dipole–quadrupole and quadrupole–quadrupole interaction, respectively. Since the factor (6.61) is close to unity the radius R_D plays the part of the effective trapping radius (see also eq. (6.35)). Equation (6.59) can also be rewritten as

$$\frac{R_D^2}{D} = \frac{1}{P_0}\left(\frac{R_D}{R_0}\right)^{2m}.$$

The left-hand side of this equation is equal to the mean time of diffusion of a particle in the region with the linear dimension R_D. The right-hand side is equal, by the order of magnitude, to the time $\tau(R) = 1/W(R)$, averaged over the sphere with the radius R_D, during which a trap captures an exciton that is at the distance R from it. Thus, we can derive a result identical to eq. (6.60) to within a numerical factor proceeding from purely qualitative considerations.

In the case of the dipole–dipole interaction between the exciton and the trap (that is, for $m = 3$) eq. (6.59) yields $R_D \sim D^{-1/4}$ so that $P_1(\infty)$ is proportional to $D^{3/4}$ as in the case of energy transfer in liquid solutions (see ch. 2). Note that a similar result was obtained earlier (de Gennes 1958, Lowe and Tse 1968, Khutsishvili 1954, 1956) in the theory of spin-lattice relaxation for the case of spin diffusion in the presence of paramagnetic impurities.

3. *Probability of exciton trapping by impurities in the Wigner–Seitz cell method*

When the mean distance between the impurity molecules is not large in comparison with the diffusion length $\lambda = \sqrt{D\tau}$ of excitons in the pure crystal the probability of exciton trapping by impurities, and hence, the quantity P given by eq. (6.26), cannot be regarded as a linear function of the concentration of traps even in the three-dimensional medium. The accompanying nonadditive effects of exciton trapping can be taken into account within the framework of the Wigner–Seitz method discussed above. We can calculate P ignoring the variation of the intensity of the exciton source and redistribution of excitons between neighbouring cells. This means that the exciton concentration c at the cell surface satisfies the boundary condition (6.30). The exciton concentration $\tilde{c}(\boldsymbol{r}, t)$ inside the cell should satisfy eq. (6.27a). For the sake of simplicity we shall analyze only the steady-state case (and assume that $I_0 = 1$ which does not limit generality since the problem is linear). Then the equation for the exciton concentration can be written as

$$D\Delta\tilde{c} - P_0\tilde{c} - W(r)\tilde{c} + 1 = 0. \tag{6.62}$$

Since under the steady-state conditions the number of generated excitons is equal to the number of disappearing excitons we have

$$P_0 \int_{v_0} \tilde{c}(\boldsymbol{r})\,\mathrm{d}\boldsymbol{r} + \int_{v_0} W(r)\tilde{c}(\boldsymbol{r})\,\mathrm{d}\boldsymbol{r} = v_0, \tag{6.63}$$

Using eq. (6.29a) we obtain now

$$P \equiv P_0 + P_1 n = \left[\frac{3}{L^3}\int_0^L \tilde{c}(r) r^2\,\mathrm{d}r\right]^{-1}. \tag{6.64}$$

We shall employ eq. (6.64) for calculating P_1 for various trapping models.

3.1. *Black sphere model: three-dimensional diffusion*

In this model eq. (6.62) can be rewritten as

$$D\Delta\tilde{c} - P_0\tilde{c} + 1 = 0, \tag{6.65}$$

Apart from eq. (6.62), the function $\tilde{c}(r)$ at the trapping sphere should satisfy the condition

$$\frac{1}{\gamma l} = \frac{1}{\tilde{c}}\frac{\mathrm{d}\tilde{c}}{\mathrm{d}r}\bigg|_{r=R}, \tag{6.66}$$

which has been discussed in sect. 2. If the diffusion length, that is, the quantity $\sqrt{D/P_0}$, is considerably greater than the mean distance between the impurities then we can omit the term $-P_0\tilde{c}$ in eq. (6.62). Now the right-hand side of eq. (6.64) determines not the entire P but only its part which is due to trapping of excitons (that is, $P - P_0$). For $P_0 = 0$ eq. (6.65) with the boundary conditions (6.30) and (6.66) has the following solution:

$$\tilde{c}(r) = -r^2/6D + A_1/r + A_2.$$

Here

$$A_1 = -\frac{L^3}{3D}, \qquad A_2 = \frac{1}{3D}\left[\frac{L^3}{R^2}(R + \gamma l) + R\left(\frac{R}{2} - \gamma l\right)\right],$$

so that according to eq. (6.64) and the above results we have

$$(P - P_0)^{-1} = \frac{3}{L^3}\left[\frac{L^5 - R^5}{30D} + \frac{A_1}{2}(L^2 - R^2) + \frac{A_2}{3}(L^3 - R^3)\right].$$

For instance, for $R/L \ll 1$ we obtain to within the terms of the order R/L inclusively

$$P - P_0 \approx \frac{4\pi DRn}{1 + \gamma l/R}\left[1 + \frac{9R(4\pi n/3)^{1/3}}{5(1 + \gamma l/R)} + \cdots\right]. \tag{6.67}$$

A nonlinear dependence of R on n can manifest itself only for $n \gtrsim n_0 = 3\pi R^3/4$. If $R \approx 30$ Å then $n_0 \approx 10^{19}$ cm^{-3}. Thus, the values of n_0 lie within the range of impurity concentrations used in experimental studies. However, we know of no experiments in which this nonlinear dependence of the exciton lifetime on the concentration of the impurity (acceptors) has been taken into consideration. It can be seen from eq. (6.67) that analysis of this effect can, in principle, yield additional data on the trapping radius R and the mean free path l.

A problem which is formally similar to the one discussed above is encountered in the analysis of the rate of capture of the impurity atoms by Frenkel pairs (see, for instance, Strieder and Aris (1973) and the references given in it). Recent results on this problem are reported by Gösele (1978), Kalnin (1981). According to Kalnin (1981), eq. (6.67) giving the correction term of the order of $n^{1/3}$ is valid only for such distributions of traps that possess, at least, short-range order. For the statistically uniform distribution of acceptors the correction term of the order of $n^{1/3}$ disappears and if we take into account the "forbidden" volume we obtain a correction term of the order of n, that is, of a higher order of smallness. This large difference between the dependences of the exciton trapping rates on the impurity concentration which is related to the differences in distributions of the impurities could, apparently, be used for analyzing these distributions proceeding from the experimental data on the dependence of $P - P_0$ on the trap concentration n.

3.2. *Long-range trapping mechanism: three-dimensional diffusion*

When trapping of localized (incoherent) excitons is due to the dipole–dipole interaction a more rigourous treatment gives $W(r) = P_0(R_0/r)^6$ in eq. (6.62) where R_0 is given by the Förster formula (see ch. 1). The results of this treatment for the steady-state case are presented by Agranovich (1968), Agranovich et al. (1973). For instance, numerical solution of the diffusion equation yielded the dimensionless function $f(R_0/\lambda, L/R_0)$ which determines

$$P = P_0/f\left(\frac{R_0}{\lambda}, \frac{L}{R_0}\right),$$

where $\lambda = \sqrt{D/P_0}$ is the diffusion length for excitons in the absence of impurities (traps). Table 6.3 gives some values of f. The analytic expression

$$f\left(\frac{R_0}{\lambda}, \frac{L}{R_0}\right) = \left[1 + 2.04\left(\frac{R_0}{L}\right)^3\left(\frac{\lambda}{R_0}\right)^{3/2}\right]^{-1},$$

found by Golubov and Konobeev (1975) gives a fairly good approximation of f for $L/R_0 \gg 1$.

Table 6.3
Function $f(R_0/\lambda, L/R_0)$.

$\frac{L}{R_0}$	R_0/λ 0.001	0.004	0.01	0.04	0.1	0.2	0.4	1
1	0.0000	0.0001	0.0004	0.0031	0.0100	0.0228	0.0471	0.0995
1.1	0.0000	0.0001	0.0006	0.0043	0.0138	0.0320	0.0670	0.1434
1.2	0.0000	0.0002	0.0008	0.0056	0.0185	0.0432	0.0912	0.1949
1.3	0.0000	0.0003	0.0010	0.0073	0.0240	0.0565	0.1194	0.2521
1.4	0.0000	0.0003	0.0012	0.0089	0.0306	0.0720	0.1515	0.3126
1.6	0.0000	0.0005	0.0019	0.0135	0.0467	0.1094	0.2247	0.4340
1.8	0.0000	0.0007	0.0027	0.0195	0.0672	0.1547	0.3056	0.5446
2.0	0.0001	0.0010	0.0037	0.0270	0.0919	0.2065	0.3883	0.6374
3.0	0.0004	0.0033	0.0126	0.0891	0.2686	0.4934	0.7120	0.8760
4.0	0.0010	0.0078	0.0298	0.1917	0.4740	0.7080	0.8620	0.9468
6.0	0.0033	0.0259	0.0945	0.4497	0.7587	0.8954	0.9569	0.9841
8.0	0.0079	0.0595	0.1991	0.6619	0.8834	0.9539	0.9817	0.9933
10.0	0.0153	0.1101	0.3275	0.7937	0.9372	0.9761	0.9906	0.9965
12.0	0.0262	0.1765	0.4575	0.8697	0.9630	0.9861	0.9946	0.9980
14.0	0.0409	0.2540	0.5729	0.9140	0.9764	0.9913	0.9966	0.9987
16.0	0.0599	0.3371	0.6672	0.9408	0.9841	0.9942	0.9977	0.9991
18.0	0.0832	0.4202	0.7408	0.9578	0.9888	0.9960	0.9984	0.9994
20.0	0.1107	0.4986	0.7969	0.9689	0.9920	0.9970	0.9988	0.9996
30.0	0.2959	0.7699	0.9303	0.9906	0.9976	0.9991	0.9996	0.9998
40.0	0.4992	0.8885	0.9695	0.9960	0.9991	0.9996	0.9998	0.9999
80.0	0.8886	0.9846	0.9961	0.9995	0.9999	0.9999	0.9999	1.0000
100.0	0.9398	0.9920	0.9980	0.9998	0.9999	0.9999	0.9999	1.0000

Since the quantum yield of the donor luminescence is $\eta_D = P_0/P = f(R_0/\lambda, L/R_0)$ and the quantum yield of the acceptor luminescence is $\eta_A = 1 - \eta_D$ (when quenching does not occur or is neglected) the above approximate expression for $R_0/L \ll 1$ (low impurity concentrations) gives

$$\eta_A = 2.04 \left(\frac{R_0}{L}\right)^3 \left(\frac{\lambda}{R_0}\right)^{3/2}.$$

We can write $L^3 = 3\pi n_A/4$ where n_A is the concentration of acceptors, and $\lambda = \sqrt{D\tau_0}$ where D and τ_0 are the diffusion coefficient and the lifetime of excitons, and thus obtain

$$\eta_A = 0.68\, 4\pi R_0^{3/2}(D\tau_0)^{3/4} n_A.$$

We have come to the relationship $\eta_A \sim D^{3/4}$ which practically does not differ from the exact result of sect. 2.3.

3.3. *Black circumference (radius R) model: two-dimensional diffusion*

In the case of two-dimensional or quasi-two-dimensional systems (see sect. 1) with low transverse diffusion coefficient $D_\perp \ll D$ (see eq. (6.24)) we can calculate P assuming that diffusion to traps in the plane plays a predominant role. Under such conditions the Wigner–Seitz cell is a plane figure formed by two concentric circumferences with the radii R and L where $L = (4\pi dn)^{-1/2}$, n is the number of the impurity molecules per unit volume, and d is the distance between the nearest planes (layers) in which the diffusion coefficient is high. In this case eq. (6.65) is replaced by

$$D\left(\frac{\mathrm{d}^2\tilde{c}}{\mathrm{d}r} + \frac{1}{r}\frac{\mathrm{d}\tilde{c}}{\mathrm{d}r}\right) - P_0\tilde{c} + 1 = 0, \tag{6.68}$$

with the boundary condition

$$\tilde{c}(R) = 0, \qquad \left.\frac{\mathrm{d}\tilde{c}}{\mathrm{d}r}\right|_L = 0. \tag{6.69}$$

(we assume here that R is much greater than the mean free path of excitons). The solution of eq. (6.68) can be expressed in terms of the Bessel functions.

To illustrate the results we shall limit ourselves to treatment of the case of a not too low impurity concentration when L is small in comparison with the exciton diffusion length $\lambda = \sqrt{D/P_0}$. Now we can omit the term proportional to P_0 in eq. (6.68) so that the solution of the simplified equation with the boundary condition (6.69) is

$$\tilde{c}(r) = \frac{L^2}{2D}\ln\frac{r}{R} - \frac{r^2 - R^2}{4D}. \tag{6.70}$$

The exciton concentration averaged over the cell is

$$\frac{1}{2L^2}\int_R^L \tilde{c}(r) r\,\mathrm{d}r,$$

and the probability of exciton disappearance due to trapping is

$$P - P_0 = \frac{8D}{R^2} f(x), \tag{6.71}$$

where

$$x = L/R, \qquad f(x) = x^2[x^4 \ln x - \tfrac{3}{4}x^4 + x^2 - \tfrac{1}{4}]^{-1}. \tag{6.71a}$$

Let us find the asymptotic expression for P in the case of low impurity concentrations when $x = L/R \gg 1$. Using eqs. (6.71) and (6.71a) we obtain for this limiting case

$$P - P_0 = \frac{64\pi D n_s}{|\ln 4\pi R^2 n_s| - 3/2}, \tag{6.72}$$

where $n_s = nd$ is the number of the impurity molecules (traps) per 1 cm^2 of the plane in which the diffusion coefficient is high.

The most important feature of eq. (6.72) is that the trapping probability is no longer an analytic function of the trap concentration n_s. In contrast to the case of three-dimensional diffusion (see eq. (6.67)), the dependence of $P - P_0$ on the trapping radius R in the two-dimensional case is logarithmic (so that it is weaker). This fact can be very significant for the analysis of the temperature dependence of the exciton diffusion coefficient in two-dimensional and quasi-two-dimensional systems in which the temperature dependence of $P - P_0$ is mainly determined by the temperature dependence of the diffusion coefficient.

Furthermore, as shown by eq. (6.72), the data on the effect of trapping on the exciton lifetime at various low impurity concentrations makes it possible to determine not only the product DR (see eq. (6.67)) but, in contrast to the case of three-dimensional diffusion, to find separately D and R. The results on trapping rates in one-dimensional and two-dimensional lattices obtained with the random walks theory are reported by Rosenstock (1970), Ryazanov (1972).

3.4. *Long-range trapping mechanism: two-dimensional diffusion*

To find P in this case we have to determine the solution of the equation

$$D\left(\frac{d^2c}{dr^2} + \frac{1}{2}\frac{dc}{dr}\right) - P_0 c - P_0\left(\frac{R_0}{r}\right)^6 c + 1 = 0, \tag{6.73}$$

within the circle of the radius L with the boundary condition $dc/dr = 0$ for $r = L$. Such a solution was found by Ivanova by means of numeric integration with a computer. These calculations first yielded the concentration averaged over the cell and then the value of

$$P = P_0/f\left(\frac{L}{R_0}, \frac{R_0}{\lambda}\right).$$

Table 6.4 presents the values of the dimensionless function f for some ratios R_0/λ and L/R_0. These data show that for large $L/R_0 \gg 1$ (that is, for low impurity concentrations) the difference $P - P_0$ is approximated by eq. (6.72) with $R^2 \approx R_0^2$.

Table 6.4
Function $f(L/R_0, R_0/\lambda)$: Two-dimensional diffusion.

	R_0/λ							
$\frac{L}{R_0}$	0.001	0.004	0.01	0.04	0.1	0.2	0.4	1
1	0.0000	0.0000	0.0000	0.0010	0.0044	0.0118	0.0282	0.0686
1.1	0.0000	0.0000	0.0000	0.0013	0.0058	0.0160	0.0391	0.0981
1.2	0.0000	0.0000	0.0000	0.0017	0.0074	0.0209	0.0521	0.1328
1.3	0.0000	0.0000	0.0000	0.0021	0.0021	0.0093	0.0265	0.1718
1.4	0.0000	0.0000	0.0001	0.0025	0.0114	0.0329	0.0839	0.2140
1.6	0.0000	0.0000	0.0002	0.0036	0.0164	0.0480	0.1227	0.3022
1.8	0.0000	0.0000	0.0003	0.0048	0.0224	0.0659	0.1680	0.3905
2.0	0.0000	0.0001	0.0005	0.0062	0.0294	0.0865	0.2151	0.4713
3.0	0.0000	0.0003	0.0014	0.0168	0.0790	0.2186	0.4589	0.7353
4.0	0.0000	0.0005	0.0026	0.0328	0.1489	0.3675	0.6400	0.8488
6.0	0.0001	0.0012	0.0067	0.0808	0.3169	0.6092	0.8244	0.9332
8.0	0.0002	0.0023	0.0129	0.1459	0.4774	0.7533	0.9001	0.9844
10.0	0.0003	0.0037	0.0208	0.2214	0.6038	0.8351	0.9537	0.9762
12.0	0.0004	0.0056	0.0310	0.3007	0.6971	0.8835	0.9553	0.9814
14.0	0.0005	0.0078	0.0431	0.3783	0.7648	0.9139	0.9672	0.9876
16.0	0.0007	0.0105	0.0571	0.4508	0.8141	0.9338	0.9749	0.99
18.0	0.0009	0.0136	0.0728	0.5163	0.8505	0.9476	0.9804	0.99
20.0	0.0015	0.0170	0.0901	0.5742	0.8816	0.9576	0.9840	0.99
30.0	0.0029	0.0402	0.1942	0.7668	0.9442	0.9811	0.9937	0.99
40.0	0.0054	0.0724	0.3107	0.8598	0.9586	0.9895	0.99	0.99
80.0	0.0237	0.2580	0.6682	0.9635	0.9921	0.99	0.99	0.99
100.0	0.0371	0.3593	0.7645	0.9766	0.9991	0.99	0.99	0.99

4. *Effect of fluctuations of trap distribution on the rate of exciton trapping*

If the only cause of exciton disappearance were trapping, that is, if the mean exciton lifetime τ in the pure crystal were infinite, then after asymptotically long periods of time some excitons would always be able to avoid trapping by penetrating such regions of the crystal were the trap concentration is much lower than the mean concentration. It is precisely this fact that provides for a significant effect of fluctuations of the trap distribution on the kinetics of exciton trapping for long times t. The primary question here is under which real conditions the exciton lifetime can be regarded as being asymptotically long. If the total number of excitons trapped by impurities at asymptotically large times is relatively small the inclusion of fluctuations does not have any significant effect on the calculations of the quantum yield of luminescence or the quantity P in the case of steady-state excitation. A possibility of the reverse situation is illustrated below for the case of a one-dimensional crystal.

We shall discuss the results reported by Balagurov and Vaks (1973) and generalize them to the case of steady-state excitation and the case of $\tau \neq \infty$. The exact solution can be found for the one-dimensional crystal and we shall treat this case first.

Assume that the acceptor concentration n is small ($na \ll 1$ where a is the lattice parameter) and $t \gg T$ where T is the hop time. Then the concentration of excitons along the distance l between two neighbouring traps satisfies the solution

$$\frac{\partial c}{\partial t} = D\frac{\partial^2 c}{\partial x^2} - \frac{c}{\tau},$$

with the boundary condition $c(0, x) = c_0$ (delta function excitation at $t = 0$). The solution of this equation in the region $x_i \leqslant x \leqslant x_{i+1}$ (where x_i and x_{i+1} are the coordinates of the neighbouring traps, $x_{i+1} - x_i = l_i$) with the boundary conditions

$$c(t, x_i) = c(t, x_{i+1}) = 0, \tag{6.74}$$

is

$$c_i(t, x) = 4c_0 \sum_{m=0}^{\infty} \exp\left(-\tfrac{1}{2}k_m^2 t - \frac{t}{\tau}\right) \frac{\sin k_m (x - x_i)}{k_m l_i}, \tag{6.75}$$

where $k_m = \pi(2m + 1)/l_i$. The mean concentration

$$\bar{c}_i(t) = \frac{1}{l_i} \int_{x_i}^{x_{i+1}} c_i(t, x)\, dx,$$

of excitons in the above interval is a random parameter which should be averaged over all possible lengths of the intervals between traps per 1 cm of the crystal length.

According to the law of large numbers we have

$$c(t) = \langle c_l(t) \rangle, \tag{6.76}$$

where the angle brackets denote averaging over the interval lengths, that is, the distances between traps. If the trap distribution is random the distribution function has the Poisson form $f(l) = n \exp(-nl)$. Using eqs. (6.75) and (6.76) we obtain

$$c(t) = \frac{4c_0}{\pi^2} \int_0^{\infty} \exp\left(-\frac{b}{\xi^2} - \frac{t}{\tau}\right) \frac{\xi\, d\xi}{\sinh \xi}, \quad b = \pi^2 n^2 Dt. \tag{6.77}$$

For $b \gg 1$ when the diffusion length $\sqrt{Dt}$ for the time t is large compared with the mean distance between the traps $\bar{l} = 1/n$ the main contribution to the integral in eq. (6.77) is made by large $\xi \gg 1$ for which $\sinh \xi \approx \frac{1}{2} \exp \xi$. Using this approximation we can easily obtain

$$c(t) = 8c_0 \left(\frac{4n^2Dt}{3\pi}\right)^{1/2} \exp\left[-3 \times 2^{-2/3}\pi^{2/3}(n^2Dt)^{1/3} - \frac{t}{\tau}\right]. \tag{6.78}$$

It is interesting to compare eq. (6.78) with the expression given by the cell method. Our approximation corresponds to replacement of l_i in eq. (6.75) with the mean value $\bar{l} = 1/n$. For large t we can retain only one term with $m = 0$ in eq. (6.75) so that we obtain

$$c(t) \sim \exp\left(-\frac{a^2}{2T}\pi^2 n^2 t - \frac{t}{\tau}\right).$$

A comparison of this expression with expression (6.78) shows that fluctuations play the predominant part in the one-dimensional case while the cell method is not applicable at all.

Let us now analyze the part played by fluctuations in the case of steady-state uniform excitation I_0. The exciton concentration here satisfies the equation

$$D\frac{d^2c}{dx^2} - \frac{c}{\tau} + I_0 = 0,$$

with the above boundary conditions for the interval (x_i, x_{i+1}). The solution of this equation is

$$c_i(x) = I_0 \frac{\Lambda^2}{D} + A\, e^{x/\Lambda} + B\, e^{-x/\Lambda},$$

where A and B are constants found from the boundary conditions and $\Lambda = \sqrt{D\tau}$. If the exciton lifetime is sufficiently long, that is, if the diffusion length $\Lambda \gg l_i$, we have

$$c_i(x) \approx \frac{I_0 x(l_i - x)}{2D}.$$

Using this expression we can obtain

$$\bar{c}_i \sim I_0 l_i^2 / 12D, \tag{6.79}$$

so that the mean exciton concentration (per unit length) is

$$\bar{c} = \frac{I_0}{12D}\int_0^\infty n\, e^{-nl} l^2\, dl = \frac{I_0}{6Dn^2}.$$

By definition, the effective probability P_1 of trapping for $\tau = \infty$ satisfies the balance equation

$$P_1\bar{c} = I_0,$$

Hence, we have

$$P_1 = 6Dn^2. \tag{6.80}$$

Thus, we see that in the one-dimensional case under the conditions of steady-state excitation P_1 exhibits a quadratic dependence on the acceptor concentration, rather than a linear one.

The cell method corresponds to the replacement of l_i in eq. (6.79) with its mean value $\bar{l} = 1/n$. This method yields $\bar{c} = I_0/12Dn^2$ so that the resulting P_1 is twice as high as that given by eq. (6.80) owing to underestimation of the contribution of the large cells with $l > 1/n$ though the character of the dependence of P_1 on n remains correct. Thus, the role played by fluctuations is significant even in the case of steady-state excitation. Is this a general result? Clearly, if we replace the boundary conditions (6.74) with less strict conditions of the type of

$$c(t, x_i) = \alpha \left|\frac{\mathrm{d}c(t, x)}{\mathrm{d}x}\right|_{x_i},$$

then for sufficiently high α values traps can play no role at all since under such conditions at $t \lesssim \tau$ all excitons will survive rather than only those that are in the region of low impurity concentrations. The role played by fluctuations, naturally, should decrease then. The role of fluctuations proves to be less significant in two-dimensional crystals and even less so in three-dimensional crystals. However, no exact solutions have been derived for these cases. In this connection, we shall discuss the analysis (Balagurov and Vaks 1973) of the asymptotic behaviour of the function $c(t)$ for $t \gg T$ which has the so-called "exponential accuracy" and is similar to the treatment by Lifshits (1964) of the density of states near the band edge. At large t excitons will survive only in sufficiently large regions free of traps. We can assume that the concentration of excitons within these regions satisfies the diffusion equation and at the boundary of the region with traps we have $c(t, \mathbf{r}) \approx 0$ (in fact, the function $c(t, \mathbf{r})$ has a "corrugated" profile at the surface with the characteristic length equal to the mean distance between traps but this, apparently, does not change the results given below). At large t the exciton concentration

$$c(t, \mathbf{r}) \sim \exp(-Dk_0^2 t), \tag{6.81}$$

where k_0 is the lowest possible wave number in the given region which is of the order of $V^{-1/3}$ (or $S^{-1/2}$ for a two-dimensional region). The mean concentration $\bar{c}(t)$ is equal to a sum of expressions of the type of eq. (6.81) over all possible fluctuation regions. Assuming that these regions have a spherical (circular) shape we can write $k_0 = \pi/R = \pi(4\pi/3V)^{1/3}$ for the sphere, and for the circle $k_0 = \mu_0/R = \mu_0\sqrt{\pi}\ S^{-1/2}$ where $\mu_0 \approx 2.405$ is the first zero of the Bessel function $J_0(\mu)$. Averaging eq. (6.81) over the Poisson distribution $f(V) = n\exp(-nV)$ we obtain for the three-dimensional system (Balagurov and Vaks 1973):

$$\bar{c}(t) \sim \exp[-\alpha_3(na^3)^{2/5}(t/T)^{3/5}], \qquad \alpha_3 = 4.7; \qquad t \gg T(na^3)^{-2/3}.$$

Similarly, for the two-dimensional system for which $f(S) = n\exp(-nS)$ we have

$$\bar{c}(t) \sim \exp[-\alpha_2(na^2t/T)^{1/2}], \qquad \alpha_2 = 4.3; \quad t \gg T/na^2.$$

Let us compare these results with those given by the cell method. This method yields for the three-dimensional system

$$\bar{c}(t) = \exp(-4\pi DRnt).$$

Since $4\pi DRnt \approx a^3nt/T$ the fluctuation mechanisms becomes predominant when

$$\alpha_3(na^3)^{2/5}(t/T)^{3/5} \ll na^3t/T,$$

that is, when

$$t/T \gg (na^3)^{-3/2}. \tag{6.82}$$

If the relative impurity concentration is, for instance, $na^3 = 10^{-4}$ inequality (6.82) is satisfied for $t/T \gg 10^6$. This means that for such impurity concentrations the fluctuation mechanisms for singlet excitons has no time to work since for singlet excitons $\tau/T \sim \max(t/T) \approx 10^3$–$10^5$. For triplet excitons, fluctuations, generally speaking, can have a significant effect on quenching kinetics for large $t \approx \tau$. Similar estimates of the contribution made by the fluctuation mechanism can, of course, be made for other impurity concentrations.

For the two-dimensional system (Balagurov and Vaks 1973) the cell method yields

$$\bar{c}(t) \sim \exp[-na^2t/T\ \ln(1/na^2)]. \tag{6.83}$$

Hence, the fluctuation mechanism becomes predominant in this case for

$$\frac{na^2t}{T\ln(1/na^3)} \gg (na^2t)^{1/2},$$

that is, for

$$t/T \gg (na^2)^{-1}\ln^2(1/na^2). \tag{6.84}$$

For instance, for $na^2 = 10^{-4}$ inequality (6.84) gives

$$t/T \gg 6 \times 10^5,$$

so that it also cannot be satisfied for singlet excitons.

In conclusion, note that we are not aware of any results on the kinetics of exciton luminescence which would definitely indicate inapplicability of the cell method and require inclusion of fluctuations of the trap distribution. However, we thought it worthwhile to discuss the effects of fluctuations here since the experimental techniques are constantly being improved and the range of systems under investigation is being extended (for instance, quasi-one-dimensional crystals are studied).

As noted in ch. 2, a different situation can occur in ternary solutions in which the concentrations of both donors and acceptors are low. The diffusion approximation can prove to be unsuitable for describing energy migration via donors in such solutions. The relevant problems are discussed in ch. 2 sect. 4 (see also Dubovskiy 1976).

5. *Kinetics of condensation of excitons in semiconductors*

At low temperatures the nonequilibrium electrons and holes in semiconductors form excitons under the effect of the Coulomb interaction. When the exciton concentrations become sufficiently high the exciton–exciton interaction starts to play a significant role. In most cases (always in cubic crystals and when the effective masses of electrons and holes are isotropic) this interaction gives rise to biexciton states for two excitons (Moskelenko 1958, Lampert 1958). The biexciton consists of four particles, and the effective masses of the electrons and holes in semiconductors are, generally, not very different. The adiabatic approximation is thus not applicable to many systems of interest, in contrast to a case such as the hydrogen molecule. Therefore, various variational methods are typically employed for calculating the bonding energy in biexcitons (for a review of methods and results see, for instance, Haken and Nikitine (1975)). If biexcitons in a system of excitons were the only possible formations with a positive bonding energy then at temperatures $T < E_{be}/k_B$ where E_{be} is the bonding energy of excitons in the biexciton the exciton gas would convert into the biexciton gas.

Keldysh (1970, 1971) has shown, however, that in some cases an exciton gas can convert into electron–hole drops, rather than biexcitons. According to Keldysh (1971), the electron–hole drops (EHD) are lumps of a new equilibrium phase in which electrons and holes form a degenerate electron–hole plasma. Therefore, the EHD should be formed at low temperatures in those cases when the bonding energy in the EHD per one electron–hole pair is greater than the respective energy in the biexciton (that is, $\frac{1}{2}E_{be}$). This is precisely the case for silicon and germanium which are the subjects of most experimental and theoretical studies in this field.

In this section we shall discuss some theoretical concepts of nucleation and growth of the EHD in the case of bulk excitations of excitons (Bagaev et al. 1976a, Silver 1975).

Variation of the number of particles in the spherical ehd of the radius R is described by the equation

$$\frac{\mathrm{d}}{\mathrm{d}t}\left(\frac{4\pi}{3}R^3 n_0\right) = I - \frac{4\pi}{3}R^3\frac{n_0}{\tau_0}, \tag{6.85}$$

where n_0 and τ_0 are the equilibrium number density and the equilibrium lifetime of the carriers in the drop, I is the total flux of excitons across the drop surface so that $I = I_1 - I_2$ where I_1 is the exciton flux flowing into the drop and I_2 is the exciton flux out of the drop. For instance, at 7 K in germanium the exciton diffusion coefficient $D = 1500\ \mathrm{cm^2\,s^{-1}}$ (Pokrovsky 1972) so that the exciton mean free path is much greater than the mean drop size (which is about 10 μm). Under such conditions (see sect. 2) we obtain

$$I_1 = \pi R^2 \gamma v c (1 + vR/4D)^{-1}, \tag{6.86}$$

where c is the exciton concentration at the site of the drop, v is the thermal velocity of excitons, and γ is the "coefficient of adhesion" of excitons to the drop. The flux I_2 of excitons leaving the drop is given by the conventional formula describing the rate of evaporation

$$I_2 = \pi R^2 \gamma v c_T(R)(1 + vR/4D)^{-1}, \tag{6.87}$$

where

$$c_T(R) = \nu\left(\frac{k_B T M}{2\pi\hbar^4}\right)^{3/2} \exp\left(-\frac{\Delta}{k_B T} + \frac{2\alpha}{n_0 R k_B T}\right) \tag{6.88}$$

is the thermodynamically equilibrium concentration of excitons over a drop of the radius R, Δ is the work function for excitons in the drop for $R \to \infty$, M is the effective exciton mass, α is the surface tension of the electron–hole liquid, and ν is the degeneracy of the exciton ground state. The term $2\alpha/n_0 R k_B T$ in the exponent of eq. (6.88) corresponds to an increase in the evaporation rate owing to the surface tension which is typically related to coalescence of drops. Under the steady-state conditions eq. (6.85) gives the exciton concentration as a function of R

$$c = c_T(R) + n_0\frac{4R}{3\gamma\tau_0 v}\left(1 + \frac{vR}{4D}\right), \tag{6.89}$$

which has a minimum at a certain R value as can be seen in fig. 6.2.

In the plane of the variable c and R in fig. 6.2 the points that are not on the curve $c(R)$ given by eq. (6.89) cannot correspond to the stationary states of the system. For the points above the curve the drop radius increases with time ($\mathrm{d}R/\mathrm{d}t > 0$) and for the points below the curve the

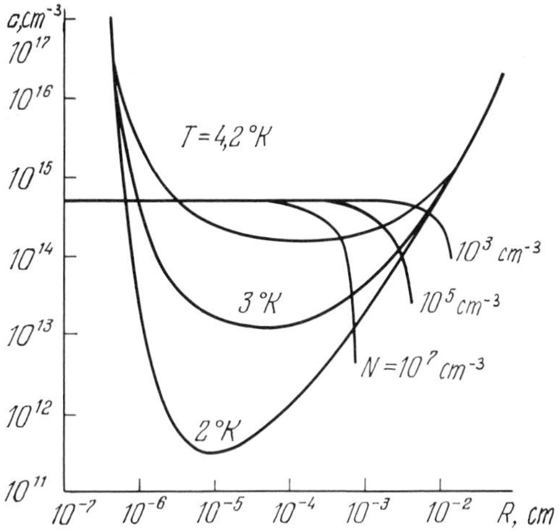

Fig. 6.2. The dependence of c on R. The curves were calculated from eq. (6.89) for three different temperatures and from eq. (6.91) for three concentrations N of the EHD for $\tau = 5\times10^{-5}$ s, $I = 10^{20}$ cm^{-3} s^{-1}, $\gamma = 1$, $D = 1500$ cm^2/s and $T = 7$ K (Pokrovsky 1972); $D \sim T^{-1/2}$ (see ch. 5), $\alpha \sim 2\times10^{-4}$ erg/cm^2, $M \approx 4\times10^{-28}$ g, $\tau_0 = 40\times10^{-6}$ s, $n_0 = 2\times10^{17}$ cm^{-3}, $\nu = 16$, $\Delta = 2.1\times10^{-3}$ eV.

radius decreases with time ($dR/dt < 0$). Therefore, only the states on the right-hand ascending branch of the $c(R)$ curve ($R > R_{min}$) are stable while for the states on the left-hand descending branch any small deviations of c or R give rise to variation of R with time so that the state moves even further from the equilibrium curve (Silver 1975). Thus, for any given temperature there is a certain minimum radius given by eq. (6.89). Interestingly, under these conditions when the exciton mean free path l is large compared with the drop radius R no depletion of the exciton concentration takes place in the region of each of the drops. Thus, for instance, in the case of spatially uniform excitation the concentration c in eq. (6.86) coincides with the mean exciton concentration and, hence, all the drops should have identical radius. To find the concentration c in eq. (6.86) we can make use of the balance equation

$$NI + c/\tau = P, \tag{6.90}$$

where τ is the exciton lifetime, N is the number of the drops, and P is the rate of exciton generation. Using eq. (6.85) we can rewrite eq. (6.90) as

$$\frac{c}{\tau} + \frac{4\pi}{3} R^3 \frac{n_0}{\tau_0} N = P. \tag{6.91}$$

The plots of the mean exciton concentration as a function of the EHD radius calculated from eq. (6.91) are also given in fig. 6.2. The intersection of these plots with the stable branch of the $c(R)$ dependence given by eq. (6.89) determine the EHD radius for a given drop concentration and the given generation rate while the intersections with the unstable branch determine the "critical" radius of the drop embryos.

It has been assumed in the analysis of Silver (1975) that under the conditions of steady-state excitation the concentration N of the EHD increases with time owing to production of new drops while c and R decrease tending to their minimum values R_{min} and $c(R_{min})$ consistent with eqs. (6.89) and (6.91) for which

$$N_{max}(P) = \frac{3}{4\pi}\frac{\tau_0}{n_0 R_{min}^3}\left(P - \frac{c(R_{min})}{\tau}\right). \tag{6.92}$$

However, Bagaev et al. (1976a) showed that this limiting equilibrium state can be obtained in germanium only at $T \approx 4$ K, that is, only at fairly high temperatures ($T > \Delta/k_B$). At lower temperatures the observed drop radius $R \leqslant 10\ \mu$m but $R > R_{min}$ and the number of drops is also much lower than the value given by eq. (6.92). It has been noted by Bagaev et al. (1976a), that the experimental data for germanium indicate, apparently, the existence of a mechanism that sharply limits the growth of the drop radius for $R \sim 10\ \mu$m. Bagaev et al. (1976a) suggest that out of many possible mechanisms (a decrease in the exciton diffusion coefficient owing to exciton–exciton collisions, overheating of the drops, etc.) the most probable one is deceleration of the exciton flux feeding the drop by the escaping flux of nonequilibrium phonons generated within the drop owing to recombination of carriers. Keldysh (1976), Bagaev et al. (1976b) has shown that this mechanism makes large drops unstable. The phonon wind gives rise to bulk forces equivalent to the Coulomb repulsion bulk forces which break large drops from within. The existence of the phonon wind significantly alters the kinetics of drop development. A wider review of these problems and new experimental results was given by Keldysh and Jeffries (1983).

An attempt to estimate the size of hypothetical dielectric drops in a molecular crystal (anthracene) is discussed by Benderskiy et al. (1977).

6. *Effects of radiative energy transfer on the exciton distribution in time and space: macroscopic characteristics of exciton luminescence*

Equation (6.1) is not applicable to crystals in which radiative energy transfer plays a significant part. If the light of exciton luminescence can be absorbed in the crystal giving rise to excitons, that is, if reabsorption can occur in the crystal, this effect should also be reflected by the equation for the function $c(\boldsymbol{r}, t)$ as was done by Agranovich and Faidysh (1956), Agranovich (1957).

For the sake of simplicity let us ignore anisotropy of the medium and assume that luminescence is not polarized. Then each volume element $\mathrm{d}v_1$ emits at the moment t $\mathrm{d}v_1 \times c(\boldsymbol{r}, t) \times \rho(\nu)\,\mathrm{d}\nu$ photons (polaritons for $\hbar\nu = E \leqslant \hbar\Omega_\perp$; see fig. 4.3(b) and (c)) per unit time in the frequency range $\mathrm{d}\nu$ owing to decay of excitons. Here $\rho(\nu)$ is the probability of generation of a photon of the frequency ν per unit time owing to exciton decay. The number of photons produced per unit time in the volume element $\mathrm{d}v_1$ and absorbed in the volume element $\mathrm{d}v$ is

$$c(\boldsymbol{r}_1, t)\rho(\nu)\,\mathrm{d}\nu\,\mathrm{d}v_1\,\mathrm{d}v \frac{\exp[-k(\nu)|\boldsymbol{r} - \boldsymbol{r}_1|]}{|\boldsymbol{r} - \boldsymbol{r}_1|^2} k(\nu), \tag{6.93}$$

where $k(\nu)$ is the absorption coefficient for light of the frequency ν in the crystal.

However, not all the photon energy absorbed in the crystal volume $\mathrm{d}v$ is converted into excitons. To take into account this fact we have to multiply eq. (6.93) by the quantity $\beta(\nu)$ equal to the fraction of the events of absorption of photons with the frequency ν resulting in excitation of electronic (excitonic) states. Integrating over all volumes $\mathrm{d}v_1$ we find that the increase in the number of excitons due to reabsorption is

$$\left.\frac{\partial c}{\partial t}\right|_{\mathrm{reabs}} = \frac{1}{4\pi} \int_0^\infty \rho(\nu)\beta(\nu)k(\nu)\,\mathrm{d}\nu \int_V c(\boldsymbol{r}_1, t) \frac{\exp[-k(\nu)|\boldsymbol{r} - \boldsymbol{r}_1|]}{|\boldsymbol{r} - \boldsymbol{r}_1|^2}\,\mathrm{d}v_1. \tag{6.94}$$

Strictly speaking, we must subtract a term of the order of $1/ck(\nu_0)$ from t in this equation where ν_0 is the frequency of the reabsorption peak, and c is the velocity of light in the crystal. However, in practically all cases this term is negligibly small in comparison with the luminescence decay time during which the variation of the exciton concentration is noticeable. Therefore, we shall ignore this retardation.

Apart from eq. (6.94) we should also take into account reabsorption of the luminescence photons reflected from the crystal surface. However, we shall neglect this effect for the sake of simplicity.

Using eqs. (6.1) and (6.94) we obtain the equation for the exciton concentration including reabsorption

$$\frac{\partial c}{\partial t} = D\Delta c - Pc + \frac{1}{4\pi} \int_0^\infty \rho(\nu)\beta(\nu)k(\nu)\,\mathrm{d}\nu \int_V c(\boldsymbol{r}_1, t) \frac{\exp[-k(\nu)|\boldsymbol{r} - \boldsymbol{r}_1|]}{|\boldsymbol{r} - \boldsymbol{r}_1|^2}\,\mathrm{d}v_1 + I_0(t)k\,\mathrm{e}^{-kx}, \tag{6.95}$$

where k is the absorption coefficient for the excitation radiation (see also, Agranovich 1957).

The quantity $\beta(\nu)$ in eq. (6.95) is directly related to the quantum yield $\eta(\nu)$ of luminescence for excitation by light of the frequency ν. Indeed, $\eta(\nu)$ is equal to the probability that absorption of a photon of the frequency ν gives rise to an exciton multiplied by the probability that decay of this exciton gives rise to a photon of luminescence. Thus, we have

$$\eta(\nu)=\beta(\nu)\frac{1}{P}\int_0^\infty \rho(\nu_1)\,\mathrm{d}\nu_1,$$

where P is the total probability of exciton decay per 1 s. Therefore, introducing the normalized luminescence spectrum

$$E(\nu)=\frac{\rho(\nu)}{\int_0^\infty \rho(\nu)\,\mathrm{d}\nu}, \qquad \int E(\nu)\,\mathrm{d}\nu=1, \tag{6.96}$$

we obtain

$$\beta(\nu)\rho(\nu)\equiv PE(\nu)\eta(\nu). \tag{6.97}$$

It should be noted that from the viewpoint of phenomenological description this problem is not specifically that of excitons in crystal. Treatment of excitons affects only the values of the parameters in the equation for $c(\boldsymbol{r}, t)$. For instance, eq. (6.95) with $D=0$ was first analyzed by Holstein (1947, 1951) (see also, Biberman 1947, Seiwert 1956, Walsh 1957), where it was used to describe the variation of the concentration of excitations in the isotropic medium with only radiation transfer. Fundamental results in radiation transfer theory were obtained by Ambartsumyan (1942), Chandrasekhar (1950) and Sobolev (1956) who studied radiation transfer in planetary atmospheres. However, they analyzed in detail only the steady-state problem when $\partial c/\partial t=0$.

The time dependence of $c(\boldsymbol{r}, t)$ is also a problem of interest for optics, for instance, in connection with the experimental studies of the kinetics of luminescence decay. Therefore, we shall discuss below the theory treated by Agranovich (1957, 1959c), Agranovich and Konobeev (1959, 1961), Konobeev (1961) which makes it possible to calculate the luminescence decay time, the luminescence spectrum and its variation with time (this theory is generalized for anisotropic (uniaxial) crystals by Agranovich et al. (1977)). The theory is based on generalization of the Ambartsumyan method for nonstationary processes. Some results in this field were also obtained by Stepanov and Samson (1960), Samson (1960a, b, 1962).

In our analysis of eq. (6.95) we shall consider the following two limiting cases (we proceed from the discussion of Agranovich (1968) with the addition of some new theoretical and experimental results).

(a) The case of strong reabsorption. The exciton diffusion coefficient is so small that the diffusion term in eq. (6.95) can be omitted.

(b) The case of weak reabsorption. The integral terms in eq. (6.95) can be omitted. Konobeev (1963) has analyzed a more complicated case when both radiative and radiationless energy transfer must be included. However, for the sake of simplicity we shall discuss the theoretical results only for the above limiting cases.

It should be borne in mind that even under the conditions of strong reabsorption when the diffusion term in eq. (6.95) can be omitted the radiationless diffusion mechanism of energy transfer can play an important part if the crystal contains impurities capable of trapping excitons. Under certain conditions exciton trapping by impurities can be limited precisely by the exciton diffusion flux towards impurities so that we still have to take into consideration exciton diffusion when we calculate P_1 (see sects. 2 and 3).

Reabsorption of luminescence photons provides for the situation when the exciton distribution in the crystal, that is, the function $c(\mathbf{r}, t)$, does not follow the distribution of the external sources of excitons even under the steady-state conditions and for $D = 0$. Determination of the function $c(\mathbf{r}, t)$ satisfying eq. (6.95) is a very difficult task even in the absence of the diffusion term. The task is simplified by the fact that we must know only a few integral parameters, rather than the full details of the function $c(\mathbf{r}, t)$, in order to find the macroscopic characteristics of luminescence, for instance the spectrum and the decay time of the luminescence light emitted from the crystal which can differ significantly from those for a single elementary (molecular) luminescence event. Indeed, let $I_0^{(\nu)}(\theta, t)$ be the number of the luminescence photons in the frequency range $d\nu$ emitted from the crystal at the angle θ to the outer normal per 1 cm^2 of the plane $x = 0$ per unit solid angle per 1 s. Clearly, we have then

$$I_0^{(\nu)}(\theta, t) = \frac{\rho(\nu)[1 - r_\nu(\cos\theta)]}{4\pi} \int_0^d c(x, t) \exp[-k(\nu)x \sec\theta]\, dx, \tag{6.98}$$

where $r_\nu(\cos\theta)$ is the reflection coefficient for the light of the frequency ν incident at the internal surface of the crystal at the angle θ. Similarly, we have for the surface $x = d$

$$I_d^{(\nu)}(\theta, t) = \frac{\rho(\nu)[1 - r_\nu(\cos\theta)]}{4\pi} \int_0^d c(x, t) \exp[-k(\nu)(d - x \sec\theta)]\, dx. \tag{6.99}$$

Equations (6.98) and (6.99) ignore the fact that the radiation escaping from the crystal at the moment t is determined by the values of the function $c(x, t)$ at earlier moments. However, even if the thickness of a plane-parallel plate is as large as 1 cm and the angle θ is not anomalously large the decay time is not more than $d/c \sim 3 \times 10^{-11}$ s which is considerably shorter than the time of significant variation of the function $c(x, t)$ (in fact, the characteristic length $d \sim 1/k(\nu)$).

Let us now discuss the time of decay of luminescence. If excitation is discontinued at the moment $t = 0$ the measured mean time $\tau_{0,d}$ or the luminescence decay time can be found from

$$\tau_{0,d}(\nu, \theta) = \frac{\int_0^d I_{0,d}^{(\nu)}(\theta, t) t \, dt}{\int_0^d I_{0,d}^{(\nu)}(\theta, t) \, dt}. \tag{6.100}$$

Moreover, let us define the time

$$\tilde{\tau}_{0,d}(\nu, \theta) = \frac{\int_0^\infty I_{0,d}^{(\nu)-}(\theta, t) \, dt}{I_{0,d}^{(\nu)}(\theta, 0)}. \tag{6.101}$$

which approximately describes decay of luminescence. This time coincides with $\tau_{0,d}(\nu, \theta)$ only if the intensity $I_{0,d}^{(\nu)}$ decreases exponentially with time. Therefore, the difference $|\tau - \tilde{\tau}|$ can be regarded as a measure of deviation of the time dependence of $I_{0,d}^{(\nu)}(\theta, t)$ from the exponential decay.

Thus, as implied by eqs. (6.98)–(6.101), to find the macroscopic characteristics of luminescence we have just to know the integral of the function $c(x, t)$ of the form

$$\int_0^d c(x, t) \, e^{-\eta x} \, dx$$

as a function of η. Hence, the analysis is simplified if such initial parameters entering into eq. (6.95) as D, P, $\eta(\nu)$, $k(\nu)$, and $E(\nu)$ can be regarded as known. In many cases, however, the procedure is reversed, that is, some of the above parameters are found by treating the experimental data on luminescence. In any case, we have to know relationships expressing the macroscopic luminescence characteristics in terms of the above parameters and geometric factors.

To analyze eq. (6.95) let us introduce the following dimensionless parameters:

$$z = k_0 x, \quad z_0 = k_0 d, \quad \tau = Pt, \quad c(z, \tau) = \frac{Pc(x, t)}{k(\nu) I_0(\nu)},$$

$$\gamma = k/k_0, \quad b(\nu) = k(\nu)/k_0, \quad a(\nu) = \tfrac{1}{2} E(\nu)\eta(\nu), \quad \lambda^2 = Dk_0^2/P. \tag{6.102}$$

Here $k_0 \equiv k(\nu_0)$ where the frequency ν_0 is, generally, arbitrary. If the function $a(\nu)$ is represented by a bell-shaped curve it is convenient to take the frequency ν_0 at the peak of the function and then k_0 is the absorption coefficient of light at the reabsorption peak. Rewriting eq. (6.95) in terms of the parameters (6.102) we obtain

$$\frac{\partial c(z,\tau)}{\partial \tau} = \lambda^2 \frac{\partial^2 c(z,\tau)}{\partial z} - c(z,t) + e^{-\gamma z}$$
$$+ \int_0^\infty a(\nu)b(\nu)\,d\nu \int_0^{z_0} c(z_1,\tau)E_1(b(\nu)|z-z_1|)\,dz_1, \tag{6.103}$$

where

$$E_1(z) = \int_1^\infty \frac{e^{-z\eta}}{\eta}\,d\eta. \tag{6.104}$$

and we have used the fact that $c(r,t) \equiv c(x,t)$. To neglect the diffusion term we must take λ equal to zero in eq. (6.103).

In conclusion, let us include possible reflection of luminescence light from the plane-parallel crystal surfaces (Agranovich and Konobeev 1959, 1961, Konobeev 1971, Doronina et al. 1970, Agranovich et al. 1981) in the equation for exciton concentration. Assume that the conditions of applicability of the diffusion equation for excitons are satisfied and analyze together the diffusion equation for the exciton concentration $c(\boldsymbol{r},t)$ and the equation for the photon distribution function $\varphi(\boldsymbol{p},\boldsymbol{r},t)$, as discussed by Agranovich et al. (1981). We ignore the polariton effects here so that the resulting system of equations has the form

$$\frac{\partial c}{\partial t} = D\Delta c - \frac{c}{\tau} + I(\boldsymbol{r},t) + \int \frac{d^3p}{(2\pi\hbar)^3}\Phi(\boldsymbol{p})\varphi(\boldsymbol{p},\boldsymbol{r},t) \tag{6.105a}$$

$$\boldsymbol{u}(\boldsymbol{p})\frac{\partial \varphi}{\partial \boldsymbol{r}} = F(\boldsymbol{p})c(\boldsymbol{r},t) - \Phi(\boldsymbol{p})\varphi(\boldsymbol{p},\boldsymbol{r},t) \tag{6.105b}$$

where $\boldsymbol{u}(\boldsymbol{p})$ is the velocity of photons with the momentum $\boldsymbol{p}$, $\Phi(\boldsymbol{p})$ is the probability (per unit time) of absorption of photon with the momentum $\boldsymbol{p}$ resulting in formation of exciton, and $F(\boldsymbol{p})$ is the probability of radiative exciton decay (per unit time) giving rise to luminescence photon with the momentum $\boldsymbol{p}$. The omission of the term $\partial\varphi/\partial t$ in eq. (6.105b) corresponds to ignoring the delay time as discussed above. Below we shall ignore crystal anisotropy and assume that the crystal is a plane-parallel plate of the thickness d $(0 \leq x \leq d)$. If the exciton source function depends only on x the distribution functions c and φ also depend only on x. Write down the function φ as the sum

$$\varphi(\boldsymbol{p},\boldsymbol{r},t) = \varphi^+(\boldsymbol{p},\boldsymbol{r},t) + \varphi^-(\boldsymbol{p},\boldsymbol{r},t),$$

where φ^- and φ^+ are the distributions of photons propagating towards the planes $x = 0$ and $x = d$, respectively. If θ is the angle between the photon velocity vector $\boldsymbol{u}$ and the axis $x (0 \leq \theta \leq \pi/2)$ then the equation for φ is equivalent to the following system of two equations:

$$u(\boldsymbol{p}) \cos\theta \frac{\partial \varphi^+}{\partial x} = F(\boldsymbol{p}) c(x,t) - \Phi(\boldsymbol{p}) \varphi^+(\boldsymbol{p}, x, t),$$

$$u(\boldsymbol{p}) \cos\theta \frac{\partial \varphi^-(\boldsymbol{p}, \boldsymbol{r}, t)}{\partial x} = \Phi(\boldsymbol{p}) \varphi^-(\boldsymbol{p}, x, t) - F(\boldsymbol{p}) c(x,t).$$

These equations yield

$$\varphi^+(\boldsymbol{p}, x, t) = \frac{F(\boldsymbol{p}) \exp(-k_p x/\cos\theta)}{u(\boldsymbol{p}) \cos\theta} \int_0^x c(x', t) \exp(k_p x'/\cos\theta)\, dx'$$

$$+ A_1(\boldsymbol{p}) \exp(-k_p x/\cos\theta),$$

$$\varphi^-(\boldsymbol{p}, x, t) = \frac{F(\boldsymbol{p}) \exp(k_p x/\cos\theta)}{u(\boldsymbol{p}) \cos\theta} \int_x^d c(x', t) \exp(-k_p x'/\cos\theta)\, dx'$$

$$+ A_2(\boldsymbol{p}) \exp(k_p x/\cos\theta),$$

where $k_p = \Phi(\boldsymbol{p})/u(\boldsymbol{p})$ is the extinction coefficient for the photons with the momentum $\boldsymbol{p}$. The quantities A_1 and A_2 can be found from the boundary conditions

$$\varphi^+(\boldsymbol{p}, 0, t) = r_p(\theta) \varphi^-(\boldsymbol{p}, 0, t),$$

$$\varphi^-(\boldsymbol{p}, d, t) = R_p(\theta) \varphi^+(\boldsymbol{p}, d, t)$$

where $r_p(\theta)$ and $R_p(\theta)$ are the regular reflection coefficients for photons with the momentum $\boldsymbol{p}$ incident at the angle θ on the crystal surfaces $x = 0$ and $x = d$, respectively. Finding A_1 and A_2 we can determine the functions φ^+ and φ^- and, hence, the function φ. Substituting the function $\varphi(\boldsymbol{p}, \boldsymbol{r}, t)$ into eq. (6.105a) we obtain the sought equation for the exciton concentration including reabsorption. It can be easily seen that the equation has the following form:

$$\frac{\partial c(x,t)}{\partial t} = D\Delta c(x,t) - \frac{c(x,t)}{\tau} + I(x,t) + \int_0^\infty \rho(\nu) k(\nu)\, d\nu \int_0^d dx_1 c(x_1, t)$$

$$\times \left\{ E_1(k(\nu)|x - x_1|) + \int_1^\infty \frac{d\eta}{\eta} \frac{1}{\exp[k(\nu) d\eta] - r(\nu,\eta) R(\nu,\eta) \exp[-k(\nu)\eta d]} \right.$$

$$\times \Big(r(\nu,\eta) \exp[k(\nu)\eta(d - x - x_1)] + R(\nu,\eta) \exp[-k(\nu)\eta(d - x - x_1)]$$

$$\left. + 2r(\nu,\eta) R(\nu,\eta) \exp[-k(\nu)\eta d] \cosh k(\nu)\eta(x - x_1) \Big) \right\}. \qquad (6.105c)$$

Here

$$\rho(\nu) = \frac{p^2}{(2\pi\hbar)^2} F(p) \frac{dp}{d\nu}.$$

When $d \to \infty$ eq. (6.105c) is transformed into the equation derived by Agranovich and Konobeev (1959, 1961). For finite values of d and $r(\nu, \eta) = R(\nu, \eta)$ eq. (6.105c) is transformed into the equation derived by Konobeev (1971).

It should be noted that if we take into account the polariton effects we must have the boundary conditions of a more general form than those given above. Such boundary conditions provide for the possibility of transferring polaritons from one spectral branch to another. As shown by Agranovich et al. (1980), this gives rise to additional bulk and surface sources of reabsorption in eq. (6.105c). A numerical solution of eq. (6.105c) was found by Doronina et al. (1970) for the steady-state conditions in the absence of exciton diffusion (that is, for $D = \lambda = 0$) and bimolecular quenching ($A = 0$). These results were used for interpreting the experimental data on the effects of electron bombardment (fast electrons produced in beta-decay of tritium) on the quantum yield of luminescence of anthracene.

Luminescence spectrum of semi-infinite crystal under the steady-state conditions. Ambartsumyan (1942) analyzed the steady-state equation for the function $c_0(z)$ in the semi-infinite medium

$$c_0(z) = e^{-\gamma z} + \frac{q}{2} \int_0^\infty c_0(z_1) E_1(|z - z_1|)\, dz_1, \tag{6.106}$$

where q is a positive number smaller than unity. This equation is a special case of the more general equation

$$c_0(z) = e^{-\gamma z} + \int_0^\infty a(\nu) b(\nu)\, d\nu \int_0^\infty c_0(z_1) E_1(b(\nu)|z - z_1|)\, dz_1, \tag{6.107}$$

which follows from eqs. (6.95) and (6.105) and corresponds to $\lambda = 0$, $A = 0$, $r_\nu = 0$ and

$$a(\nu) = \frac{q}{2} \delta(\nu - \nu_0). \tag{6.108}$$

Discussing the results of Ambartsumyan (1942) we shall use the more general equation (6.107). Equation (6.107) shows that its solution depends on γ. Therefore, we can rewrite eq. (6.107) as

$$c_0(z, \gamma) = e^{-\gamma z} + \int_0^\infty a(\nu) b(\nu)\, d\nu \int_0^\infty c_0(z_1, \gamma) E_1(b(\nu)|z - z_1|)\, dz_1. \tag{6.109}$$

Differentiation of both sides of this equation with respect to z yields

$$c_0'(z,\gamma) = -\gamma\, e^{-\gamma z} + c_0(0,\gamma)\int_0^\infty a(\nu)b(\nu)\,d\nu \int_1^\infty \frac{\exp(-\eta b(\nu)z)}{\eta}\,d\eta$$

$$+\int_0^\infty a(\nu)b(\nu)\,d\nu \int_0^\infty c_0'(z_1,\gamma)E_1(b(\nu)|z-z_1|)\,dz_1. \tag{6.110}$$

Equation (6.110) for the function $c_0'(z,\gamma)$ differs from eq. (6.109) for the function $c_0(z,\gamma)$ only in the exciton sources. Using the linearity of these equations and the fact that eq. (6.109) without the free term has only trivial solution we can obtain

$$c_0'(z,\gamma) = -\gamma c_0(z,\gamma) + c_0(0,\gamma)\int_0^\infty a(\nu)b(\nu)\,d\nu \int_1^\infty \frac{1}{\eta}\,c_0(z,\eta b(\nu))\,d\eta. \tag{6.111}$$

Introduce a new function

$$R_0(\xi,\eta) = \int_0^\infty e^{-\xi z} c_0(z,\eta)\,dz, \tag{6.112}$$

which fully determines $I_0^{(\nu)}(\theta)$ in the steady-state case (see eq. (6.98)). To find the function $R_0(\xi,\eta)$ multiply both sides of eq. (6.111) by $\exp(-\xi z)$ and integrate the results between the limits $z=0$ and $z=\infty$. We obtain

$$\int_0^\infty e^{-\xi z} c_0'(z,\gamma)\,dz = -\gamma R_0(\xi,\gamma) + c_0(0,\gamma)$$

$$\times\int_0^\infty a(\nu)b(\nu)\,d\nu \int_1^\infty \frac{1}{\eta}\,R_0(\xi,\eta b(\nu))\,d\eta.$$

Integration by parts yields

$$\int_0^\infty e^{-\xi z} c_0'(z,\gamma)\,dz = -c_0(0,\gamma) + \xi R_0(\xi,\gamma).$$

Hence, we have

$$(\xi+\gamma)R_0(\xi,\gamma) = c_0(0,\gamma)\left[1+\int_0^\infty a(\nu)b(\nu)\,d\nu \int_1^\infty \frac{1}{\eta}\,R_0(\xi,\eta b(\nu))\,d\eta\right]. \tag{6.113}$$

Since $R(\xi, \eta) = R(\eta, \xi)$ as it follows from eq. (6.115), eq. (6.109) for $z = 0$ yields

$$c_0(0, \xi) = 1 + \int_0^\infty a(\nu)b(\nu)\,d\nu \int_0^\infty \frac{1}{\eta} R_0(\xi, \eta b(\nu))\,d\eta. \tag{6.114}$$

Equation (6.114) is valid for arbitrary ξ. Comparing it with eq. (6.113) we can see that

$$R_0(\xi, \gamma) = \frac{c_0(0, \gamma)c_0(0, \xi)}{\xi + \gamma}. \tag{6.115}$$

Thus, the function $R_0(\xi, \gamma)$ is fully determined by the function $c_0(z, \gamma)$ for $z = 0$. To find $c_0(0,\gamma)$ substitute eq. (6.115) into eq. (6.114)..Introducing a new function $\varphi(x) = c_0(0, \gamma)$ where $x = 1/\gamma$ we find that this function is the solution of the nonlinear integral equation

$$\varphi(x) = 1 + x\varphi(x) \int_0^\infty a(\nu)b(\nu)\,d\nu \int_0^{1/b(\nu)} \frac{\varphi(y)}{x + y}\,dy. \tag{6.116}$$

If we know the function $\varphi(x)$ we can find the luminescence spectrum determined by eq. (6.98). Indeed, using eqs. (6.102), (6.112) and (6.115) we find

$$I_0^{(\nu)}(\theta) = \frac{\gamma I_0 \rho(\nu)[I - r_\nu(\cos\theta)]}{4\pi P[\gamma + b(\nu)\sec\theta)]} \varphi\left(\frac{1}{\gamma}\right) \varphi\left(\frac{\cos\theta}{b(\nu)}\right). \tag{6.117}$$

Equations (6.117) demonstrates that the shape of the spectrum of luminescence differs significantly from the shape of the spectrum of an elementary emission event described by the function $\rho(\nu)$. Moreover, eq. (6.117) implies a strong dependence of $I_0^{(\nu)}$ on the angle θ.

The function $\varphi(x)$ is not easy to find since different crystals have different functions $a(\nu)$ and $b(\nu)$. The problem is simplified only if the function $a(\nu)$ is practically different from zero in a certain frequency range $\nu \approx \nu_0$ so that the function

$$b(\nu) \int_0^{1/b(\nu)} \frac{\varphi(y)\,dy}{x + y}$$

in eq. (6.116) can be taken outside the integral by taking $\nu = \nu_0$. In this so-called q approximation the equation for the function $\varphi(x)$ is simplified to

$$\varphi(x) = 1 + \frac{q}{2} x\varphi(x) \int_0^1 \frac{\varphi(y)}{x + y}\,dy, \tag{6.118}$$

where

$$q = 2\int_0^\infty a(\nu)\,\mathrm{d}\nu. \tag{6.119}$$

Equation (6.102) yields $q \leqslant 1$. If $q < 0.4$ we can expand $\varphi(x)$ in the series

$$\varphi(x) = 1 + \frac{q}{2}x\ln\left(1+\frac{1}{x}\right) + \frac{q^2}{4}x^2\ln^2\left(1+\frac{1}{x}\right) + \cdots \tag{6.120}$$

Table 6.5 gives the values of the function $\varphi(x)$ for $q \geqslant 0.4$. If the function $a(\nu)$ has a more complicated structure the accuracy of the q approximation can prove to be insufficient. Then we can obtain sufficiently accurate solutions of eq. (6.116) by using the values of $\varphi(x)$ found in the q

Table 6.5
Function $\varphi(x)$.

	q						
x	0.4	0.5	0.6	0.7	0.8	0.9	0.95
0	1.00	1.00	1.00	1.00	1.00	1.00	1.00
0.1	1.07	1.07	1.09	1.11	1.14	1.17	1.19
0.2	1.09	1.11	1.15	1.18	1.23	1.29	1.34
0.3	1.11	1.14	1.19	1.24	1.30	1.39	1.46
0.4	1.13	1.17	1.22	1.28	1.36	1.48	1.57
0.5	1.14	1.19	1.25	1.32	1.41	1.56	1.67
0.6	1.15	1.20	1.27	1.35	1.46	1.63	1.76
0.7	1.16	1.22	1.29	1.38	1.50	1.69	1.85
0.8	1.17	1.23	1.31	1.40	1.54	1.75	1.93
0.9	1.18	1.24	1.32	1.42	1.57	1.80	2.01
1.0	1.18	1.25	1.34	1.44	1.60	1.85	2.08
2.0	1.22	1.31	1.40	1.57	1.79	2.19	2.60
3.0	1.24	1.34	1.46	1.63	1.89	2.38	2.93
4.0	1.25	1.35	1.48	1.67	1.96	2.52	3.16
5.0	1.26	1.36	1.50	1.69	2.00	2.63	3.34
6.0	1.26	1.37	1.51	1.71	2.03	2.69	3.47
7.0	1.27	1.37	1.52	1.73	2.06	2.74	3.57
8.0	1.27	1.38	1.53	1.74	2.08	2.78	3.65
9.0	1.27	1.38	1.53	1.74	2.09	2.82	3.74
10.0	1.28	1.39	1.54	1.75	2.10	2.84	3.78
30.0	1.29	1.40	1.57	1.80	2.18	3.05	4.20
50.0	1.29	1.41	1.57	1.81	2.20	3.09	4.29
100.0	1.29	1.41	1.58	1.83	2.21	3.12	4.36
∞	1.29	1.41	1.58	1.83	2.24	3.16	4.46

approximation and performing one or two successive iterations. Bendersky (1971) attempted analysis outside the framework of the q approximation for the case of anthracene assuming exponential dependence of the coefficient of light absorption in the region of the long-wavelength edge (Urbach's rule; see ch. 4 sect. 9). In particular, he treated the case of sinusoidally modulated illumination.

7. *Luminescence decay time for semi-infinite crystal*

If excitation of luminescence is discontinued at the moment $t = 0$ (it is assumed that at $t < 0$ luminescence was excited by constant-intensity light of the frequency ν and was stationary) then at subsequent moments the exciton concentration satisfies the equation

$$\frac{\partial c(z, \gamma, \tau)}{\partial \tau} = -c(z, \gamma, \tau) + \int_0^\infty a(\nu)b(\nu)\,\mathrm{d}\nu \int_0^\infty c(z_1, \gamma, \tau)E_1(b(\nu)|z - z_1|)\,\mathrm{d}z_1, \tag{6.121}$$

with the starting condition

$$c(z, \gamma, \tau)|_{\tau=0} = c_0(z, \gamma), \tag{6.121a}$$

Equations (6.100) and (6.101) for the luminescence decay times imply that to find these quantities we have to know just the following integral expressions:

$$\tilde{\varphi}(z, \gamma) = \int_0^\infty c(z, \gamma, \tau)\,\mathrm{d}\tau, \tag{6.122}$$

and

$$\tilde{\psi}(z, \gamma) = \int_0^\infty c(z, \gamma, \tau)\tau\,\mathrm{d}\tau. \tag{6.123}$$

At first, let us find the form of the equation that is satisfied by the function (6.122). Integrate both parts of eq. (6.121) over τ between zero and infinity. Using the starting condition we obtain

$$\tilde{\varphi}(z, \gamma) = c_0(z, \gamma) + \int_0^\infty a(\nu)b(\nu)\,\mathrm{d}\nu \int_0^\infty \tilde{\varphi}(z_1, \gamma)E_1(b(\nu)|z - z_1|)\,\mathrm{d}z_1. \tag{6.124}$$

We see that eq. (6.124) differs from eq. (6.109) in that it contains, instead of the exponent, the function $c_0(z, \gamma)$ which itself satisfies eq. (6.111).

Finally, we have

$$\tilde{\varphi}'(z,\gamma) = -\gamma c_0(z,\gamma) + c_0(0,\gamma)\int_0^\infty a(\nu)b(\nu)\,d\nu \int_1^\infty \frac{1}{\eta} c_0(z,\eta b(\nu))\,d\eta$$

$$+ \tilde{\varphi}(0,\gamma)\int_0^\infty a(\nu)b(\nu)\,d\nu \int_1^\infty \frac{1}{\eta} e^{-z\eta b(\nu)}\,d\eta$$

$$+ \int_0^\infty a(\nu)b(\nu)\,d\nu \int_0^\infty \tilde{\varphi}'(z,\gamma)E_1(b(\nu)|z-z_1|)\,dz_1. \tag{6.125}$$

Thus, the function $\tilde{\varphi}'(z,\gamma)$ satisfies the equation with the same kernel as that of eqs. (6.109) and (6.124). All these equations have the form

$$\Phi(z) = f(z) + \int_0^\infty a(\nu)b(\nu)\,d\nu \int_0^\infty \Phi(z_1)E_1(b(\nu)|z-z_1|)\,dz_1,$$

where

$$\Phi(z) = \begin{cases} c_0(z_1,\gamma), & \text{if } f(z) = e^{-\gamma z}, \\ \tilde{\varphi}(z,\gamma), & \text{if } f(z) = c_0(z,\gamma). \end{cases}$$

Since these equations are linear we obtain

$$\tilde{\varphi}'(z,\gamma) = -\gamma\tilde{\varphi}(z,\gamma) + c_0(0,\gamma)\int_0^\infty a(\nu)b(\nu)\,d\nu \int_1^\infty \frac{1}{\eta}\tilde{\varphi}(z,\eta b(\nu))\,d\eta$$

$$+ \tilde{\varphi}(0,\gamma)\int_0^\infty a(\nu)b(\nu)\,d\nu \int_1^\infty \frac{1}{\eta} c_0(z,\eta b(\nu))\,d\eta. \tag{6.126}$$

Below we shall make use of the function

$$R(\xi,\gamma) = \int_0^\infty e^{-\xi z}\tilde{\varphi}(z,\gamma)\,dz. \tag{6.127}$$

To find this function multiply both sides of eq. (6.126) by $\exp(-\xi z)$ and integrate the results over z between zero and unity obtaining

$$(\xi+\gamma)R(\xi,\gamma) = \tilde{\varphi}(0,\gamma)\left[\int_0^\infty a(\nu)b(\nu)\,d\nu \int_1^\infty \frac{1}{\eta} R_0(\xi,\eta b(\nu))\,d\eta\right]$$

$$+ c_0(0,\gamma)\int_0^\infty a(\nu)b(\nu)\,d\nu \int_1^\infty \frac{1}{\eta} R(\xi,\eta b(\nu))\,d\eta. \tag{6.128}$$

This equation can be considerably simplified by making use of eq. (6.114) and eq. (6.124) for $z = 0$ which now has the form

$$\varphi(0, \gamma) = c_0(0, \gamma) + \int_0^\infty a(\nu)b(\nu)\,d\nu \int_1^\infty \frac{1}{\eta} R(\gamma, \eta b(\nu))\,d\eta. \tag{6.129}$$

Proceeding from the above we transform eq. (6.128) into

$$(\xi + \gamma)R(\xi, \gamma) = \tilde{\varphi}(0, \gamma)c_0(0, \xi) + \tilde{\varphi}(0, \xi)c_0(0, \gamma) - c_0(0, \xi)c_0(0, \gamma). \tag{6.130}$$

Thus, the function $R(\xi, \gamma)$ is expressed in terms of the functions $\varphi(z, \gamma)$ and $c_0(z, \gamma)$ for $z = 0$. The equation for $\varphi(0, \gamma)$ can be found by substituting eq. (6.130) into eq. (6.129) and using the equation satisfied by the function $c_0(0, \gamma)$. Simple transformations yield the sought equation:

$$\frac{\tilde{\varphi}(0, \gamma)}{c_0(0, \gamma)} = 1 + c_0(0, \gamma) \int_0^\infty a(\nu)b(\nu)\,d\nu \int_1^\infty \frac{1}{\eta} \tilde{\varphi}(0, \eta b(\nu))\,d\eta.$$

If we introduce the function $\chi(x) \equiv \tilde{\varphi}(0, \gamma)$, $x = 1/\gamma$, this equation can be rewritten as

$$\frac{\chi(x)}{\varphi(x)} = 1 + x\varphi(x) \int_0^\infty a(\nu)b(\nu)\,d\nu \int_0^{1/b(\nu)} \frac{\chi(y)}{x + y}\,dy. \tag{6.131}$$

It is clear, however, that we can also write

$$\chi(x) = 1 + x \int_0^\infty a(\nu)b(\nu)\,d\nu \int_0^{1/b(\nu)} \frac{\varphi(z)\chi(x) + \varphi(x)\chi(z)}{x + z}\,dz. \tag{6.131a}$$

But it is more convenient to use the function $\theta(x) \equiv \chi(x)/\varphi(x)$. According to eqs. (6.131) we have

$$\theta(x) = 1 + x\varphi(x) \int_0^\infty a(\nu)b(\nu)\,d\nu \int_0^{1/b(\nu)} \frac{\theta(y)\varphi(y)}{x + y}\,dy. \tag{6.132}$$

It is sufficient to know the function $\theta(x)$ to find the approximate decay time $\tau_0(\nu, \theta)$. According to the definition (6.101) we have

$$\tilde{\tau}_0(\theta, \nu) = \frac{R(\gamma, b(\nu) \sec \theta)}{R_0(\gamma, b(\nu) \sec \theta)}. \tag{6.133}$$

Therefore, using eqs. (6.130) and (6.115) we find

$$\tilde{\tau}_0(\nu, \theta) = \theta\left(\frac{1}{\gamma}\right) + \theta\left(\frac{\cos \theta}{b(\nu)}\right) - 1. \tag{6.134}$$

In that spectral region of radiation in which the absorption coefficient is very small, $b(\nu) \ll 1$, the quantity $\tau_0(\nu, \theta)$ does not depend on either the frequency ν or the angle θ at which luminescence is observed. In this spectral region we have

$$\tilde{\tau}_0 = \theta(1/\gamma) + \theta(\infty) - 1, \tag{6.135}$$

that is, $\tilde{\tau}_0$ is fully determined by the absorption coefficient for the light exciting luminescence.

All the remarks made above concerning the solution of eq. (6.116) are valid for the function $\theta(x)$.

In the q approximation eq. (6.132) has the form

$$\theta(x) = 1 + \frac{q}{2} x\varphi(x) \int_0^1 \frac{\theta(y)\varphi(y)}{x+y} \, dy. \tag{6.136}$$

Table 6.6
Function $\theta(x)$.

x	q = 0.4	0.5	0.6	0.7	0.8	0.9	0.95
0	1.00	1.00	1.00	1.00	1.00	1.00	1.00
0.1	1.07	1.09	1.12	1.15	1.21	1.32	1.47
0.2	1.10	1.13	1.18	1.24	1.34	1.54	1.78
0.3	1.12	1.17	1.23	1.32	1.45	1.74	2.10
0.4	1.14	1.20	1.27	1.38	1.55	1.92	2.30
0.5	1.16	1.22	1.31	1.43	1.63	2.07	2.65
0.6	1.17	1.24	1.34	1.48	1.71	2.22	2.90
0.7	1.19	1.26	1.36	1.52	1.77	2.35	3.14
0.8	1.19	1.28	1.38	1.55	1.83	2.47	3.35
0.9	1.20	1.29	1.40	1.58	1.88	2.61	3.55
1.0	1.21	1.30	1.42	1.61	1.94	2.73	3.76
2.0	1.25	1.35	1.53	1.76	2.25	3.45	5.21
3.0	1.28	1.40	1.59	1.85	2.43	3.86	6.13
4.0	1.28	1.42	1.62	1.94	2.54	4.13	6.74
5.0	1.30	1.43	1.64	1.97	2.61	4.35	7.25
6.0	1.30	1.44	1.66	2.00	2.67	4.49	7.62
7.0	1.31	1.45	1.67	2.03	2.71	4.60	7.92
8.0	1.31	1.46	1.68	2.04	2.73	4.68	8.12
9.0	1.31	1.46	1.69	2.05	2.76	4.74	8.37
10.0	1.31	1.46	1.69	2.06	2.78	4.82	8.51
30.0	1.33	1.49	1.73	2.08	2.91	5.26	9.67
50.0	1.33	1.49	1.74	2.14	2.94	5.35	9.92
100.0	1.33	1.49	1.74	2.15	2.96	5.43	10.13
∞	1.33	1.50	1.75	2.17	3.00	5.50	10.50

If $q < 0.4$ the function $\theta(x)$ is equal to the function $\varphi(x)$ to within 3%. Table 6.6 presents the values of the function $\theta(x)$ for large values of the parameter q. Table 6.7 gives the values of $\tilde{\tau}_0$ found from eq. (6.135) for the same q values.

Now let us analyze the function $\tilde{\psi}(z, \gamma)$ which is needed to calculate $\tau_0(\nu, \theta)$.

Multiplying eq. (6.121) by τ and integrating the result between zero and infinity we obtain

$$\tilde{\psi}(z, \gamma) = \tilde{\varphi}(z, \gamma) + \int_0^\infty a(\nu) b(\nu)\, d\nu \int_0^\infty \tilde{\psi}(z_1, \gamma) E_1(b(\nu)|z - z_1|)\, dz_1. \qquad (6.137)$$

This equation is similar to eq. (6.124). Therefore, we can employ the same

Table 6.7
Function $\tilde{\tau}_0(\gamma)$, $x = 1/\gamma$.

x	q = 0.4	0.5	0.6	0.7	0.8	0.9	0.95
0	1.33	1.50	1.75	2.17	3.00	5.50	10.50
0.1	1.40	1.59	1.87	2.32	3.17	5.82	10.97
0.2	1.43	1.63	1.93	2.41	3.30	6.04	11.28
0.3	1.45	1.67	1.98	2.49	3.41	6.24	11.60
0.4	1.47	1.70	2.02	2.55	3.51	6.42	11.89
0.5	1.49	1.72	2.06	2.60	3.59	6.57	12.15
0.6	1.50	1.74	2.09	2.65	3.67	6.72	12.40
0.7	1.52	1.76	2.11	2.69	3.73	6.85	12.64
0.8	1.52	1.78	2.13	2.72	3.79	6.97	12.85
0.9	1.53	1.79	2.15	2.75	3.84	7.11	13.05
1.0	1.54	1.80	2.17	2.78	3.90	7.23	13.26
2.0	1.58	1.85	2.28	2.93	4.22	7.95	14.71
3.0	1.61	1.90	2.34	3.02	4.39	8.36	15.63
4.0	1.61	1.92	2.37	3.11	4.50	8.63	16.24
5.0	1.63	1.93	2.39	3.14	4.57	8.85	16.75
6.0	1.63	1.94	2.41	3.17	4.63	8.99	17.12
7.0	1.64	1.95	2.42	3.20	4.67	9.10	17.42
8.0	1.64	1.96	2.43	3.21	4.69	9.18	17.62
9.0	1.64	1.96	2.44	3.22	4.72	9.24	17.87
10.0	1.65	1.96	2.44	3.23	4.74	9.32	18.01
30.0	1.66	1.99	2.48	3.25	4.87	9.76	19.17
50.0	1.66	1.99	2.49	3.31	4.90	9.85	19.42
100.0	1.66	1.99	2.49	3.32	4.92	9.93	19.63
∞	1.67	2.00	2.50	3.34	5.00	10.00	20.00

procedure as that used above and express the function

$$T(\eta, \gamma) = \int_0^\infty e^{-\eta z} \tilde{\psi}(z, \gamma)\, dz \tag{6.138}$$

in terms of the functions $\tilde{\psi}(z, \gamma)$, $\tilde{\varphi}(z, \gamma)$ and $c_0(z, \gamma)$ for $z = 0$. We obtain

$$T(\xi, \gamma) = \tilde{\psi}(0, \gamma)c_0(0, \xi) + \tilde{\psi}(0, \xi)c_0(0, \gamma)$$
$$+ \tilde{\varphi}(0, \gamma)\tilde{\varphi}(0, \xi) - c_0(0, \gamma)\tilde{\varphi}(0, \xi) - c_0(0, \xi)\tilde{\varphi}(0, \gamma). \tag{6.139}$$

Introducing the function

$$\omega(x) = \tilde{\psi}(0, \gamma)/c_0(0, \gamma), \qquad x = 1/\gamma,$$

as it was done above we can easily show that the function $\omega(x)$ satisfies the

Table 6.8
Function $\omega(x)$.

	q						
x	0.4	0.5	0.6	0.7	0.8	0.9	0.95
0	1.00	1.00	1.00	1.00	1.00	1.00	1.00
0.1	1.15	1.20	1.29	1.39	1.61	2.25	3.81
0.2	1.22	1.31	1.45	1.66	2.07	2.35	6.42
0.3	1.27	1.40	1.59	1.90	2.49	4.44	0.09
0.4	1.32	1.48	1.71	2.10	2.88	5.46	11.71
0.5	1.36	1.54	1.82	2.28	3.22	6.41	14.27
0.6	1.39	1.59	1.91	2.45	3.56	7.36	16.80
0.7	1.43	1.64	1.98	2.59	3.83	8.17	19.35
0.8	1.45	1.69	2.05	2.71	4.11	9.08	21.74
0.9	1.47	1.72	2.11	2.82	4.35	10.00	24.12
1.0	1.49	1.76	2.17	2.94	4.62	10.86	26.56
2.0	1.61	1.94	2.56	3.55	6.29	17.04	46.35
3.0	1.68	2.07	2.75	3.97	7.26	20.99	61.18
4.0	1.69	2.12	2.86	4.29	7.92	23.83	72.13
5.0	1.73	2.17	2.94	4.45	8.35	26.20	81.61
6.0	1.74	2.19	3.00	4.58	8.72	27.75	88.85
7.0	1.76	2.22	3.04	4.67	8.98	29.01	94.85
8.0	1.77	2.23	3.07	4.75	9.14	29.95	99.18
9.0	1.77	2.25	3.10	4.80	9.32	30.73	104.4
10.0	1.77	2.26	3.17	4.86	9.45	31.58	107.4
30.0	1.82	2.33	3.26	5.03	10.33	37.03	134.1
50.0	1.82	2.34	3.30	5.23	10.54	38.20	140.3
∞	1.83	2.37	3.34	5.38	11.00	40.40	155.3

equation

$$\omega(x) = 1 + \theta^2(x) - \theta(x) + x\varphi(x) \int_0^\infty a(\nu)b(\nu)\,\mathrm{d}\nu \int_0^{1/b(\nu)} \frac{\omega(y)\varphi(y)}{x+y}\,\mathrm{d}y, \quad (6.140)$$

or in the q approximation

$$\omega(x) = 1 + \theta^2(x) - \theta(x) + \frac{q}{2} x\varphi(x) \int_0^1 \frac{\omega(y)\varphi(y)}{x+y}\,\mathrm{d}y. \quad (6.141)$$

We can show that $\omega(x) \approx 2\varphi(x) - 1$ for $q < 0.4$. The values of the function $\omega(x)$ for $q > 0.4$ are given in table 6.8.

According to eq. (6.100) the luminescence decay time

$$\tau_0(\nu, \theta) = \frac{T(\gamma, b(\nu)\sec\theta)}{R(\gamma, b(\nu)\sec\theta)}. \quad (6.142)$$

Hence, using eqs. (6.139) and (6.130) we obtain

$$\tau_0(\nu, \theta) = \frac{\omega(1/\gamma) + \omega(\cos\theta/b(\nu)) + \theta(1/\gamma)\theta(\cos\theta/b(\nu)) - 1}{\theta(1/\gamma) + \theta(\cos\theta/b(\nu)) - 1} - 1. \quad (6.143)$$

In the spectral region in which we can ignore absorption of luminescence light, that is, in which $b(\nu) \ll 1$, we have

$$\tau_0 = \frac{\omega(1/\gamma) + \omega(\infty) + \theta(1/\gamma)\theta(\infty) - 1}{\theta(1/\gamma) + \theta(\infty) - 1} - 1, \quad (6.144)$$

that is, τ_0 is independent of either the frequency ν or the angle θ at which luminescence is observed. Table 6.9 presents the values of τ_0 given by eq. (6.144) for $q > 0.4$.

A comparison of τ_0 and $\tilde{\tau}_0$ indicates whether luminescence decay is close to exponential. As noted above, if the decay is exponential τ_0 and $\tilde{\tau}_0$ must coincide. A coincidence within 2–3% is, indeed, found for $q < 0.4$. In this range of q values we have

$$\tau_0(\nu, \theta) \approx \tilde{\tau}_0(\nu, \theta) = 1 + \frac{1}{\gamma} \int_0^\infty a(\nu_1)b(\nu_1)\,\mathrm{d}\nu_1 \ln\left(1 + \frac{\gamma}{b(\nu_1)}\right)$$

$$+ \frac{\cos\theta}{b(\nu)} \int_0^\infty a(\nu_1)b(\nu_1)\,\mathrm{d}\nu_1 \ln\left(1 + \frac{b(\nu)}{b(\nu_1)\cos\theta}\right). \quad (6.145)$$

In the spectral range in which $b(\nu) \ll 1$ eq. (6.145) is transformed into the following expression derived by Galanin (1950):

$$\tau_0(\gamma) \approx \tilde{\tau}_0(\gamma) = 1 + \frac{1}{2} \int_0^\infty E(\nu)\eta(\nu)\,\mathrm{d}\nu \left[1 + \frac{b(\nu)}{\gamma} \ln\left(1 + \frac{\gamma}{b(\nu)}\right)\right]. \quad (6.146)$$

Table 6.9
Function $\tau_0(x)$, $x = 1/\gamma$.

	q						
x	0.4	0.5	0.6	0.7	0.8	0.9	0.95
0.0	1.37	1.58	1.91	2.48	3.67	7.33	14.78
0.1	1.43	1.64	1.98	2.58	3.69	7.37	14.80
0.2	1.45	1.68	2.03	2.63	3.76	7.44	14.82
0.3	1.48	1.71	2.07	2.68	3.83	7.52	14.90
0.4	1.49	1.74	2.10	2.72	3.89	7.60	15.00
0.5	1.51	1.76	2.13	2.76	3.94	7.67	15.08
0.6	1.52	1.77	2.16	2.79	3.99	7.74	15.18
0.7	1.53	1.79	2.18	2.82	4.03	7.80	15.27
0.8	1.54	1.80	2.19	2.85	4.07	7.88	15.36
0.9	1.54	1.81	2.21	2.87	4.10	7.94	15.45
1.0	1.55	1.82	2.22	2.89	4.14	8.00	15.54
2.0	1.59	1.88	2.32	3.01	4.37	8.46	16.29
3.0	1.61	1.91	2.36	3.10	4.50	8.74	16.84
4.0	1.62	1.93	2.39	3.14	4.58	8.93	17.24
5.0	1.63	1.94	2.41	3.18	4.64	9.09	17.57
6.0	1.64	1.96	2.42	3.20	4.68	9.19	17.82
7.0	1.64	1.96	2.43	3.21	4.71	9.27	18.02
8.0	1.64	1.96	2.44	3.23	4.74	9.33	18.17
9.0	1.65	1.96	2.44	3.23	4.76	9.39	18.34
10.0	1.65	1.97	2.45	3.25	4.77	9.44	18.44
30.0	1.66	1.98	2.48	3.29	4.88	9.77	19.29
50.0	1.66	1.99	2.49	3.31	4.90	9.90	19.48
100.0	1.66	2.00	2.50	3.34	4.92	9.93	19.63
∞	1.67	2.00	2.50	3.34	5.00	10.00	20.00

Though the above approach provides information on the characteristics of the radiation emitted from the medium it does not furnish us with an effective procedure for calculating the exciton distribution perpendicular to the crystal surface. The knowledge of this distribution can prove to be essential for analyzing some effects involving excitons (see sect. 8).

There are many various methods for solving the transport equations which are satisfied by the function $c(z, \gamma)$. These methods are reviewed in the theory of reactors (see, e.g., Davison 1957). Here we shall discuss only one simple and convenient method used by Agranovich (1957). According to this method, the kernel of the integral equation (6.121) taken in the q approximation, that is, the function $E_1(|z - z_1|)$, is replaced with a simpler kernel of the type of $\mu \exp(-\mu|z - z_1|)$ where the condition of minimization of the standard deviation yields $\mu = 3$ (this method was also used by Galanin et al. (1962) for describing decay of luminescence of anthracene in the case of gamma excitation). After this replacement we can find exact solutions of eqs. (6.121) and (6.107). The mean "exponential" decay time for the spectral region in which absorption of luminescence light is insignificant is found to be

$$\tilde{\tau}_0 = \frac{\gamma(1-\frac{1}{2}q) + 3\sqrt{1-q}}{(1-q)(\gamma + 3\sqrt{1-q})}. \tag{6.147}$$

If $\gamma = k/k_0 \gg 1$ we have

$$\tilde{\tau}_0 = (1-\tfrac{1}{2}q)/(1-q). \tag{6.148}$$

If $\gamma \ll 1$ we have

$$\tilde{\tau}_0 = 1/(1-q). \tag{6.149}$$

Equation (6.149) was first derived by Galanin (1950) but without the applicability criterion ($\gamma \ll 1$).

In the analysis of Galanin (1950) reabsorption was regarded as a multistep process in which the instantaneous external source generates excited molecules whose number n_1 decays as $\exp(-t/\tau)$, these molecules excite other molecules and so on.

Thus, the number of excited molecules of "successive generations" can be expressed by the following system of equations (q describes the relative fraction of reabsorbed light which is the same for all the steps of the process if $\gamma \ll 1$):

$$\frac{dn_1}{dt} = -\frac{n_1}{\tau}, \qquad \frac{dn_2}{dt} = -\frac{n_2}{\tau} + \frac{q}{\tau} n_1, \qquad \frac{dn_3}{dt} = -\frac{n_3}{\tau} + \frac{q}{\tau} n_2,$$

and so on.

The solution has the form

$$n_i = \frac{n_0}{(i-1)!}\left(\frac{qt}{\tau}\right)^{i-1} \exp(-t/\tau).$$

Hence, we obtain

$$n(t) = \sum n_i(t) = n_0 \exp[-(1-q)t/\tau],$$

which corresponds to eq. (6.149).

Now we shall derive the explicit expression for the time dependence (for $t \geq 0$) of the luminescence intensity normalized to unity for $t = 0$ assuming the conditions discussed above. According to Galanin et al. (1962) this quantity given by

$$B_\gamma(t) = \frac{\int_0^\infty c(z,\gamma,t)\,dz}{\int_0^\infty c_0(z,\gamma)\,dz}.$$

can be transformed by means of the above replacement of the integral equation kernel into

$$B_\gamma(t) = \frac{1}{\pi}\frac{\gamma s}{\gamma - s/2}\, e^{-\tau}\, [F_{s/2}(qt) - F_{\gamma - s/2}(qt)],$$

where

$$s = 2\sqrt{1-q}, \qquad F_\gamma(z) = \int_0^\infty \frac{\exp[-z/(1+x^2)]}{\gamma^2 + x^2}\, dx.$$

In the limiting case of $\gamma \gg \sqrt{1-q}$ we have

$$B_\gamma(t) \approx B_\infty(t) = \frac{s}{\pi}\, e^{-t} F_{s/2}(qt).$$

For $\gamma \ll \sqrt{1-q}$ we obtain the expression

$$B_\gamma(t) \approx B_0(\tau) = \exp[-(1-q)t],$$

which corresponds to eq. (6.149) and agrees with the conclusion made below that in the case of excitation by low-absorption light luminescence decay has an exponential nature.

Konobeev (1971) (see below) has shown that although eqs. (6.148) and (6.149) were derived with an approximate method they are accurate even in the case when the q approximation is not used. As for eq. (6.147), it provides a convenient extrapolation for the intermediate values of γ and its accuracy seems to be not lower than that of the available experimental techniques for determining the luminescence decay time.

To estimate how luminescence decay deviates from the exponential behaviour under the conditions of strong reabsorption (Konobeev 1971) let us compare the above limiting values of $\tilde{\tau}_0$ with the respective limiting values of τ_0 (definitions of these parameters are given by eqs. (6.100) and (6.101)). Equation (6.134) indicates that, as noted above, $\tilde{\tau}_0$ does not depend on the angle of observation of luminescence in the frequency range for which $k(\nu) \ll k_0$ (that is, $b(\nu) \ll 1$) and (see eq. (6.135))

$$\tau_0 = 2\theta(\infty) - 1, \tag{6.150}$$

if the absorption coefficient for the excitation radiation is $k \ll k_0$ (that is, if $\gamma \ll 1$) or

$$\tilde{\tau}_0 = \theta(\infty), \tag{6.151}$$

if $k \gg k_0$ (that is, if $\gamma \gg 1$). For the same conditions of observation eq. (6.144) yields

$$\tau_0 = \frac{2\omega(\infty) + \theta^2(\infty) - 2\theta(\infty)}{2\theta(\infty) - 1}, \quad \gamma \ll 1, \tag{6.152}$$

and

$$\tau_0 = \omega(\infty)/\theta(\infty), \quad \gamma \gg 1. \tag{6.153}$$

Thus, to find all the above limiting values of $\tilde{\tau}_0$ and τ_0 we have to know the functions $\theta(x)$ and $\omega(x)$ for $x \to \infty$.

Equation (6.132) shows that

$$\theta(\infty) = 1 + \varphi(\infty)\chi_0, \tag{6.154}$$

where χ_0 is the zero moment of the function $\chi(x)$,

$$\chi_m = \int_0^\infty a(\nu)b(\nu)\,\mathrm{d}\nu \int_0^{1/b(\nu)} \chi(x)x^m\,\mathrm{d}x.$$

At the same time, eq. (6.116) yields

$$\varphi(\infty) = 1 + \varphi(\infty)\varphi_0 = 1/(1-\varphi_0),$$

and, moreover,

$$\varphi_0 = \tfrac{1}{2}q + \tfrac{1}{2}\varphi_0^2.$$

Finding φ_0 from this equation and noting that $\varphi(\infty) > 0$ we obtain

$$\varphi(\infty) = 1/\sqrt{1-q}, \qquad \varphi_0 = 1 - \sqrt{1-q}. \tag{6.155}$$

Similarly, using eq. (6.131) we obtain

$$\chi_0 = q/2(1-\varphi_0). \tag{6.156}$$

Finally, eqs. (6.154)–(5.156) yield

$$\theta(\infty) = (1 - \tfrac{1}{2}q)/(1-q). \tag{6.157}$$

We see that $\tilde{\tau}_0$ (see eqs. (6.150) and (6.151)) is, indeed, described by eqs. (6.148) and (6.149) in the limiting cases of strong and weak absorption of the excitation radiation. Let us now find the respective expressions for τ_0.

The parameter $\omega(\infty)$ entering into eqs. (6.152) and (6.153) can be derived from eq. (6.140) by using a similar procedure. Simple transformations which we omit here yield

$$\omega(\infty) = 1 + \frac{q}{(1-q)^2}\left(1 - \frac{5q}{8}\right). \tag{6.158}$$

Making use of eqs. (6.152) and (6.157) we obtain now

$$\tau_0 = \begin{cases} 1/(1-q), & \gamma \ll 1, \\ (1-\tfrac{1}{2}q)/(1-q) + q^2/8(1-q)(1-\tfrac{1}{2}q), & \gamma \gg 1. \end{cases}$$

Hence, the times τ_0 and $\tilde{\tau}_0$ coincide in the case of weak absorption of the excitation radiation and this fact indicates exponential decay of the luminescence intensity irrespective of the value of the parameter q. In the case of strong absorption of the excitation radiation we have $\tau_0 \neq \tilde{\tau}_0$ which indicates some deviation from exponential decay. This deviation can be approximately characterized by the ratio $\Delta = (\tau_0 - \tilde{\tau}_0)/\tilde{\tau}_0$ the maximum value of which is $\Delta = q^2/8(1-q/2)^2$ when $\gamma \gg 1$ and increases with increasing q. For $q = 0.72$ (anthracene) Δ is only about 16% while for $q \approx 1$ we have $\Delta \approx 0.5$.

8. *Spectrum and decay time of luminescence for crystals of finite thickness: annihilation of excitons and contribution of reabsorption*

A method similar to that used in the above sections can be applied to crystals of finite thickness (Agranovich and Konobeev 1959, 1961). The function (6.112) is replaced now with

$$R_0(\xi, \gamma) = \int_0^{z_0} e^{-\xi z} c_0(z, \gamma)\, dz. \tag{6.159}$$

This function can be expressed in terms of the function $c_0(z, \gamma)$ for $z = 0$ and $z = z_0$ as

$$R_0(\xi, \gamma) = \frac{c_0(0, \gamma)c_0(0, \xi) - c_0(z_0, \gamma)c_0(z_0, \xi)}{\xi + \gamma}, \tag{6.160}$$

where the functions $\varphi(x) \equiv c_0(0, \gamma)$ and $\tilde{\chi}(x) \equiv c_0(z_0, \gamma)$ satisfy the following system of two nonlinear equations:

$$\varphi(x) = 1 + x\int_0^\infty a(\nu)b(\nu)\, d\nu \int_0^{1/b(\nu)} \frac{\varphi(x)\varphi(y) - \tilde{\chi}(x)\tilde{\chi}(y)}{x+y}\, dy,$$

$$\tilde{\chi}(x) = \exp(-z_0/x) + x\int_0^\infty a(\nu)b(\nu)\, d\nu \int_0^{1/b(\nu)} \frac{\varphi(x)\tilde{\chi}(y) - \varphi(y)\tilde{\chi}(x)}{y-x}\, dy. \tag{6.161}$$

In the q approximation these equations coincide with those derived by Chandrasekhar (1950) for scattering of light in planetary atmospheres. If the dimensionless thickness of the crystal is $z_0 \to \infty$ then $\tilde{\chi}(x) \to 0$ and the equation for $\varphi(x)$ reduces to eq. (6.116).

As in the above treatment, to find τ_0 and $\tilde{\tau}_0$ in the nonstationary case we need only the quantities:

$$R(\xi, \gamma) = \int_0^{z_0} e^{-\xi z} \tilde{\varphi}(z, \gamma)\, dz, \tag{6.162}$$

$$T(\xi, \gamma) = \int_0^{z_0} e^{-\xi z} \tilde{\psi}(z, \gamma)\, dz, \tag{6.163}$$

where the functions $\tilde{\varphi}(z, \gamma)$ and $\tilde{\psi}(z, \gamma)$ are given by eqs. (6.122) and (6.123). It can be shown (Agranovich and Konobeev 1959, 1961) that

$$\begin{aligned} R(\xi, \gamma) = \frac{1}{\xi + \gamma} [&\tilde{\varphi}(0, \gamma) c_0(0, \xi) + \tilde{\varphi}(0, \xi) c_0(0, \gamma) \\ &- \tilde{\varphi}(z_0, \gamma) c_0(z_0, \xi) - \tilde{\varphi}(z_0, \xi) c_0(z_0, \gamma)] - R_0(\xi, \gamma), \end{aligned} \tag{6.164}$$

$$\begin{aligned} T(\xi, \gamma) = \frac{1}{\xi + \gamma} [&\tilde{\psi}(0, \gamma) c_0(0, \xi) + \tilde{\psi}(0, \xi) c_0(0, \gamma) \\ &- c_0(z_0, \gamma) \tilde{\psi}(z_0, \xi) - c_0(z_0, \xi) \tilde{\psi}(z_0, \gamma) + \tilde{\varphi}(0, \gamma) \tilde{\varphi}(0, \xi) \\ &- \tilde{\varphi}(z_0, \xi) \tilde{\varphi}(z_0, \gamma)] - R(\xi, \gamma) - R_0(\xi, \gamma). \end{aligned} \tag{6.165}$$

We see that $R(\xi, \gamma)$ is expressed in terms of the quantities $\tilde{\varphi}(0, \gamma) \equiv M(x)$ and $\tilde{\varphi}(z_0, \gamma) \equiv N(x)$ while $T(\xi, \gamma)$ is expressed in terms of these quantities and, in addition, $\tilde{\psi}(0, \gamma) \equiv K(x)$ and $\tilde{\psi}(z_0, \gamma) \equiv L(x)$. The pairs of the functions $M(x)$ and $N(x)$ and $K(x)$ and $L(x)$ satisfy the following systems of linear equations:

$$\begin{aligned} M(x) = 1 + x \int_0^{\infty} a(\nu) b(\nu)\, d\nu \int_0^{1/b(\nu)} &\left[\frac{\varphi(z) M(x) + \varphi(x) M(z)}{z + x} \right. \\ &\left. - \frac{\tilde{\chi}(x) N(z) + \tilde{\chi}(z) N(x)}{z + x} \right] dz, \end{aligned} \tag{6.166a}$$

$$\begin{aligned} N(x) = \exp(-z_0/x) + x \int_0^{\infty} a(\nu) b(\nu)\, d\nu \int_0^{1/b(\nu)} &\left[\frac{M(x) \tilde{\chi}(z) - \tilde{\chi}(x) M(z)}{z - x} \right. \\ &\left. + \frac{\varphi(x) N(z) - N(x) \varphi(z)}{z - x} \right] dz, \end{aligned} \tag{6.166b}$$

$$\begin{aligned} K(x) = 1 + x \int_0^{\infty} a(\nu) b(\nu)\, d\nu \int_0^{1/b(\nu)} &\left[\frac{M(x) M(z) + N(x) N(z)}{z + x} \right. \\ &\left. + \frac{\varphi(x) K(z) + \varphi(z) K(x)}{x + z} - \frac{\tilde{\chi}(x) L(z) + \tilde{\chi}(z) L(x)}{z + x} \right] dz, \end{aligned} \tag{6.167a}$$

$$L(x) = \exp(-z_0/x) + x \int_0^\infty a(\nu)b(\nu)\,d\nu \int_0^{1/b(\nu)} \left[\frac{M(x)N(z) - M(z)N(x)}{z - x} + \frac{\varphi(x)L(z) - \varphi(z)L(x)}{z - x} - \frac{\tilde{\chi}(x)K(z) - \tilde{\chi}(z)K(x)}{z - x}\right] dz. \quad (6.167b)$$

The functions $N(x)$ and $L(x) \to 0$ for $z_0 \to \infty$ while $M(x) \to \chi(x)$ and $K(x) \to \varphi(x)\omega(x)$ as shown in sect. 7.

The results of numerical calculations of the functions $\varphi(x)$, $\tilde{\chi}(x)$, $M(x)$, $N(x)$, $K(x)$ and $L(x)$ in the q approximation for $q = 0.5$, 0.8 and 0.9 for various z_0 values are reported by Agranovich and Konobeev (1959, 1961). The decay times are also calculated by Agranovich and Konobeev (1959, 1961). For $q < 0.4$ these functions can be found from eqs. (6.166) and (6.167) by successive approximations of expansions in powers of q.

Samson (1960) has suggested a very effective method for finding approximate analytical expressions for the functions $\varphi(x)$ and $\tilde{\chi}(x)$ in the q approximation. The method is based on the fact that the integrands in eqs. (6.161) are slowly varying functions of y. Therefore, they can be approximated with their values at a certain intermediate point which is largely arbitrary and taken outside the integral. Samson has shown that if this point is $y = \frac{1}{2}$ the difference between the resulting solutions for φ and $\tilde{\chi}$ and the exact ones is 2–3%. Using this method we obtain for eqs. (6.161) in the q approximation

$$\varphi(x) = 1 + \frac{qx}{2x+1}[\varphi(x)\varphi(\tfrac{1}{2}) - \tilde{\chi}(x)\tilde{\chi}(\tfrac{1}{2})], \quad (6.168a)$$

$$\tilde{\chi}(x) = \exp(-z_0/x) + \frac{q(x)}{2x-1}[\varphi(\tfrac{1}{2})\tilde{\chi}(x) - \varphi(x)\tilde{\chi}(\tfrac{1}{2})]. \quad (6.168b)$$

A solution of this system is

$$\varphi(x) = \frac{1 - \dfrac{qx}{2x-1}\varphi(\tfrac{1}{2}) - \dfrac{qx}{2x+1}\tilde{\chi}(\tfrac{1}{2})\exp(-z_0/x)}{4(1-q)x^2 - 1}(4x^2 - 1). \quad (6.169)$$

Since the function $\varphi(x)$ is finite for all x eq. (6.169) implies that not only the denominator but also numerator vanishes for $x = x_0 \equiv \frac{1}{2}(1-q)^{-1/2}$. Hence, we have

$$1 - \frac{qx_0}{2x_0 - 1}\varphi(\tfrac{1}{2}) = \frac{qx_0}{2x_0+1}\tilde{\chi}(\tfrac{1}{2})\exp(-z_0/x_0). \quad (6.170)$$

Taking $x = \frac{1}{2}$ in the right-hand side of eq. (6.168a) we can obtain the second equation relating $\varphi(\frac{1}{2})$ to $\tilde{\chi}(\frac{1}{2})$. Using this equation and eq. (6.170) we obtain

$$\varphi(\tfrac{1}{2}) = 4x_0 \frac{(2x_0+1)-(2x_0-1)\exp(-2z_0/x_0)}{(2x_0+1)^2-(2x_0-1)^2\exp(-2z_0/x_0)},$$

$$\tilde{\chi}(\tfrac{1}{2}) = \frac{8x_0\exp(-z_0/x_0)}{(2x_0+1)^2-(2x_0-1)^2\exp(-2z_0/x_0)}. \tag{6.171}$$

Equations (6.171) and (6.168) fully determine the functions $\varphi(x)$ and $\tilde{\chi}(x)$ in this approximation.

Konobeev (1969) has used this method for finding the functions $M(x)$ and $N(x)$ and thus approximately calculated the dependence of the mean "exponential" time of luminescence decay on the q value and the thickness of the crystal.

In the frequency range in which the absorption of luminescence light can be ignored eqs. (6.133), (6.160) and (6.164) yield

$$\tilde{\tau}_0 = \frac{R(\gamma,0)}{R_0(\gamma,0)} = \frac{M(x)-N(x)}{\varphi(x)-\tilde{\chi}(x)} + \frac{M(\infty)}{\varphi(\infty)} - 1.$$

Using this expression Konobeev derived the following explicit dependence assuming $\gamma \gg 1$:

$$\tilde{\tau}_0(\gamma) = \frac{1-\frac{1}{2}q}{1-q} - 2qx_0^2 \frac{x_0(2x_0+1)^2-(2x_0-1)^2[z_0(2x_0+1)+1]}{x_0(2x_0+1)^2-(2x_0-1)^2[z_0(2x_0+1)-1]\exp(-2z_0/x_0)} \times \exp(-z_0/x_0). \tag{6.172}$$

In many experiments luminescence is excited with light whose absorption coefficient is much higher than the absorption coefficient for the light corresponding to maximum reabsorption. Therefore, we shall limit our analysis to eq. (6.172) for $\gamma \gg 1$ though a formula for a more general case can be obtained in a similar way.

In conclusion, note that Stepanov and Samson (1960), Samson (1960) have derived a number of convenient formulas describing the luminescence decay time in the cases of pulsed excitation, transmission experiments, etc. These authors have also developed a method based on inclusion of terms of different orders of ratiation (that is, the method of successive approximations) which is effective enough only for small q values. The results of these studies are reviewed by Samson (1962) which contains a more detailed bibliography.

Bimolecular annihilation of excitons and the contribution of reabsorption. We have already discussed various processes occurring with exciton–exciton collisions in ch. 5 sect. 8. The frequency of collisions is proportional to the squared exciton concentration and, hence, it should be particularly sensitive to the exciton distribution in depth of the crystal.

Since we are concerned here with high exciton concentrations we must, generally speaking, solve eq. (6.105).

If we know the exciton distribution in the crystal at any moment we can find the parameters measured in experiments, namely, the quantum yield B of luminescence and the decay time τ. We can derive (Agranovich et al. 1969) the relationship

$$\tau/\tau_0 = B/B_0, \tag{6.173}$$

where B_0 and τ_0 are the quantum yield and the decay time at low intensities of the excitation light when we can take $A = 0$ in eq. (6.105). Therefore, we shall calculate B/B_0 as a function of the parameter A or, which is the same, as a function of the excitation intensity (see eq. (6.105a)).

Since the parameter A depends on the rate constant γ of bimolecular annihilation of excitons we can find the numerical value of γ if we can predict variation of B/B_0 as a function of A for a given value of J_0.

In our discussion of the singlet–singlet annihilation we shall assume that the luminescence spectrum of the crystal corresponds essentially to its transparency region. In this case the luminescence intensity is proportional to the integral of the exciton concentration taken over the total depth of the crystal (which is assumed to be a plane-parallel plate of the thickness d). If we ignore diffusion of excitons and reabsorption of luminescence the solution of eq. (6.105) under the steady-state conditions has the form

$$c(x) = \frac{1}{2A}(\sqrt{1 + 4A\,\mathrm{e}^{-kx}} - 1).$$

For a sufficiently thick crystal we have

$$B \sim \int_0^\infty c(x)\,\mathrm{d}x,$$

and we obtain (Tolstoi and Abramov 1967)

$$\frac{B}{B_0} = \frac{1}{A}\left(\sqrt{1+4A} - 1 - \ln\frac{\sqrt{1+4A}+1}{2}\right). \tag{6.174}$$

To illustrate the role played by reabsorption we shall ignore exciton diffusion and reflection of luminescence light from the crystal boundaries in eq. (6.105). Then the dimensionless concentration $c(x, \tau)$ of excitons (see eq. (6.102)) satisfies the equation

$$\frac{\partial c}{\partial \tau} = \mathrm{e}^{-kx} - c(x,\tau) - Ac^2(x,\tau) + \int_0^\infty a(\nu)k(\nu)\,\mathrm{d}\nu \int_0^{z_0} E_1(k(\nu)|x - x_1|)c(x,\tau)\,\mathrm{d}x_1 = 0. \tag{6.175}$$

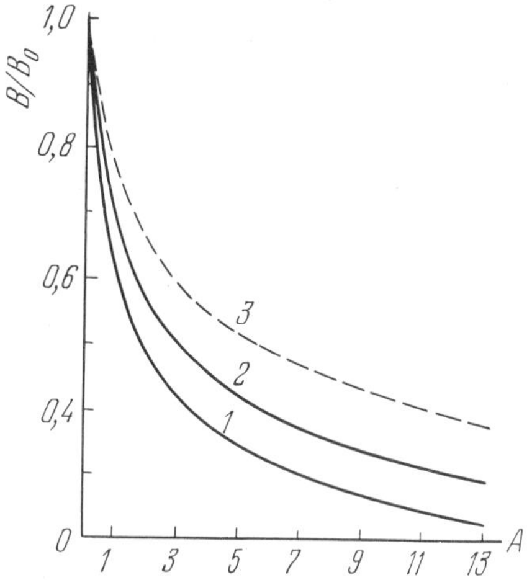

Fig. 6.3. The relative yield of luminescence as a function of the excitation intensity: (1) $d = 50\ \mu\text{m}$, (2) $d = 10\ \mu\text{m}$, (3) thick crystal, no reabsorption.

If $z_0 \gg 1$ we can ignore exciton diffusion under the conditions of strong reabsorption (as, for instance, in anthracene) only when the dimensionless parameter $\lambda^2 = (Lk_0)^2$ is small in comparison with unity (here $L = \sqrt{D\tau}$ is the diffusion length for excitons; for anthracene $Lk_0 \approx 10^{-2}$) and exciton quenching at the crystal surface is insignificant (this effect is analyzed by Babenko et al. (1971)). Below we shall assume that these conditions are satisfied.

The ratio B/B_0 as a function of A for the anthracene crystal was calculated by Agranovich et al. (1969) where light reabsorption was taken into account and the absorption spectrum and the luminescence spectrum given in ch. 7 were used. The function $c(x, 0)$ satisfying eq. (6.175) was calculated by iterations and the integral term in eq. (6.175) was regarded as a perturbation. The calculated results for $k = 5 \times 10^4\ \text{cm}^{-1}$ are given in fig. 6.3 (curves 1 and 2). The dashed line in fig. 6.3 shows the function (6.174). The plots in fig. 6.3 indicate that inclusion of reabsorption has a significant effect on γ.

For instance, in anthracene $B/B_0 = 0.4$ (Tolstoi and Abramov 1967) $J_0 = 5 \times 10^{19}$ quantums per $1\ \text{cm}^2$ per 1 s and $d = 50\ \mu\text{m}$. This corresponds to $A \approx 3$ (see fig. 6.3). Since in anthracene $1/P_0 = 10^{-8}$ s this value of $A = \gamma k J_0 / P_0^2$ corresponds to $\gamma \approx 1.1 \times 10^{-8}\ \text{cm}^3\ \text{s}^{-1}$. If we neglect reabsorption the value of γ is overestimated by a factor of 3–4 (see fig. 6.3).

In fact, the above theory of reabsorption assumes that the polariton effects discussed in ch. 4 sect. 11 are not significant. This assumption is valid for the regions of exciton transitions with low oscillator strengths or at high temperatures when $k_T \gg k_0$. If these conditions are not satisfied we must use the Boltzmann equations for polaritons (weak exciton–phonon coupling) taking into consideration not only the interband transitions caused by the inelastic polariton–phonon processes but also possible reciprocal transformations of prolaritons at the same value of $E = \hbar\omega$ when they are reflected from the crystal boundaries. Under these conditions the surface polaritons should also be taken into consideration (Agranovich and Leskova 1979).

All these factors tend to complicate the mathematics of the theory of exciton luminescence but make it more interesting. The theory will, most probably, develop in precisely this direction in the nearest future (Agranovich et al. 1980).

CHAPTER 7

Experimental Studies of Kinetic Parameters Determining Electronic Excitation Energy Transfer in Crystals

1. *Introduction*; *cooperative luminescence*

A large number of experimental studies of electronic excitation energy transfer in crystals have been reported (they are reviewed by Terenin (1956), Terenin and Ermolaev (1956, 1960), Rozman and Kilin (1959), Ganguly and Chanudhury (1959), Hochstrasser (1962), Wolf (1967), Widsor (1965), Rice and Jortner (1967), Avakian and Merrifield (1968); Avakian (1974), Powell and Soos (1975). Energy transfer in biological materials and structures has also been intensely studied.*

These studies have yielded enormous amounts of data. It is, however, fairly difficult to analyze these data since relatively few experimental results have been obtained under the conditions making possible a more or less convincing comparison with the theoretical results.

Therefore, we shall only note the main trends in the current experimental studies of energy transfer in crystals and the respective experimental techniques. However, we shall pay considerable attention to the fundamental experimental results which are important for estimating relative contributions of various mechanisms of energy transfer (radiative transfer, exciton transfer, etc.), determining the conditions for realization of coherent and incoherent excitons, analyzing the mechanisms of interaction between excitons and impurities and lattice defects, and studying the exciton–exciton collisions and, in general, the properties of excitons at high concentrations. Most such experiments have been done with molecular crystals which serve as very convenient model systems. But in recent years the demand for new laser materials has stimulated intense studies of energy transfer in doped inorganic crystals, particularly, in crystals doped with rare-earth ions. The cooperative luminescence processes due to cumulation of the energy of several guest rare-earth ions at one of them have been first observed in such crystals (Ovsyankin and Feofilov 1966, 1971a, b).

The theory of cooperative luminescence was developed by Tulub and Patzer (1968), Patzer (1971), Ivliev (1970, 1971), Miyakawa and Dexter

*See, Energy Conversion by Photosynthetic Apparatus, Report of Symposium held June 6–9, 1966. Brookhaven Symposia, Brookhaven National Laboratory.

(1970), Kiriy and Shekhtman (1972). For instance, transfer of the excitation energy of a pair of identical ions to a third ion of another species was analyzed by Miyakawa and Dexter (1970). (As shown by Altarelli and Dexter (1973) (see, also, Kaplan and Jortner 1977) a significant role can sometimes be played by single events consisting of absorption of photon by an ion and transfer to it of the energy of an earlier excited neighbouring ion.) Typically, the processes analyzed by Miyakawa and Dexter (1970) are distinguished by large resonance defects so that multiphonon processes have been taken into account in this study.

In contrast to molecular crystals where the exciton mechanism of energy transfer is of the greatest interest, in the studies of laser crystals, such as ruby, YAG-Nd^{3+}, etc., the attention is focussed, typically, on resonance energy transfer in disordered systems of impurity rare-earth ions. The relevant problems have been discussed in chs. 1, 2, 4 and 5. Here we shall be concerned only with the exciton transfer mechanism.

The following trends can be identified in this field at present.

(1) Studies of radiative energy transfer (the processes: exciton–photon–exciton). These experiments yield the photon lifetime in the medium as a function of its energy and the size and shape of the crystal. The main experimental approach is to study the kinetics of fluorescence decay for various excitation techniques.

(2) Studies of exciton mobility aimed at determining the exciton diffusion coefficient D and its temperature dependence. There are many experimental techniques for finding D (see sect. 3). The temperature dependence of D indicates whether the exciton motion has a coherent or incoherent character (see ch. 5).

(3) Studies of physical processes accompanying the interaction between excitons and impurity molecules. These studies aim at determining the cross section of exciton trapping by impurities (or the effective trapping radius R, see ch. 6) and its temperature dependence. The basic approach here is the study of the sensitized luminescence.

(4) Studies of the exciton–exciton collisions yielding various constants of exciton–exciton annihilation (singlet+singlet, singlet+triplet, etc.). These results are discussed in more detail in sect. 3.

(5) Studies of the collective properties of excitons. This is a new field. The studies aim at finding possibilities of formation of the Frenkel biexcitons and observation of phase transitions in exciton systems (exciton drops, Bose–Einstein condensation of excitons, structural phase transitions in crystals with high exciton concentrations, electron–exciton complexes (see ch. 5 sect. 8), laser effects, etc.). Some of these problems are discussed by Agranovich and Toshich (1967). Experimental studies of these phenomena attempted in recent years are reported by Galanin et al. (1972,

1973), Benderskiy et al. (1975, 1977) (see also a theoretical treatment by Kaplan and Ruvinskiy (1976)).

2. *Radiative energy transfer: comparison with the reabsorption theory*

If the emission spectrum and the absorption spectrum overlap radiative energy transfer has a significant effect on the characteristics of luminescence and, in particular, it increases considerably the photon lifetime in the crystal.

To compare the predictions of the theory of luminescence reabsorption discussed in ch. 6 with the experimental data we must know such spectral characteristics as the "molecular" emission spectrum $\rho(\nu)$, the absorption coefficient $k(\nu)$ for the excitation light, and the quantum yield $\eta(\nu)$ of luminescence for excitation by the light of the frequency ν. The characteristics $\rho(\nu)$ and $\eta(\nu)$ must be independently determined in experiments with very thin crystals in which reabsorption is insignificant.

The respective measurements for anthracene at room temperature are given by Galanin and Chizhikova (1954). Figure 7.1 presents the data of Galanin and Chizhikova (1954): (1) – the luminescence spectrum of a very thin crystal (practically without reabsorption); (2) – the absorption spectrum $k(\lambda)$, and (3) – the quantum yield. Substitution of these data into eq. (6.119) yields $q = 0.72$. The luminescence decay time for anthracene was measured by Wright (1955). Using the light in the strong absorption region (3000–3900 Å) he found the following decay time for the thick crystal: $t_{cryst} = \tau_0/P_0 = (18.0 \pm 0.5) \times 10^{-9}$ s. On the other hand, according to

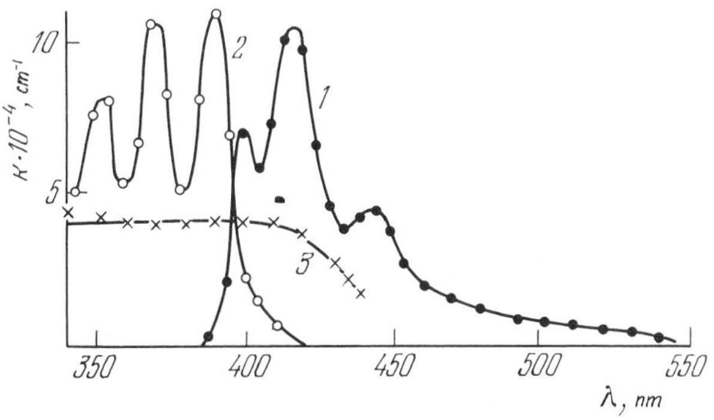

Fig. 7.1. Spectrum of an anthracene crystal (Galanin and Chishikova 1954).

his measurements the decay time under the conditions when reabsorption is insignificant (thin films) is $t_{mol} = 1/P_0 = (6.4 \pm 0.2) \times 10^{-9}$ s so that the ratio $t_{cryst}/t_{mol} = \tau_0 = 2.8 \pm 0.1$.

The data of fig. 7.1 indicate that in the region of strong reabsorption $k = k_0 = 0.5 \times 10^4\ \text{cm}^{-1}$. At the same time, in the wavelength interval $3000\ \text{Å} < \lambda < 3900\ \text{Å}$ the absorption coefficient $k \geqslant 5 \times 10^4\ \text{cm}^{-1}$. Thus, in this wavelength interval $\gamma = k/k_0$ is approximately 10. From the data given in table 6.9 we find that for $x = 1/\gamma = 0.1$ and $q = 0.72$ the value of τ_0 (that is, the time of luminescence decay expressed in terms of the units $t_{mol} = 1/P_0$) is about 2.8 which is in a good agreement with the experimental results.

The above discussion concerned the decay time τ_0 for the long-wavelength spectral region in which absorption of luminescence light can be neglected (see eq. (6.144)). But if we make use of the more general expression (6.143) we can find the dependence of the decay time τ_0 on the frequency ν. Equation (6.143) indicates that in the luminescence region in which

$$b(\nu) \equiv k(\nu)/k_0 \gg 1, \qquad \tau(\nu) \approx \omega(1/\gamma)/\theta(1/\gamma),$$

so that for $\gamma = k/k_0 \gg 1$ (remember that k here is the absorption coefficient for the excitation light) we have $\tau_0(\nu) \approx \omega(0)/\theta(0) = 1$. Thus, when we increase the luminescence frequency, that is, go over from the spectral region in which absorption of the luminescence light is weak to the region with strong absorption, the decay time of luminescence must decrease by a factor of about 2.8.

This result is in a fairly good agreement with the data reported by

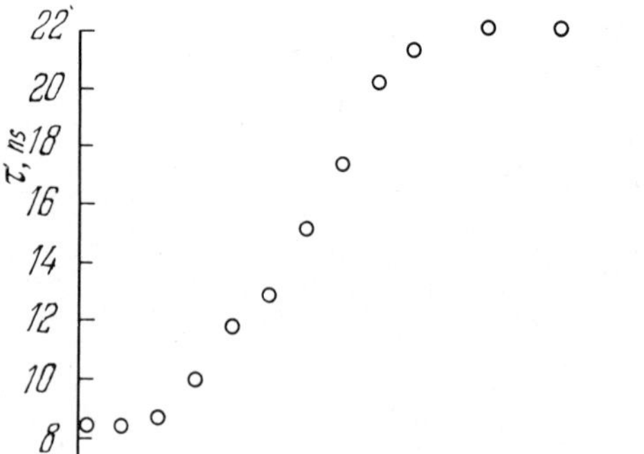

Fig. 7.2. Time of decay of anthracene fluorescence as a function of the wavelength in the fluorescence spectrum (excitation by UV light; see Gammill and Powell (1974)).

Gammill and Powell (1974) (see fig. 7.2 and Galanin et al. (1974a)) according to which $\tau_{max}/\tau_{min} \approx 22/8 \approx 2.75$.

Konobeev (1969) compared theoretical predictions with experimental results for the case when the finite thickness of the crystal should be taken into account. He derived eq. (6.172) which yields $\tilde{\tau}_0 = 1$ for $z_0 \to 0$ and is reduced to eq. (6.148) for $z_0 \to \infty$. Using eq. (6.172) he has shown that it describes the experimental data of Kareh and Wolf (1966), Hammer et al. (1967) on the decay time of luminescence of naphthalene as a function of the crystal thickness with a high accuracy if $q \approx 0.73$. This is not an arbitrary value. It is found from the data of table 6.9 and the experimental results of Kareh and Wolf (1966), Hammer et al. (1967) (see fig. 7.3) according to which the decay time of luminescence produced in naphthalene by strongly absorbed light ($k/k_0 \approx 10$) is $t_{mol} = 1/P_0 = 4 \times 10^{-8}$ s for very thin crystals and $t_{cryst} = 11.5 \times 10^{-8}$ s for very thick crystals so that their ratio is $\tau_0 = 11.5/4 \approx 2.9$. For the given value of q the best agreement with the experimental results (deviation of the order of 1–2%) is obtained if we take $k_0 = 3 \times 10^3\ \mathrm{cm}^{-1}$ for the absorption coefficient of light at the reabsorption maximum which determines $z_0 = dk_0$ entering into eq. (6.102). An independent measurement of this parameter would be of interest.

In connection with light reabsorption in condensed medium which we treated in ch. 6 using integral equations of radiative transfer, let us discuss the results of semi-empirical analysis including reabsorption reported by Birks (1964), Birks and Munro (1967). The analysis proceeds from the simplest kinetic equation for the number of excited molecules $-\mathrm{d}n/\mathrm{d}t = (k_f + k_i)n$ where k_f and k_i are the probabilities of emission of photon and quenching per unit time. Light reabsorption is taken into account by introducing a parameter $a (1 > a > 0)$ and the equation for the number of excited molecules is written as

$$-\frac{\mathrm{d}n}{\mathrm{d}t} = [(1-a)k_f + k_i]n.$$

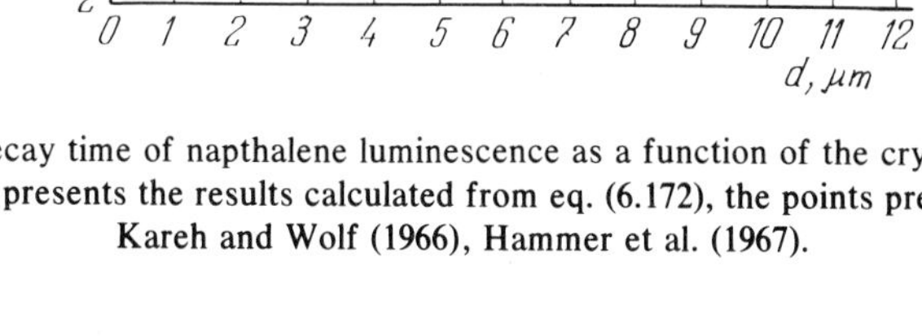

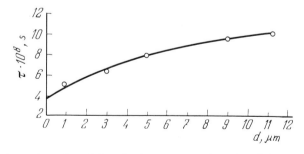

Fig. 7.3. The decay time of napthalene luminescence as a function of the crystal thickness d. The solid curve presents the results calculated from eq. (6.172), the points present the data of Kareh and Wolf (1966), Hammer et al. (1967).

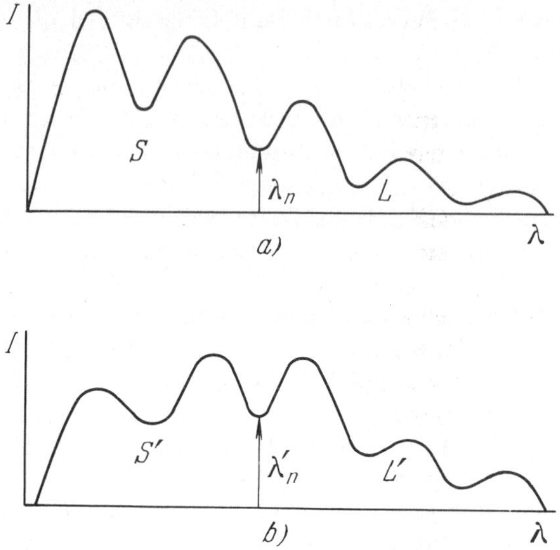

Fig. 7.4. Calculation of the empirical reabsorption factor a from the spectral areas (Birks 1964, Birks and Munro 1967, Munro et al. 1972). It is assumed that reabsorption occurs at $\lambda < \lambda_n$ and that $\lambda_n = \lambda_{n'}$. After normalization $L + S = L' + S' = 1$ so that $a = 1 - L/L'$. (a) Dilute solution, no reabsorption, (b) crystal.

This equation yields the luminescence decay time

$$t = \frac{1}{(1-a)k_f + k_i} = \frac{t_0}{1 - aQ}, \tag{7.1}$$

where $t_0 = 1/(k_i + k_f)$ is the decay time under the conditions when there is no reabsorption, and $Q = k_f/(k_i + k_f)$ is the quantum yield of luminescence. We have shown in ch. 6 that for uniform excitation eq. (7.1) indeed follows from the consistent theory of reabsorption (see eq. (6.149)) and the product aQ is equal to q given by eq. (6.119). Birks (1964), Birks and Munro (1967), Munro et al. (1972) determined q from so-called normalized areas of emission spectra. In any spectrum we can always identify a wavelength λ_n separating the regions of strong ($\lambda < \lambda_n$) and weak ($\lambda > \lambda_n$) reabsorption. Birks (1964), Birks and Munro (1967), Munro et al. (1972) assumed that a entering into eq. (7.1) can be found as $a = 1 - L/L'$ where L and L' are the areas under the emission spectral regions corresponding to $\lambda > \lambda_n$ and normalized to unity found for sufficiently thin (no reabsorption) and thick (with reabsorption) crystals. The procedure for finding a is illustrated in fig. 7.4. This approach has proved to be successful in some cases (Munro et al. 1972).

3. Experimental studies of exciton diffusion in molecular crystals

Let us now discuss the experimental results on the kinetic parameters describing radiationless exciton mechanisms of electronic excitation energy transfer in molecular crystals.

The first four reports in this field were published in 1956–1957. Three of them described application of different experimental techniques to anthracene crystals.

The results on sensitized luminescence of naphthacene (tetracene) dissolved in the anthracene crystal were reported by Galanin and Chizhikova (1956). The bright green luminescence of naphthacene in anthracene has been repeatedly studied earlier (Terenin 1956). As it was first shown by Winterstein et al. (1934), a very low concentration of naphthacene (of the order of 10^{-6}–10^{-5} mole/mole) results in a sharp drop in the intensity of anthracene luminescence and gives rise to a strong naphthacene luminescence. For such low impurity concentrations the excitation light is absorbed mostly by anthracene and this fact directly indicates transfer of the excitation energy. The accompanying decrease in the lifetime of the excited state of the anthracene molecules suggests that this transfer of the excitation energy is radiationless resonance transfer, rather than absorption of the light of anthracene luminescence by naphthacene.

According to eqs. (6.26) and (6.35), when the impurity concentration is low the exciton lifetime $\tau = 1/P$ in the host material is given by

$$\tau/\tau_0 = 1 - 4\pi DRn\tau_0, \tag{7.2}$$

where $\tau_0 = 1/P_0$ is the lifetime of excitations in the absence of impurities (eqs. (6.26) and (6.35) yield $\tau/\tau_0 \equiv P_0/P = 1/(1+4\pi DRn\tau_0)$ so that eq. (7.2) is valid only for $4\pi DRn\tau_0 \ll 1$). The quantities τ and τ_0 were determined from experimental data yielding $DR = 0.7 \times 10^{-10}\ \mathrm{cm^3/s}$.

As noted in ch. 6 sect. 2, eq. (6.35), and hence eq. (7.2), are valid only if the mean free path of excitons is small in comparison with the trapping radius R. However, to find the correct estimate for R we have to know the type of excitons which transfer energy from the host material to the impurity. If, as assumed by Galanin and Chishikova (1956), energy is transferred by incoherent excitons we see that eq. (7.2) is valid and eqs. (2.52) or (6.59) yield the estimate

$$R = 0.676 R_0^{3/2} (D\tau_0)^{1/4},$$

where R_0 is the Förster radius (1.34).

Then the Förster formula gives $R_0 = 32$ Å (Galanin and Chizhikova 1956). Hence we have $\sqrt{D\tau_0} = 230$ Å and $D \approx 1.3 \times 10^{-3}\ \mathrm{cm^2/s}$ (Galanin and Chizhikova (1956) assumed that the capture radius $R \sim 0.5\,R_0$ and thus obtained somewhat lower values).

The experimental technique for measuring the diffusion length for excitons suggested by Galanin and Chizhikova (1956) was repeatedly used in later studies (Wolf 1967). The main difficulty associated with this technique is the need to use a relationship of the type of eq. (7.2) between the variation of the luminescence decay time τ and the exciton diffusion coefficient D. It has been noted in ch. 6 that this relationship is far from being universal. To say nothing of the fact that this relationship ignores reabsorption of the luminescence light (this can be done for sufficiently thin crystals) formulation of relationships of the type of eq. (7.2) always implies some *a priori* assumptions on the mechanism of exciton trapping by impurities and on the relation between the effective trapping radius R and the mean free path l of excitons (for incoherent excitons l is of the order of the lattice parameter; as noted above, eq. (7.2) is valid for $R \gg l$ and other cases have been discussed in ch. 6).

In this sense, the method for finding the exciton diffusion coefficient used by Agranovich et al. (1956, 1957) is less troublesome. In this method the exciton diffusion length is found by comparing the theoretical predictions based on the diffusion equation (6.1) with the experimentally measured quantum yields for the host material (anthracene) and the impurity (naphthacene) as functions of the absorption coefficient for the excitation light. According to these results, the exciton diffusion length is about 0.15 μm which is considerably higher than the value reported by Galanin and Chizhikova (1956). It is true, this method yields, in fact, just a dimensionless quantity, namely, the product of the exciton diffusion length by the light absorption coefficient, and this coefficient was, apparently, underestimated by a factor of 2–3 by Agranovich et al. (1956, 1957) when the estimate for the diffusion coefficient was found (Faidysh, private communication). Nevertheless, even the corrected value of the diffusion length is still higher than that reported by Galanin and Chizhikova (1956).

To make use of the above experimental methods for determining the diffusion length we must know not only such parameters as the light absorption coefficient but also the impurity concentration. Now we shall discuss the method suggested by Simpson (1957) which seems to be more direct and accurate than the two above methods.

Assume that a pure anthracene crystal (see fig. 7.5) has the shape of a plane-parallel plate of thickness d whose surface is perpendicular to the x-axis. If one of the crystal surfaces, for instance, the surface $x = d$ is coated with a layer of exciton detector, for instance, anthracene with a high concentration of luminescent impurity (naphthacene) and the other surface is illuminated with light whose frequency is in the main absorption band of anthracene then we can directly determine the exciton mobility

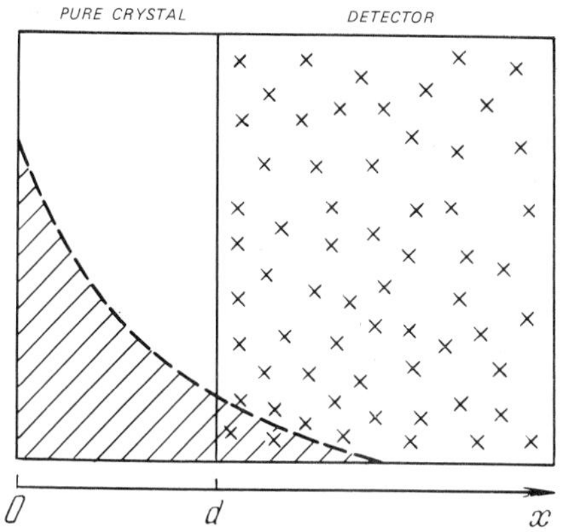

Fig. 7.5. Schematic of the Simpson technique for determining the diffusion coefficient (Simpson 1957). The dashed curve for $I(x) = \exp(-kx)$.

from the intensity of the detector luminescence if the crystal thickness d is greater than the penetration depth for the excitation light. Solving the diffusion equation (6.1) with the boundary conditions

$$\left.\frac{\mathrm{d}c}{\mathrm{d}x}\right|_{x=0} = 0, \qquad c(x = d) = 0, \tag{7.3}$$

we can readily see that under the steady-state conditions ($\mathrm{d}c/\mathrm{d}t = 0$) the exciton flux at the detector is

$$S = -D\left.\frac{\mathrm{d}c}{\mathrm{d}x}\right|_{x=d} = \frac{Ik^2\lambda^2}{k^2\lambda^2 - 1}\left\{\operatorname{sech}\frac{d}{\lambda} - \mathrm{e}^{-kd}\left[1 + \frac{\tanh(d/\lambda)}{k\lambda}\right]\right\}. \tag{7.4}$$

If each exciton in the detector gives rise to a quantum of naphthacene luminescence then we can find the exciton diffusion length $\lambda = \sqrt{D/P_0}$ by comparing the experimental results on the intensity of naphthacene luminescence as a function of the thickness d with eq. (7.4). Simpson has assumed that the absorption coefficient in the main absorption band of anthracene is $k = 8.2 \times 10^3\ \mathrm{cm}^{-1}$ obtaining $\lambda = 0.046\ \mu\mathrm{m}$ which is close to the result of Agranovich et al. (1956, 1957) (if the corrected value of k is used). The crystals studied by Simpson had the thickness of not more than $0.6\ \mu\mathrm{m}$. Therefore, reabsorption of the anthracene luminescence light under these conditions could hardly have any significant effect on the intensity of energy transfer from the host material to the impurity. Using the diffusion length found by Simpson and the exciton lifetime $1/P_0 = 6 \times 10^{-9}\ \mathrm{s}$ (Wright 1955) we find the exciton diffusion coefficient $D \approx 3.5 \times 10^{-3}\ \mathrm{cm}^2/\mathrm{s}$. When we compare the results of Simpson and the results reported by Agranovich et al. (1956, 1957) we should bear in mind that the boundary conditions at the illuminated crystal surface in these studies were different.

The form of the boundary condition depends on the state of the crystal surface. The condition of total reflection of excitons from the crystal surface used by Simpson is applicable if the crystal surface is not contaminated and transitions of bulk excitons into surface excitons can be ignored. At present it is difficult to evaluate the correctness of the boundary conditions by Agranovich et al. (1956, 1957), Simpson (1957). We shall only note that this important problem must be specially analyzed since for $k = 8.2 \times 10^3\,\text{cm}^{-1}$ most excitons in the crystal are generated near the illuminated surface (at a distance of the order of diffusion length).

Borisov and Vishnevsky (1957) have attempted to determine the exciton diffusion length in naphthalene from the quantum yield of luminescence of the anthracene impurity as a function of the absorption coefficient of the excitation light. They have assumed that the intensity of impurity luminescence decreases with increasing coefficient k owing to photo-oxidation of anthracene at the crystal surface. Unfortunately, their crystals were very thick ($d = 0.5$ cm) and reabsorption led to some increase in the transfer efficiency. Moreover, they have ignored the fact that photo-oxidation of anthracene occurs at some depth in the crystal. Thus, their result for the exciton diffusion length was very large: $\lambda = 0.2$–$0.3\,\mu$m.

A modification of the Simpson method suggested by Khan-Magometova (1965, 1972) consists in replacing the transfer to a luminescent detector with quenching of excitons at the defects produced in a thin layer by bombardment with electrons. The diffusion length found with this technique is about $0.13\,\mu$m which is close to the result of Agranovich et al. (1956, 1957). The same diffusion length was found by the Simpson method by Kurik and Piryatinskiy (1971) where it was shown that λ decreases down to $0.09\,\mu$m at 77 K. The study of pulse propagation in a two-layer Simpson system reported by Takahashi and Tomura (1971) yielded a diffusion length of $0.85\,\mu$m which seems to be overestimated.

We have noted a considerable difference between the diffusion coefficients of singlet excitons in anthracene found in the experiments discussed above. In part, this can be attributed to difficulties in measurement procedures. However, we should bear in mind that, according to Galanin et al. (1974a, 1975) (see also sect. 4) exciton diffusion in anthracene crystals is strongly affected by crystal defects which serve as shallow exciton traps. Therefore, the differences between the results may be explained by the differences between the crystals used in the experiments. (An original technique for measuring the diffusion coefficient of triplet excitons was used by Ern et al. (1966). The result for anthracene is $D_T = 2 \times 10^{-4}\,\text{cm}^2/\text{s}$.)

The results on the diffusion length of singlet excitons in a phenanthrene crystal obtained with the Simpson method are reported by Gallus and Wolf (1966). When the boundary condition at the illuminated surface was $dc/dx = 0$ for $x = 0$ the result was $\lambda = 78 \pm 25$ Å. For the boundary condition $c(0) = 0$ the diffusion length was $\lambda = 152 \pm 48$ Å. Gallus and Wolf (1966) failed to discuss the applicability of one or another of these boundary conditions in their experiments.

The diffusion length of excitons can be found also from the intensity of metallic quenching of excitons as a function of the absorption coefficient of the excitation light (these experiments are described in ch. 8) and from the data on the rate of bimolecular quenching of excitons. Below, we shall discuss the results of these studies in more detail and, in particular, relate them the data on the temperature dependence of the exciton diffusion coefficient.

As it was repeatedly noted in chs. 4 and 5, the temperature dependence of the exciton diffusion coefficient provides evidence on the character of exciton motion (indicating if excitons are coherent or incoherent and so on). In recent years interesting results in this field have been reported by Japanese (Inoue et al. 1972, Yoshihara et al. 1973, Inoue 1975) and Soviet (Vol et al. 1971, 1974) authors. But the first results on the temperature dependence of the exciton mobility are presented by Kazzaz and Zahlan (1961), Hammer and Wolf (1968). The authors of these studies determined the temperature dependence of the ratio between the luminescence intensity B_1 of the impurity and the luminescence intensity B_0 of the host material in the naphthalene crystals with anthracene impurity under the conditions when the excitation light was absorbed only by the host material. The impurity concentration in a paper by Kazzaz and Zahlan (1961) varied in a wide range – from $n = 4 \times 10^{-8} N_0$ to $n = 4.8 \times 10^{-3} N_0$ where N_0 is the number of molecules of the host material per unit volume. For high impurity concentrations the intensity ratio B_1/B_0 was found to be practically independent of temperature in the temperature range from 125 to 300 K while for a low impurity concentration $n = 4 \times 10^{-8} N_0$ at low temperatures this ratio was found to decrease to a few tenths of its value with increasing temperature.

For $n = 4 \times 10^{-8} N_0$ the mean distance between impurity molecules is about 300 lattice parameters. Since the effective radius of exciton trapping by impurities and the exciton mean free path in this temperature range are both definitely smaller than the distance between impurities the rate of exciton trapping should, indeed, depend on the exciton mobility.

This result has been confirmed even more convincingly by Hammer and Wolf (1968) who have shown that in the temperature range from 6 to 100 K the ratio B_1/B_0 varies as $1/\sqrt{T}$ which corresponds to the temperature dependence of the diffusion coefficient of coherent excitons (see ch. 5 sect. 4).

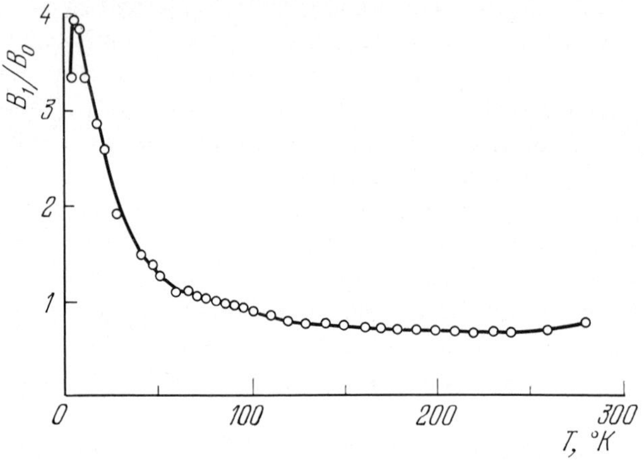

Fig. 7.6. The rate of energy transfer in naphthalene as a function of temperature (Hammer and Wolf 1968).

According to eq. (6.42a), we can write

$$\frac{B_1}{B_0} = \frac{\eta}{\eta_0} \frac{4\pi DRn}{1 + \gamma l/R} \tau_0, \tag{7.5}$$

where $\tau_0 = (P_0 + P_T)^{-1}$ is the exciton lifetime in the host material, and $\eta_0 = P_0/(P_0 + P_T)$ is the "molecular" quantum yield of luminescence of the host material in the absence of impurities.

As demonstrated by independent measurements, variation of η, η_0 and τ_0 in the temperature range between 6 and 100 K is very small (a few tenths). Therefore, the observed variation of the ratio B_1/B_0 in this temperature range can be attributed only to the temperature dependences of D, R and l.

The results of Hammer and Wolf (1968) suggest that $l \leqslant R$ in this temperature range. Indeed, if we had $l > R$ then the ratio $4\pi DR/(1 + \gamma l/R)$ would be equal to $v\pi R^2$, where v is the mean (thermal) velocity of excitons, that is, it would be equal to the probability of exciton trapping by impurities calculated without taking into account the effect of "attrition" of the exciton flux in the impurity vicinity. At low temperatures the temperature dependence of this probability should be weak since at low temperatures exciton trapping is accompanied mostly with emission of phonons, rather than absorption of phonons. Moreover, this probability should grow with increasing temperature while the ratio determined in experiments decreases (the difference between the energy levels of the impurity molecule and the host molecule in this case is more or about 0.5 eV so that anthracene in naphthalene is a deep trap for excitons).

Furthermore, the results reported by Broude et al. (1961) seem to indicate that the energy minimum in the lowest singlet exciton band of naphthalene is close to the point $k = 0$. However, we must know the effective exciton mass (entering into eq. (5.29)) in order to evaluate the applicability of the relationship $D \sim 1/\sqrt{T}$ in this temperature range.

For instance, in the anthracene crystal the effective exciton mass in the lowest exciton band (the Davydov splitting $\Delta = 250\ \mathrm{cm}^{-1}$) is $m^* = 4.2m_0$ (in the direction 010) and $m^* = 7.8m_0$ (in the direction 100) in the *ab* plane (here m_0 is the mass of the free electron; see Matsui and Oeda (1973)). In naphthalene the Davydov splitting is $\Delta = 150\ \mathrm{cm}^{-1}$ and we can assume $m^* \approx 10m_0$. Then the inequality (5.29) is satisfied practically in the whole temperature range covered by the study reported by Hammer and Wolf (1968) so that eq. (5.31) can be assumed to be valid.

Thus, the experimental results reported by Hammer and Wolf (1968) seem to indicate the coherent character of exciton motion in naphthalene in the temperature range between 6 and 100 K. It remains unclear why the mean free path l is smaller than the trap radius R. For coherent excitons the mean free path is much greater than the lattice parameter. Therefore, the condition $l < R$ suggests the existence of a long-rang mechanism. Most probably, this mechanism can be attributed to the fact that the vicinity of the impurity contains a shallow trap, apart from the deep trap, and it corresponds to a local exciton state (see ch. 5 sect. 6). The lattice deformation produced by the impurity may play a part in this mechanism. Possibly, both factors affect simultaneously the rate of exciton trapping (the $D(T)$ dependence in naphthalene is discussed also by Schmid et al. (1977)).

In this connection, note that such states are indeed observed in low-temperature spectra of crystals with impurities. Local excitons were reported by Propstl and Wolf (1963), Ostapenko and Shpak (1969), for instance, in naphthalene crystals with thionaphthene, indole and benzofuran impurities. The depth of these local states varies in a wide range from several to tens of reciprocal centimetres for various impurity clusters and depending on the position of a state in the lattice cell (Ostapenko et al. 1973). More shallow local states can serve as traps at sufficiently low temperatures. With growing temperature increasingly deeper traps start to operate while the effective value of R should decrease. However, the mean free path also decreases with increasing temperature so that the inequality $R > l$ can, generally, hold not only at low temperatures but also at "moderately" high temperatures (we mean here the region of existence of coherent excitons; the inequality $R > l$ can be satisfied for incoherent excitons, too, if only the Förster radius $R_0 \approx R$ is large in comparison with the lattice parameter; see ch. 6).

However, we should bear in mind that the above mechanism of exciton trapping by impurities, if it is, indeed, realized, must not only provide for sufficiently high effective values of $R > l$ but, to conform to the experimental results, these values of R must not vary significantly in that range of low temperatures in which $RD \sim 1/\sqrt{T}$. In this connection, note that our interpretation of large R values and the trapping mechanism on the whole could be changed fundamentally if the density of dislocations were observed to affect the rate of transfer of exciton energy to impurities. Such an observation would indicate that exciton trapping by impurities is preceded by exciton trapping by dislocation and subsequent transfer of exciton energy to the impurity along the dislocation. Further, even more careful exerimental studies are needed to answer these questions. The important factors to be studied are the effects of the size distributions of the impurity clusters on energy transfer, the relationship of these distributions to the dislocation structure, and so on.

In connection to the above discussion, note that the results reported by Hammer and Wolf (1968) indicate the need for not only qualitative analysis of the interaction between excitons and impurities but also quantitative calculations of various trapping mechanisms. In such calculations it may prove to be important to include the anisotropy of the exciton bands, lattice deformations and many other factors (for instance, various local and quasilocal relaxation processes; see Khachaturyan and Estrin (1971)). The need for such theory is demonstrated by the results on the temperature dependence of the exciton diffusion coefficient found from the measurements of the rate of exciton bimolecular annihilation.

We can assume that a relationship of the type of eq. (5.92) holds for incoherent excitons in the singlet-singlet and singlet–triplet collisions so that, according to the results of ch. 6 sect. 2.3 (see also eq. (2.51)) we obtain $\gamma \sim D^{3/4}$. Since for the diffusion coefficient of incoherent excitons we have eq. (5.15) we should expect a dependence of the type

$$\gamma = \gamma_0 \, e^{-\Delta E/kT}, \qquad \Delta E = \tfrac{3}{4}\Delta E_a, \tag{7.6}$$

where ΔE_a is the activation energy of exciton diffusion.

The relationship (5.92) is not valid for coherent excitons. Here we encounter problems similar to those in the analysis of trapping of coherent excitons by deep traps. One difference is that in this case under certain conditions biexcitons of the type of those discussed in ch. 4 sect. 6 can be formed, instead of local excitons. If biexciton states are formed and their energy is lower than the energy of two free excitons then the annihilation radius R entering into eq. (5.88) should be related to the cross section of binding of two excitons into the biexciton. Further similarities can be found between the process discussed here and trapping by impurity. However, the experimental results are of much greater interest for us. Table 7.1 presents the data on the temperature dependence $\gamma(T)$ reported by Inoue et al. (1972), Yoshihara et al. (1973) and the result on exciton

Table 7.1
Temperature dependence of the rate of singlet-singlet exciton annihilation (Inoue et al. 1972, Yoshihara et al. 1973).

Material	Temperature dependence
Anthracene	$1/\sqrt{T}$ (5–250 K)
Naphthalene	$1/\sqrt{T}$ (5–50 K)
TCNB	$\exp(-\Delta E/k_B T)$, $\Delta E \approx 160\ \mathrm{cm}^{-1}$ (120–300 K)
Pyrene	Constant (5 – 100 K)
	$\exp(-\Delta E/k_B T)$, $E \approx 330\ \mathrm{cm}^{-1}$ (120–300 K)
Naphthalene (Hammer and Wolf 1968)	$1/\sqrt{T}$ (6–100 K)

trapping in naphthalene with anthracene impurity reported by Hammer and Wolf (1968).

The data of the experiments on exciton capture in pyrene by impurities also yielded an exponential dependence with the activation energy $\Delta_a E \approx 440\ \mathrm{cm}^{-1}$ for the ratio B_1/B_0 in the range of high temperatures (Klöffer et al. 1972). Since the activation energy for $\gamma(T)$ in pyrene reported by Inoue et al. (1972), Yoshihara et al. (1973) is $\frac{3}{4}$ of $\Delta E_a = 440\ \mathrm{cm}^{-1}$ (see eq. (7.6)) we can come to the conclusion that independent experimental results for pyrene agree. The results indicate that the effective radius of exciton trapping by impurities (the black sphere radius) is practically independent of the diffusion rate and that the activation energy for exciton diffusion $\Delta E_a = 440\ \mathrm{cm}^{-1}$.

Figure 7.7 illustrates the spread of experimental results for $\gamma(T)$ determinations (Yoshihara et al. 1973).

Independent experiments employing fundamentally different techniques for determining the rate of exciton trapping by impurities (Hammer and Wolf 1968) and the rate of bimolecular annihilation of excitons (Inoue et al. 1972, Yoshihara et al. 1973) in naphthalene in the range of low temperatures have yielded the same temperature dependence of the exciton diffusion coefficient which coincides with the theoretically predicted dependence $D = D(T)$. This fact, apparently, evidences independence of the respective trapping radii from temperature; a similar conclusion, most probably, could be made in the case of anthracene for which we also have $\gamma \sim 1/\sqrt{T}$ according to the data of Yoshihara et al. (1973).

However, in the range of high temperatures (incoherent excitons) the conditions of bimolecular annihilation with large Förster radii are different as, apparently, evidenced by the above results for pyrene for which $\Delta E = 3\Delta E_a/4$.

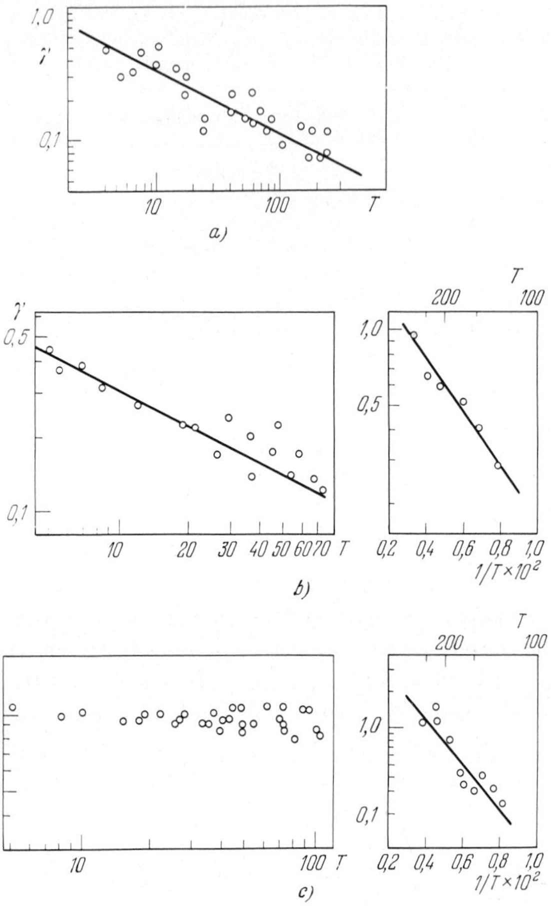

Fig. 7.7. The bimolecular quenching rate constant as a function of temperature (Yashihara et al. 1973). (a) Anthracene crystal, the slope of the log-log plot is −0.5. (b) TCNB naphthalene crystal (at the left the slope of the log–log plot is −0.5; at the right log γ as a function of $1/T$. (c) Pyrene crystal (as in (b)), at the right activation energy $\Delta E = 330\ \mathrm{cm}^{-1}$).

The experimental studies of the rate of bimolecular annihilation as a function of various factors (crystal structure, impurities, temperature) are of essential significance and must be continued.

At very high exciton concentrations when the mean distance between excitons is approximately equal to the radius R of bimolecular annihilation the exciton diffusion should become insignificant (this possibility was noted by Inoue (1975)). Then bimolecular annihilation can increase and this effect can be related to some observed features of the rate of bimolecular annihilation as a function of the excitation intensity (Inoue 1975) which are found in anthracene for $n \geq 3 \times 10^{18}\,\text{cm}^{-3}$. The condition $4\pi R^3 n/3 \approx 1$ then yields $R \leq 40$ Å which is close to the value reported by Rahman and Knox (1973) (see ch. 5. sect. 8).

In conclusion, note that above we discussed the experimental data on exciton trapping by impurities and on bimolecular annihilation using expressions for steady-state trapping probabilities. It has been repeatedly questioned whether these expressions are applicable to various rapid processes of exciton diffusion and exciton trapping by impurities. However, only in the recent years the improvement in the experimental techniques in the studies of luminescence kinetics made it possible to give a sufficiently definite answer to such questions, at least, for some crystals. These results are discussed in sect. 4. No respective experiments have been done for bimolecular annihilation of excitons and its theoretical study has just started (see ch. 5 sect. 8).

4. *Studies of electronic excitation energy transfer in crystals with time-resolved spectroscopy*

Starting from 1969 Powell and co-workers published the results of a series of experimental studies of a number of crystals making use of time-resolved spectroscopy (Powell and Soos 1975, Powell and Kepler 1969, 1970, Powell 1970, 1971, 1973, Gammil and Powell 1974). This method consists in tracing the evolution with time of the donor and acceptor luminescence produced by pulsed excitation which makes it possible to determine the time dependence of the rate of transfer of energy from the host material to the impurity. Since different transfer mechanisms, generally speaking, lead to different time dependence of the energy transfer rate the experimental investigation of it is very important.

Typical kinetic curves shown in fig. 7.8 present the data for the anthracene crystal with 10^{-6} concentration of tetracene (Powell 1970, 1971, 1973, Gammil and Powell 1974). To derive the information on the elementary processes from curves of this type we have to interpret them using the kinetic equations describing variations with time of the concentration c_1 of the donor excitations and the concentration c_2 of the acceptor excitations.

If excitation is uniform and we ignore bimolecular annihilation of excitons we have the linear equations

$$\frac{dc_1}{dt} = G(t) - \frac{1}{\tau_S} c_1(t) - P_1(t) c_1(t) N_A,$$

$$\frac{dc_2}{dt} = P_1(t) N_A c_1(t) - \frac{1}{\tau_A} c_2(t), \tag{7.7}$$

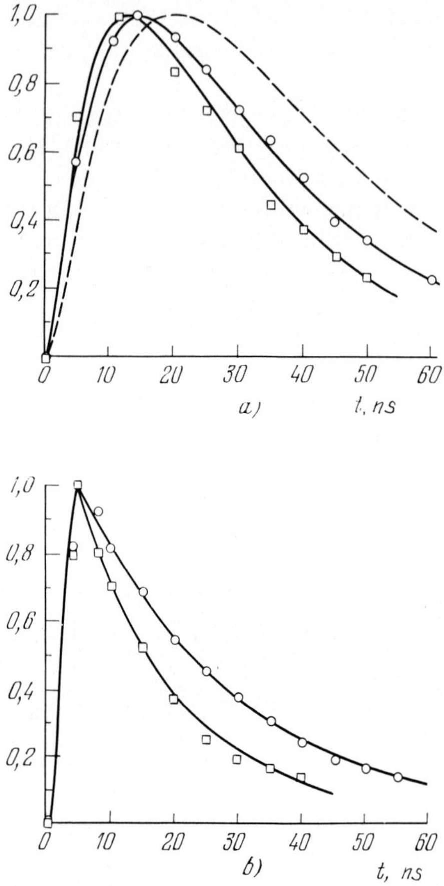

Fig. 7.8. Kinetic curves according to the data of Powell (1970, 1971, 1973); Gammil and Powell (1974): (a) for tetracene, (b) for anthracene. The dashed curve is the theoretical function for $P_1 = \text{const}$. The tetracene concentration: ○ – 10^{-6}, □ – 12×10^{-6} (Powell and Kepler 1969, 1970, Powell 1970, 1971, 1973, Gammil and Powell 1974).

where N_A is the acceptor concentration, τ_S is the lifetime of excitations in the pure crystal, τ_A is the acceptor excitation lifetime in the matrix, $P_1(t)N_A$ is the rate of energy transfer from the donor (matrix) to the impurity, and $G(t)$ is the shape of the excitation pulse. Numerical calculations show that the experimental results reported by Powell and Soos (1975), Powell and Kepler (1969, 1970), Powell (1970, 1971, 1973), Gammil and Powell (1974), are described by the solutions of eqs. (7.7) if we assume

$$P_1(t) = (a + bt^{-1/2}), \tag{7.8}$$

where a and b are constants (see figs. 7.9 and 7.10).

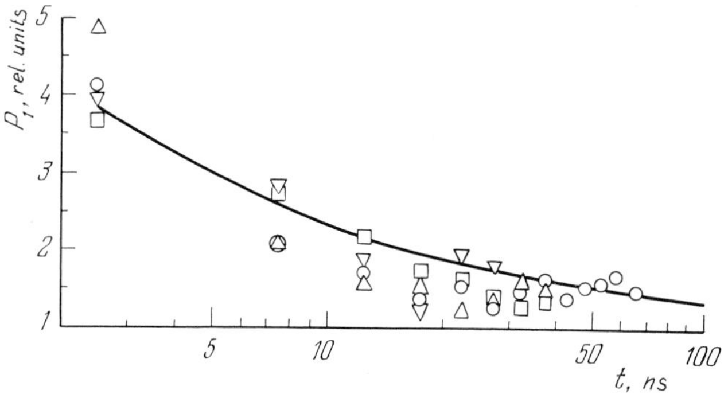

Fig. 7.9. Experimental dependence $P_1(t)$ for anthracene.

In the black sphere approximation (see ch. 6) we have

$$P_1(t) = 4\pi DR\left(1 + \frac{R}{\sqrt{\pi Dt}}\right). \tag{7.9}$$

Therefore, the results on the nonstationary trapping rate, generally speaking, make it possible to find not only the product DR in the approximation of eq. (7.9) using the parameter a entering into eq. (7.8) but also R and D (the results for R in the reports under discussion were unusually high (for instance, $R = 108$ Å for tetracene in anthracene) which led to a stormy discussion; see Golubov and Konobeev (1977) and the reports cited there). If $D = 0$ and the Förster mechanism is operational we have $P_1(t) \sim 1/\sqrt{t}$ (see eq. (2.12)).

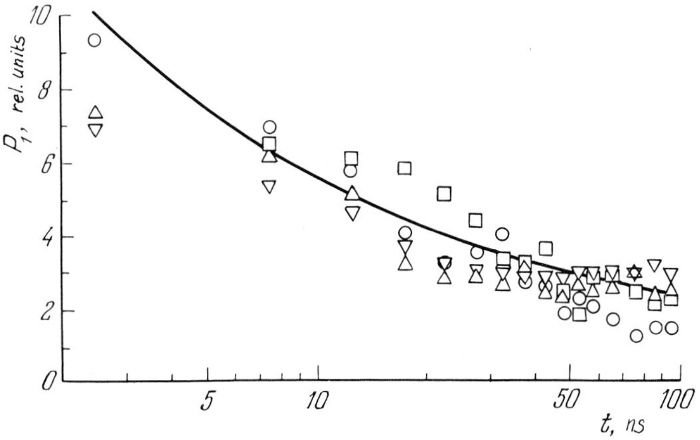

Fig. 7.10. Experimental dependence $P_1(t)$ for naphthalene.

To compare the results with eq. (7.9) let us now analyze the rate of energy transfer from the host material to the impurities due to reabsorption of fluorescent light. By analogy to eq. (6.94) we have

$$\left.\frac{\partial c_2}{\partial t}\right|_{\text{reabs}} = \frac{1}{4\pi}\int \rho(\nu)\beta_A(\nu)k_A(\nu)\,d\nu \int_V c_1(r_1, t)\frac{\exp[-k(\nu)|r - r_1|]}{|r - r_1|^2}\,dv_1, \tag{7.10}$$

where $k(\nu) = k_0(\nu) + k_A(\nu)$, $k_0(\nu)$, and $k_A(\nu)$ are the coefficients of absorption of light with frequency ν by the host molecules and the acceptor molecules, respectively, $k_A(\nu) = k'_A(\nu)N_A$, $\beta_A^{(\nu)}$ is the fraction of the events of absorption of light with frequency ν by the acceptor molecule which produce its electronic excitation, and $\rho(\nu)$ is the probability of generation of the fluorescence photon with the frequency ν owing to exciton decay per unit time.

To estimate the highest rate of radiative energy transfer to impurities we shall assume that $c_1(r_1, t)$ only slightly depends on r_1 (spatially uniform excitation, thick crystals). Then we have

$$\left.\frac{\partial c_2}{\partial t}\right|_{\text{reabs}} = P'c_1(t)N_A, \tag{7.11}$$

where

$$P' = \int_0^\infty \rho(\nu)\beta_A(\nu)\frac{k'_A(\nu)}{k(\nu)}\,d\nu. \tag{7.12}$$

Since we have

$$\rho(\nu) = P_0E(\nu) = \frac{1}{\tau_0}E(\nu), \tag{7.13}$$

where $E(\nu)$ is the exciton radiation spectrum normalized to unity and $P_0 = 1/\tau_0$ is the total probability of exciton decay in the crystal without impurities we obtain

$$P' = \frac{1}{\tau_0}\int_0^\infty E(\nu)\beta_A(\nu)\frac{k'_A(\nu)}{k(\nu)}\,d\nu. \tag{7.14}$$

Thus, under the conditions of spatially uniform excitation the radiative transfer of energy to the impurities results in a transfer rate P' independent of time. The contribution of this transfer mechanism can be significant only if the ratio $P'/P_1 \gtrsim 1$, that is, if (see eq. (7.12))

$$\frac{1}{4\pi DR\tau_0}\int_0^\infty E(\nu)\beta_A(\nu)\frac{k'_A(\nu)}{k(\nu)}\,d\nu \gtrsim 1. \tag{7.15}$$

The main contribution to eq. (7.14) is made by the frequency range corresponding to the maximum overlapping of $\rho(\nu)$ and $k_A(\nu)$. Therefore, eq. (7.15), as eqs. (7.12)–(7.14), is valid only for the crystals whose thickness d in the given spectral range satisfies the condition

$$dk(\nu) \gg 1. \qquad (7.16)$$

For very low N_A values we can assume $k(\nu) \approx k_0(\nu)$ in the integral in eq. (7.14). The value of $k(\nu)$ increases with increasing impurity concentration N_A while the contribution of the radiative energy transfer to the impurities declines as can be seen from eq. (7.14).

If we take into consideration the radiative transfer rate P' which is independent of time the meaning of the parameter a found from the experimental data (see eq. (7.8)) is changed. Then eq. (7.9) yields lower values for the new effective R and D.

If we cannot assume that the function $c(\mathbf{r}_1, t)$ in eq. (7.10) is independent of $\mathbf{r}_1$ the trapping rate P' proves to be dependent on t. Then not only the parameter a but also the parameter b in eq. (7.8) change their meaning (of course, if we continue to extrapolate the experimental data with an expression of the type of eq. (7.8)). The character of the function $P'(t)$ under such conditions should depend on the shape and size of the crystals under study and cannot be analyzed in a general form.

We have discussed a possible contribution of radiative transfer in detail because the crystals studied by Powell and Kepler (1969, 1970), Powell (1970, 1971, 1973), Gammil and Powell (1974) were, apparently, fairly thick.

Similar measurements of the kinetics of luminescence of the host material and the impurity (anthracene doped with tetracene) were performed for very thin crystals with thicknesses of about 0.5 μm and less produced by sublimation (the tetracene concentrations were 10^{-6}–10^{-5} mole/mole; Galanin et al. (1974a, b, 1975). The statistical one-photon method was used to obtain the rise-up and decay curves for luminescence of the host material and impurity in the same crystal (the time resolution of the apparatus was 0.2 ns). The measurements were made at room temperatures when excitons in anthracene are incoherent (see sect. 3) and at a low temperature (4.2 K) when excitons are coherent. At first, let us discuss the results obtained at room temperature and make a few remarks on treatment of the results reported by Galanin et al. (1974a, b, 1975).

If the quantity P_1 in eqs. (7.7) is independent of time then for pulsed excitation of the matrix (the starting concentration $c(0) = c_0$) eqs. (7.7) have the following solutions:

$$c_2(t) = \frac{P_1 c_0 N_A}{1/\tau_A - P_1 N_A - 1/\tau_S}\left[\exp\left(-\frac{1}{\tau_A}t\right) - \exp\left(-\left(P_1 N_A + \frac{1}{\tau_S}\right)t\right)\right], \qquad (7.17)$$

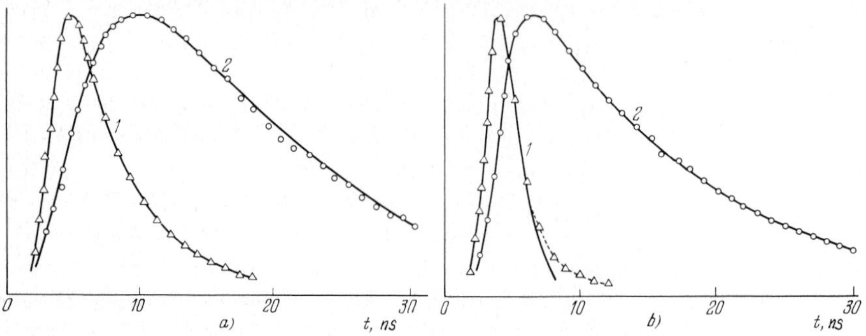

Fig. 7.11. Kinetic curves of luminescence for the anthracene–tetracene crystals at room temperature. (a) Crystals with low concentration of shallow exciton traps; $\tau_A = 12$ ns, $\tau_l = 4.5$ ns (the tetracene concentration is 10^{-6} g/g). (b) Crystals with high concentration of traps, $\tau_A = 12$ ns, $\tau_l = 1$ ns (the tetracene concentration is 10^{-5} g/g). Curves: (1) – anthracene luminescence, (2) – tetracene luminescence. Solid curves were calculated as the convolutions of the exponential functions (7.17) with the excitation pulse. Points present experimental results.

Therefore, the experimental curves obtained by Galanin et al. (1974a, b, 1975) were approximated with convolution of exponential functions and the excitation pulse. According to eq. (7.17) the luminescence curve for tetracene was approximated with the difference $\exp(-t/\tau_2) - \exp(-t/\tau_1)$ where $\tau_1 = 1/(PN_A + 1/\tau_S)$ is the lifetime of excitons (that is, the time of decay of the anthracene luminescence for the given tetracene concentration) and $\tau_2 \equiv \tau_A$ is the lifetime of the impurity excitation. This approximation is successful if we can ignore the time-dependent term in eq. (7.9). The comparison between experimental and calculated curves in fig. 7.11 shows that this approximation has a good accuracy (only for thin crystals). This means that for anthracene-tetracene crystals the energy transfer rate is practically independent of time under the conditions when the contribution of radiative transfer from the host to the impurities is not significant (this experimental result contradicts the explanation of the results of Powell and Soos (1975) given by Golubov and Konobeev (1977)).

Let us now discuss the luminescence results for 4.2 K. First, it was shown that at low temperatures the luminescence spectrum of pure anthracene crystals evidences two types of excitations with different lifetimes, namely, excitons with the lifetime about 1 ns which disappear converting into other excitations with the lifetime about 3.5 ns (at 4.2 K). These results directly lead to the question what is the nature of these excitations and which of them are responsible for energy transfer to impurities.

Though this question arose long ago (Galanin and Chizhikova 1956) it has been answered only for few crystals. We have already mentioned the

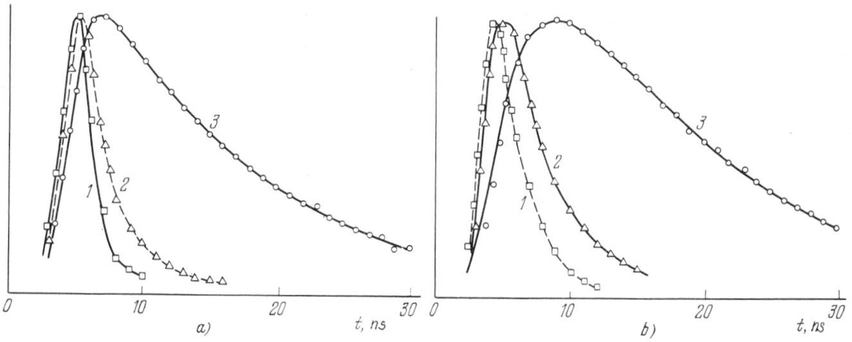

Fig. 7.12. Kinetic curves of luminescence for anthracene–tetracene mixed crystals at liquid helium temperatures. (a) Crystals with low concentration of shallow exciton traps, $\tau_A = 12$ ns, $\tau_1' = 1$ ns, $\tau_1'' = 1.5$ ns (the tetracene concentration is 10^{-6} g/g). (b) Crystals with high concentration of traps, $\tau_A = 12$ ns, $\tau' = 1$ ns, $\tau_1'' = 4$ ns (the tetracene concentration is 10^{-5} g/g). Curves: (1) and (2) – anthracene luminescence, (3) – tetracene luminescence. Solid curves were calculated as the convolutions of the exponential functions (7.17) with the excitation pulse.

results of Hammer and Wolf (1968) according to which $P_1(T) \sim 1/\sqrt{T}$ in the range of low temperatures (6–60 K) for the anthracene impurity in naphthalene. As noted above, these results imply that in this system energy transfer to impurities is carried out by coherent excitons. However, the situation in anthracene has proved to be more complicated in certain respect according to the results of Galanin et al. (1974a, b, 1975). When the rise-up and decay curves for the impurity (tetracene) luminescence were processed the authors substituted into the approximation convolution the difference of two exponential functions with $\tau_1 = \tau_1'$ or $\tau_1 = \tau_1''$ and $\tau_2 = 12$ s which corresponded to the lifetime of tetracene excitations. It came out that the peculiar features of luminescence kinetics were determined by the concentration of the shallow exciton traps (*X* centres according to Wolf (1967); see fig. 7.12). When their concentration is low (crystals of the type I) coherent excitons are practically not trapped by the X centres and under these conditions they are responsible for energy transfer to tetracene (the experimental data correspond to $\tau_1 = \tau_1'$ and the temperature dependence of energy transfer to the impurities is relatively weak). When the concentration of the X centres is high (crystals of the type II) they trap coherent excitons. In this case the experimental data correspond to $\tau_1 = \tau_1''$ indicating energy transfer to tetracene from excitons localized at the *X* centres (localized excitons).

Under such conditions the efficiency of energy transfer to the impurities is observed to increase strongly with increasing temperature. This fact seems to suggest that in the process of energy transfer an exciton localized at the *X* centre is "freed" and then migrates to tetracene molecules as a free and coherent exciton. In this connection, very interesting results on the kinetics of energy transfer by excitons in tetracene crystals doped with pentacene are reported by Campillo et al. (1977). The use of picosecond excitation pulses in this study made it possible to obtain data which suggested that the rate of energy transfer to the impurities was independent of time as it was done for anthracene by Galanin et al. (1974a, b, 1975). Further studies in this field will be of great interest.

CHAPTER 8

Metallic Quenching of Excitons, Exciton Rearrangements and Exciton Reactions at the Metal Boundary

1. *Introduction*: *energy transfer to the surface*

In the above discussion of energy transfer by excitons we have taken the crystal boundary into account only once, namely, in ch. 6 for the formulation of boundary conditions for the exciton concentration. In the diffusion approximation the effect of the boundary is described by the rate of the surface annihilation of excitons v (see eq. (6.1a)) which is a phenomenological quantity and can be calculated only outside the framework of the diffusion approximation. But irrespective of this fact, the most interesting problem related to the effect of the crystal surface (or the interface) on migration and lifetime of excitons is the strong dependence of v on the electronic structure of the boundary layer of the molecular crystal and the electronic structure of the medium adjoining the crystal (the effect of the crystal boundary on the impurity spectrum was reported by Brillante and Craig (1974)). In this connection it is especially interesting to study various physical phenomena at the interfaces of molecular crystals and metals or narrow-band semiconductors. In this case excitons can decay in the region of the interface interacting with the continuum of the states of the conduction electrons. The analysis of the resulting effects is not interesting only for physics of excitons but has a wider importance in connection with some of current trends in modern materials science. The recent advances of this science made it possible to produce fairly complex inhomogeneous systems consisting of metallic and dielectric components.

For instance, in the search for high-temperature superconductors such systems are manufactured by implanting complex organic molecules into metal matrices* of dichalcogenides (Gamble et al. 1971), by simultaneous condensation of metal vapours and organic molecules on cooled substrates (Alekseevskiy et al. 1971), by placing metal films between dielectric plates (Little 1969), and by growing crystals consisting of polymer chains with one-dimensional conductivity and polarizable groups (Little 1974).

The physical properties of organometallic systems of the above types are

*Optical properties of such systems are discussed by Agranovich and Mekhtiev (1971), Mal'shukov (1974).

being intensely analyzed at present. In a certain respect, the metal–molecular crystal contact region is a convenient model system which can be used for analyzing some phenomena (for instance, those related to hybridization of molecular states and electronic states in the conduction band) occurring in homogeneous periodic organometallic structures (Gamble et al. 1971, Alekseevskiy et al. 1971, Little 1969, 1974).

Returning to the discussion of the interaction between excitons and conduction electrons in the metal substrate, note that, as it will be shown below (Agranovich et al. 1972) just the presence of the substrate produces an essential effect on the spectrum of surface excitons. If excitons are not coherent the presence of metal or any other strongly polarizable substrate gives rise to forces promoting exciton migration towards the substrate (Efremov and Mal'shukov 1975).

Under these conditions the exciton lifetime decreases noticeably since there appears a new mechanism of their decay accompanied with generation of surface plasmons and electron–hole pairs in the metal (metallic quenching of excitons; see Agranovich et al. (1968, 1972), Efremov and Mal'shukov (1975), Kuhn (1970)). As observed in experiments (Kallmann et al. 1971, Kallmann 1971, Vaubel et al. 1971, Killesreiter and Baessler 1971, 1972, Kurczewska and Baessler 1977, Heisel et al. 1978) metallic quenching of excitons is especially intense for singlet excitons for which this process is determined by the dipole–dipole mechanism of radiationless energy transfer from excitons to the metal.

Apart from metallic quenching, experiments have evidenced exciton dissociation processes which result in injection of electron or hole into the metal (Singh and Baessler 1974). For singlet excitons the efficiency of this process is much lower than for metallic quenching but it is precisely the exciton dissociation at the metal boundary which is the cause of intrinsic photoconductivity of a molecular crystal under the conditions when the crystal is illuminated by light with high absorption coefficient through a semitransparent metal contact. Dissociation is the main decay mechanism for triplet excitons since radiationless energy transfer to the metal has a considerably lower rate for them.

In our discussion we shall assume the absence of the effects of hybridization of the molecular electronic states and the electronic states in metal. Then the molecules in the boundary layer of the crystal remain neutral and their interaction with the metal electrons is purely electrodynamic (hybridization has been taken into consideration by Agranovich and Mal'shukov (1976), Mal'shukov (1974)). However, even in this approximation the presence of a metallic substrate gives rise to a number of effects. Apart from metallic quenching discussed above, these include the very important effect of the metallic substrate on the spectrum of surface excitons in the molecular crystal.

The surface exciton states which lie below the band of bulk excitons can act as traps for bulk excitons at sufficiently low temperatures and this should lead to an increase in the rate of surface annihilation of excitons under such conditions. Therefore, the theory of the interaction between excitons and a boundary between phases must include analysis of the spectrum of surface excitons.

The qualitative mechanism of the effect of metal contact on the spectrum of surface excitons (Agranovich et al. 1972) can be readily understood from the model of a dielectric whose lattice sites contain dipole oscillators with the frequency ω which is small in comparison with the frequency ω_p of plasma oscillations in the metal ($\hbar\omega_p \approx 10$ eV).

In the first approximation for $\omega \ll \omega_p$ we can assume that the electrons in the metal adiabatically follow the distribution of charges in the dielectric and thus we can employ the quasistatic approximation.

If $\boldsymbol{P}_n$ is the dipole moment of the oscillator on the site n of the dielectric at the distance z_n from the boundary with the metal then the electrical image forces produce a decrease $\delta U = -\alpha P_n^2/(2z_n)^3$ in the potential energy of the oscillator where α is a positive factor depending on the dipole orientation (of course, we mean the energy of the interaction of the dipole P_n with its electrostatic image). Since $P_n = ex_n$ where x_n is the displacement of the charge in the oscillator this potential energy decrease $\delta U \sim -x_n^2$ and it results in a decrease in the natural oscillator frequency ω. In terms of quantum mechanics we can say that the electrical image forces at the boundary with the metal produce a potential well. $U(z) = -A/z^3$, $(A > 0)$ for the exciton with the energy $E = \hbar\omega$ in which, generally speaking, surface exciton states can be formed.

The quantitative theory of the surface state at the boundary with metal (Agranovich et al. 1972) takes into consideration the interaction of the vibrations of charges in a molecule with the electrical images of other molecules. The theory includes penetration of the electric field into the metal (going outside the framework of the electrostatic approximation) which (in the language of this approximation) somewhat blurs the image and, moreover, leads to metallic quenching of excitons. These effects are discussed in sect. 2.

Since we are interested in both long-wavelength and short-wavelength excitations of the dielectric we shall treat them within the framework of the microscopic theory. However, we shall describe the field in the metal with the phenomenological Maxwell equations. This method is applicable here since the mean distance between the conduction electrons in the metal (about 1 Å) is much smaller than the exciton wavelength (the minimum exciton wavelength $\lambda_{\min} = 2a = (10\text{–}20)$ Å where a is the lattice parameter).

2. *Theory of small-radius surface excitons taking into account field penetration into metal. General relations*

The total Hamiltonian of the system molecular crystal + metal can be written as

$$\hat{H} = \hat{H}_1 + \hat{H}_2 + \hat{H}^{(1)}_{\text{int}} + H^{(2)}_{\text{int}}, \tag{8.1}$$

where $\hat{H}_1$ is the Hamiltonian of the metal, $\hat{H}_2$ is the Hamiltonian of the system of noninteracting molecules, $\hat{H}^{(1)}_{\text{int}}$ is the Hamiltonian of the interaction between the molecules and the metal, and $\hat{H}^{(2)}_{\text{int}}$ is the Hamiltonian of the intermolecular interaction. If $\varphi(\boldsymbol{r})$ is the potential of the electric field at the point $\boldsymbol{r}$ of the dielectric ($z > 0$) produced by the metal occupying the half-space $z < 0$ then the Hamiltonian of the interaction between the metal and the dielectric is

$$\hat{H}^{(1)}_{\text{int}} = \sum_n \int \varphi(\boldsymbol{r})\hat{\rho}_n(\boldsymbol{r})\, \mathrm{d}\boldsymbol{r}, \tag{8.2}$$

where $\hat{\rho}_n(\boldsymbol{r})$ is the operator of the charge density of the molecule on the nth site of the crystal lattice. At the same time the operator $\hat{H}^{(2)}_{\text{int}}$ can be written as

$$\hat{H}^{(2)}_{\text{int}} = \tfrac{1}{2} \sum_{n \neq m} \hat{V}_{n,m}, \tag{8.3}$$

where $\hat{V}_{n,m}$ is the operator of the interaction between the molecules on the sites n and m.

Under the effect of the metal the operator (8.2) gives rise to the intermolecular interaction due to the exchange of the virtual excitations of the metal, apart from the conventional intermolecular interaction in the system. It is precisely the interaction of this type that gives rise to the Van der Waals forces at the boundary which are determined mainly by renormalization of the ground state of the system at low temperatures. Moreover, as noted above, the interaction of molecules with the metal leads to metallic quenching of excitons.

Since the metal is assumed to be neutral the mean value of the potential produced by it, that is, $\varphi(\boldsymbol{r})$, is zero. Therefore, the operator (8.2) makes a contribution to the energy of the interaction between molecules and the metal only in the second and higher orders of the perturbation theory.

Below we shall be interested in the excited states of the dielectric whose energies are close to the energy of one of the excited states f of an isolated molecule. Since both the intermolecular interaction and the interaction between molecules and the metal are weak in comparison with the intramolecular interaction we can find the exciton states regarding the operators (8.2) and (8.3) as perturbations as is typically done for the operator $\hat{H}^{(2)}_{\text{int}}$ in the theory of small-radius excitons. It can be shown (Agranovich et al. 1972) that in this approximation the Hamiltonian determining the excited states of the crystal can be written as (compare with eq. (4.34))

$$\Delta\hat{H} = \sum_n [\Delta\epsilon_f + D_f(n)]B_n^+B_n + \sum_{n\neq m}{}' \tilde{M}^f_{nm}B^+_mB_n, \tag{8.4}$$

where

$$\tilde{M}^f_{nm} = M^f_{nm} + \int [K(\mathbf{r}', \mathbf{r}, -\omega_{0f}) + K(\mathbf{r}, \mathbf{r}', \omega_{0f})]\rho^{f0}_n(\mathbf{r}')\rho^{0f}_m(\mathbf{r})\, \mathrm{d}\mathbf{r}\, \mathrm{d}\mathbf{r}', \tag{8.5a}$$

$$D_f(n) = \sum_{i=1}^{4} D^{(i)}_f(n), \tag{8.5b}$$

$$D^{(1)}_f(n) = \sum_m{}' \{\langle 0f|\hat{V}_{nm}|0f|\rangle - \langle 00|\hat{V}_{nm}|00\rangle\}, \tag{8.5c}$$

$$D^{(2)}_f(n) = \sum_m{}' \int [K(\mathbf{r}', \mathbf{r}, 0) + K(\mathbf{r}, \mathbf{r}', 0)][\rho^{ff}_n(\mathbf{r}') - \rho^{00}_n(\mathbf{r}')]\rho^{00}_m(\mathbf{r})\, \mathrm{d}\mathbf{r}\, \mathrm{d}\mathbf{r}', \tag{8.5d}$$

$$D^{(3)}_f(n) = \int K(\mathbf{r}, \mathbf{r}', 0)[\rho^{ff}_n(\mathbf{r}')\rho^{ff}_n(\mathbf{r}) - \rho^{00}_n(\mathbf{r})\rho^{00}_n(\mathbf{r}')]\, \mathrm{d}\mathbf{r}\, \mathrm{d}\mathbf{r}', \tag{8.5e}$$

$$D^{(4)}_f(n) = \sum_{f'\neq f} \int K(\mathbf{r}', \mathbf{r}, -\omega_{f'f})\rho^{ff'}_n(\mathbf{r}')\rho^{f'f}_n(\mathbf{r})\, \mathrm{d}\mathbf{r}\, \mathrm{d}\mathbf{r}' - \sum_{f'\neq 0} \int K(\mathbf{r}', \mathbf{r}, -\omega_{f'0})\rho^{0f'}_n(\mathbf{r}')\rho^{f'0}_n(\mathbf{r})\, \mathrm{d}\mathbf{r}\, \mathrm{d}\mathbf{r}'. \tag{8.5f}$$

In these expressions we have

$$\rho^{ff'}_n(\mathbf{r}) \equiv \langle f|\hat{\rho}_n(\mathbf{r})|f'\rangle, \qquad \omega_{ff'} = \frac{1}{\hbar}(\epsilon_f - \epsilon_{f'}), \tag{8.6}$$

and the correlation function

$$K(\mathbf{r}, \mathbf{r}', \omega) = -\mathrm{i}\int_0^\infty \langle \varphi(\mathbf{r}, t)\varphi(\mathbf{r}', 0)\rangle\, \mathrm{e}^{\mathrm{i}\omega t}\, \mathrm{d}t,$$

where the operator in the integral is averaged over the ground state of the metal.

If we take $K = 0$ in eq. (8.5) we obtain the Hamiltonian for the semi-infinite crystal bordering on vacuum ($\tilde{M} \to M$, $D_f \to D^{(1)}_f$) and in this case $D^{(1)}_f$ is also dependent on n (this is not so in the infinite crystal; see eq. (4.22a))*.

*The theory of surface excitons for this case is presented by Schipper (1975) where the references to earlier studies are given also.

As shown by Agranovich et al. (1972), the correlation function can be expressed in terms of the dielectric constant of the metal as

$$K(\boldsymbol{r}, \boldsymbol{r}', \omega) = \frac{1}{\pi} \int_0^\infty \frac{\operatorname{Im} D(\boldsymbol{r}, \boldsymbol{r}', \omega')\,\mathrm{d}\omega'}{\omega - \omega' + \mathrm{i}\delta}, \tag{8.7}$$

where $\delta \to +0$ and

$$\operatorname{Re} D(\boldsymbol{r}, \boldsymbol{r}', \omega) = \frac{1}{\pi} \int_{-\infty}^{\infty} \frac{\operatorname{Im} D(\boldsymbol{r}, \boldsymbol{r}', \omega')}{\omega' - \omega}\,\mathrm{d}\omega', \tag{8.7a}$$

$$D(\boldsymbol{r}', \boldsymbol{r}, \omega) = \frac{1}{2\pi} \int \frac{\mathrm{d}\boldsymbol{k}_\perp}{k_\perp} \left(\frac{1 - f(\omega, \boldsymbol{k}_\perp)}{1 + f(\omega, \boldsymbol{k}_\perp)}\right) \exp[\mathrm{i}\boldsymbol{k}_\perp(\boldsymbol{r} - \boldsymbol{r}') - |\boldsymbol{k}_\perp|(z + z')], \tag{8.8}$$

$$f(\omega, \boldsymbol{k}_\perp) = \frac{\omega^2}{\pi c^2 k_\perp} \int_{-\infty}^{\infty} \frac{\mathrm{d}k_3}{k^2} \left[\frac{k_\perp^2}{(\omega^2/c^2)\epsilon_\ell(\omega, \boldsymbol{k})} + \frac{k_3^2}{(\omega^2/c^2)\epsilon_\mathrm{t}(\omega, \boldsymbol{k}) - k^2}\right], \tag{8.8a}$$

Here $\epsilon_\ell(\omega, \boldsymbol{k})$ and $\epsilon_\mathrm{t}(\omega, \boldsymbol{k})$ are the "longitudinal" and "transverse" dielectric constants of the metal.

The following notation for vectors is used in the above expressions: $\boldsymbol{r} \equiv (x, y, z)$, $\boldsymbol{k} \equiv (k_1, k_2, k_3) \equiv (\boldsymbol{k}_\perp, k_3)$; as noted above, the z-axis is perpendicular to the boundary.

The wave function of the excited state of the crystal with the energy $E(E \approx \Delta\epsilon_f)$ can be written as

$$\Psi = \sum_n U_n B_n^+ |0\rangle, \tag{8.9}$$

where the coefficients U_n satisfy the system of equations

$$[E - \Delta\epsilon_f - D_f(n)] U_n = \sum_m{}' \tilde{M}_{nm}^f U_m. \tag{8.10}$$

Since the function $K(\boldsymbol{r}, \boldsymbol{r}', \omega)$ (see eq. (8.7)) determining the effective intermolecular interaction is complex the exciton states in the dielectric are decaying (metallic quenching). But the exciton level width Γ corresponding to this decay is small compared with the exciton energy. Therefore, we can ignore this decay solving eqs. (8.10) in the first approximation. However, in the next approximation we obtain $E \to E + \mathrm{i}\Gamma$ from eqs. (8.10) and

$$\Gamma(E) = \sum_n \operatorname{Im} D_f(n) |U_n|^2 + \sum_{np}{}' U_n^* U_p \operatorname{Im} \tilde{M}_{np}^f. \tag{8.11}$$

Equation (8.7) yields

$$\operatorname{Re} K(\boldsymbol{r}, \boldsymbol{r}', \omega) = \frac{1}{\pi} \int_0^\infty \frac{\operatorname{Im} D(\boldsymbol{r}, \boldsymbol{r}', \omega')}{\omega - \omega'}\,\mathrm{d}\omega', \tag{8.12}$$

while

$$\operatorname{Im} K(\boldsymbol{r}, \boldsymbol{r}', \omega) = \begin{cases} -\operatorname{Im} D(\boldsymbol{r}, \boldsymbol{r}', \omega) & \text{for} \quad \omega > 0, \\ 0 & \text{for} \quad \omega \leq 0. \end{cases} \tag{8.13}$$

Thus, we obtain from eqs. (8.5) and (8.13)

$$\operatorname{Im} D_f(n) = \operatorname{Im} D_f^4(n)$$

$$= -\int \operatorname{Im} D(\boldsymbol{r}', \boldsymbol{r}, \omega_{f0}) \rho_n^{f0}(\boldsymbol{r}') \rho_n^{0f}(\boldsymbol{r})\, \mathrm{d}\boldsymbol{r}\, \mathrm{d}\boldsymbol{r}', \tag{8.14}$$

$$\operatorname{Im} \tilde{M}_{nm}^f = -\int \operatorname{Im} D(\boldsymbol{r}', \boldsymbol{r}, \omega_{f0}) \rho_n^{f0}(\boldsymbol{r}') \rho_m^{0f}(\boldsymbol{r})\, \mathrm{d}\boldsymbol{r}\, \mathrm{d}\boldsymbol{r}',$$

so that

$$\Gamma(E) = -\int \operatorname{Im} D(\boldsymbol{r}', \boldsymbol{r}, \omega_{f0}) \Phi(\boldsymbol{r}', \boldsymbol{r})\, \mathrm{d}\boldsymbol{r}\, \mathrm{d}\boldsymbol{r}', \tag{8.15}$$

where

$$\Phi(\boldsymbol{r}', \boldsymbol{r}) = \sum_{nm} U_n^* U_m \rho_n^{f0}(\boldsymbol{r}') \rho_m^{0f}(\boldsymbol{r}). \tag{8.15a}$$

we see that $\Gamma(E)$ is fully determined by the imaginary part of the function $D(\boldsymbol{r}, \boldsymbol{r}', \omega)$.

Since the frequencies of the surface plasmons in the metal at the boundary with vacuum satisfy the equation (see, Romanov 1964, Richie and Marusak 1966)

$$1 + f(\omega, \boldsymbol{k}_\perp) = 0,$$

we see from eq. (8.8) that $\operatorname{Im} D(\boldsymbol{r}, \boldsymbol{r}', \omega)$ has a sharp peak at the surface plasmon frequency $\omega \approx \omega_\mathrm{p}$. Therefore, at $\omega \ll \omega_\mathrm{p}$ we can neglect the dependence of $\operatorname{Re} K(\boldsymbol{r}, \boldsymbol{r}', \omega)$ on ω taking $\omega = 0$. In this approximation (see also eq. (8.7a)) we obtain

$$\operatorname{Re} K(\boldsymbol{r}, \boldsymbol{r}', \omega) \approx \operatorname{Re} K(\boldsymbol{r}, \boldsymbol{r}', 0) = K(\boldsymbol{r}, \boldsymbol{r}', 0)$$

$$= -\tfrac{1}{2} \operatorname{Re} D(\boldsymbol{r}, \boldsymbol{r}', 0). \tag{8.16}$$

To find $\operatorname{Re} D(\boldsymbol{r}, \boldsymbol{r}', 0)$ we shall use the relationship $\epsilon_1(0, \boldsymbol{k}) = 1 + k_\mathrm{D}^2/k^2$, where $\hbar k_\mathrm{D}$ is the Debye momentum ($k_\mathrm{D} \approx 10^8\ \mathrm{cm}^{-1}$) for electrons of the metal. Then, eq. (8.8a) yields $f(0, \boldsymbol{k}_\perp) \approx 2k_\perp/k_\mathrm{D} \ll 1$ for $k_\perp \ll k_\mathrm{D}$. Hence, according to eq. (8.8) we have

$$\operatorname{Re} D(\boldsymbol{r}', \boldsymbol{r}, 0) = D(\boldsymbol{r}', \boldsymbol{r}, 0)$$

$$\approx \frac{1}{2\pi} \int \frac{\mathrm{d}\boldsymbol{k}_\perp}{k_\perp} \exp[\mathrm{i}\boldsymbol{k}_\perp(\boldsymbol{r} - \boldsymbol{r}') - |\boldsymbol{k}_\perp|(z + z')] = \frac{1}{|\boldsymbol{r} - \tilde{\boldsymbol{r}}'|}, \tag{8.17}$$

where the vector $\tilde{r}' \equiv (x', y', -z')$ determines the position of the mirror image of the point r'.

When we calculate $\operatorname{Im} D(r, r', \omega)$ we cannot ignore the frequency dependence of $f(\omega, k_\perp)$. But we have $|f(\omega, k_\perp)| < 1$ for $\omega < \omega_p$. Therefore, according to eq. (8.8) we obtain

$$D(r', r, \omega) \approx \frac{1}{2\pi} \int \frac{dk_\perp}{k_\perp} [1 - 2f(\omega, k)] \exp[ik_\perp(r - r') - |k_\perp|(z + z')],$$

so that

$$\operatorname{Im} D(r', r, \omega) \approx -\frac{1}{\pi} \int \frac{dk_\perp}{k_\perp} \operatorname{Im} f(\omega, k_\perp) \exp[ik_\perp(r - r') - |k_\perp|(z + z')],$$

or, finally,

$$\operatorname{Im} D(r, r', \omega) = \frac{1}{\pi^2} \int \left\{ \frac{\epsilon_l''(\omega, k)}{|\epsilon_l(\omega, k)|^2} + \frac{k_3^2}{k_\perp^2} \frac{\epsilon_t''(\omega, k)}{|\epsilon_t(\omega, k) - c^2k^2/\omega^2|^2} \right\}$$
$$\times \exp[ik_\perp(r - r') - |k_\perp|(z + z')] \frac{dk}{k^2}. \tag{8.18}$$

Note that the mechanism of decay of the exciton states discussed here is due to the Joule losses produced by penetration of the excitonic electric field into the metal and, thus, it is quite similar to the mechanism discussed in ch. 2 sect. 2 and ch. 5 sect. 9 (where decay of the excited states of individual molecules was analyzed).

It is particularly easy to calculate $\Gamma(E)$ for incoherent excitons. If an incoherent exciton is, for instance, on the site n_0 then $U_n = \delta_{nn_0}$ and, according to eq. (8.15) we have

$$\Gamma_{n_0}(E) = \int \operatorname{Im} D(r', r, \omega_{f0}) \rho_{n_0}^{f0}(r') \rho_{n_0}^{0f}(r) \, dr \, dr'. \tag{8.19}$$

The calculations of decay for coherent (both surface and volume) excitons are, generally, more complicated since they necessitate solving eqs. (8.10) with inclusion of the matrix elements for resonance energy transfer between molecules, $\tilde{M}_{nm}^f$. In this connection, in sect. 3 we shall be concerned with the determination of the form of the wave functions describing coherent surface excitons (that is, the coefficients U_n) and the structure of their spectrum. Metallic quenching of such excitons and some associated problems will be discussed in the subsequent sections.

3. *Surface excitons in the range of frequencies of intramolecular transitions*

Note that, apart from the dipole–dipole interaction, the intermolecular multipole interactions also make a significant contribution to the exciton energy. Therefore, to simplify the treatment of excited states of the dielectric in the region of the spectrum of electronic transitions, we shall assume that the dielectric molecules have an inversion centre. The dipole moment operator for such molecules can have only nondiagonal nonzero matrix elements so that the quantities $\langle 0f|\hat{V}_{nm}|0f\rangle$ and $\langle 00|\hat{V}_{nm}|00\rangle$ (as well as $\rho_n^{00}(\boldsymbol{r})$ and $\rho_n^{ff}(\boldsymbol{r})$) depend only on the quadrupole moment and higher moments. Therefore, these diagonal matrix elements of the operator $\hat{V}_{nm}$ rapidly decrease with increasing $|\boldsymbol{n}-\boldsymbol{m}|$ and it proves to be sufficient to take into account the interaction between the nearest neighbours in order to calculate their contribution to the exciton energy. Assuming that the crystal has the simplest cubic lattice with the parameter a and that molecular layers with $n_3 = 1, 2, 3, \ldots$ are parallel to the metal–molecular crystal boundary (see fig. 8.1) we find that in this approximation $D_f^{(1)}(\boldsymbol{n})$ (see eq. (8.5c)) is independent of n_3 even for $n_3 \geqslant 2$. According to eqs. (8.16) and (8.17), $D_f^{(2)}(\boldsymbol{n})$ given by eq. (8.5d) is determined by the energy of the Coulomb interaction of the charges with the density $\Delta\rho_n(\boldsymbol{r}) = \rho_n^{ff}(\boldsymbol{r}) - \rho_n^{00}(\boldsymbol{r})$ with the electrostatic images of the charges with the density $\rho_m^{00}(\boldsymbol{r})$ for $m \neq n$. Therefore, in the approximation of the nearest neighbours $D_f^{(2)}(\boldsymbol{n})$ can be completely ignored. For $n_3 = 1$ the quantity $D_f^{(1)}(\boldsymbol{n})$ differs from its "bulk" value (that is, the value for $n_3 \geqslant 2$) in the energy of the interaction with one (absent) nearest neighbour. Instead of this interaction, eq. (8.5b) contains the term $D_f^{(3)}(\boldsymbol{n})$ which is determined by the variation of the energy of interaction of the molecule with its electrostatic image in the

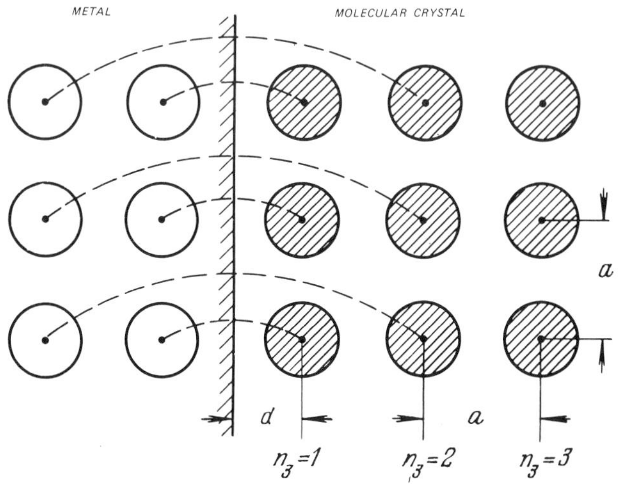

Fig. 8.1. The metal–molecular crystal contact region. The vertical line is the metal boundary. The shaded circles (molecules of the crystal) are connected by dashed lines with the open circles (the mirror images of the molecules). Only the molecules in the layers $n_3 = 1$ and $n_3 = 2$ are shown.

metal in the transition $0 \to f$. If the distance $2d$ were equal to a the sum $D_f^{(1)}(n_3 = 1) + D_f^{(3)}(n_3 = 1)$ in the approximation of the nearest neighbours would be equal to the "bulk" value of $D_f^{(1)}(n)$. It is most likely, however, that even under the conditions of close contact $2d > a$ so that the above equality does not hold.

In our case of a plane boundary it is convenient to write for the quantities U_n entering into eqs. (8.10)

$$U_n = U(n_3)\exp(\mathrm{i}\boldsymbol{k}_\perp \boldsymbol{n}). \tag{8.20}$$

Then we can rewrite eqs. (8.10) as

$$[E - \Delta\epsilon_f - D^{(0)} - \Gamma_{k_\perp}(0) - D^{(1)}\delta_{1n_3}]U(n_3) - \sum_{m_3}{}' \Gamma_{k_\perp}(n_3 - m_3)U(m_3)$$
$$- \sum_{m_3} \tilde{\Gamma}_{k_\perp}(n_3 + m_3)U(m_3) = A(n_3)U(n_3). \tag{8.21}$$

Here $D^{(0)}$ is the "bulk" value of $D_f^{(1)}(n)$ while

$$D^{(1)} = \int \frac{\rho_n^{00}(\boldsymbol{r})\rho_n^{00}(\boldsymbol{r}') - \rho_n^{ff}(\boldsymbol{r})\rho_n^{ff}(\boldsymbol{r}')}{|\boldsymbol{r} - \tilde{\boldsymbol{r}}'|}\,\mathrm{d}\boldsymbol{r}\,\mathrm{d}\boldsymbol{r}'$$
$$+ \int \frac{\rho_n^{00}(\boldsymbol{r}) - \rho_n^{ff}(\boldsymbol{r})}{|\boldsymbol{r} - \boldsymbol{r}'|}\,\rho_m^{00}(\boldsymbol{r}')\,\mathrm{d}\boldsymbol{r}\,\mathrm{d}\boldsymbol{r}', \tag{8.22}$$

where $\boldsymbol{n} \equiv (n_1, n_2, 1)$, $m \equiv (n_1, n_2, 2)$, and $\tilde{\boldsymbol{r}}' = (x', y', -z')$. In contrast to $D^{(0)}$ and $D^{(1)}$, the quantities $A(n_3)$, $\Gamma_{k_\perp}$ and $\tilde{\Gamma}_{k_\perp}$ are expressed in terms of the nondiagonal matrix elements of the charge density operator $\hat{\rho}_n(\boldsymbol{r})$ and do not vanish even in the dipole approximation. Since $p_n^{ff'}(\boldsymbol{r}) = -p^{ff'}\nabla\delta(\boldsymbol{r} - \boldsymbol{n})$ where $\boldsymbol{p}^{ff'}$ is the dipole moment of the transition $f \to f'$ (this relationship is derived below) we obtain in the dipole approximation

$$A(n_3) = D_f^{(4)}(n) - \int D(\boldsymbol{r}, \boldsymbol{r}', \omega_{f0})\rho_n^{0f}(\boldsymbol{r})\rho_n^{f0}(\boldsymbol{r}')\,\mathrm{d}\boldsymbol{r}\,\mathrm{d}\boldsymbol{r}',$$

$$\Gamma_{k_\perp}(m_3 - n_3) = \sum_{m_\perp} W_{nm}^{0ff0} \exp[\mathrm{i}\boldsymbol{k}_\perp(\boldsymbol{m} - \boldsymbol{n})]. \tag{8.23}$$

$$\tilde{\Gamma}_{k_\perp}(n_3 + m_3) = \sum_{m_\perp} \int D(\boldsymbol{r}', \boldsymbol{r}, \omega_{0f})\rho_n^{f0}(\boldsymbol{r}')\rho_m^{0f}(\boldsymbol{r})\exp[\mathrm{i}\boldsymbol{k}_\perp(\boldsymbol{m} - \boldsymbol{n})]\,\mathrm{d}\boldsymbol{r}\,\mathrm{d}\boldsymbol{r}'.$$

As shown by Agranovich et al. (1972), the last two expressions can be rewritten as

$$\Gamma_{k_\perp}(m_3 - n_3) = \Gamma^*_{k_\perp}(n_3 - m_3) = \frac{2\pi}{a^2}\sum_{g_\perp} \frac{|(\boldsymbol{p}^{0f}, \boldsymbol{g}_\perp + \boldsymbol{k}_\perp) + \mathrm{i}p_3^{0f}|\boldsymbol{g}_\perp + \boldsymbol{k}_\perp||^2}{|\boldsymbol{g}_\perp + \boldsymbol{k}_\perp|}$$
$$\times \exp[-|\boldsymbol{g}_\perp + \boldsymbol{k}_\perp||m_3 - n_3|a], \qquad m_3 > n_3, \tag{8.24}$$

$$\tilde{\Gamma}_{k_\perp}(n_3+m_3) = -\frac{2\pi}{a^2}\sum_{g_\perp}\frac{(\boldsymbol{p}^{0f}, \boldsymbol{g}_\perp+\boldsymbol{k}_\perp)^2+(p_3^{0f})^2|\boldsymbol{g}_\perp+\boldsymbol{k}_\perp|^2}{|\boldsymbol{g}_\perp+\boldsymbol{k}_\perp|}$$

$$\times\frac{1-f(\boldsymbol{g}_\perp+\boldsymbol{k}_\perp, \omega_{f0})}{1+f(\boldsymbol{g}_\perp+\boldsymbol{k}_\perp, \omega_{f0})}\exp\{-|\boldsymbol{g}_\perp+\boldsymbol{k}_\perp|[(n_3+m_3)a+2(d-a)]\}. \tag{8.25}$$

In eqs. (8.24) and (8.25) summation is performed over all possible projections of the vectors **g** of the reciprocal lattice on the plane (x, y) and d is equal to the distance from the plane $n_3 = 1$ to the metal surface (see fig. 8.1). For a good metal–dielectric contact we can, probably, assume $d \approx a/2$.

Expression (8.24) gives the energy of interaction of the dipole $\boldsymbol{p}^{0,f}$ on the site $\boldsymbol{n}$ with the plane network of dipoles $\boldsymbol{p}^{0f}\exp(i\boldsymbol{k}_\perp\boldsymbol{m})$ located on the lattice sites with the constant m_3. Expression (8.25) gives the energy of the interaction between the same dipole and the mirror image of the dipole network with $m_3 = \text{const}$ (here $m_3 \neq n_3$) which takes into account penetration of the field into the metal. If $m_3 = n_3$ then eq. (8.25) takes into account only part of the total energy of the interaction of the molecular transition moment with the mirror image of the dipole network. This is due to the fact that the energy of the interaction of the transition dipole with its mirror image (the quantity $D_f^{(4)}(n)$ given by eq. (8.5f)) is determined by the totality of the excited states of the molecule and is, generally, different from the energy of the interaction with a foreign mirror image (the second term in eq. (8.5a) on the right-hand side) even if we take $m = n$ in this term. It is precisely this fact that gives rise to the term with $A(n_3)$ taking into account the part of the interaction energy not included in $\tilde{\Gamma}_{k_\perp}(n_3+m_3)$ in eqs. (8.21). Equation (8.23) implies that $A(n_3)$ decreases with increasing n_3 according to the relation $A(n_3) \approx (2z_n)^{-3}$ where $z_n = n_3 a + d - a$.

Before solving the system of equations (8.21) let us analyze the surface states in the dielectric in the range of the fundamental tone frequency $(f = 1)$ of the dipole-active nondegenerate intramolecular vibrations of the nuclei. For excitations of this type the matrix elements $\rho_n^{ff'}(\boldsymbol{r})$ differ from zero only if $f' = f \pm 1$ (the selection rules for the oscillator) and

$$\rho_n^{f+1,f}(\boldsymbol{r}) = \rho_n^{f,f+1}(\boldsymbol{r}) = -\sqrt{f+1}\,(\boldsymbol{p}^{01}\nabla\delta(\boldsymbol{r}-\boldsymbol{n})). \tag{8.26}$$

Using eqs. (8.16) and (8.23) we obtain

$$A(n_3) = \int K(\boldsymbol{r}', \boldsymbol{r}, 0)\Big\{\sum_{p\neq 1}\rho_n^{1p}(\boldsymbol{r}')\rho_n^{p1}(\boldsymbol{r})$$
$$-\sum_{p\neq 0}\rho_n^{0p}(\boldsymbol{r}')\rho_n^{p0}(\boldsymbol{r}) - 2\rho_n^{01}(\boldsymbol{r}')\rho_n^{10}(\boldsymbol{r})\Big\}\,\mathrm{d}\boldsymbol{r}\,\mathrm{d}\boldsymbol{r}'$$

is identically zero if we take into consideration eq. (8.26).

Equations (8.24) and (8.25) indicate that for $\boldsymbol{k}_\perp = 0$ and for $|\boldsymbol{k}_\perp| \lessdot \pi/a$ the quantities $\Gamma_{k_\perp}(z)$ and $\tilde{\Gamma}_{k_\perp}(z)$ exponentially decrease with increasing z. Therefore, we can use the nearest neighbour approximation for analyzing the system of equations (8.21) with $k_\perp = 0$ and $k_\perp \leqslant \pi/a$. In this approximation the wave function and the energy of the respective surface states are given by

$$u(n_3) = \exp(-\kappa n_3), \qquad E(\boldsymbol{k}_\perp) = \hbar\omega_{f0} + D^{(0)} + \Gamma_{\boldsymbol{k}_\perp}(0) + 2\Gamma_{\boldsymbol{k}_\perp}(1)\cosh\kappa,$$

$$\mathrm{e}^{-\kappa} = \frac{\Gamma_{\boldsymbol{k}_\perp}(1)}{D^{(1)} + \tilde{\Gamma}_{\boldsymbol{k}_\perp}(2)}. \tag{8.27}$$

According to eqs. (8.24) and (8.25), the quantities $\Gamma_{\boldsymbol{k}_\perp}$ and $\tilde{\Gamma}_{\boldsymbol{k}_\perp}$ depend on the direction of the dipole $\boldsymbol{p}^{0f}$. Therefore, the surface states of this type can occur only for such orientations of the molecules when $|\mathrm{e}^{-\kappa}| < 1$. Since for the same values of $\boldsymbol{k}_\perp$ for the bulk states the energy is

$$E(\boldsymbol{k}_\perp, k_3) = \hbar\omega_{f0} + D^{(0)} + \Gamma_{\boldsymbol{k}_\perp}(0) + 2\Gamma_{\boldsymbol{k}_\perp}(1)\cos k_3 a,$$

the depth of the surface levels, that is, $\Delta \equiv E(\boldsymbol{k}_\perp, k_3) - E(\boldsymbol{k}_\perp)$, is

$$\Delta = \frac{[D^{(1)} + \tilde{\Gamma}_{\boldsymbol{k}_\perp}(2) - \Gamma_{\boldsymbol{k}_\perp}(1)]^2}{|D^{(1)} + \tilde{\Gamma}_{\boldsymbol{k}_\perp}(2)|}. \tag{8.28}$$

In those cases when $\mathrm{Re}\,\kappa \ll 1$ the surface states with $\boldsymbol{k}_\perp = 0$ are macroscopic. As shown by Agranovich et al. (1972b), these states can be found in the framework of the phenomenological Maxwell equations taking into account spatial dispersion though additional boundary conditions are needed for that.

Moreover, note that for small $|D^{(1)}|$ the total width of the energy band of the surface states (that is, the energies for arbitrary $\boldsymbol{k}$) can be greater than Δ. In this case the surface exciton states found above lie below the energy band of all the bulk states only for such dipole orientations for which the absolute minimum of the bulk exciton energies corresponds either to $\boldsymbol{k}_\perp = 0$ or to $|\boldsymbol{k}_\perp| \leqslant \pi/a$.

Now let us consider the surface states in the frequency range of the intramolecular electronic transitions. In this case eqs. (8.26) are no longer applicable and $A(n_3)$, generally, does not vanish. To find $A(n_3)$ we have to know the behaviour of the function $K(\boldsymbol{r}, \boldsymbol{r}', \omega)$ in a wide frequency range since the whole spectrum of the molecular excitations which now has no oscillator character makes the contribution to $D_f^{(4)}(\boldsymbol{n})$. But the dispersion relationships (8.7) show that to find the correlation function K we have to know just the dependence on ω of the imaginary part of the function $D(\boldsymbol{r}, \boldsymbol{r}', \omega)$ given by eq. (8.8). In this connection, we should bear in mind that the imaginary part of the integrand in eq. (8.8) has a maximum at the frequency $\omega_s(\boldsymbol{k}_\perp)$ of the surface plasmon. In the range of small $k_\perp \ll k_F$ the surface plasmons are well-defined elementary excitations and this maximum is a sharp peak of a small width. Since small $k_\perp \leqslant (z + z')^{-1}$ make the contribution to the integrand in eq. (8.8) we have approximately

$$\operatorname{Im} \frac{1 - f(\omega, \boldsymbol{k}_\perp)}{1 + f(\omega, \boldsymbol{k}_\perp)} = \tilde{A}\delta(\omega - \omega_s(\boldsymbol{k}_\perp)), \qquad ((8.29)$$

where $\tilde{A}$ is a constant.

For very small $k_\perp \lesssim \omega_p/c$ retardation effects become significant and the frequency $\omega_s(\boldsymbol{k}_\perp)$ rapidly tends to zero. Since we are interested in the region $z, z' \ll c/\omega_p \approx 300$ Å where ω_p is the bulk plasmon frequency we can ignore the region of small $k_\perp \lesssim \omega_p/c$ where retardation is significant since the main region of integration in eq. (8.8) is much wider in this case and the integrand does not have singularities. Here $\omega_s \approx \omega_p/\sqrt{2}$ so that we have

$$\operatorname{Im} D(\boldsymbol{r}, \boldsymbol{r}', \omega) = \frac{\tilde{A}\delta(\omega - \omega_p/\sqrt{2})}{|\boldsymbol{r} - \tilde{\boldsymbol{r}}'|}. \qquad (8.30)$$

The quantity $\tilde{A}$ can be found from the dispersion relation (8.7a) if we take $\omega = 0$ and use eq. (8.17) for $D(\boldsymbol{r}, \boldsymbol{r}', 0)$.

Substitution of eq. (8.30) into eq. (8.7) yields (see also eqs (8.16) and (8.17)):

$$\tilde{A} = \frac{\pi}{2} \frac{\omega_p}{\sqrt{2}}. \qquad (8.31)$$

Now, using eqs. (8.23), (8.30), (8.31) and (8.7) we obtain

$$\operatorname{Re} A(n_3) = R^f/16Z_n^3, \qquad (8.32)$$

where

$$R^f = \sum_{f' \neq 0, f} \left\{ \frac{1}{2\sqrt{2}} \frac{\omega_p}{\omega_{ff'} - \omega_p/\sqrt{2}} [|\boldsymbol{p}^{ff'}|^2 + (p_3^{ff'})^2] - \frac{\omega_p}{2\sqrt{2}} \frac{|\boldsymbol{p}^{0f'}|^2 + (p_3^{0f'})^2}{\omega_{0f} - \omega_p/\sqrt{2}} \right\} - \frac{\omega_p}{\sqrt{2}} \frac{|\boldsymbol{p}^{0f}|^2 + (p_3^{0f})^2}{\omega_{0f} - \omega_p/\sqrt{2}}. \qquad (8.33)$$

The sign of R^f significantly depends on the structure of the spectrum of electronic excitatons of molecule. If we take into consideration only two molecular states 0 and f then $R^f > 0$ (we assume that $\omega_{0f} < \omega_p/\sqrt{2}$). Inclusion of other molecular excitations can also lead to $R^f < 0$ (for the oscillator spectrum $R^f \equiv 0$).

If $|R^f| \ll |\boldsymbol{p}^{0f}|^2$ we can omit the term with $A(n_3)$ in the system of equations (8.21). Here we obtain the situation which has been analyzed above for the oscillatory surface excitons.

The most interesting features of the spectrum of the electronic surface states appear when

$$|R^f| \gg |\boldsymbol{p}^{0f}|^2,$$

and the depth of the potential well (or the height of the potential barrier for $\operatorname{Re} A(n_3) > 0$) is large in comparison with the width of the energy band of the bulk excitons. Under such conditions for small n_3 and, at least for $k_\perp \approx 0$ and for $k_\perp \leqslant \pi/a$ (when the quantities Γ and $\tilde{\Gamma}$ are exponentially small; see eqs. (8.27) and (8.28)) the following inequalities are satisfied:

$$|\mathrm{Re}\, A(n_3)| \gg |\Gamma_{k_\perp}(n_3 - m_3)|, \qquad |\tilde{\Gamma}_{k_\perp}(n_3 + m_3)|.$$

Therefore, for these values of n_3 in the first approximation we can omit the terms with $\Gamma_{k_\perp}$ and $\tilde{\Gamma}_{k_\perp}$ in the system of equations (8.21). The presence of the term with $A(n_3)$ gives rise to the surface states with the energies

$$E_{n_3}(\boldsymbol{k}_\perp) = \hbar\omega_{f0} + D^{(0)} + \delta_{n_3 1} D^{(1)} + \mathrm{Re}\, A(n_3), \tag{8.34a}$$

and the wave functions

$$\Psi(\boldsymbol{k}_\perp, n_3) = \frac{1}{\sqrt{N_1 N_2}} \sum_{n_1 n_2} \exp(\mathrm{i}\boldsymbol{k}_\perp \boldsymbol{n}) \chi^f_{n_1 n_2 n_3}. \tag{8.34b}$$

Each of such states is related practically only to one crystal plane $n_3 =$ const. However, the number of these states is small and even for $|R^f| = 10|\boldsymbol{p}^{0f}|^2$ is not more than 2 (with $n_3 = 1$ and with $n_3 = 2$).

Note that apart from the above surface states of the type described by eq. (8.34) there also can occur the "macroscopic" surface states in a large number of subsurface molecular planes. Their properties correspond to those of the states treated by Agranovich et al. (1972b) within the framework of the phenomenological approach.

In conclusion, let us discuss once more the difference between the surface states of the dielectric corresponding to the electronic and vibrational intramolecular excitations. As noted above, these surface states are due to the interaction of the intramolecular excitations with the metal surface. The energy of this interaction is equal to the difference between the shifts of the terms of the ground and excited states of the molecule; naturally, this energy depends on the structure of the molecular spectrum and the wave functions of the stationary states as is indicated by eqs. (8.5). At the same time, the surface states discussed above lie in the energy range of the lowest molecular excitations where the quantum properties are most markedly manifested. Therefore, it is not surprising that intramolecular vibrations with different spectral character are related to surface states with different structures. The contribution of the electronic surface states to the mechanism of the interaction between the bulk excitons and the surface has been noted above and will be discussed in sects. 5 and 6.

4. *Metallic quenching of incoherent excitons: the role of the "dead" zone*

For an incoherent exciton on the site n (this case has been experimentally studied by Kallmann et al. (1971), Kallmann (1971), Vaubel et al. (1971), Killesreiter and Baessler (1971, 1972)) the partial broadening of the molecular term due to metallic quenching is determined by eq. (8.19). Since the operator of the electron density in the molecule n is

$$\hat{\rho}_n(\boldsymbol{r}) = -e \sum_{\nu} \delta(\boldsymbol{r} - \boldsymbol{n} - \boldsymbol{x}_\nu),$$

where $\boldsymbol{x}_\nu$ is the radius vector determining the position of the νth electron in the molecule n with respect to the lattice site $\boldsymbol{n}$ then in the approximation linear in $\boldsymbol{x}_\nu$ we have

$$\delta(\boldsymbol{r} - \boldsymbol{n} - \boldsymbol{x}_\nu) = -\boldsymbol{x}_\nu \nabla \delta(\boldsymbol{r} - \boldsymbol{n}), \tag{8.35}$$

so that the matrix element of the operator of the electron density is

$$\rho_n^{0f}(\boldsymbol{r}) = -\boldsymbol{p}_n^{0f} \nabla \delta(\boldsymbol{r} - \boldsymbol{n}), \tag{8.36}$$

where $\boldsymbol{p}_n^{0f}$ is the matrix element of the dipole moment operator $\boldsymbol{p}_n = -\Sigma_\nu e\boldsymbol{x}_\nu$ corresponding to the transition $0 \to f$.

Equation (8.36) is written in the framework of the dipole approximation and to extend its scope we must include in eq. (8.35) the higher-order terms in $\boldsymbol{x}_\nu$. Then eq. (8.25) will contain terms with higher multipoles.

Now, using eq. (8.19) we obtain

$$\Gamma_n = \Gamma_n^\ell + \Gamma_n^{\mathrm{t}}, \tag{8.37}$$

$$\Gamma_n^\ell = \frac{1}{\pi^2} \int \frac{\mathrm{d}\boldsymbol{k}}{k^2} \frac{\epsilon_\ell''(\omega, \boldsymbol{k})}{|\epsilon_\ell(\omega, \boldsymbol{k})|^2} |\boldsymbol{p}^{0f}\boldsymbol{K}|^2 \exp(-2k_\perp z_n), \tag{8.37a}$$

$$\Gamma_n^{\mathrm{t}} = \frac{1}{\pi^2} \int \frac{\mathrm{d}\boldsymbol{k}}{k^2} \frac{k_3^2}{k_\perp^2} \frac{\epsilon_{\mathrm{t}}''(\omega, \boldsymbol{k})}{|\epsilon_{\mathrm{t}}(\omega, \boldsymbol{k}) - k^2c^2/\omega^2|^2} |\boldsymbol{p}^{0f}\boldsymbol{K}|^2 \exp(-2k_\perp z_n), \tag{8.37b}$$

where vector $\boldsymbol{K} \equiv (\boldsymbol{k}_\perp, i\boldsymbol{k}_\perp)$, $\omega \equiv \omega_{f0}$, and z_n is the distance from the molecule n to the metal surface. The quantities Γ_n^ℓ and Γ_n^{t} are due to the longitudinal and transverse forced vibrations excited in the metal; $\Gamma_n^{\mathrm{t}} \to 0$ if we ignore retardation (that is, if $c \to \infty$).

Equations (8.37) imply that the contributions of the pole $k^2 = 0$ in the sum $\Gamma_n = \Gamma_n^\ell + \Gamma_n^{\mathrm{t}}$ are compensated. Therefore, we shall ignore this pole in the calculations of Γ_n^ℓ and Γ_n^{t}.

The calculated results for Γ_n^ℓ and Γ_n^{t} significantly depend on the relative importance of taking into account the dependence of the dielectric constant of the metal on the wave vector $\boldsymbol{k}$ in the given frequency range. To illustrate this importance, we shall consider at first the case of normal skin effect when the dielectric constant of the metal can be regarded as independent of $\boldsymbol{k}$. We have

$$\epsilon_\ell(\omega, \boldsymbol{k}) = \epsilon_{\mathrm{t}}(\omega, \boldsymbol{k}) = \epsilon(\omega), \tag{8.38}$$

and after simple integration (8.37a) yields the result (reported by Kuhn (1970)):

$$\Gamma_n^\ell = A/z_n^3, \tag{8.39}$$

where

$$A = \frac{1+\cos^2\theta}{4\pi} \frac{\epsilon''(\omega)}{|\epsilon(\omega)|^2} |p^{0f}|^2, \tag{8.39a}$$

and θ is the angle between the vector p^{0f} and the z-axis. Since the radiation lifetime of the excited state f is given by

$$\frac{1}{\tau_0} = \frac{4\omega_{0f}^3}{3\hbar c^3} |p^{0f}|^2, \tag{8.39b}$$

we can write

$$A = \frac{3\hbar}{\tau_0} \frac{1+\cos^2\theta}{\pi} \left(\frac{c}{2\omega_{0f}}\right)^3 \frac{n\kappa}{(n^2+\kappa^2)^2}, \tag{8.39c}$$

where n and κ are the indexes of absorption and refraction for the metal at the frequency $\omega = \omega_{0f}$; we have $\epsilon'' = 2n\kappa$ and $|\epsilon|^2 = (n^2+\kappa^2)^2$.

The reverse limiting case (a strong dependence of the dielectric constant on $\boldsymbol{k}$) is realized for the metals which satisfy the inequality $\omega \ll v_F k$ where v_F is the electron velocity on the Fermi surface and $\boldsymbol{k}$ describes the characteristic wave vector values (see below). Under such conditions the longitudinal dielectric constant provides for Debye screening ($k \lesssim k_D$)

$$\epsilon'_\ell \approx 1 + k_D^2/k^2, \tag{8.40}$$

where $k_D \approx k_F \equiv m v_F/\hbar$. Equation (8.37a) indicates that though $k_\perp \approx 1/2z_n \ll k_D$ is relatively small k_3 determining the main contribution to eq. (8.37a) varies in a wide range $k_3 \sim k_D$ so that $\omega_{f0} \ll v_F k_D \sim \omega_p$. Thus, for the "good" metals we must take into account the dependence of ϵ_ℓ on $\boldsymbol{k}$ but for $z_n \gg 1/k_D$ we can replace the function $\epsilon''_\ell/|\epsilon_\ell|^2$ in the integral in eq. (8.37a) with its value for $k_\perp = 0$. In this approximation (Agranovich et al. 1972) we have

$$\Gamma_n^\ell = B/z_n^4, \tag{8.41}$$

where

$$B = \frac{3}{8\pi} [|p^{0f}|^2 + (p_3^{0f})^2] \int_{-\infty}^{\infty} \frac{\epsilon''_\ell(\omega, k_3, \boldsymbol{k}_\perp = 0)\, dk_3}{k_3^2 |\epsilon_\ell(\omega, k_3, \boldsymbol{k}_\perp = 0)|^2}. \tag{8.41a}$$

Now if we use the approximation of dense electron gas for the metal (Hubbard 1958) then, apart from eq. (8.40) we can take

$$\epsilon''_\ell = \frac{e^2 \omega k_D^4}{2k^3 E_F^2} \quad \text{for} \quad k < 2k_F, \qquad \omega < \frac{k(2k_F - k)}{k_F^2} E_F, \tag{8.40a}$$

and $\epsilon''_\ell = 0$ in all other cases so that eq. (8.37a) yields

$$B \approx \frac{e^2\hbar\omega}{30E_F^2}|p^{0f}|^2 \ln\frac{E_F}{\hbar\omega}, \tag{8.41b}$$

where E_F is the Fermi energy for the metal.

A comparison of eqs. (8.39) and (8.41) shows that owing to the Debye character of screening metallic quenching of incoherent excitons has a stronger dependence on their distance from the metallic surface at the boundary with a "good" metal (the mean free path $l \gg 1/k_D$; eqs. (8.40) and (8.40a) are applicable only in this case). But the quantitative difference in Γ_n^ℓ for the minimum $z_n \approx a$ is insignificant.

Let $p^{0f} = 1e \times 1$ Å, $\hbar\omega = 3$ eV, $E_F = 10$ eV, $z_n = 5$ Å (e is the electron charge). Then eq. (8.41) yields an order-of-magnitude estimate $\Gamma_n^\ell = 3\ \text{cm}^{-1}$. A similar estimate from eq. (8.39) for $n \approx 1$ and $\kappa = 5$ (typical values for dirty metals) is also $\Gamma_n^\ell \simeq 3\ \text{cm}^{-1}$.

For large z_n (for instance, $z_n = 100$ Å) the difference between the results obtained from eqs. (8.39) and (8.41) is significant. For instance, for the above values of the parameters in eqs. (8.39) and (8.41) the difference is about 20 times. It should be borne in mind that owing to the Debye character of screening of the longitudinal field in metals a relationship of the type of eq. (8.41) must be valid not only for the contact between a molecular crystal and a thick metal but also for the contact between a molecular crystal and a metal film with a thickness $L_0 \gg 1/k_D$. Since in metals the Debye screening length $1/k_D \approx 1$ Å this condition is satisfied for films with metallic conductivity even for the thicknesses as small as $L_0 \gtrsim$ 10 Å.

In contrast to eq. (8.41), eq. (8.39) is valid only for L_0 much greater than the depth of the skin layer. Otherwise, as shown by Kuhn (1970), we have $\Gamma_n^\ell \sim z_n^{-4}$ as in the case of strong spatial dispersion in the metal (see eq. (8.41)):

$$\Gamma_n^\ell = \frac{h}{\tau_0}\left(\frac{2\omega_{0f}\kappa}{c}\right)\left(\frac{c}{2\omega_{0f}nz_n}\right)^4 \varphi(\theta)L_0, \tag{8.42}$$

Here $\varphi(\theta)$ is a factor of the order of unity depending on the dipole orientation. Now we shall estimate Γ_n^t ignoring the dependence of ϵ_t on $\boldsymbol{k}$ (as in the case of normal skin effect; see eq. (8.38)).

In this approximation eq. (8.37b) yields for $z_n \ll 1/\delta$

$$\Gamma_n^t \approx \frac{\epsilon''(\omega)}{4|\epsilon(\omega)|^2}\frac{1}{z_n^3}[|\boldsymbol{p}^{0f}|^2 + (p_3^{0f})^2] + O(\delta z_n), \tag{8.43}$$

where $\delta = (\omega/c)\sqrt{|\epsilon'_t(\omega)|}$ is the wave vector of the light wave in the metal. If $z_n \gg 1/\delta$ (but still $z_n < c/\omega$) we have

$$\Gamma_n^t \approx \frac{\delta^3\epsilon''(\omega)}{8|\epsilon(\omega)|^2}[|\boldsymbol{p}^{0f}|^2 + (p_3^{0f})^2](\delta z_n)^{-2}. \tag{8.44}$$

For $\lambda = 2\pi c/\omega = 5 \times 10^3$ Å, for instance, for Ag, $|\epsilon| \approx 10$, $\epsilon'' \approx 0.3$, $|p^{0f}| = e$ Å, and $z_n < 1/\delta$ we have $\Gamma_n^t \approx 0.5\ \text{cm}^{-1}$.

Of course, it should be borne in mind that this estimate is valid only for sufficiently "dirty" metals or at high enough temperatures. Otherwise, the conditions of the anomalous skin effect can be realized with eq. (8.43) is not valid.

Nevertheless, we can expect that even under the conditions of the anomalous skin effect the quantitative results for Γ_n^t at $z_n \ll 1/\delta$ will not contradict the inequality $\Gamma_n^t \ll \Gamma_n^\ell$ found from the above estimates.

According to the above results we can ignore Γ_n^t in the calculations of the rate of surface annihilation as we shall do in sect. 5.

The number of excitons disappearing per 1 s owing to metallic quenching is

$$N = \sum_n \frac{1}{2\pi\hbar} \Gamma_n^\ell c_n, \tag{8.45}$$

where c_n is the number of excitons on the site n and we have ignored Γ_n^t.

When excitons are generated by light the function c_n can be assumed to depend only on the distance from the crystal surface, that is, we can assume $c_n \equiv c_{n_3}$. If the variation of c_n in the surface layer is generally weak (it is enough if the variation of c_n is small at the distance of significant variation of the function Γ_n^ℓ) then the number of disappearing excitons per unit area of the boundary surface is $vc(0)$ where $c(0) = c_0/a^3$ is the exciton concentration at the boundary (that is, for $z = 0$), and

$$v = \frac{a}{2\pi\hbar} \sum_{n_3=1}^{\infty} \Gamma_{n_3}^\ell \approx \frac{a\Gamma_1^\ell}{2\pi\hbar}. \tag{8.46}$$

Under the conditions of metallic quenching the variation of the function c_n in the surface layer is small only if the quenching is a slow process (the case of weak metallic quenching), that is, if the quantity $\Gamma_1^\ell/2\pi\hbar$ is in comparison with the rate of the $n_3 \to n_3 \pm 1$ hops of excitons. But if the intensity of metallic quenching is sufficiently high the variations of c_{n_3} can be strong owing to the rapid decrease of Γ_n^ℓ for small n_3. Under such conditions eq. (8.46) can yield only rather rough estimates of the rate v of surface annihilation. If the intensity of metallic quenching is high the rate v can be found only outside the framework of the diffusion approximation. These results are discussed in sect. 5.

Here we shall estimate v from eq. (8.46). Taking, for instance, $\Gamma_1^\ell \approx 3\ \text{cm}^{-1}$ and $a \approx 5$ Å we obtain $v \approx 5000$ cm/s which agrees by the order of magnitude with the experimental results on metallic quenching in anthracene crystals reported by Kallmann et al. (1971), Kallmann (1971), Vanbel et al. (1971).

In conclusion, it should be recalled that in the above discussion we ignored possible violation of the neutrality of the molecules which are in contact with the metal (the molecules with $n_3 = 1$, see fig. 8.1). This phenomenon occurs always under the effect of hybridization of the states of the molecules and the states of the electrons of the metal. In those cases when this hybridization is considerable (that is, in fact, under the conditions of chemisorption) the molecular levels are broadened (the respective width $\Gamma \approx 1\,\text{eV}$) and shifted (the shift $\Delta \approx 1\,\text{eV}$, too). Owing to this effect the molecules in the contact region acquire a charge and the opposite charge appears on the metal so that in the metal–molecular crystal contact plane there is formed a kind of a two-dimensional ionic crystal (Agranovich 1972). New branches of elementary excitations appear under such conditions, namely, the dipole-active optical phonons corresponding to the motion of the charged molecules with respect to the contact with the metal as well as the charge transfer (CT) excitons corresponding, for instance, to the transition of an electron from the conduction band of the metal to the free molecular levels. All these features of the spectrum of the contact have been discussed by Agranovich (1972) in connection with the effect of chemisorption on the critical temperature of the superconducting transition in the thin metal film (see also Agranovich et al. (1980); Burstein et al. (1979) included these states in the analysis of the gigantic Raman scattering of light by the molecules adsorbed on the metal). It is, of course, clear that chemisorption in the contact region can have a significant effect on the interaction between the molecular excitons and the metal. Since the electron excitation energy for the molecular ion is, generally, considerably different from the respective energy for the neutral molecule the molecules of the layer with $n_3 = 1$ (see fig. 8.1) cease to be in resonance with the bulk molecules owing to chemisorption. If the energy of ion excitation is lower than the exciton energy, surface traps for excitons are formed. If, on the contrary, the excitation energy of the molecular ion is higher than the exciton energy a so-called dead zone for excitons is formed – excitons will be repulsed from the contact region and the rate of metallic quenching must decline. These problems are discussed in more detail by Agranovich and Mal'shukov (1976), in particular, in connection with the effect of chemisorption on the rate of exciton autoionization at the contact with the metal.

5. *Quenching and adsorption of small-radius incoherent excitons at the boundary between molecular crystal and metal or semiconductor*

We have noted in sect. 4 that to find the rate of surface annihilation of incoherent excitons we have to abandon the diffusion approximation. Efremov and Malshukov (1975) have used the random walk equation to calculate the rate of surface annihilation (see chs. 5 and 6). They have also taken into account the dependence of the energy of incoherent excitons on the distance from the boundary surface. The dependence is obtained from the system of equations (8.10) if we take in them $\tilde{M}_{nm} = 0$ and thus ignore their right-hand side which contains the matrix elements of the resonance intermolecular interaction.

In this approximation the energy of the exciton on the nth site is

$$E_n = \Delta\epsilon_f + D_f(n), \tag{8.47}$$

where $D_f(n)$ is given by eq. (8.5b). The imaginary part of $D_f(n)$ determining the rate of metallic quenching of incoherent excitons was treated in sect. 4. According to the results of sect. 3, the real part of $D_f(n)$ is

$$\operatorname{Re} D_f(n) = D^{(0)} + D^{(1)}\delta_{1n_3} + \operatorname{Re} D_f^{(4)}(n). \tag{8.48}$$

If we employ the harmonic oscillator model of the molecule and ignore penetration of the field into the metal then, as noted in sect. 1, we obtain

$$\operatorname{Re} D_f^{(4)}(n) = -E_0 a^3/z_n^3, \tag{8.49}$$

where $E_0 = |\boldsymbol{p}^{01}|^2\alpha(\theta)/a^3$, $\alpha(\theta) \sim 1$, and a is the lattice parameter corresponding to the crystallographic direction along the z-axis. However, as noted above, $\operatorname{Re} D_f^{(4)}(n)$ calculated for the electronic states f significantly depends on the structure of the spectrum of electronic excitations and can lead to a potential well or a potential barrier for excitons. Of course, if the depth of such a well (or the height of the barrier) is not very small compared with the width of the exciton energy level the presence of the well must affect the kinetics of metallic quenching of incoherent excitons. These effects (see sect. 3), of course, play a part for coherent excitons, too, since they can give rise to surface levels. This opens an additional channel of exciton quenching due to bulk excitons going over to the surface levels.

The substrate affects incoherent excitons via a different mechanism. The potential well (or barrier) gives rise to the forces acting on incoherent excitons attracting them to (or repulsing them from) the metal surface. The most interesting case is that of exciton attraction to the substrate, of course. This attraction (for coherent excitons it means trapping by a surface level) results in an increase in the exciton concentration near the boundary. On the other hand, metallic quenching decreases the concentration. The relationship between these competing processes can vary depending on the properties of the substrate and excitons. We shall discuss such effects for incoherent excitons following the results of Efremov and Mal'shukov (1975). In particular, we shall calculate the rate of surface annihilation of excitons as a function of the force of attraction of excitons to the boundary.

Assume that the intensity I of exciton pumping is constant and spatially uniform and that the molecular crystal is in the region $z \geqslant 0$ while the metal is in the region $z < d$ (here d is the gap width). The occupancy c_n of a lattice site here depends only on n_3 so that the random walk equation for incoherent excitations is one-dimensional (the occupancy will be denoted by c_n, $n \equiv n_3$). Including only the hops between the nearest neighbours we can write down this equation as

$$-(W^n_{n-1} + W^n_{n+1})c_n + W^{n-1}_n c_{n-1} + W^{n+1}_n c_{n+1} - \left(\frac{1}{\tau_n} + \frac{1}{\tau}\right) c_n + I = 0, \qquad (8.50)$$

where τ is the exciton lifetime far from the boundary between the crystal and the metal, $1/\tau_n = \Gamma_n/2\pi\hbar$ is the probability of metallic quenching at the site n, and $W^{n'}_n$ is the probability of the hop $n' \to n$.

We shall describe the hopping probability by eq. (4.117) bearing in mind crystals of the type of anthracene or naphthalene excitons in which at sufficiently high temperatures are incoherent (see ch. 7) and the temperature dependence of their diffusion coefficient is given by eq. (5.15). In this case we can rewrite eq. (4.117) as

$$W^{n'}_n = 2\hbar^{-1}|M_{nn'}|^2 \sqrt{\frac{\pi}{B}} \exp\left[-\frac{(E_n - E_{n'} + A)^2}{B}\right], \qquad (8.51)$$

where eqs. (8.47) and (8.50) give the energy E_n (we neglect here the effect produced by correlation of displacements on the hopping probability and therefore write A and B instead of $\tilde{A}$ and $\tilde{B}$ (see ch. 4 sect. 10)).

In the infinite crystal $E_n = E_{n'}$ and if we include only the hops to the neighbouring sites (see eq. (8.5)) and assume that the variation of c_n at distances of the order of the lattice parameter a is small we find that the diffusion coefficient $D_{zz} \equiv D$ can be written as

$$D = a^2 W^n_{n+1} \equiv a^2/t = D_0 \exp(-E_a/k_B T),$$

where $E_a = k_B T A^2/B$ is the height of the activation barrier (see eq. (5.15)), T is the tempeature, and $t = (W^n_{n+1})^{-1}$ is the mean time of hops. In the diffusion approximation used here the occupance c_n which is a function of a discrete argument is replaced with the exciton concentration $c(z) = c_n/a^3$ where a^3 is the volume of the unit cell.

It can be readily seen that the function $c(z)$ satisfies the equation

$$Dc''(z) - \frac{1}{\tau} c(z) + \frac{I}{a^3} = 0. \qquad (8.52)$$

If we take into consideration the metallic substrate eq. (8.52) can also be used in the size of the crystal region which is in contact with the metal and where we must take account of the difference of Γ_n and $E_n - E_{n'}$, from zero, is much smaller than the characteristic dimensions of the problem as the exciton diffusion length and the thickness of the molecular crystal. Under such conditions, as it was repeatedly noted above, eq. (8.52) should be solved with the boundary condition (6.1a).

Equation (8.50) indicates that the total number of excitons disappearing per unit time (per 1 cm^2 of the boundary area) is given by

$$\frac{1}{2\pi\hbar a^3}\sum_n \Gamma_n c_n + \frac{1}{\tau a^3}\sum_n c_n = \frac{I}{a^3}. \tag{8.53}$$

A similar balance equation with the function $c(z)$ has the form

$$vc(0) + \frac{1}{\tau}\sum_n c(na) = \frac{I}{a^3}. \tag{8.54}$$

Hence, the rate of surface annihilation must satisfy the condition

$$vc(0) = \frac{1}{2\pi\hbar a^3}\sum_n \Gamma_n c_n + \frac{1}{\tau}\sum_n \left[\frac{c_n}{a^3} - c(na)\right]. \tag{8.55}$$

The function $c(z)$ satisfying eq. (8.52) with the boundary condition (6.1a) depends on the parameter v:

$$c(z) = \frac{I\tau}{a^3}\left[1 - \frac{v\tau}{\tau v + L}\,\mathrm{e}^{-z/L}\right]. \tag{8.56}$$

Thus, if we have solved the system of equations (8.50) we can use eq. (8.55) to find the rate of surface annihilation as a function of the intensity of metallic quenching and to take into account the effects of the electrostatic interaction between excitons and the metal surface.

A numerical solution of eq. (8.50) was found by Efremov and Mal'shukov (1975) discussed above. The exciton energy (compare with eq. (8.49)) was written there as

$$E_n = \Delta\epsilon_f + D^0 - \frac{E}{(\gamma + n)^3}, \qquad n = 0, 1, 2, \ldots, \tag{8.57}$$

where the energy $E > 0$ corresponding to attraction of excitons to the boundary with the metal. The intensity of quenching (see eq. (8.41)) was

Table 8.1
The rate of metallic quenching for $E = 0$ and various values of Γ and the diffusion length $L(\gamma = 1)$.

$\tau\Gamma$	750	1500			3000	6000
L/a	100.0	40.0	100.0	200.0	100.0	100.0
$v_0\tau/L$	8.11	40.5	16.2	8.11	32.4	64.9
$v\tau/L$	8.23	61.0	17.0	8.45	35.6	86.1

Table 8.2
The dependence of μ on the attraction parameter E, the distance to the metal, d, and the diffusion length L. $\Gamma = 1500/\tau$. The rate of metallic quenching $v = \mu L/\tau$.

$\gamma \equiv d/a$	L/a	$\tau v_0(\gamma)/L$	$E = 0$	$E = E_0$	$E = 3E_0$
1	40	40.5	61.0	475.0	–
1	100	16.2	17.0	28.5	131.0
1	200	8.11	8.235	12.7	49.3
3	40	0.743	0.783	0.852	–
3	100	0.297	0.301	0.299	0.12
3	200	0.149	0.149	0.146	–

taken to be

$$\Gamma_n = \Gamma_n^l = 2\pi\hbar\Gamma/(n+\gamma)^4. \tag{8.58}$$

The parameter γ entering into eqs. (8.57) and (8.58) determines the gap width $d = \gamma a$ between the metal and the molecular crystal (a is the lattice constant of the crystal).

Tables 8.1 and 8.2 present the calculated rate v of surface annihilation as a function of the above parameters. Table 8.1 gives the calculated results for v ignoring the effects of the electrostatic interaction between excitons and the metal (that is, for $E = 0$, see eq. (8.57)) but for different values of the parameter $\xi = \Gamma t/\hbar$ determining the ratio of the hop time t and the time of metallic quenching of the exciton localized at the molecule of the crystal nearest to the metal (that is, for $n = 0$ and $\gamma = 1$; see eq. (8.58); $\xi = \tau\Gamma(a/L)^2$, $L = \sqrt{D\tau}$).

The following crystal parameters were used in the calculations: $a = 5$ Å, $E_0 = |p^{0f}|^2/8a^3 = 100\ \mathrm{cm}^{-1}$ ($p^{0f} = e$ Å), and the height of the activation barrier $E_a = 300\ \mathrm{cm}^{-1}$.

When we substitute eq. (8.58) into eq. (8.46) we can find the rate of surface annihilation under the conditions of weak metallic quenching (see sect. 4). This rate will be denoted below by v_0 to distinguish it from the more correct value v given by eq. (8.55). According to eqs. (8.46) and (8.58), we have

$$v_0 = a\Gamma\varphi(\gamma), \tag{8.59}$$

where

$$\varphi(\gamma) = \sum_{n=0}^{\infty} \frac{1}{(n+\gamma)^4}. \tag{8.59a}$$

Before discussing the calculated results in tables 8.1 and 8.2 note that for large $\gamma \gg 1$ we can replace the sum in eq. (8.59a) with an integral which can be readily calculated. Then, for $\gamma \gg 1$ we obtain

$$\varphi(\gamma) = \int_0^\infty \frac{dn}{(n+\gamma)^4} = \frac{1}{3}\frac{1}{\gamma^3}. \tag{8.60}$$

Thus, for sufficiently wide gaps between the metal and the dielectric when quenching is definitely weak its rate $v_0 \approx v$ is inversely proportional to the third power of the gap width as it is observed in experiments (see below).

The data of table 8.1 imply, as it should be expected, that v and v_0 are close when the parameter $\Gamma t \equiv \tau\Gamma(a/L)^2 \ll 1$. If this condition is not satisfied then $v > v_0$.

The data presented in table 8.2 illustrate the effect of exciton attraction to the metal on metallic quenching of excitons. This effect is particularly strong for small gaps between the metal and the dielectric (that is, for small γ). Here we also see a large difference between the exact (v) and approximate (v_0) rates of surface annihilation. This difference is due to the fact that strong attraction of excitons to the metal results in a sharp variation of the exciton concentration at the boundary when eq. (8.46) becomes inapplicable. But as the width of the gap between the metal and the dielectric increases the effect of exciton attraction on metallic quenching rapidly diminishes so that eq. (8.46) becomes sufficiently accurate.

Experimental studies of metallic quenching of excitons were started about 10 years ago but still available results are few (Kallmann et al. 1971, Kallmann 1971, Vaubel et al. 1971, Killesreiter and Baessler 1971, 1972, Singh and Baessler 1974, Kurczewska and Baessler 1977, Heisel et al. 1978). For instance, Kallmann et al. (1971), Kallmann (1971) report on the effect of metallic quenching of excitons in anthracene on the intensity of luminescence excited by the light with varying absorption coefficient $k = 1/l_a$ where l_a is the penetration depth. Under the stationary conditions the exciton concentration $c(z)$ satisfies the equation

$$Dc''(z) - \frac{1}{\tau}c(z) + I_0 k\, e^{-kz} = 0.$$

Solving this equation with the boundary condition (6.1a) we can easily find the function $F(v)$ which determines the intensity of luminescence in the presence of metallic quenching,

$$F(v) = \frac{1}{\tau}\int_0^\infty c(z)\, dz,$$

and the quantity $F(0)$ for the conditions when metallic quenching does not occur. The relative variation of intensity can be conveniently described by

$$\eta = F(0)/[F(0) - F(v)],$$

which is equal to

$$\eta = \frac{l_a + L}{L}\left(1 + \frac{L}{v\tau}\right), \qquad L = \sqrt{D\tau}. \tag{8.61}$$

In the study reported by Kallmann et al. (1971), Kallmann (1971) one of the surfaces of a single anthracene crystal grown as a plane-parallel plate was coated with a thin aluminium film which transmitted 10% of the light exciting luminescence. This crystal was then used to study the dependence of the luminescence intensity $F(v)$ on the penetration depth l_a of the exciting light. Then a similar dependence was studied for another surface of the crystal which was not coated with metal and so metallic quenching did not occur (that is, for $v = 0$). The data on $F(v)$ and $F(0)$ were used to find the function $\eta = \eta(l_a)$ plotted in fig. 8.2. We see that the experimental results agree with eq. (8.61) since η is, approximately, a linear function of l_a the plot of which intersects the vertical axis ($l_a \approx 0$) for $\eta = 1 + L/v\tau = 1 \pm 0.1$ so that $l/v\tau \leqslant 0.1$ or $v \geqslant 10L/\tau$. On the other hand, according to eq. (8.61), the slope of the plot of $\eta = \eta(l_a)$ is $\alpha = (1/L)(1 + L/v\tau)$ or, from the above results, $\alpha = (1 \pm 0.1)/L$. A comparison of this expression with the measured slope of the plot yielded the diffusion length $L = 420 \pm 20$ Å which is in a good agreement with the earlier results (see ch. 7). The knowledge of L made it possible to find the lower estimate for $v \geqslant 10L/\tau$ which for $\tau \approx 10^{-8}$ s is $v \geqslant 4 \times 10^3$ cm/s.

In similar experiments reported by Vaubel et al. (1971) a layer of transparent dielectric was introduced between the metal film and anthracene (preparations of pure fatty acids with C_{14}, C_{18}, C_{20} and C_{22} were used; see also Kurczewska and Baessler (1977)).

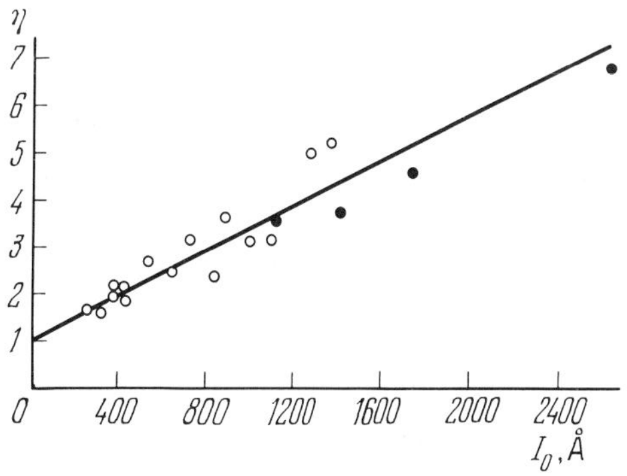

Fig. 8.2. The parameter $\eta = F(0)/[F(0) - F(v)]$ as a function of the depth of penetration of the excitation light (the anthracene crystal is coated with a semitransparent silver film).

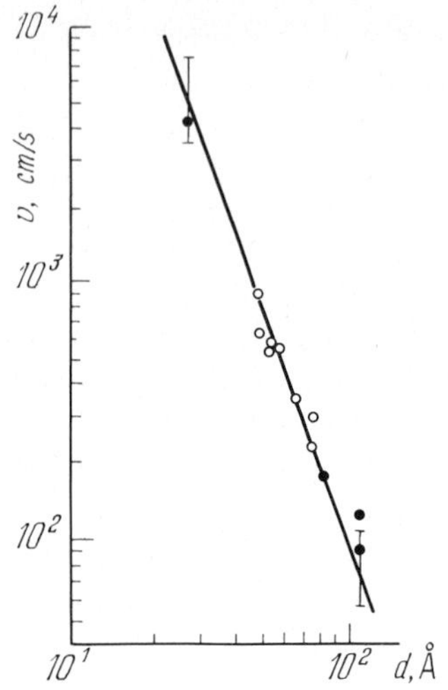

Fig. 8.3. The rate v of metallic quenching of incoherent excitons as a function of the gap width.

The thickness of the transparent layer was varied which made it possible to find the rate of surface annihilation as a function of the width d of the gap between the metal and the molecular crystal.

It has been noted above that for $\gamma = d/a \gg 1$ quenching is weak so that $v \approx v_0$. Therefore, using eqs. (8.61), (8.59) and (8.60) we can write

$$\eta = \frac{l_a + L}{L}\left(1 + \frac{d^3}{d_0^3}\right), \tag{8.61a}$$

where $d_0 = (a^4 \tau \Gamma / 3L)^{1/3}$. The experimental results reported by Vaubel et al. (1971) agree with eq. (8.61a) and, thus, confirm the dependence $v_0 \sim d^{-3}$ (see fig. 8.3). The parameter d_0 in eq. (8.61a) was found to be about 62 ± 4 Å. Since in anthracene the lattice parameter perpendicular to the plane ab is $a \approx 10$ Å and the ratio $L/a\tau \approx 4 \times 10^9$ s then using the above value of d_0 we obtain $\Gamma \approx 4.5 \times 10^{11}\ \mathrm{s}^{-1}$. For $\gamma = 2$ (Vaubel et al. 1971; the minimum width $d_{\mathrm{min}} \approx 20$ Å $= 2a$) the quenching probability is $\Gamma/24 \approx 2 \times 10^{10}\ \mathrm{s}^{-1}$, according to eqs. (8.59) and (8.60). If, according to Vaubel et al. (1971), we take $\Gamma = 4.5 \times 10^{11}\ \mathrm{s}^{-1}$ then eq. (8.59) yields the rate $v_0 \approx 4.5 \times 10^4$ cm/s for $\gamma = 1$. This result is higher than the lower estimate of v found by Kallmann et al. (1971), Kallmann (1971) by an order of magnitude. This spread of results must be specially analyzed if we eliminate the possible effects due to subtle differences in the preparation of crystals by Kallmann et al. (1971), Kallmann (1971), Vaubel et al. (1971).

The experimental results described above give a qualitative confirmation of the theory of metallic quenching of incoherent excitons; apart from observation of quenching itself the most important experimental result is, of course, the measured dependence $\eta = \eta(d)$.

An interesting suggestion is to study attraction of excitons to the boundary between a molecular crystal and a semiconductor (rather than a metal) whose forbidden gap width Δ is somewhat larger than the exciton energy $\hbar\omega_{0f}$ (Efremov and Mal'shukov 1975). This system provides for weak quenching but sufficiently strong attraction of excitons to the boundary if the dielectric constant of the semiconductor at the frequency $\omega = \omega_{0f}$ is sufficiently high.

Numerical solution of eq. (8.50) for $\tau_n = \infty$ using eq. (8.57) gave the following results (Efremov and Mal'shukov 1975). For $E = 30E_0$, $a = 5$ Å and $L = 100a$ the concentration of incoherent excitons in the crystal plane nearest to the semiconductor was 100 times the bulk concentration of excitons; for $E = 10E_0$ the ratio was 70, and for $E = 3E_0$ the ratio was 5 (we are talking about the number of excitons per 1 cm^2 of the monomolecular layer parallel to the boundary).

The accumulation effect is interesting also in connection with new prospects it opens for studying the collective effects in systems of two-dimensional coherent excitons. We have noted in sect. 5 that the presence of sufficiently deep surface levels due to the effect of a substrate with a sufficiently high polarizability results in trapping of bulk excitons. At low enough temperatures this trapping is irreversible and can lead to increasing concentration of surface excitons.

Indeed, for $v = \infty$ eq. (8.56) yields that I_s excitons are trapped at the surface per 1 s where

$$I_s = -D \frac{\mathrm{d}c}{\mathrm{d}z}\bigg|_0 = \frac{D}{L} c_0, \qquad c_0 = \frac{I\tau}{a_3},$$

and c_0 is the concentration of bulk excitons for $z \gg L$. Now, if we assume that the lifetime of surface excitons is about the lifetime of bulk excitons, that is, far from the boundary (as it is the case at the boundary between molecular crystal and semiconductor, see above), then under stationary conditions the concentration of surface excitons is $c_s = I_s\tau = c_0L$.

At a large distance from the crystal boundary there are c_0a excitons per 1 cm^2 of the monomolecular layer so that the concentration of surface excitons is higher than the concentration of bulk excitons by a factor of L/a. For anthracene this means a 50-fold increase in concentration ($L \approx 500$ Å, $a \approx 10$ Å) and, for instance, for $c_0 \approx 10^{18}$ cm^{-3} we have $c_s \approx 5 \times 10^{12}$ cm^{-2}.

6. Metallic quenching of coherent excitons, effects of spatial dispersion and additional boundary conditions

When we analyze decay of the surface states in a dielectric with the penetration depth smaller than the depth of the skin we can omit the terms depending on $\epsilon_t(\omega, \boldsymbol{k})$ in the expression for the function $f(\omega, \boldsymbol{k}_\perp)$; see eq. (8.8a). In this approximation the use of eq. (8.15) leads to the following expression for the level width due to metallic quenching of the "microscopic" surface excitons (see eq. (8.34)):

$$\Gamma(\boldsymbol{k}_\perp, n_3) = \frac{1}{a^2} \sum_{\boldsymbol{g}_\perp} \int \frac{\epsilon_l''(\omega, k_3, \boldsymbol{k}_\perp + \boldsymbol{g}_\perp)|\boldsymbol{p}^{0f}\boldsymbol{K}(\boldsymbol{g}_\perp)|^2 \, \mathrm{d}k_3}{[k_3^2 + (\boldsymbol{g}_\perp + \boldsymbol{k}_\perp)^2]|\epsilon_\ell(\omega, k_3, \boldsymbol{k}_\perp + \boldsymbol{g}_\perp)|^2} \exp[-2z_n|\boldsymbol{k}_\perp + \boldsymbol{g}_\perp|], \tag{8.62}$$

Here $\boldsymbol{K}(\boldsymbol{g}_\perp) \equiv (\boldsymbol{k}_\perp + \boldsymbol{g}_\perp, \mathrm{i}|\boldsymbol{k}_\perp + \boldsymbol{g}_\perp|)$.

Thus, when n_3 and the width d of the gap between the metal and the molecular crystal increase the quantity Γ for coherent surface excitons (see eq. (8.34)) decreases exponentially, rather than as a power function as is the case for incoherent excitons.

Metallic quenching of macroscopic surface excitons and, in a more general form, metallic quenching of surface polaritons can be treated in the framework of the phenomenological Maxwell equations with the impedance boundary conditions (see Agranovich et al. (1972), Agranovich (1975), Bryksin et al. (1974) where the references to experimental studies are given).

We shall not discuss here these results (reviewed by Agranovich (1975), Bryksin et al. (1974)) but shall calculate the rate of surface annihilation of coherent bulk excitons and bulk polaritons (Agranovich et al. (1968); see ch. 4 sect. 11).

Assume that there is a gap of width d between a metal and a semi-infinite isotropic crystal. The axis $0z$ is perpendicular to the boundary. The calculation of the rate of surface annihilation is considerably facilitated by the fact that the size of the region near the boundary where intense annihilation of excitons (polaritons) occurs is considerably smaller than their mean free path (the results reported below have been obtained in cooperation with A. Malshukov and E. Glushko).

Under such conditions the rate v can be easily related to the coefficient of internal reflection of excitons (polaritons) from the crystal boundary by

$$vc(0) = \sum_{\boldsymbol{k}} (1 - R_{\boldsymbol{k}}) I_z(\boldsymbol{k}), \tag{8.63}$$

where $R_{\boldsymbol{k}}$ is the coefficient of reflection of the wave with the wave vector $\boldsymbol{k}$, $I_z(\boldsymbol{k})$ is the exciton flux incident at the boundary, and $c(0)$ is the exciton concentration at the boundary. Clearly, the right-hand and the left-hand sides of eq. (8.63) give the number of excitons annihilated per unit time per unit area of the boundary.

Assume that the incident excitons are at thermodynamic equilibrium. Furthermore, the number of the waves whose electric field vector $\boldsymbol{E}$ is in the incidence plane (p polarization) is taken to be equal to the number of the waves whose vector $\boldsymbol{E}$ is perpendicular to this plane (s polarization). In the effective mass approximation we have

$$I_z(\boldsymbol{k}) = \frac{\hbar k_z}{M} A \exp(-\hbar^2 k^2/2MT), \tag{8.64}$$

where A is the normalization constant, and M is the effective mass of excitons. When we calculate the reflection coefficient we should bear in mind that our frequencies lie in the region of the exciton spectrum and therefore it is important to include additional waves due to reflection (Agranovich and Ginzburg 1983). This leads to the problem of the additional boundary condition for the field components. A fairly general additional boundary condition (Agranovich and Ginzburg 1983) has the form

$$\left(\boldsymbol{P} + \alpha \frac{\partial \boldsymbol{P}}{\partial z}\right)\bigg|_{z=0} = \gamma E(0), \tag{8.65}$$

where $\boldsymbol{P}$ is the exciton part of polarization. We shall use below only a special form of this condition with $\gamma = \alpha = 0$. Other boundary conditions are the conventional boundary conditions which can be conveniently written as

$$\frac{E_y^{(s)}}{H_x^{(s)}}\bigg|_{z=0} = -Z_s(\omega, k_x), \qquad \frac{E_x^{(p)}}{H_y^{(p)}}\bigg|_{z=0} = Z_p(\omega, k_x), \tag{8.66}$$

where H is the vector of magnetic field strength and the superscripts s and p denote polarization of the incident waves. The x-axis lies in the incidence plane. The quantities $Z(\omega, k_x)$ can be easily related to the surface impedance $\xi(\omega, k_x)$ of the metal. Since the width of the gap between the crystal and the metal is d the Maxwell equations yield

$$Z(\omega, k_x) = Z_0 \frac{Z_0 + \xi(\omega, k_x) \coth \kappa d}{\xi(\omega, k_x) + Z_0 \coth \kappa d}, \tag{8.67}$$

where

$$Z_{0p} = -\frac{ic}{\omega\epsilon}\sqrt{k_x^2 - \epsilon\frac{\omega^2}{c^2}}, \qquad Z_{0s} = \frac{i\omega}{c\sqrt{k_x^2 - \epsilon(\omega^2/c^2)}}, \qquad \kappa = \sqrt{k_x^2 - \epsilon\frac{\omega^2}{c^2}},$$

and ϵ is the dielectric constant of the material in the gap. In the case of mirror reflection of electrons from the boundary the surface impedance of the metal is

$$\xi_s(\omega, k_x) = -\frac{i\omega}{\pi c}\int_{-\infty}^{+\infty}\frac{dk_z}{(\omega^2/c^2)\epsilon_\perp - k^2},$$

$$\xi_p(\omega, k_x) = -\frac{i\omega}{\pi c}\int_{-\infty}^{+\infty} dk_z\left[\frac{k_z^2}{k^2[(\omega^2/c^2)\epsilon_t - k^2]} + \frac{(c^2/\omega^2)k_x^2}{k^2\epsilon_\ell(\omega, k_x)}\right], \tag{8.68}$$

where ϵ_t and ϵ_ℓ are the transverse and longitudinal dielectric constants of the metal.

In the presence of additional waves the electric field in the crystal is the result of superposition of several waves, namely, the fields of the incident and reflected waves $\boldsymbol{E}_0$ and $\boldsymbol{E}_1$ and the fields of two additional waves, $\boldsymbol{E}_2$ (transverse) and $\boldsymbol{E}_3$ (longitudinal). For each of these waves we have

$$\boldsymbol{E}_i(\boldsymbol{r}) = \boldsymbol{E}_i \exp(ik_x x + ik_z z). \tag{8.69}$$

The refractive indexes $n_i^2 = k_i^2 c^2/\omega^2$ for the waves satisfy the dispersion relations:

$$\epsilon(\omega, n_i) = n_i^2, \qquad i = 0, 1, 2, \qquad \epsilon(\omega, n_3) = 0. \tag{8.70}$$

In the case of s polarization the two additional waves reduce to one transverse wave (n_2) and the boundary conditions have the form

$$\sum_{i=0}^{2} E_{yi} + \sum_{i=0}^{2} H_{xi}Z_s(\omega, k_x) = -E_{0y} - H_{0x}Z_s,$$

$$\sum_{i=0}^{2} E_{yi}(n_i^2 - \epsilon_\infty) = -E_{0y}(n_0^2 - \epsilon_\infty). \tag{8.71}$$

In the case of p polarization both additional waves should be included and the boundary conditions now have the form

$$\sum_{i=0}^{3} E_{xi} - \sum_{i=0}^{3} H_{yi}Z_p(\omega, k_x) = H_{0y}Z_p(\omega, k_x) - E_{0x},$$

$$\sum_{i=0}^{2} E_{xi}(n_i^2 - \epsilon_\infty) - E_{x3}\epsilon_\infty = -E_{0x}(n_0^2 - \epsilon_\infty),$$

$$\sum_{i=0}^{2} E_{zi}(n_i^2 - \epsilon_\infty) - E_{z3}\epsilon_\infty = -E_{0z}(n_0^2 - \epsilon_\infty). \tag{8.72}$$

Equations (8.71) and (8.72) yield the reflection coefficients R_s and R_p:

$$R = |E_1/E_0|^2.$$

When we calculate the metal impedance (8.68) we can use the dielectric constants of the metal found in the hydrodynamic approximation (Pines and Noziere 1966):

$$\epsilon_\ell(\omega, k) = 1 - \frac{\omega_p^2}{\omega^2 - \frac{1}{3}k^2 v_F^2 + i\omega/\tau'}, \qquad \epsilon_t(\omega, k) = 1 - \frac{\omega_p^2}{\omega^2 + i\omega/\tau'},$$

where τ' is the mean free time of electrons in the metal.

The rate of metallic quenching estimated from eq. (8.63) for $\omega_p = 6\times 10^{16}\ s^{-1}$, $\tau' = 2\times 10^{-15}$ s for excitons in CdS (see ch. 4 sect. 11) is $v \approx 1000$ cm/s for $T = 100$ K and the gap width $d = 0$.

Since the intensity of metallic quenching of excitons depends on the form of the additional boundary conditions experimental and theoretical studies of this type of quenching open up new ways for determining the form of the additional boundary conditions (the most interesting results are obtained from the analysis of the dependences on temperature and the gap width d).

If $d \gtrsim 1/k_T$ where $k_T = (1/\hbar)(3Mk_BT)^{1/2}$ we have a situation which is similar to that in the studies of the surface waves with the ATR technique (Agranovich 1975, Bryksin et al. 1974). When $d \ll 1/k_T$ the bulk polaritons do not excite surface waves. If, however, $d \gtrsim 1/k_T$ a new mechanism of metallic quenching is activated since under these conditions, as in the ATR technique, polariton can decay giving rise to a surface plasmon with the same energy in the metal (if we take into account retardation the spectrum of the surface plasmons in the metal ranges from $\omega = 0$ to $\omega = \omega_p/\sqrt{2}$; see Agranovich (1975), Bryksin et al. (1974).

These problems have been analyzed by Philpott (1975) Morawitz and Philpott (1974) where references to other theoretical studies in the field are given; see also interesting experimental data reported by Drexhage (1966), Drexhage et al. (1968), Tews (1973), Chance et al. (1974) who have discussed the effect of the metal surface on the lifetime and the channels of decay of the excited state of an individual molecule taking into consideration the mechanism of excitation of the surface plasmons.

Excitation of surface plasmons becomes possible because the presence of one molecule over the metal surface totally violates the translational symmetry (the case of a radiator (antenna) over a conducting surface which was discussed by Sommerfield back at the beginning of the century; see Agranovich (1975), Bryksin et al. (1974)). But if we have a molecular crystal over a metal which is in a close contact with it ($d = 0$) then its bulk excitations (bulk polaritons) cannot excite surface waves under the conditions of ideal contact, as we have noted above. Nevertheless, if we generate in some way a nonstationary excited state corresponding to electronic excitation of one molecule then decay of this state will give rise to the whole series of resonance ordinary waves and, in particular, surface waves. As shown by Gerbshtein et al. (1975) these surface waves can be registered in the fluorescence spectrum by means of an ATR prism. The spectrum of these waves should differ from the spectrum of the surface plasmons of the metal in the effect of the dielectric layer. In the study reported by Gerbshtein et al. (1975) the excited states of individual molecules in the crystal were generated owing to radiationless transitions from the higher excited electronic states.

No experimental studies of metallic quenching of coherent excitons have been performed. Clearly, such studies will open new prospects for analysis of exciton properties of the metal–dielectric contacts (particularly the electronic structure of the boundary layer; it has an especially strong effect on surface dissociaton of exciton giving rise to electron and hole (Agranovich and Mal'shukov 1976)).

In conclusion, note that the metal–dielectric contacts have attracted considerable attention recently in connection with Raman scattering of light by molecules adsorbed on the metal (Burstein et al. 1979). The experimental results show enormous increases in the scattering cross section (up to 10^5–10^6 times). Theoretical analysis of this interesting effect gives rise to problems similar to those encountered in the analysis of the metal–dielectric contact and discussed in this chapter. This can have a significant influence on the interpretation of the effect of thin metal films on light scattering by the dielectric crystal surfaces.

It has been recently shown by Malshukov (1981) that to explain the gigantic Raman effect we should take into account the hybridization effects (chemisorption) which Mal'shukov discussed in (1974). These effects result in adiabatic transfer of charge from the molecule to the metal and back owing to the intramolecular vibrations. This effect allows us to explain the anomalous broadening of the vibrational spectrum of the molecules adsorbed on the metal (Person and Persson 1980). Apparently, this effect must be included in the analysis of the electron–phonon interaction for molecular crystals in contact with metals and semiconductors.

REFERENCES

Abeles, F., 1972, Optical Properties of Metal, in: Optical Properties of Solids, ed., F. Abeles (North-Holland, Amsterdam-London) p. 93.

Agabekyan, A.S. and A.O. Melyakyan, 1970, in: Energy transfer in condensed media, Proc. All-Union Seminar on radiationless energy transfer, Erevan, pp. 5–16 (in Russian).

Agranovich, V.M., 1957, Opt. i Spektr. **3**, 29, 84.

Agranovich, V.M., 1958, Opt. i Spektr. **4**, 586.

Agranovich, V.M., 1959a, Zh. ETF **37**, 430.

Agranovich, V.M., 1959b, Role of defects in exciton luminescence of molecular crystals, X All-Union Conf. on luminescence, Leningrad, October, 1959.

Agranovich, V.M., 1959c, Izv. AN SSSR, Ser. Fiz. **23**, 40.

Agranovich, V.M., 1960, Opt. i Spektr. **9**, 113, 798.

Agranovich, V.M., 1960, Usp. Fiz. Nauk, **71**, 141.

Agranovich, V.M., 1961, Fiz. Tverd. Tela, **3**, 811.

Agranovich, V.M., 1968, Theory of excitons (Nauka, Moscow) (in Russian).

Agranovich, V.M., 1970, Fiz. Tverd. Tela, **12**, 562.

Agranovich, V.M., 1972, Fiz. Tverd. Tela, **14**, 3684.

Agranovich, V.M., 1974, Usp. Fiz. Nauk, **112**, 143.

Agranovich, V.M., 1975, Usp. Fiz. Nauk, **115**, 199.

Agranovich, V.M., 1983, in Spectroscopy and Excitation Dynamics of Condensed Molecular Systems, ed., R. Hochstrasser and V.M. Agranovich (North-Holland, Amsterdam) (in press).

Agranovich, V.M. and O.A. Dubovskiy, 1966, Pis'ma Zh. ETF **3**, 223.

Agranovich, V.M. and O.A. Dubovsky, 1970, Fiz. Tverd. Tela, **12**, 2055.

Agranovich, V.M. and N.A. Efremov, 1980, Fiz. Tverd. Tela, **22**, 998.

Agranovich, V.M. and A.N. Faidysh, 1956, Opt. i. Spektr. **1**, 885.

Agranovich, V.M. and V. Ginzburg, 1983, Crystal optics with spatial dispersion and the theory of excitons (Springer, Berlin) (in press).

Agranovich, V.M. and Yu.V. Konobeev, 1959, Opt. i Spektr. **6**, 242, 648.

Agranovich, V.M. and Yu.V. Konobeev, 1961, Opt. i Spektr. **11**, 369.

Agranovich, V.M. and Yu.V. Konobeev, 1961, Fiz. Tverd. Tela, **3**, 360.

Agranovich, V.M. and Yu.V. Konobeev, 1968, Phys. Status Solidi **27**, 435.

Agranovich, V.M. and T.A. Leskova, 1979, Pis'ma Zh. TEF **29**, 151.

Agranovich, V.M. and A.G. Mal'shukov, 1976, Chem. Phys. Lett. **43**, 221.

Agranovich, V.M. and M.A. Mekhtiev, 1971, Fiz. Tverd. Tela, **13**, 2732.

Agranovich, V.M. and A.A. Rukhadze, 1959, Fiz. Tverd. Tela, sb. statei II, 235.

Agranovich, V.M. and B.S. Toshich, 1967, Zh. ETF **53**, 149.

Agranovich, V.M. and A.A. Zakhidov, 1977, Chem. Phys. Lett. **50**, 278.

Agranovich, V.M. and A.A. Zakhidov, 1979, Chem. Phys. Lett. **68**, 86.

Agranovich, V.M., I.Ya. Kucherov and A.N. Faidysh, 1956, Uchenye zapiski Kievskogo gos. universiteta No. 1, 27.

Agranovich, V.M., I.Ya. Kucherov and A.N. Faidysh, 1957, Ukr. Fiz. Zh. **2**, 61.

Agranovich, V.M., Yu.V. Konobeev and M.A. Mekhtiev, 1968, Fiz. Tverd. Tela, **10**, 1754.
Agranovich, V.M., V.I. Doronina and Yu.V. Konobeev, 1969a, Fiz. Tverd. Tela, **11**, 2607.
Agranovich, V.M., Yu.V. Konobeev and N.E. Kamenogradskiy, 1969b, Fiz. Tverd. Tela, **11**, 1445.
Agranovich, V.M., N.A. Efremov and E.P. Kaminskaya, 1971, Opt. Commun. **3**, 387.
Agranovich, V.M., A.G. Mal'shukov and M.A. Mekhtiev, 1972a, Zh. ETF **63**, 2274.
Agranovich, V.M., A.G. Mal'shukov and M.A. Mekhtiev, 1972b, Fiz. Tverd. Tela, **14**, 849.
Agranovich, V.M., E.P. Ivanova and Sh.S. Nikolaishvili, 1973, Fiz. Tverd. Tela, **15**, 2701.
Agranovich, V.M., N.A. Efremov and I.K. Kobozev, 1976a, Fiz. Tverd. Tela 18, 3421.
Agranovich, V.M., B.P. Antonyuk, E.P. Ivanova and A.G. Malshukov, 1976b, Zh. ETF Pis'ma **23**, 492.
Agranovich, V.M., N.A. Efremov and E.P. Ivanova, 1977a, Opt. i Spektr. **42**, 933.
Agranovich, V.M., B.P. Antonyuk, E.P. Ivanova, and A.G. Malshukov, 1977b, Zh. ETF **72**, 616.
Agranovich, V.M., O.A. Dubovskiy and K.Ts. Stoichev, 1979, Fiz. Tverd. Tela, **21**, 3012.
Agranovich, V.M., S.A. Darmanyan and V.I. Rupasov, 1980a, Zh. ETF **78**, 656.
Agranovich, V.M., N.A. Efremov and V.V. Kirsanov, 1980b, Fiz. Tverd. Tela, **22**, 2118.
Agranovich, V.M., A.A. Zakhidov and N.A. Efremov, 1980c, Izv. AN SSSR, Ser. Fiz. **44**, 759.
Agranovich, V.M., V.E. Kravtsov and A.G. Malshukov, 1980d, Solid State Commun. **33**, 137.
Agranovich, V.M., S.A. Darmanyan and V.I. Rupasov, 1981, Optika i Spektr **51**, 364.
Agrest, M.M., S.F. Kilin, M.M. Rikenglaz and I.M. Rozman, 1936, Dokl. Akad. Nauk SSSR **9**, 93.
Agrest, M.M., S.F. Kilin, M.M. Rikenglaz and I.M. Rozman, 1969, Opt. i Spektr. **27**, 946.
Agrest, M.M., E.A. Andreeshchev, S.F. Kilin, M.M. Rikenglaz and I.M. Rozman, 1970, Izv. AN SSSR, Ser. Fiz. **34**, 625.
Alekseevskiy, N.E., V.I. Tsebro, and E.I. Fillipovich, 1971, Pis'ma Zh. ETF **13**, 247.
Allen, L. and J. Eberly, 1975, Optical resonance and two-level atoms (Wiley, New York).
Al'shits, E.I., E.D. Godyaev and R.I. Personov, 1972, Fiz. Tverd. Tela, **14**, 1605.
Altarelli, M. and D.L. Dexter, 1973, Phys. Rev. B7, 5335.
Ambartsumyan, V.A., 1942, Astron. Zh. **19**, 30.
Anderson, P.W., 1958, Phys. Rev. **109**, 1492.
Anderson, P.W., 1970, Comments Solid State Phys. **2**, 193.
Andreeshchev, E.A., E.E. Baroni, V.S. Viktorova, K.A. Kovyrzina, I.M. Rozman and V.M. Shoniya, 1963a, Opt. i Spektr. "Luminescence", p. 128.
Andreeshchev, E.A., E.E. Baroni, I.M. Rozman, and V.M. Shoniya, 1963b, Opt i Spektr., "Luminescence", p. 131.
Andreeshchev, E.A., S.F. Kilin, I.M. Rozman, and V.I. Shirokov, 1963, Izv. AN SSSR, Ser. Fiz. **27**, 533.
Ansel'm, A., 1962, Introduction to theory of semiconductors (Fizmatgiz, Moscow) (in Russian).
Ansel'm, A.I. and Yu.A. Firsov, 1955, Zh. ETF **28**, 151.
Ansel'm, A.I. and Yu.A. Firsov, 1956, Zh. ETF **30**, 719.
Antonov-Romanovsky, V.V., 1966, Kinetics of photoluminescence of phosphors (Nauka, Moscow) (in Russian).
Apker, K. and E. Taft, 1951, Phys. Rev. **81**, 678.
Appel, J., 1968, Solid State Phys. **21**, 193.
Artamonova, M.V., Ch.M. Briskina, A.I. Burshtein, L.D. Zusman and A.G. Skleznev, 1972, Zh. ETF **62**, 863.
Avakian, P., 1974, Pure Appl. Chem. **37**, 1.
Avakian, P. and R.E. Merrifield, 1964, Phys. Rev. Lett. **13**, 541.
Avakian, P. and R.E. Merrifield, 1968, Mol. Cryst. **5**, 37.
Avakian, P., E. Abramson, R.A. Kepler and J.C. Caris, 1963, J. Chem. Phys. **39**, 1127.
Avery, J.S., 1975, Theor. Chim. Acta (Berl.) **39**, 281.

Axe, J.D. and P.E. Weller, 1964, J. Chem. Phys. **40**, 3066.
Babenko, S.D., V.A. Benderskii, V.I. Goldanskii, A.G. Lavrushko and V.P. Tuchinskii, 1971, Chem. Phys. Lett. **8**, 598.
Babenko, S.D., V.A. Benderskii, V.I. Goldanskii, A.G. Lavrushko and V.P. Tychinskii, 1971, Phys. Status Solidi (b) **45**, 91.
Bagaev, V.S., N.V. Zamkovets, L.V. Keldysh, N.N. Sibel'din and V.A. Tsvetkov, 1976a, Zh. ETF **70**, 1501.
Bagaev, V.S., L.V. Keldysh, N.N. Sibel'din and V.A. Tsvetkov, 1976b, Zh. ETF **70**, 702.
Bagdasar'yan, Kh.S. and A.L. Muler, 1965, Opt. i Spektr. **18**, 990.
Bakhshiev, N., 1972, Spectroscopy of intermolecular interactions (Nauka, Leningrad) (in Russian).
Balagurov, B.Ya. and V.G. Vaks, 1973, Zh. ETF **65**, 1939.
Basiev, T.T., Yu.K. Voron'ko and I.A. Shcherbakov, 1974, Zh. ETF **66**, 2118.
Basiev, T.T., T.G. Mamedov, and I.A. Shcherbakov, 1975, Kvantovaya electronika **2**, 1269.
Bates, O.R. and G. Poots, 1953, Proc. Roy. Soc. A**66**, 784.
Benderskiy, V., D.Sc. Thesis, 1971, Inst. of Chemical Physics AN SSSR, Moscow.
Benderskiy, V.A., V.Kh. Brikenshtein, V.L. Broude and I.I. Tartakovskiy, 1975, Pis'ma Zh. ETF **22**, 332.
Benderskiy, V.A., B.Kh. Brikenshtein, M.A. Kozhushner, I.A. Kuznetsova and P.G. Filippov, 1976a, Zh. ETF **70**, 521.
Benderskiy, V.A., V.Kh. Brikenshtein, M.A. Kozhushner and P.G. Filippov, 1976b, Zh. ETF, **70**, 521.
Benderskiy, V.A., V.M. Beskrovnyi, V.Kh. Brikenshtein, V.L. Broude, A.G. Lavrushko and A.A. Ovchinnikov, 1977, Zh. ETF **72**, 106.
Benderskiy, V.A., V.Kh. Brikenshtein, P.G. Filippov and A.V. Yatsenko, 1978, Opt. i Spektr., **45**, 512.
Benoit a la Guillaume, C. and A. Bonnot, 1970, Phys. Rev. Lett. **24**, 1235.
Bergman, A., M. Levine and J. Jortner, 1967, Phys. Rev. Lett. **18**, 593.
Berlman, I.B., 1965, Handbook of Fluorescence Spectra of Aromatic Molecules (Academic Press, New York, London) p. 123.
Biberman, L.M., 1947, Zh. ETF **17**, 416; **19**, 585.
Birks, J.B., 1964, The Theory and Practice of Scintillation Counting (Pergamon Press, Oxford).
Birks, J.B., 1967, Nature **214**, No. 5094, 1187.
Birks, J.B., 1968, J. Phys. B (Proc. Phys. Soc.) **1**, 946.
Birks, J.B. and S. Georghiou, 1968, J. Phys. B (Proc. Phys. Soc.) **1**, 958.
Birks, J.B. and I.H. Munro, 1967, Prog. Reaction Kinetics, **4**, 239.
Birks, J.B., M. Salete and S.C.P. Leite, 1970, J. Phys. B**3**, 513.
Bisti, V.E., 1976, Fiz. Tverd. Tela, **18**, 1056.
Blumen, A. and J. Manz, 1979, Journ. Chem. Phys. **71**, 4694.
Bodunov, E.N., 1971, Opt. i Spekt. **31**, 410, 949.
Bodunov, E.N., 1972, Izv. AN SSSR, Ser. Fiz. **36**, 996.
Bodunov, E.N., 1973, Opt. i. Spektr. **34**, 490.
Bodunov, E.N., 1976, Opt. i. Spektr. **41**, 990.
Bodunov, E.N., 1977, Zh. Prikl. Spektr. **26**, 1123.
Bodunov, E.N., V. Ermolaev, E. Sveshnikova and T. Shakhverdov, 1977, Radiationless transfer of electronic excitation energy (Nauka, Leningrad) (in Russian).
Bogani, F., 1978, J. Phys. C: Solid State Phys. **11**, 1283, 1297.
Bojarski, C., 1968, Acta Phys. Polonica, **34**, 853.
Bojarski, C., 1972, J. Lumin. **5**, 413.

Borisov, M.D. and V.N. Vishnevskiy, 1957, Ukr. Fiz. Zh. **1**, 371.
Borisova, O.F., L. Gorachek, G.V. Gurskiy, E.E. Minyat and V.G. Tumanyan, 1968, Mol. Biol. **2**, 475.
Borisova, O.F., L. Gorachek, G.V. Gurskiy, E.E. Minyat and V.G. Tumanyan, 1968, Izv. AN SSSR, Ser. Fiz. 32, 1317.
Born, M. and K. Huang, 1954, Dynamical theory of crystal lattices (Clarendon Press, Oxford).
Boron'ko, Yu.K. T.G. Mamedov, V.V. Osiko, M.I. Timoshechkin and I.A. Shcherbakov, 1973, Zh. ETF **65**, 1141.
Bouckaert, L.P., R. Smoluchowski, E.P. Wigner, 1936, Phys. Rev. **50**, 58.
Bowen, E.J., 1938, Nature, **142**, 1081.
Bowen, E.J. and E. Mikiewicz, 1947, Nature, **159**, 706.
Brillante, A. and D.P. Craig, 1974, Chem. Phys. Lett. **29**, 17.
Brodin, M.S. and M.G. Matsko, 1979, Pis'ma Zh. TEF, **30**, 571.
Brodin, M.S., M.A. Dubinskiy, S.V. Marisova and E.N. Myasnikov, 1975, Proc. Int. Conf. Lumin. Tokyo, p. 32.
Broude, V.L. and E.I. Rashba, 1961, Fiz. Tverd. Tela, **3**, 1941.
Broude, V.L., E.I. Rashba and E.F. Sheka, 1961, Dokl. Akad. Nauk SSSR **139**, 1085.
Brown, M.R., J.S. Whiting and W.A. Shand, 1965, J. Chem. Phys. **43**, 1.
Bryskin, V.V., D.N. Mirlin and Yu.A. Firsov, 1974, Usp. Fiz. Nauk **113**, 29.
Budo, A. and I. Ketskemety, 1962, Acta Phys. Hung. **14**, 167.
Buetner, A.V., B.B. Snavely and O.G. Peterson, 1969, in: Molecular luminescence, ed., E.C. Lim (Benjamin, New York) p. 416.
Burshtein, A.I., 1965, Teor. i Eksp. Khimiya, **1**, 563.
Burshtein, A.I., 1972, Zh. ETF **62**, 1695.
Burshtein, K.Ya. and M.A. Kozhushner, 1971, Fiz. Tverd. Tela, **15**, 504.
Burstein, E., 1969, Comments on Solid State Phys. **1**, 202.
Burstein, E. and D.L. Mills, 1969, Comments on Solid State Phys. **2**, 93, 111.
Burstein. E., and D.L. Mills, 1970, Comments on Solid State Phys. **3**, 12.
Burstein, E., C.Y. Chen and S. Lundquist, 1979, in: Light Scattering in Solids, eds. J.L. Birman, H.Z. Cummins, K.K. Rebane (Plenum Press, New York, London) p. 479.
Campillo, A.J., R.C. Hyer, S.I. Shapiro and C.E. Swenberg, 1977a, Chem. Phys. Lett. **48**, 495.
Campillo, A.J., S.L. Shapiro and C.F. Swenberg, 1977b, Chem. Phys. Lett. **52**, 11.
Chabr, W. and D.F. Williams, 1979, Phys. Rev. B**19**, 5206.
Chaiken, R.E. and D.R. Kearns, 1968, J. Chem. Phys. **49**, 2846.
Chance, R., A. Prock and R. Silbey, 1974, J. Chem. Phys. **60**, 2184, 2744.
Chandrasekhar, S., 1943, Rev. Mod. Phys. **15**, 1.
Chandrasekhar, S., 1950, Radiative Transfer (Clarendon Press, Oxford).
Choi, S.I., 1967, Phys. Rev. Lett. **19**, 358.
Choi, S.I. and S.A. Rice, 1962, Phys. Rev. Lett. **80**, 410.
Choi, S.I. and S.A. Rice, 1963, J. Chem. Phys. **38**, 366.
Clark, R.H. and R.M. Hochstrasser, 1967, J. Chem. Phys. **46**, 4532.
Craig, D.P., 1974, Electronic Energy levels of Molecular solids, in: Orbital theories of Molecules and solids, ed., N.H. March (Clarendon Press, Oxford) p. 344.
Craig, D.P. and S.H. Walmsley,1968, Excitons in Molecular crystals. Theory and Applications (Benjamin, New York, Amsterdam).
Craver, F.W. and R.S. Knox, 1971, Mol. Phys. **22**, 385.
Croxton, C.A., 1975, Introduction to Liquid State Physics (Wiley, London).
Cummins, P.G., D.A. Dunmur and R.W. Munn, 1973, Chem. Phys. Lett. **22**, 519.
Cummings, P.G., D.A. Dunmur, R.W. Munn and R.J. Newham, 1974, Acta Crystallogr. A**32**, 847, 854.
Davison, B., 1957, Neutron transport theory (Clarendon Press, Oxford).

Davydov, A., 1971, Theory of molecular excitons (Plenum Press, New York).
Davydov, A.S. and N.I. Kislukha, 1973, Phys. Status Solidi (b) **59**, 465.
Davydov, A.S. and N.I. Kislukha, 1976, Zh. ETF **71**, 293.
Davydov, A.S. and A.A. Serikov, 1972, Phys. Status Solidi, **51**, 57.
Demidenko, A.A., 1963, Fiz. Tverd. Tela, **5**, 489.
Dexter, D.L., 1953, J. Chem. Phys. **21**, 836.
Dexter, D.L., 1972, Phys. Status Solidi (b) **51**, 571.
Dexter, D.L. and R.S. Knox, 1965, Excitons (Interscience Publishers).
Dexter, D.L., Th. Förster and R.S. Knox, 1969, Phys. Status Solidi, **34**, K159.
Doronina, V.I., Yu.V. Konobeev and Sh.D. Khan-Magometova, 1970, Opt. i Spektr. **28**, 811.
Dow, J.D., 1968, Phys. Rev. **174**, 962.
Drexhage, K.H., 1966, Habilitations Schrift, Marburg.
Drexhage, K.H., H. Kuhn and F.P. Schaefer, Ber. Bunsendes, 1968, Phys. Chem. **72**, 329.
Dubovskiy, O.A., 1976, Fiz. Tverd. Tela, **18**, 2301.
Dubovskiy, O.A. and Yu.V. Konobeev, 1964, Fiz. Tverd. Tela, **6**, 2599.
Dubovskiy, O.A. and Yu.V. Konobeev, 1965, Fiz. Tverd. Tela, **7**, 945.
Dubovskiy, O.A. and Yu.V. Konobeev, 1970, Fiz. Tverd. Tela, **12**, 406.
Dunmur, D.A., 1971, Chem. Phys. Lett. **10**, 49.
Dunmur, D.A., 1972, Mol. Phys. **23**, 109.
Dunmur, D.A. and R.W. Munn, 1975, Chem. Phys. **11**, 297.
Efremov, N.A., 1975, Fiz. Tverd. Tela, **17**, 1895.
Efremov, N.A. and E.P. Kaminskaya, 1972, Fiz. Tverd. Tela, **14**, 1185.
Efremov, N.A. and E.P. Kaminskaya, 1973, Fiz. Tverd. Tela, **15**, 3338.
Efremov, N.A. and A.G. Mal'shukov, 1975, Fiz. Tverd. Tela, **17**, 2239.
Elkana, Y., J. Feitelson and E. Katchalski, 1968, J. Chem. Phys. **48**, 2399.
El'yashevich, M., 1962, Atomic and molecular spectroscopy (Fizmatgiz, Moscow) (in Russian).
Ern, V., P. Avakian, and R.E. Merrifield, 1966, Phys. Rev. **148**, 862.
Ern, V., A. Suna, Y. Tomkiewicz, P. Avakian and R.P. Groff, 1972, Phys. Rev. **B5**, 3222.
Faidysh, A.N., 1955, Dokl. Akad. Nauk Ukr. SSR **6**, 215.
Fano, U., 1956, Phys. Rev. **103**, 1202.
Fano, U., 1958, in: Comparative Effects of Radiation, eds., M. Burton, J.S. Kirby-Smith, J.L. Magee (Wiley, New York) p. 14.
Fano, U., 1961, Phys. Rev. **15**, 1866.
Fayer, M.D. and C.B. Harris, 1974, Chem. Phys. Lett. **25**, 149.
Fayer, M.D. and C.B. Harris, 1974, Phys. Rev. **B9**, 748.
Feofilov, P., 1959, Polarized luminescence of atoms, molecules and crystals (Fitzmatgiz, Moscow) Sec. IV. 8 (in Russian).
Feofilov, P.P. and B.Ya. Sveshnikov, 1940, Zh. ETF **10**, 1372.
Ferguson, J., 1977, Chem. Phys. Lett. **36**, 316.
Firsov, Yu.A., 1968, Fiz. Tverd. Tela, **10**, 1950.
Firsov, Yu.A., ed, 1975, Polarons (Nauka, Moscow).
Firsov, Yu.A. and E.K. Kudinov, 1964, Zh. ETF **47**, 601.
Firsov, Yu.A. and E.K. Kudinov, 1965, Fiz. Tverd. Tela, **7**, 546.
Flytzanis, C., Treatise in Quantum Electronics, eds., H. Rabin, C.L. Tang, 1A (Academic, New York) p. 111.
Förster, Th., 1948, Ann. Phys. **2**, 55.
Förster, Th., 1949, Zs. Naturf. **4a**, 321.
Förster, Th., 1951, Fluoreszenz organischer Verbindungen (Vandenhoek und Ruprecht Göttingen).
Förster, Th., 1959, Discussion Faraday Soc. **27**, 7.

Förster, Th., 1965, in: Mod. Quantum Chem., p. 111, 93, L., N.Y.
Fourny, J., G. Delacote and M. Schott, 1968, Phys. Rev. Lett. **21**, 1085.
Frank J. and E. Teller, 1938, J. Chem. Phys. **6**, 861.
Freedhoff, H. and J. van Kranendonk, 1967, Can. J. Phys. **45**, 1833.
Frenkel', Ya.I., 1931, Phys. Rev. **37**, 17, 1276.
Fröhlich, H., 1937, Proc. Roy. Soc. **160**, 230.
Fugol', I.Ya., E.V. Savchenko and A.G. Belov, 1975, Fiz. Nizk. Temp. **1**, 750.
Fulton, R.L., 1974, J. Chem. Phys, **61**, 4141.
Galanin, M.D., 1950, Trudy FIAN SSSR **5**, 341.
Galanin, M.D., 1951, Zh. ETF **21**, 126.
Galanin, M.D., 1955, Zh. ETF **28**, 485.
Galanin, A.D., 1957, Theory of nuclear thermal-neutron reactors (Atomizdat, Moscow) (in Russian).
Galanin, M.D., 1960, Trudy FIAN SSSR **12**, 3.
Galanin, M.D., 1977, Acta Phys. Chem. (Hungaria) **23**, 83.
Galanin, M.D. and Z.A. Chizhikova, 1954, Zh. ETF, **26**, 624.
Galanin, M.D. and Z.A. Chizhikova, 1956, Opt. i Spektr. **1**, 175.
Galanin, M.D. and I.M. Frank, 1951, Zh. ETF **21**, 114.
Galanin, M.D., Yu.V. Konobeev and Z.A. Chizhikova, 1962, Opt. i Spektr. **13**, 386.
Galanin, M.D., Sh.D. Khan-Magometova and Z.A. Chizhikova, 1972, Pis'ma Zh. ETF **16**, 141.
Galanin, M.D., Sh.D. Khan-Magometova and Z.A. Chizhikova, 1973, Izv. AN SSSR, Ser. Fiz. **37**, 298.
Galanin, M.D., Sh.D. Khan-Magometova, Z.A. Chizhikova, M.I. Demchuk and A.F. Chernyavskii, 1974a, J. Lumin. **9**, 459.
Galanin, M.D., M.I. Demchuk, Sh.D. Khan-Magometova, A.F. Chernyavskiy and Z.A. Chizhikova, 1974b, Pis'ma Zh. ETF **20**, 260.
Galanin, M.D., Sh.D. Khan-Magometova and Z.A. Chizhikova, 1975, Izv. AN SSSR, Ser. Fiz. **39**, 1807.
Galanin, M.D., Sh.D. Khan-Magometova and E.N. Myasnikov, 1980, Mol. Cryst. Liq. Crist. **57**, 119.
Gale, G.M. and A. Mysyrowicz, 1974, Phys. Rev. Lett. **32**, 727.
Gallus, G. and H.C. Wolf, 1966, Phys. Status Solidi, **16**, 277.
Gamble, F.G., J.H. Osiecki and F.J. Di Slavo, 1971, J. Chem. Phys. **55**, 3525.
Gammill, L.S. and R.C. Rowell, 1974, Mol. Cryst. Liq. Cryst. **25**, 123.
Gamurar', V.Ya., Yu.E., Perlin and B.S. Tsukerblat, 1969, Fiz. Tverd. Tela, **11**, 1193.
Ganguly, S.C. and N.K. Chanudhury, 1959, Rev. Mod. Phys. **31**, 990.
Gennes, P.G., de, 1958, J. Phys. Chem. Sol. **7**, 345.
Gerbshtein, Yu.M., D.N. Mirlin and I.A. Merkulov, 1975, Pis'ma Zh. ETF **22**, 80.
Goldhirsch, I., E. Levich and V. Yakhot, 1979, Phys. Rev. B**19**, 4780.
Golubkov, G.V., F.I. Dalidchik and G.K. Ivanov, 1980, Zh. ETF **78**, 1423.
Golubov, S.I. and Yu.V. Konobeev, 1971, Fix. Tverd. Tela, **13**, 3185.
Golubov, S.I. and Yu.V. Konobeev, 1973, Phys. Status Solidi, **56**, 69.
Golubov, S.I. and Yu.V. Konobeev, 1975, Phys. Status Solidi (b) **70**, 373; **71**, 777.
Golubov, S.I. and Yu.V. Konobeev, 1977, Phys. Status Solidi (b) **79**, 79.
Gorokhovskiy, A.A., R.K. Kaarli and L.A. Rebane, 1974, Pis'ma Zh. ETF **20**, 474.
Gösele, U., 1978, J. Nucl. Mat. **78**, 83.
Gösele, U., M. Hauser, U.K.A. Klein and R. Frey, 1975, Chem. Phys. Lett. **34**, 519.
Gösele, U., M. Hauser, U.K.A. Klein and R. Frey, 1976, Chem. Phys. Lett. **41**, 139.
Greenwood, D.A., 1958, Proc. Phys. Soc. (London) **71**, 585.
Greer, W.L., 1974, J. Chem. Phys. **60**, 744.
Gross, E.F., S.A. Permogorov, V.V. Travnikov and A.V. Sel'kin, 1971, Fiz. Tverd. Tela, **13**, 699.

Gross, E., S. Permogorov, V. Travnikov and A. Selkin, 1972, Solid State Commun. **10**, 1071.
Grover, M. and R. Silbey, 1971, J. Chem. **54**, 4843.
Gurari, M.L. and M.A. Kozhushner, 1970, Zh. ETF **58**, 1967.
Gurevich, L., 1940, Fundamentals of physical kinetics (Gostekhizdat, Moscow) (in Russian).
Gurskiy, G.V., A.S. Zasedatelev and M.V. Vol'kenshtein, 1973, Mol. Biol. **7**, 49.
Haarer, D., 1977, J. Chem. Phys. **67**, 4076.
Haarer, D., 1979, J. Lumin. 18/19 (Part 1), 453.
Haken, H., 1958, Bull. Soc. Franc. Phys. **20**, 60.
Haken, H., 1958, Fortschr. Phys. **6**, 271.
Haken, H. and S. Nikitine, eds, 1975, Excitons at High Density, Springer Tracts in Modern Physics.
Haken, H., and P. Reineker, 1972, Z. Phys. **249**, 253.
Haken, H. and E. Schwarzer, 1974, Chem. Phys. Lett. **27**, 41.
Haken, H. and G. Strobl, 1967, in: Triplet State, ed., Zahlan A. (Cambridge University Press) 311.
Haken, H. and G. Strobl, 1973, Z. Phys. **262**, 135.
Hammer, A. and H.C. Wolf, 1968, Mol. Cryst. **4**, 191.
Hammer, A., T.B. El. Kareh and H.C. Wolf, 1967, Mol. Cryst. **3**, 160.
Hanamura, E., 1973, Solid State Commun. **12**, 951.
Hanson, D.M., 1973, Crit. Rev. Solid State Sci., p. 243.
Haugland, R.P., J. Yguerabide and L. Stryer, 1969, Crit. Rev. Solid State Sci. **63**, 23.
Hayes, J.M. and G.J. Small, 1978, Chem. Phys., **27**, 151.
Heim, U. and P. Weisner, 1975, Phys. Rev. Lett. **30**, 1205.
Heisel, F., J.A. Miehe, M. Schott, and B. Sipp, 1978, Mol. Cryst. Liq. Cryst. Lett. **41**, 251.
Heitler, W., 1954, The quantum theory of radiation (Clarendon Press, Oxford).
Hemenger, R.P., Katja Lakatos-Lindenberg and R.M. Pearlstein, 1974, J. Chem. Phys. **60**, 3271.
Himsel, F.J., J. Jortner, E.E. Koch, Z. Ophir, D. Pudewill, B. Raz, V. Saile, N. Schwenter and M. Skibowski, 1975, Seventh Molecular Crystal Symposium, Nikko, Japan, p. 3.
Hochstrasser, R.M., 1962, Rev. Mod. Phys. **34**, 531.
Hochstrasser, R.M., 1972, J. Chem. Phys. **56**, 1.
Hochstrasser, R.M., 1973, Acc. Chem. Res. **6**, No. 8.
Hochstrasser, R.M. and L.J. Noe, 1969, J. Chem. Phys. **50**, 1684.
Hochstrasser, R.M., L.W. Johnson and T.Y. Li, 1975, Seventh Molecular Crystal Symposium, Nikko, Japan, p. 47.
Holstein, I., S.K. Lyo and R. Orbach, 1976, Phys. Rev. Lett. **36**, 891.
Holstein, T., 1947, Phys. Rev. **72**, 1212.
Holstein, T., 1951, Phys. Rev. **83**, 1159.
Hong, H.K. and G.W. Robinson, 1970, J. Chem. Phys. **52**, 825.
Hopfield, J.J., 1958, Phys. Rev. **112**, 1555.
Hopfield, J.J., 1966, J. Phys. Soc. Japan, Suppl. **21**, 77.
Hoshen, J. and J. Jortner, 1970, Chem. Phys. Lett. **5**, 351.
Hsu, C. and R. Powell, 1975, Phys. Rev. Lett. **35**, 734.
Huang, K., 1951, Proc. Roy. Soc. A**208**, 352.
Hubbard, J., 1958, Proc. Roy. Soc. A**243**, 336.
Imbush, G.F., 1967, Phys. Rev. **153**, 326.
Inoue, A., 1975, J. Phys. Soc. Japan, **39**, 467.
Inoue, A., K. Yoshiara and S. Nagakura, 1972, Bull. Chem. Soc. Japan, **45**, 720.
Ivanov, G.K. and M.A. Kozhushner, 1980, Phys. Status Solidi, (b) **97**, 329.
Ivanova, E.P., 1980, J. Lumin., **21**, 373.
Ivanova, E.P., 1981, Fiz. Tverd. Tela, **22 23**, 547.
Ivliev, Yu.A., 1970, Opt. i Spektr. **28**, 206.

Ivliev, Yu.A., 1971, Fiz. Tverd. Tela, **13**, 205.
Jablonski, A., 1955, Acta Phys. Polonica, **14**, 295.
Jablonski, A., 1958, Acta Phys. Polonica, **17**, 481.
Janke-Emde-Losch, 1960, Tafeln höherer Funktionen (B.G. Teubner Verlagsgeselschaft. Stutgart).
Jemenger, R.P. and R.M. Pearlstein, 1973, J. Chem. Phys. **59**, 4064.
Jortner, J., 1968, Phys. Rev. Lett. **20**, 244.
Jortner, J. and S.I. Choi, 1965, J. Chem. Phys. **42**, 309.
Jortner, J., S.I. Choi, J.L. Katz and S. Rice, 1963, Phys. Rev. Lett. **11**, 323.
Kallmann, H., 1971, Zs. Naturf. **26**a, 799.
Kallmann, H., G. Vaubel and H. Baessler, 1971, Phys. Status Solidi (b) **44**, 813.
Kalnin, Yu.H., 1981, in: Defects in Insulating Crystals, Proc. Int. Conf. Riga, May 1981, eds., V.M. Tuchkevich and K.K. Shvarts (Springer, Riga, Berlin, New York) p. 150.
Kamenogradskiy, N.E. and Yu.V. Konobeev, 1969, Fiz. Tverd, Tela, **11**, 2357.
Kamenogradskiy, N.E. and Yu.V. Konobeev, 1970, Phys. Status Solidi, **37**, 29.
Kamke, E., 1959, Differentialbleichungen; I, Gewöhnliche, differentialbleichungen, Leipzig.
Kapkhanin, Yu.I. and V.E. Lashkarev, 1955, DAN SSSR **101**, 829.
Kaplan, I. and J. Jortner, 1977, Chem. Phys. Lett. **51**, 1.
Kaplan, I.G. and M.A. Ruvinskiy, 1976, Zh. ETF **71**, 2142.
Kareh, El. T.B. and H.C. Wolf, 1966, Report at symposium on organic scintillators, Chicago.
Katal'nikov, V.V., 1968, Fiz. Tverd. Tela, **2**, 1392.
Katal'nikov, V.V., 1974, Fiz. Tverd. Tela, **16**, 3498.
Katal'nikov, V.V. and O.S. Rudenko, 1977, Ukr. Fiz. Zh. **22**, 1362.
Kawski, A., 1963, Zs. Naturf. **18**a, 961.
Kawski, A., 1963, Bull. Acad. Pol. Sci., Ser. Sci. Math. Astron. Phys. **11**, 567.
Kawski, A. and J. Kaminski, 1970, Akta Phy. Polonica, A**37**, 591.
Kawski, A. and J. Kaminski, 1974, Zs. Naturf. **29**a, 452.
Kazzaz, A.A. and A.B. Zahlan, 1961, Phys. Rev. **124**, 90.
Kearns, D.R., 1963, J. Chem. Phys. **39**, 2697.
Keldysh, L., 1971, in: Excitons in semiconductors (Nauka, Moscow) p. 5. (in Russian).
Keldysh, L.V. 1976, Pig'ma Zh. ETF **23**, 2.
Keldysh, L.V., 1970, UFN **100**, 514.
Keldysh, L.V. and C.D. Jeffries, eds., 1983, Electron–hole droplets in semiconductors. (North-Holland, Amsterdam) (in press).
Keller, R.A., 1969, Chem. Phys. Lett. **3**, 27.
Kenkre, V.M. 1975, Phys. Rev. B**11**, 1741.
Kenkre, V.M. and R.S. Knox, 1974, Phys. Rev. B**9**, 5279.
Khachaturyan, A.G. and Yu.Z. Estrin, 1971, Zh. ETF **61**, 2607.
Khan-Magometova, Sh.D., 1965, Izv. AN SSSR, Ser. Fiz. **29**, 1321.
Khan-Magometova, Sh.D., 1972, Trudy FIAN SSSR **59**, 236.
Kharlamov, B.M., R.I. Personov and L.A. Bykovskaya, 1974, Optics Commun. **12**, 191.
Kharlamov, B.M., R.I. Personov and L.A. Bykovskaya, 1975, Optika i Spektr. **39**, 240.
Khizhnyakov, V.V., 1972, Phys. Status Solidi (b) **51**, K17.
Khokhlov, Yu.K., 1972, Trudy FIAN SSSR **59**, 221.
Khutsishvili, G.P., 1954, Trudy In-ta fiziki AN Gruz. SSR **2**, 115.
Khutsishvili, G.P., 1956, Trudy In-ta fiziki AN Gruz. SSR **4**, 3.
Kilin, S.F., M.S. Mikhelashvili and I.M. Rozman, 1964, Opt. i Spektr. **16**, 1063.
Killesreiter, H. and H. Baessler, 1971, Chem. Phys. Lett. **11**, 411.
Killesreiter, H. and H. Baessler, 1972, Phys. Status Solidi (b) **51**, 657.
Kink, P.A., G.G. Liyd'ya, CH.B. Lushchik and G.A. Soovik, 1969, Trudy In-ta fisiki AN EST. SSR **36**, 5.
Kiriy, V.B. and V.L. Shekhtman, 1972, Fiz. Tverd. Tela, **14**, 2980.

Kirkwood, J.G. and E.M. Boggs, 1942, J. Chem. Phys. **10**, 394.
Klein, G. and R. Voltz, 1975, Intern. J. Radiat. Phys. Chem. **7**, 155.
Klinger, M., 1979, Problems of linear electron (polaron) transport theory (Pergamon Press, Oxford).
Klöffer, W., H. Bauser, F. Dolezalek and G. Naundorf, 1972, Mol. Cryst. Liq. Cryst. **16**, 229.
Knox, R.S., 1963, Theory of Excitons (Academic Press, New York).
Knox, R.S., 1968, Physica, **39**, 361.
Kogan, V.D., 1975, Fiz. Tverd. Tela, **17**, 2578.
Komponeets, A.S., 1970, (private communication).
Konobeev, Yu.V., 1961, Opt. i Spektr. **11**, 504.
Konobeev, Yu.V., 1962, Fiz. Tverd. Tela, **4**, 3634.
Konobeev, Yu.V., 1963, Opt. i Spektr., sb. statei, I, "Luminescence", 135.
Konobeev, Yu.V., 1969, Opt. i Spektr., sb. "Spetroskopiya tverdogo tela", 208.
Konobeev, Yu., 1971, D.Sc. Thesis, Obninsk.
Konyshev, V.P. and A.I. Burshtein, 1968, Teor. i Eksp. Khimiya **4**, 192.
Koo, J., L.R. Walker and S. Geschwind, 1975, Phys. Rev. Lett. **35**, 1669.
Kozhushner, M.A., 1971, Fiz. Tverd. Tela, **13**, 2601.
Kozhushner, M.A., 1971, Zh. ETF **60**, 220.
Kubo, R., 1957, J. Phys. Soc. Japan, **12**, 570.
Kubo, R., 1962, J. Phys. Soc. Japan, **17**, 1100.
Kuhn, H., 1970, J. Chem. Phys. **53**, 101.
Kukushkin, L.S., 1963, Opt. i Spektr. **15**, 371.
Kurczewska, H. and H. Baessler, 1977, J. Lumin. **15**, 261.
Kurik, M.V., 1971, Phys. Status Solidi (a) **8**, 9.
Kurik, M.V. and Yu.P. Piryatinskiy, 1971, Fiz. Tverd. Tela, **13**, 2877.
Kurskiy, Yu.A. and A.S. Selivanenko, 1960, Opt. i Spektr. **8**, 643.
Kustov, E.F. and L.I. Surogin, 1970, in: Energy Transfer in Condensed Media, Proc. All-Union Seminar, Erevan, p. 17 (in Russian).
Kuusman, I.L., G.G. Liyd'ya and Ch.B. Lushchik, 1976, Trudy In-ta Fiziki AN Est. SSR **46**, 5.
Lalov, I.I., 1974, Fiz. Tverd. Tela, **16**, 2476.
Lalov, I.I., 1975, Phys. Status Solidi (b) **68**, 319, 681.
Lalovic, D.J., B.S. Tosic and R.B. Zakula, 1969, Phys. Rev. **178**, 1472.
Lampert, M.A., 1958, Phys. Rev. Lett. **1**, 450.
Landau, L. and E. Lifshits, 1962, Field Theory (Fizmatgiz, Moscow) (in Russian).
Landau, L.D. and E.M. Lifshits, 1965, Quantum mechanics (Addison-Wesley, Reading, Mass.)
Landau, L. and E. Lifshits, 1966, Electrodynamics of continuous media (Addison-Wesley, Reading, Mass).
Lang, I.G. and Yu.A. Firsov, 1962, Zh. ETF **43**, 1843.
Lang, I.G. and Yu.A. Firsov, 1963, Zh. ETF **45**, 378.
Lang, I.G. and Yu.A. Firsov, 1967, Fiz. Tverd. Tela, **9**, 3422.
Levshin, V.L. and E.G. Baranova, 1959, Opt. i Spektr. **6**, 55.
Levshin, V.L. and Yu.I. Grineva, 1968, Acta physica polonica, **34**, 791.
Lifshits, I.M., 1942, Zh. ETF **12**, 117, 137.
Lifshits, I.M., 1947, Zh. ETF **17**, 1017, 1076.
Lifshits, I.M., 1948, Zh. ETF **18**, 293.
Lifshits, I.M., 1956, Nuovo Cimento, Suppl. **3** (x), 716.
Lifshits, I.M. 1964, Adv. Phys. **13**, 483; Usp. Fiz. Nauk, 83, 617.
Linschitz, H. and K. Sarkanen, 1958, J. Amer. Soc. **80**, 4826,
Little, W.A. ed., 1969, Proc. Intern. Conf. Organic Superconductors, Honolulu.
Little, W.A., 1974, Sixth Molecular Crystal Symposium, Shloss Elmau, Germany.
Lowe, I.J. and D. Tse, 1968, Phys. Rev. **166** 279.
Lushchik, Ch.B., I.K. Vitol and M.A. Elango, 1977, UFN **122**, 223.

Lyons, L.E., 1957, Austr. J. Chem. **10**, 365.
Mahan, G.D., 1967, Phys. Rev. **153**, 983.
Mahan, G.D., 1975, in: Electronic Structure of Polymers and Molecular Crystals, eds., Jean-Marie Andre and Janos Ladik (Plenum Press, New York, London) p. 79.
Mahan, G.D. and R.M. Mazo, 1968, Phys. Rev. **175**, 1191.
Maksimov, M.Z. and I.M. Rozman, 1962, Opt. i Spektr. **12**, 606.
Malshukov, A.G., 1974, Fiz. Tverd. Tela, **16**, 2274.
Malshukov, A.G., 1981, Solid State Commun. **38**, 907.
Maradudin, A. 1966, Theoretical and experimental aspects of the effects of point defects and disorder on the vibrations in crystals (Academic Press, New York).
Maradudin, A.A., E.W. Montroll, G.H. Weiss, R. Herman and H.W. Milnes, 1960, Acad. Roy. Belg. Classe. Sci. Men. Collection in 4° 2, No. 47, 14.
Marinkovic, M.M., 1975, Phys. Status Solidi (b) **69**, 291.
Marinkovic, M.M. and B.S. Tosic, 1975, Phys. Status Solidi (b) **67**, 435.
Matsui, A. and Y. Oeda, 1973, Sixth Molecular Crystal Symposium (Schloss Elmau, Germany) p.41.
Menzel, D., ed., 1955, Fundamental formulas of physics, (Prentice-Hall, New York).
Mikhelashvili, M.S., I.M. Rozman and G.S. Tsukaya, 1974, Opt. i Spektr. **36**, 352.
Mitchell, A. and M. Zemansky, 1934, 1961, Resonance radiation and excited atoms (Cambridge, University Press, London).
Miyakawa, T. and D.L. Dexter, 1970, Phys. Rev. 1B, 70.
Montroll, E.W., 1964, Proc. Symp. Appl. Math. Am. Math. Soc. **16**, 193.
Montroll, E.W., 1965, J. Math. Phys. **6**, 167.
Montroll, E.W., 1969, J. Math. Phys. **10**, 753.
Morawitz, H. and M.R. Philpott, 1974, Phys. Rev. B**10**, 4863.
Moskalenko, S.A., 1958, Opt. i Spektr. 5, 147.
Mott, N.F. and E.A. Davis, 1971, Electronic Processes in Non-Crystalline Materials (Clarendon Press, Oxford).
Munn, R.W. and W. Siebrand, 1970, J. Chem. Phys. **52**, 47.
Munro, I.H., L.M. Logan, F.D. Blair, F.R. Lipsett and D.E. Williams, 1972, Mol. Cryst. Liq. Cryst. **15**, 311.
Nienhuis, G. and J.M. Deutch, 1972, J. Chem. Phys. **56**, 235, 1819, 5511.
Northrop, D.C., O. Simpson, 1958, Proc. Roy. Soc. A**244**, 377.
Obreimov, I., 1945, Applications of Fresnel diffraction for physical and technical measurements, AN SSSR, (in Russian).
Onipko, A.A., 1976, Phys. Status Solidi (b) **73**, 699.
Onipko, A.I. and V.I. Sugakov, 1974, Opt. i Spektr. **35**, 185.
Onodera, Y. and Y. Toyozawa, 1968, J. Phys. Soc. Japan, **24**, 341.
Ophir, Z., B. Raz, J. Jortner, V. Saile, N. Schwentner, E-E. Koch, M. Skibowski, W. Steinmann, 1975, J. Chem. Phys. **62**, 650.
Orbach, R. and M. Tachiki, 1967, Phys. Rev. **158**, 524.
Ore, A., 1959, J. Chem. Phys. **31**, 442.
Osaca, T., Y. Imai and Y. Takeuti, 1968, J. Phys. Soc. Japan, **24**, 236.
Ostapenko, N.I. and M.T. Shpak, 1969, Phys. Status Solidi **31**, 531; **36**, 515.
Ostapenko, N.I., V.I. Sugakov and M.T. Shpak, 1973, in: Excitons in molecular crystals (Naukova dumka, Kiev) p. 92 (in Russian).
Ovchinnikov, A.A., 1969, Zh. ETF **57**, 263.
Ovchinnikov, A.A. and N.S. Erikhman, 1974, Zh. ETF **67**, 1474.
Ovsyankin, V.V. and P.P. Feofilov, 1966, Pis'ma Zh. ETF **3**, 494.
Ovsyankin, V.V. and P.P. Feofilov, 1971, Pis'ma Zh. ETF **14**, 548.
Ovsyankin, V.V. and P.P. Feofilov, 1971, Opt. i Spektr. **31**, 944.
Pantell, R. and H. Puthoff, 1969, Fundamentals of Quantum Electronics (Wiley, New York).

Patzer, K., 1971, Opt. i Spektr. **27**, 954.
Peierls, R., 1932, Ann. Phys. **13** (5), 905.
Peierls, R., 1955, Quantum theory of solids (Clarendon Press, Oxford).
Perlin, Yu. and B. Tsukerblat, 1974, Effects of electron vibrational interaction in optical spectra of impurity paramagnetic ions, Shtiintsa, Kishinev, (in Russian).
Perrin, F., 1929, Ann. Phys. 10th serie, **12**, 169.
Perrin, J., 1924, 2-me Conceil de Chimie Solvay, Bruxelles.
Person, B.N.J. and M. Persson, 1980, Solid State Commun., **33**, 713.
Personov, R.I., I.S. Osad'ko, E.D. Gidyaev and E.I. Al'shits, 1971, Fiz. Tverd. Tela, **13**, 2653.
Personov, R.I., E.M. Al'shits, L.A. Bykovskaya and B.M. Kharlamov, 1973, Zh. ETF **65**, 1825.
Phillips, J.C., 1964, Phys. Rev. Lett. **12**, 447.
Philpott, M.R., 1973, Adv. Chem. Phys. **23**, 227.
Philpott, M.R., 1975, J. Chem. Phys. **62**, 1812; **63**, 485.
Philpott, M.R. and G.D. Mahan, 1973, J. Chem. Phys. **59**, 445.
Pines, D. and Ph. Noziere, 1966, The Theory of Quantum Liquids (Benjamin, New York, Amsterdam).
Pitaevskiy, L.P., 1976, Zh. ETF **70**, 738.
Pokrovsky, Ya.E., 1972, Phys. Status Solidi (a) **11**, 386.
Pokrovsky, Ya.E., A. Kaminskii and K. Svistunova, 1970, Proc. X Internat. Conf. Phys. Semicond. USAEC, 504.
Polivanov, Yu.N., 1978, UFN, **126**, 185.
Polivanov, Yu.N., 1979, Pis'ma Zh. ETF, **30**, 415.
Polya, G., 1921, Math. Ann. **84**, 149.
Porter, G. and M.W. Windsor, 1958, Proc. Roy. Soc. A**245**, 238.
Powell, R.C., 1970, Phys. Rev. B**2**, 1159, 2090.
Powell, R.C., 1971, Phys. Rev. B**4**, 628.
Powell, R.C., 1973, J. Chem. Phys. **58**, 920.
Powell, R.C., 1973, J. Lumin. **6**, 285.
Powell, R.C. and R.G. Kepler, 1969, Phys. Rev. Lett. **22**, 636, 1232.
Powell, R.C. and R.G. Kepler, 1970, J. Lumin. **12**, 254.
Powell, R.C. and R.G. Kepler, 1970, Mol. Cryst. Liq. Cryst. **11**, 349.
Powell, R.C. and Z.G. Soos, 1975, J. Lumin. **11**, 1.
Propstl, A. and H.C. Wolf, 1963, Zs. Naturf. **18a**, 724.
Rahman, T.S. and R.S. Knox, 1973, Phys. Status Solidi (b) **58**, 715.
Rashba, E.I., 1972, in: Physics of impurity centres in crystals, Proc. International Seminar, Tallin, p. 427.
Rebane, K., 1968, Elementary theory of the vibrational structure of the spectra of impurity centres in crystals (Nauka, Moscow) (in Russian).
Rebane, K.K., I.Y. Tehver and V.V. Hizhnyakov, 1975, in: The Theory of Light Scattering in Solid, Proc. of the First USSR-USA Symposium, May 1975, eds., V.M. Agranovich, J.L. Birman (Publ. House Nauka).
Reineker, P., 1974, Zs. Naturf. **29a**, 282.
Rice, S.A. and J. Jortner, 1967, Physics and Chemistry of the Organic Solid State, Vol. 3, eds., D. Fox, M.M. Labes, A. Weissberger (Interscience, New York) p. 199.
Richie, R.H. and A.L. Marusak, 1966, Surf. Sci. **4**, 234.
Rikenglaz, M.M. and I.M. Rozman, 1974, Opt. i Spektr. **36**, 100.
Robinson, G.W. and R.P. Frosch, 1962, J. Chem. Phys. **37**, 1962.
Robinson, G.W. and R.P. Frosch, 1963, J. Chem. Phys. **38**, 1187.
Romanov, Yu.A., 1964, Radiofizika, **7**, 2.
Rosenstock, H.B., 1970, J. Math. Phys. **11**, 487.
Rozman, I.M., 1958, Opt. i Spektr. **4**, 536.
Rozman, I.M., 1961, Opt. i Spektr. **10**, 354.

Rozman, I.M., 1973, Izv. AN SSSR, Ser. Fiz. **37**, 502.
Rozman, I.M. and S.F. Kilin, 1959, Usp. Fiz. Nauk, **69**, 459.
Rozman, I.M. and R.Sh. Sichinava, 1975, Izv. AN SSSR, Ser. Fizich. **39**, 1863.
Rudoy, Yu. and Yu. Tserkovnikov, 1975, Suppl. to Tyablikov (1975).
Ryazanov, G.V., 1972, Teor. Mat. Fiz. **10**, 271.
Safaryan, F.I., 1977, Fiz. Tverd. Tela, **19**, 1947.
Saito, H. and S. Shionoya, 1974, J. Phys. Soc. Japan, **37**, 423.
Sakun, V.P., 1972, Fiz. Tverd. Tela, **14**, 2199.
Samoilovich, A.G., I.Ya. Korenblit, I.V. Dakhovskiy and Iskra, V.D., 1961, Fiz. Tverd. Tela, **3**, 2939.
Samson, A.M., 1960, Opt. i Spektr. **8**, 89.
Samson, A.M., 1960, Izv. AN SSSR, Ser. Fiz. **24**, 496.
Samson, A.M., 1962, Opt. i Spektr. **13**, 511.
Samson, A., 1962, Spectroscopy of light-scattering media, Collection of reports, AN BSSR, Minsk, p. 53 (in Russian).
Samson, A.M., 1964, Opt. i Spektr. **16**, 697.
Sanche, L., 1979, Chem. Phys. Lett. **65**, 61.
Sanche, L., 1979, J. Chem. Phys. **71**, 4860.
Choi, Sang-il, S.A. Rice, 1963, J. Chem. Phys. **38**, 366.
Sarzhevsky, A. and A. Sevchenko, 1971, Anisotropy of absorption and emission of light by molecules, Minsk, Ch. V (in Russian).
Savchenko, E.V. and I.Ya. Fugol', 1975, Fiz. Nizk. Temp. **1**, 1438.
Schipper, P.E., 1975, Mol. Phys. **29**, 501.
Schmid, D., H. Auwer, A. Braun, U. Mayer and H. Pfisterer, 1977, Eighth Molecular Cryst. Symp. Santa Barbara, USA, Abstracts, p. 332.
Segall, B. and G.D. Mahan, 1968, Phys. Rev. **171**, 935.
Seiwert, R., 1956, Ann. Physik, **17**, 371.
Sell, D.D., S.E. Stokowski, R. Dingle and J.V. Dilorenzo, 1973, Phys. Rev. **7**, 4568.
Selzer, P.M. and D.S. Hamilton, W.M. Yen, 1977, Phys. Rev. Lett. **38**, 858.
Shakhverdov, T.A. and E.N. Bodunov, 1973, Opt. i Spektr. **34**, 1112.
Sharma, R.D., 1967, J. Chem. Phys. **46**, 3475.
Sheka, E.F., 1971, Usp. Fiz. Nauk, **104**, 593.
Shekhtman, V.L., 1972, Opt i Spektr. **33**, 284, 776.
Shibatani, A. and Y. Toyozawa, 1968, J. Phys. Soc. Japan, **25**, 335.
Shirokov, V.I., 1964, Opt. i Spektr. **16**, 696.
Silbey, R., 1976, Ann. Rev. Phys. Chem. **27**, 203.
Silver, M., D. Olness, M. Swicord and R.C. Jarnagin, 1963, Phys. Rev. Lett. **10**, 12.
Silver, R.N., 1975, Phys. Rev. B**11**, 1569.
Simpson, O., 1957, Proc. Roy. Soc. A238, 402.
Singh, J. and H. Baessler, 1974, Phys. Status Solidi (b) **62**, 147.
Smith, D.Y. and D.L. Dexter, 1972, Prog. Opt., X, 165.
Smoluchowski, M., 1917, Zs. Phys. Chem. **92**, 129.
Sobolev, V., 1956, Transfer of radiation energy in atmospheres of stars and planets (Gostekhizdat, Moscow) (in Russian).
Soos, Z.G., 1974, Ann. Rev. Phys. **25**, 121.
Soos, Z. and R.C. Powell, 1972, Phys. Rev. B**6**, 4035.
Soules, T.F. and C.B. Duke, 1971, Phys. Rev. B3, 262.
Spitzer, F., 1964, Principles of Random Walks (Van Nostrand, New Jersey).
Stepanov, B.I. and A.M. Samson, 1960, Izv. AN SSSR, Ser. Fiz. **24**, 502.
Strickler, S.G. and R.A. Berg, 1962, J. Chem. Phys. **37**, 814.
Strieder, W. and R. Aris, 1973, Variational Methods Applied to Problems of Diffusion and Reaction (Springer, Berlin).

Stryer, L. and R.P. Haugland, 1967, Proc. Nat. Ac. Sci. **58**, 719.
Sumi, H., 1975, Solid State Commun. **17**, 701.
Suna, A., 1970, Phys. Rev. **B1**, 1716.
Sveshnikov, B.Ya., 1935, Acta Physiocochimica URSS **3**, 257.
Sveshnikov, B.Ya., 1937, Acta Physiocochimica URSS **7**, 755.
Sveshnikov, B.Ya., 1938, Trudy GOI, vol. XII, p. 108.
Sveshnikov, B.Ya. and V.I. Shirokov, 1962, Opt. i Spektr. **12**, 576.
Szent-Györgyi, A., 1957, Bioenergetics (Academic Press, New York).
Tait, W.C. and R.L. Weiher, 1968, Phys. Rev. **166**, 769.
Tait, W.C. and R.L. Weiher, 1969, Phys. Rev. **178**, 1404.
Tait, W.C., D.A. Campbell, J.R. Packard and R.L. Weiher, 1967, Bull, Am. Phys. Soc. **12**, 384.
Takahashi, Y. and M. Tomura, 1971, J. Phys. Soc. Japan, **31**, 1100.
Tanaka, M. and J. Tanaka, 1973, Theoret. Chim. Acta (Berl.) **30**, 81.
Tekhver, I.Yu. and V.V. Khizhnyakov, 1974, Pis'ma Zh. TEF **19**, 338.
Tekhver, I.Yu. and V.V. Khizhnyakov, 1974, Zh. ETF **69**, 599.
Terenin, A.N., 1956, Usp. Fiz. Nauk, **43**, 347.
Terenin, A.N. and V.L. Ermolaev, 1956, Usp. Fiz. Nauk, **58**, 37.
Terenin, A.N. and V.L. Ermolaev, 1960, Usp. Fiz. Nauk, **71**, 133.
Terskoi, Ya.A., B.G. Brudz' and O.N. Korol'kova, 1968, Opt. i Spektr. **25**, 516.
Tews, K.H., 1973, Ann. Phys., N.Y., **29**, 97.
Tolpygo, K.B., 1950, Zh. ETF **20**, 497.
Tolstoi, N.A. and A.P. Abramov, 1967, Fiz. Tverd. Tela, **9**, 340.
Tomura, M., E. Ishiguro and N. Mataga, 1968, J. Phys. Soc. Japan, **25**, 1439.
Tosic, B.S., 1971, Phys. Status Solidi (b) **48**, K129.
Tosic, B.S. and J.B. Vujaklija, 1971, Phys. Status Solidi (b) **45**, K113.
Toyozawa, J., 1959, Suppl. of Prog. Theor. Phys. **12**, No. 12, 111.
Toyozawa, Y., 1975, Resonance and Relaxation in Light Scattering, Techn. Rep. of ISSP, Ser. A, 739.
Toyozawa, Y., 1976, J. Phys. Soc. Japan, **41**, 400.
Trifonov, E.D., 1971, Izv. AN SSSR, Ser. Fiz. **35**, 1330.
Trifonov, E.D. and V.L. Shekhtman, 1970, Phys. Status Solidi, **41**, 855.
Trifonov, E.D. and V.L. Shekhtman, 1972, in: Physics of impurity centres in crystals, pp. 585–596, Proc. International Seminar, 21–26 Sept., Tallin (in Russian).
Trlifai, M., 1956, Czech. J. Phys. **6**, 533.
Trlifai, M., 1958, Czech. J. Phys. **8**, 510.
Trlifai, M., 1963, Czech. J. Phys. **13**, 644.
Trubitsin, S.I., A.N. Curobaya and O.F. Borisova, 1971, Molk. Biol. **5**, 419.
Tulub, V. and K. Patzer, 1968, Phys. Status Solidi, **26**, 693.
Tunitskiy, N.N. and Kh.S. Bagdasar'yan, 1963, Opt. i Spektr. 15, 100.
Tyablikov, S., 1975, Methods in the quantum theory of magnetism (Nauka, Moscow) (in Russian).
Van Kranendonk, J., 1959, Physica, **25**, 1080.
Vasil'ev, I.N., B.P. Kirsanov and V.A. Krongauz, 1964, Kinetika i Kataliz. **5**, 792.
Vaubel, G., H. Baessler, and D. Mobius, 1971, Chem. Phys. Lett. **10**, 334.
Vavilov, S.I., 1925, Zs. Phys. **31**, 750.
Vavilov, S.I., 1928, Zs. Phys. **50**, 52.
Vavilov, S.I., 1929, Zs. Phys. **53**, 665.
Vavilov, S.I., 1936, Acta Phys. Polonica, **5**, 417.
Vavilov, S.I., 1937, Dokl. Akad. Nauk, SSSR **16**, 263.
Vavilov, S.I., 1942, Dokl. Akad. Nauk, SSSR **35**, 110.
Vavilov, S.I., 1943, Zh. ETF **13**, 13.
Vavilov, S.I., 1947, J. Phys. USSR **7**, 141.

Vavilov, S.I., 1952, Sobr. Soch. Moscow, Vol. 2, p. 22.
Vavilov, S.I., 1954, Sobr. Soch. Moscow, Vol. I, p. 165, 267, 275, 438.
Vavilov, S.I., 1954 II, p. 116, 122, 152, Moscow.
Vavilov, S.I. and P.P. Feofilov, 1942, Dokl. Akad. Nauk SSSR **34**, 243.
Vavilov, S.I. and M.D. Galanin, 1949, Dokl. Akad. Nauk SSSR, **67**, 811.
Vineyard, G.H., 1963, J. Math. Phys. **4**, 1191.
Vinogradov, A.V., 1970, Fiz. Tverd. Tela, **12**, 3081.
Vol, E.D., V.A. Goloyadov, L.S. Kukushkin, Yu.V. Naboykin and N.S. Silaeva, 1971, Phys. Status Solidi (b) **47**, 685.
Vol, E.D., P.V. Zinov'ev, Yu.V. Naboykin and N.B. Silaeva, 1974, Fiz. Tverd. Tela, 1522.
Vol'kenstein, M., 1951, Molecular optics (Gostekhizdat, Moscow) (in Russian).
Vu Duy Phach, A. Bivas, B. Hönerlage and J.B. Grun, 1978, Phys. Status Solidi (b) **86**, 159.
Walsh, P.J., 1957, Phys. Rev. **107**, 338.
Watson, G.N., 1939, Quar. J. Math. Oxford, Ser. 10, 266.
Weber, M.J., 1971, Phys. Rev. B**4**, 2932.
Weil, G. and M. Calvin, 1963, Biopolymers, **1**, 401.
Widsor, M.W., 1965, Physics and Chemistry of the Organic Solid State, Vol. 2, eds., D. Fox, M.M. Labes, A. Weissberger (Interscience, New York) p. 343.
Winterstein, A., K. Schoen and H. Wetter, 1934, Naturwiss. **22**, 237.
Wolf, H.C. 1967, Advances in Atomic and Molecular Physics, Vol. 3, eds., D.R. Bates and I. Estermann (Academic Press, New York) p. 119.
Wright, G.T., 1955, Proc. Phys. Soc. B**68**, 241.
Wünshe, A., 1975a, Ann. Phys. **32**, 7, Folge, Heft **2**, 122.
Wünshe, A., 1975b, Experimentale Technik der Physik XXIII, Heft **3**, 223.
Yokota, M. and O. Tanimoto, 1967, J. Phys. Soc. Japan, **22**, 779.
Yoshihara, K., A. Inoue and S. Nagakura, 1973, Sixth Molecular Crystal Symposium, Schloss Elmau, Germany, p. 16.
Zel'dovich, Ya.B. and A.A. Ovchinnikov, 1971, Pis'ma Zh. ETF 13, 636.
Zewail, A.H. and C.B. Harris, 1975, Phys. Rev. B**11**, 952.
Zgierski, M.Z., 1973, Phys. Status Solidi (b) **59**, 589.
Zhizhin, G., 1975, D.Sc. Thesis, Inst. of Chemical Physics (AN SSSR, Moscow).
Zubarev, D.N., 1960, Usp. Fiz. Nauk, **71**, 71.
Zverev, G.M., I.I. Kuratev and A.M. Onischenko, 1975, Kvantovaya elektronika, **2**, 469.

AUTHOR INDEX

SUBJECT INDEX

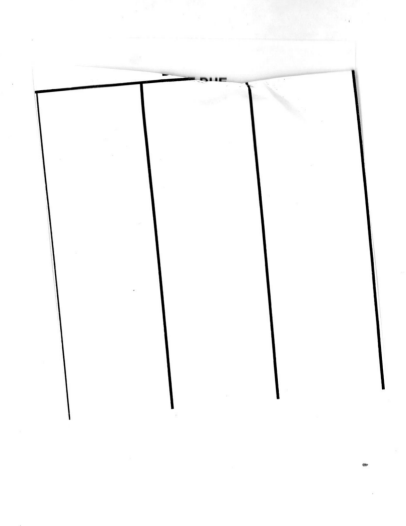